## Principle 10

An important, vigorous new business venture can emerge from a large firm when afforded the appropriate balance of independence, resources, and people to respond to the opportunity.

## Principle 11

The intellectual property and name of a new venture can provide a proprietary advantage leading to success in the marketplace.

## Principle 12

Effective leaders coupled with a good organizational plan, a collaborative, performance-based culture, and a sound compensation scheme can help align every participant with the goals and objectives of the new firm.

## Principle 13

Effective new ventures use their persuasion skills, credibility, and location advantages to secure the required resources for their firm in order to build a well-coordinated mix of outsourced and internal functions.

## Principle 14

The design and management of an efficient, real-time set of production, logistical, and business processes can become a sustainable competitive advantage for a new enterprise.

## Principle 15

All new technology business ventures should formulate a clear acquisition and global strategy.

## Principle 16

A new firm with a powerful revenue and profit engine can achieve strong but manageable growth leading to a favorable harvest of wealth for the owners.

## Principle 17

A sound financial plan demonstrates the potential for growth and profitability for a new venture and is based on the most accurate and reliable assumptions available.

## Principle 18

Many kinds of sources for investment capital for a new and growing enterprise exist and should be compared and managed carefully.

## Principle 19

The presentation of a compelling story about a venture and the resulting skillful negotiations to close a deal with investors are critical to all new enterprises.

## Principle 20

The ability to continuously and ethically execute a business plan and adapt to changing conditions provides long-term success.

# Technology Ventures
## From Idea to Enterprise

**Thomas H. Byers**
*Stanford University*

**Richard C. Dorf**
*University of California, Davis*

**Andrew J. Nelson**
*University of Oregon*

Mc
Graw
Hill
Education

TECHNOLOGY VENTURES: FROM IDEA TO ENTERPRISE, FOURTH EDITION

ISBN 978–0–07–352342–2
MHID 0–07–352342–9

Senior Vice President, Products & Markets: *Kurt L. Strand*
Vice President, General Manager, Products & Markets: *Marty Lange*
Vice President, Content Production & Technology Services: *Kimberly Meriwether David*
Managing Director: *Thomas Timp*
Brand Manager: *Raghothaman Srinivasan*
Editorial Coordinator: *Samantha Donisi-Hamm*
Marketing Manager: *Heather Wagner*
Director, Content Production: *Terri Schiesl*
Content Project Manager: *Melissa M. Leick*
Buyer: *Susan K. Culbertson*
Media Project Manager: *Sandy Schnee*
Cover Image Steel structure: *Getty Images; Binary Code: Glow Images; Test Tubes: Comstock Images/Jupiterimages; Android Tablet: McGraw-Hill Education, Mark Dierker, photographer.*
Compositor: *Laserwords Private Limited*
Typeface: *10.5/12 Times New Roman*
Printer: *R.R. Donnelley*

**Library of Congress Cataloging-in-Publication Data**

Dorf, Richard C.
  Technology ventures : from idea to enterprise / Thomas H. Byers, Stanford University, Richard C. Dorf, University of California, Davis, Andrew J. Nelson, University of Oregon.—Fourth edition. pages cm
  Dorf's name appears first on the earlier editions.
  ISBN-13: 978-0-07-352342-2 (alk. paper)
  ISBN-10: 0-07-352342-9 (alk. paper)
 1. Information technology. 2. Entrepreneurship. 3. New business enterprises. I. Byers, Thomas (Thomas H.) II. Nelson, Andrew J. III. Title.
  HC79.I55D674 2014
  658.1'1—dc23
                                                                      2013033982

# DEDICATION

For our wonderful families.
We warmly recognize their love and commitment to this
publication that will help others create important
enterprises for the benefit of all.

THOMAS H. BYERS, RICHARD C. DORF, ANDREW J. NELSON

# ABOUT THE AUTHORS

**Thomas H. Byers** is professor of management science and engineering at Stanford University and the founding faculty director of the Stanford Technology Ventures Program, which is dedicated to accelerating technology entrepreneurship education around the globe. He is the first person to hold the Entrepreneurship Professorship endowed chair in the School of Engineering at Stanford. He also is a Bass University Fellow in Undergraduate Education. He is a principal investigator and the director of the NSF's Engineering Pathways to Innovation Center (Epicenter), which seeks to spread entrepreneurship and innovation education across all undergraduate schools. After receiving his B.S., MBA, and Ph.D. from the University of California, Berkeley, Dr. Byers held leadership positions in technology ventures including Symantec Corporation. His teaching awards include Stanford University's highest honor (Gores Award) and the Gordon Prize from the National Academy of Engineering.

**Richard C. Dorf** is professor of electrical and computer engineering and professor of management at the University of California, Davis. He is a Fellow of the American Society for Engineering Education (ASEE) in recognition of his outstanding contributions to the society, as well as a Fellow of the Institute of Electrical and Electronic Engineering (IEEE). The best-selling author of *Introduction to Electric Circuits* (9th ed.), *Modern Control Systems* (12th ed.), *Handbook of Electrical Engineering* (4th ed.), *Handbook of Engineering* (2nd ed.), and *Handbook of Technology Management,* Dr. Dorf is cofounder of seven technology firms.

**Andrew J. Nelson** is assistant professor of management and Bramsen Faculty Fellow in innovation, entrepreneurship, and sustainability at the University of Oregon's Lundquist College of Business. Dr. Nelson holds a Ph.D. and a dual B.A. from Stanford University, and an M.Sc. from Oxford University. He has received numerous academic awards, including recognition from the Kauffman Foundation, INFORMS, the NCIIA, and the Industry Studies Association. At the University of Oregon, he is the four-time recipient of the James E. Reinmuth MBA Teaching Excellence Award as well as the Business Advisory Council Undergraduate Teaching Award.

# BRIEF CONTENTS

v

# CONTENTS

# FOREWORD

**by John L. Hennessy, President of Stanford University**

I am delighted to introduce this book on technology entrepreneurship by Professors Byers, Dorf, and Nelson. Technology and similar high-growth enterprises play a key role in the development of the global economy and offer many young entrepreneurs a chance to realize their dreams.

Unfortunately, there have been few complete and analytical books on technology entrepreneurship. Professors Byers, Dorf, and Nelson bring years of experience in teaching and direct background as entrepreneurs to this book, and it shows. Their connections and involvement with start-ups—ranging from established companies like Facebook and Genentech to new ventures delivering their first products—add real-world insights and relevance.

One of the most impressive aspects of this book is its broad coverage of the challenges involved in technology entrepreneurship. Part I talks about the core issues around deciding to pursue an entrepreneurial vision and the characteristics vital to success. Key topics include building and maintaining a competitive advantage and market timing. As recent history has shown, it is easy to lose sight of these principles. Although market trends in technology are ever shifting, entrepreneurs are rewarded when they maintain a consistent focus on having a sustainable advantage, creating a significant barrier to entry, and leading when both the market and the technology are ready. The material in these chapters will help entrepreneurs and investors respond in an informed and thoughtful manner.

Part II examines the major strategic decisions with which every entrepreneur grapples: how to balance risk and return, what entrepreneurial structure to pursue, and how to develop innovative products and services for the right users and customers. It is not uncommon for start-ups led by a technologist to question the role of sales and marketing. Sometimes, you hear a remark like: "We have great technology and that will bring us customers; nothing else matters!" But without sales, there is no revenue, and without marketing, sales will be diminished. It is important to understand these vital aspects of any successful business. These are challenges faced by every company, and the leadership in any organization must regularly examine them.

Part III discusses operational and organizational issues as well as the vital topic in technology-intensive enterprises of intellectual property. Similar matters of building the organization, thinking about acquisitions, and managing operations are also critical. If you fail to address them, it will not matter how good your technology is.

Finally, Part IV talks about putting together a solid financial plan for the enterprise including exit and funding strategies. Such topics are crucial, and they are often the dominant topics of "how-to" books on entrepreneurship. While the financing and the choice of investors are key, unless the challenges discussed in the preceding sections are overcome, it is unlikely that a new venture, even if well financed, will be successful.

In looking through this book, my reaction was, "I wish I had read a book like this before I started my first company (MIPS Technologies in 1984)." Unfortunately, I had to learn much of this in real-time, often making mistakes on the first attempt. In my experience, the challenges discussed in the earlier sections are the real minefields. Yes, it is helpful to know how to negotiate a good deal and to structure the right mix of financing sources, especially so employees can retain as much equity as possible. However, if you fail to create a sustainable advantage or lack a solid sales and marketing plan, the employees' equity will not be worth much.

Those of us who work at Stanford and live near Silicon Valley are in the heartland of technology entrepreneurship. We see firsthand the tenacity and intelligence of some of the world's most innovative entrepreneurs. With this book, many others will have the opportunity to tap into this experience. Exposure to the extensive and deep insights of Professors Byers, Dorf, and Nelson will help build tomorrow's enterprises and business leaders.

Entrepreneurship is a vital source of change in all facets of society, empowering individuals to seek opportunities where others see insurmountable problems. For the past century, entrepreneurs have created many great enterprises that subsequently led to job creation, improved productivity, increased prosperity, and a higher quality of life. Entrepreneurship is now playing a vital role in finding solutions to the huge challenges facing civilization, including health, communications, security, infrastructure, education, energy, and the environment.

Many books have been written to help educate others about entrepreneurship. Our textbook was the first to thoroughly examine a global phenomenon known as "technology entrepreneurship." Technology entrepreneurship is a style of business leadership that involves identifying high-potential, technology-intensive commercial opportunities, gathering resources such as talent and capital, and managing rapid growth and significant risks using principled decision-making skills. Technology ventures exploit breakthrough advancements in science and engineering to develop better products and services for customers. The leaders of technology ventures demonstrate focus, passion, and an unrelenting will to succeed.

Why is technology so important? The technology sector represents a significant portion of the economy of every industrialized nation. In the United States, more than one-third of the gross national product and about half of private-sector spending on capital goods are related to technology. It is clear that national and global economic growth depends on the health and contributions of technology businesses.

Technology has also become ubiquitous in modern society. Note the proliferation of smartphones, personal computers, tablets, and the Internet in the past two decades and their subsequent integration into everyday commerce and our personal lives. When we refer to "high-technology" ventures, we include information technology enterprises, biotechnology and medical businesses, energy and sustainability companies, and those service firms where technology is critical to their missions. At the beginning of the twenty first century, many technologies show tremendous promise, including computational systems, Internet advancements, mobile communications platforms, networks and sensors, medical devices and biotechnology, artificial intelligence, robotics, 3D manufacturing, nanotechnology, and clean energy. The intersection of these technologies may indeed enable the most promising opportunities.

The drive to understand technology venturing has frequently been associated with boom times. Certainly, the often-dramatic fluctuations of economic cycles can foster periods of extreme optimism as well as fear with respect to entrepreneurship. However, some of the most important technology companies have been founded during recessions (e.g., Intel, Cisco, and Amgen). This book's principles endure regardless of the state of the economy.

## APPROACH

Just as entrepreneurs innovate by recombining existing ideas and concepts, we integrate the most valuable entrepreneurship and technology management theories from the world's leading scholars to create a fresh look at entrepreneurship. We also provide an action-oriented approach to the subject through the use of examples, exercises, and lists. By striking a balance between theory and practice, our readers gain from both perspectives.

Our comprehensive collection of concepts and applications provides the tools necessary for success in starting and growing a technology enterprise. We show the critical differences between scientific ideas and true business opportunities. Readers benefit from the book's integrated set of cases, examples, business plans, and recommended sources for more information.

We illustrate the book's concepts with examples from the early stages of high-technology firms (e.g., Apple, Google, and Genentech) and traditional firms that use technology strategically (e.g., FedEx and Wal-Mart). How did they develop enterprises that have had such positive impact, sustainable performance, and longevity? In fact, the book's major principles are applicable to any growth-oriented, high-potential venture, including high-impact nonprofit enterprises such as Conservation International and the Gates Foundation.

## AUDIENCE

This book is designed for students in colleges and universities, as well as others in industry and public service, who seek to learn the essentials of technology and high-growth entrepreneurship. No prerequisite knowledge is necessary, although an understanding of basic accounting principles will prove useful.

Entrepreneurship was traditionally taught only to business majors. Because entrepreneurship education opportunities now span the entire campus, we wrote this book to be approachable by students of all majors and levels, including undergraduate, graduate, and executive education. Our primary focus is on science and engineering majors enrolled in entrepreneurship and innovation courses, but the book is also valuable to business students and others with a particular interest in high-growth ventures.

For example, our courses at Stanford University, the University of Oregon, and the University of California, Davis, based on this textbook regularly attract students from majors as diverse as computer science, product design, political science, economics, pre-med, electrical engineering, history, biology, and business. Although the focus is on technology entrepreneurship, these students find this material applicable to the pursuit of a wide variety of endeavors. Entrepreneurship education is a wonderful way to teach universal leadership skills, which include being comfortable with constant change, contributing to an innovative team, and demonstrating passion in any effort. Anyone can learn entrepreneurial thinking and leadership. We particularly encourage instructors to design courses in which the students form study teams early in the term and learn to work together effectively on group assignments.

## WHAT'S NEW

Based upon feedback from readers and new developments in the field of technology entrepreneurship, numerous enhancements appear in this fourth edition. Recent compelling academic theories and practitioner insights in entrepreneurship from leading scholarly journals, trade books, and popular blogs and press are included in the text. Special attention is given to business model development and measurement, lean start-ups, design thinking, intellectual property, and marketing and sales. All examples and exercises were reviewed to place even more emphasis on exciting technology ventures around the globe.

Chapters 1 and 2 have been extensively revised to better introduce the art and science of venturing. Chapter 3 now contains the latest techniques on business model development and lean start-up methodologies. The concept story and business plan development materials and tools are consolidated and improved in a new Chapter 6 to start Part II. Similarly, a new Chapter 8 consolidates and expands content regarding creativity and product development. Chapter 11 is solely focused on the vital topic of intellectual property. Chapter 12 now has all key material on teams and organizational learning in one place. Two new full-length cases regarding clean technology and sustainability are included in the appendix. The AgraQuest sequential case in each chapter has been replaced with an examination of an exemplary enterprise relevant to that material. Cases no longer in use from previous versions are available on our websites. Some reordering of sections within chapters streamlines the remaining content.

## FEATURES

The book is organized in a modular format to allow for both systematic learning and random access of the material to suit the needs of any reader seeking to learn how to grow successful technology ventures. Readers focused on business plan and model development should consider placing a higher priority on Chapters 3, 6, 9, 11, 12, 17, 18, and 19. Regardless of the immediate learning goals, the book is a handy reference and companion tool for future use. We deploy the following wide variety of methods and features to achieve this goal, and we welcome feedback and comments.

*Principles and Chapter Previews*—A set of 20 fundamental principles is developed and defined throughout the book. They are listed in the inside front cover as well. Each chapter opens with a key question and outlines its content and objectives.

*Examples and Exercises*—Examples of cutting-edge technologies illustrate concepts in a shaded-box format. Information technology is chosen for many examples because students are familiar with its products and services. Exercises are offered at the end of each chapter to test comprehension of the concepts.

*Sequential Exercise and Spotlights*—A special exercise called the "venture challenge" guides readers through a chapter-by-chapter formation of a new

**TABLE P1  Overview of cases.**

| Cases in appendix B | Synopsis | Issues |
|---|---|---|
| Method | A start-up contemplates a new product line | Opportunities, vision and the business model, marketing and sales |
| Method products | A product development effort runs into problems | Innovation strategies, creativity, and product development |
| Biodiesel | Three founders consider an opportunity in the energy industry | Opportunity identification and evaluation, business model |
| Yahoo! | Two founders face a decision on financing that forces them to confront their vision | Vision and business model, sources of capital, business plan |
| Barbara's Options | A soon-to-be graduate weighs two job offers | Stock options, finance |
| Artemis Images | A promising image management company runs into trouble | Competitive strategy, business model, team, finance |
| Sirtris Pharmaceuticals | A life sciences firm faces major decisions about its future | Alliances, licensing, market strategy |
| Cooliris | A young entrepreneur struggles to hire a team | Hiring process, scaling issues |

enterprise. At the end of each chapter's narrative, a successful enterprise is profiled in a special "spotlight" section.

*Business Plans*—Methods and tools for the development of a business plan are gathered into one special chapter, which includes a thoroughly annotated table of contents. A sample business plan is provided in appendix A.

*Cases*—Eight comprehensive cases are included in appendix B. A short description of each case is provided in Table P1. Additional cases from Harvard and ECCH are recommended on this textbook's websites.

*References and Glossary*—References are indicated in brackets [Smith, 2001] and are listed as a complete set in the back of the book. This is followed by a comprehensive glossary.

*Chapter Sequence*—The chapter sequence represents our best effort to organize the material in a format that can be used in various types of entrepreneurship courses. The chapters follow the four-part layout shown in Figure P1. Courses focused on creating business plans and models can reorder the chapters with emphasis on Chapters 3, 6, 9, 11, 12, 17, 18, and 19.

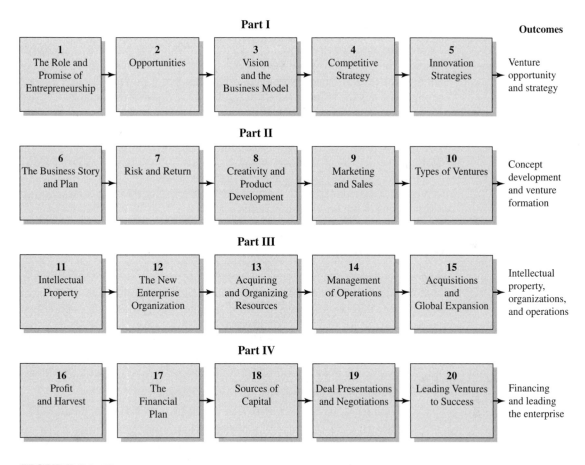

**Part I**

| 1 The Role and Promise of Entrepreneurship | 2 Opportunities | 3 Vision and the Business Model | 4 Competitive Strategy | 5 Innovation Strategies | Outcomes: Venture opportunity and strategy |

**Part II**

| 6 The Business Story and Plan | 7 Risk and Return | 8 Creativity and Product Development | 9 Marketing and Sales | 10 Types of Ventures | Concept development and venture formation |

**Part III**

| 11 Intellectual Property | 12 The New Enterprise Organization | 13 Acquiring and Organizing Resources | 14 Management of Operations | 15 Acquisitions and Global Expansion | Intellectual property, organizations, and operations |

**Part IV**

| 16 Profit and Harvest | 17 The Financial Plan | 18 Sources of Capital | 19 Deal Presentations and Negotiations | 20 Leading Ventures to Success | Financing and leading the enterprise |

**FIGURE P1**  Chapter sequence.

*Video Clips*—A collection of suggested videos from world-class entrepreneurs, investors, and teachers is listed at the end of each chapter and provided on this textbook's websites. More free videos clips and podcasts are available at Stanford's Entrepreurship Corner website (see http://ecorner.stanford.edu).

*Websites and Social Networking*—Please visit websites for this book at both McGraw-Hill Higher Education (http://www.mhhe.com/byersdorf) and Stanford University (http://techventures.stanford.edu) for supplemental information applicable to educators, students, and professionals. For example, a complete syllabus for an introductory course on technology entrepreneurship and additional learning resources for each chapter are provided for instructors.

## ELECTRONIC TEXTBOOK OPTIONS

E-books are an innovative way for students to save money and create a greener environment at the same time. An e-book can save students about half the cost of a traditional textbook and offers unique features like a powerful search engine, highlighting, and the ability to share notes with classmates using e-books.

McGraw-Hill offers this text as a CourseSmart ebook. With the CourseSmart eTextbook version of this title, students can save over half off the cost of a print book, reduce their impact on the environment, and access powerful Web tools for learning. Faculty can also review and compare the full text online without having to wait for a print desk copy. CourseSmart is an online eTextbook, which means users need to be connected to the Internet in order to access. Students can also print sections of the book for maximum portability.

## ACKNOWLEDGMENTS

Many people have made this book possible. Our editors at McGraw-Hill were Katie Neubeaur Carney and Samantha Donisi-Hamm. We thank all of them for their insights and dedication. We also thank the McGraw-Hill production and marketing teams for their diligent efforts. Our colleagues at Stanford University, the University of Oregon, and the University of California, Davis, were helpful in numerous ways. We are indebted to them for all of their great ideas and support. Lastly, we remain grateful for the continued support of educators, students, and other readers of previous editions.

**Thomas H. Byers, Stanford University, tbyers@stanford.edu**
**Richard C. Dorf, University of California, Davis, rcdorf@ucdavis.com**
**Andrew J. Nelson, University of Oregon, ajnelson@uoregon.edu**

# MEDIA SUPPLEMENTS FOR STUDENTS AND INSTRUCTORS

The 4th edition is supplemented by two websites, collectively bringing students and instructors the most extensive resources available for technology and high-growth entrepreneurship courses. Visitors to either website can link to the authors' social networking sites in order to interact with the authors and other readers.

## McGraw-Hill Website
## www.mhhe.com/byersdorf

Accessed with a password, the McGraw-Hill website for instructors features:

- Answers to end-of-chapter exercises
- Teaching notes in Word and PDF format for the cases in appendix B
- Extensive sample presentations based on the text

**Sample presentations** provide instructors with a framework for organizing their lectures, and reference topic-related videos on the textbook's websites.

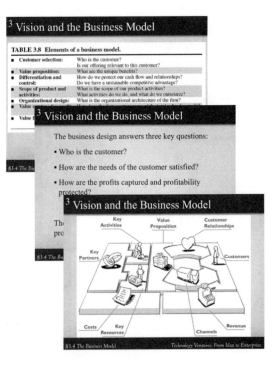

# MEDIA SUPPLEMENTS FOR STUDENTS AND INSTRUCTORS

## Stanford University Website
## http://techventures.stanford.edu

Rich with content, the author-created
Stanford website provides relevant media
for each chapter in *Technology Ventures,*
including:

- Video clips and podcasts of entrepreneurial leaders including founders,
  CEOs, venture capitalists, authors, educators, and policy makers.

- Suggested case studies from Harvard
  Business School and other universities
  around the globe.

- Resources on how to best integrate the
  book's business plans and case studies
  into entrepreneurship courses.

- Links to compelling resources on
  entrepreneurship.

- Additional sample business plans to
  augment the executive summary in
  appendix A.

- Sample syllabi including one from an
  actual Stanford University course for
  students of all majors.

- A collection of the videos listed in the
  "Video Resources" section at the end
  of each chapter in this textbook.

# PART I

# Venture Opportunity and Strategy

Entrepreneurs create new businesses that fuel progress in societies worldwide. They use innovation and technology to foster positive impact and activity in all facets of life. Entrepreneurs identify, develop, and communicate the essence of an opportunity that has attractive potential to become a successful venture. They describe the valuable contributions of the venture, and they design a business model that can adapt to changing circumstances. The venture team creates a road map (strategy) that can, with good chance, effectively lead to the commercialization of the new product or service in the marketplace with a sustainable competitive advantage. ■

# CHAPTER 1

# The Role and Promise of Entrepreneurship

*There are risks and costs to a program of action. But they are far less than the long-range risks and costs of comfortable inaction.*
**John F. Kennedy**

## What drives global entrepreneurship?

Entrepreneurs strive to make a difference in our world and to contribute to its betterment. They identify opportunities, mobilize resources, and relentlessly execute on their visions. In this chapter, we describe how entrepreneurs act to create new enterprises. We identify firms as key structures in the economy and the role of entrepreneurship as the engine of economic growth. New technologies form the basis of many important ventures where scientists and engineers combine their technical knowledge with sound business practices to foster innovation. Entrepreneurs are the critical people at the center of all of these activities. ■

## 1.1 Entrepreneurship in Context

From environmental sustainability to security, from information management to health care, from transportation to communication, the opportunities for people to create a significant positive impact in today's world are enormous. **Entrepreneurs** are people who identify and pursue solutions among problems, possibilities among needs, and opportunities among challenges.

Entrepreneurship is more than the creation of a business and the wealth associated with it. It is focused on the creation of a new enterprise that serves society and makes a positive change. Entrepreneurs can create great and reputable firms that exhibit performance, leadership, and longevity. In Table 1.1, look at the examples of successful entrepreneurs and the enterprises they created. What contributions have these people and organizations made? What organization would you add to the list? What organization do you wish you had created or been a part of during its formative years? What organization might you create in the future?

**TABLE 1.1  Selected entrepreneurs and the enterprises they started.**

| Entrepreneur | Enterprise started | Age of entrepreneur at time of start | Year of start |
| --- | --- | --- | --- |
| Benioff, Mark | Salesforce.com (USA) | 35 | 1999 |
| Bezos, Jeff | Amazon.com (USA) | 31 | 1995 |
| Brin, Sergey | Google (USA) | 27 | 1998 |
| Dell, Michael | Dell Computer (USA) | 19 | 1984 |
| Dorsey, Jack | Twitter, Square (USA) | 30 | 2006 |
| Goyanechea, Rosalia | Zara (Spain) | 31 | 1975 |
| Greene, Diane | VMWare (USA) | 42 | 1998 |
| Huateng, Ma | Tencent Inc. (China) | 27 | 1998 |
| Ibrahim, Mo | Celtel (Africa) | 42 | 1998 |
| Lerner, Sandra | Cisco (USA) | 29 | 1984 |
| Li, Robin | Baidu (China) | 32 | 2000 |
| Ma, Jack | Alibaba.com (China) | 35 | 1999 |
| Plattner, Hasso | SAP (Germany) | 28 | 1972 |
| Rottenberg, Linda | Endeavor (Chile, Argentina) | 28 | 1997 |
| Shwed, Gil | Check Point (Israel) | 25 | 1993 |
| Tanti, Tulsi | Suzlon Energy (India) | 37 | 1995 |
| Yunus, Muhammed | Grameen Bank (India) | 36 | 1976 |
| Zennstrom, Nikalas | Skype, Kazaa (Sweden) | 37 | 2003 |
| Zuckerberg, Mark | Facebook (USA) | 20 | 2004 |

Entrepreneurs seek to achieve a certain goal by starting an organization that will address the needs of society and the marketplace. They are prepared to respond to a challenge to overcome obstacles and build a business. As Martin Luther King, Jr. (1963), said, "The ultimate measure of a man is not where he stands in moments of comfort and convenience, but where he stands at times of challenge and controversy."

For an entrepreneur, a **challenge** is a call to respond to a difficult task and the commitment to undertake the required enterprise. Richard Branson, the creator of Virgin Group, reported [Garrett, 1992]: "Ever since I was a teenager, if something was a challenge, I did it and learned it. That's what interests me about life—setting myself tests and trying to prove that I can do it."

Thus, entrepreneurs are resilient people who pounce on challenging problems, determined to find a solution. They combine important capabilities and skills with interests, passions, and commitment. Over nearly a decade, Fred Smith worked on perfecting a solution to what he viewed as a growing problem of organizations to find ways to rapidly ship products to customers. To address this challenge, Smith saw an opportunity to build a freight-only airline that would fly packages to a huge airport and then sort, transfer, and fly them to their destinations overnight. He turned in his paper describing this plan to his Yale University professor, who gave it an average grade, said to be a C. After he graduated, Smith served four years as a U.S. Marine Corps officer and pilot. Following his military service, he spent a few years in the aviation industry building up his experience and knowledge of the industry. Then, he prepared a fully developed business plan for an overnight freight service. By 1972, he had secured financial backing, and Federal Express took to the air in 1973. Federal Express became a new way of shipping goods that revolutionized the cargo shipping business worldwide.

Smith and other entrepreneurs recognize a change in society and its needs, and then, based on their knowledge and skill, they respond with a new way of doing things, typically by recombining people, concepts, and technologies into an original solution. Smith saw that the combination of dedicated cargo airplanes, computer-assisted tracking systems, and overnight delivery would serve a new market that required just-in-time delivery of critically important parts, documents, and other valuable items. Smith adapted computer technology to manage the complex task of tracking and moving packages. More fundamentally, Smith matched his passions and skills as a person with a good opportunity.

An **opportunity** is a favorable juncture of circumstances with a good chance for success or progress. Attractive opportunities combine good timing with realistic solutions that address important problems in favorable contexts. It is the job of the entrepreneur to locate new ideas, to determine whether they are actual opportunities, and, if so, to put them into action. Thus, **entrepreneurship** may be described as the nexus of enterprising individuals and promising opportunities [Shane and Venkataraman, 2000]. As illustrated in Figure 1.1, the "sweet spot" exists where an individual's or team's passions and capabilities intersect with an attractive opportunity.

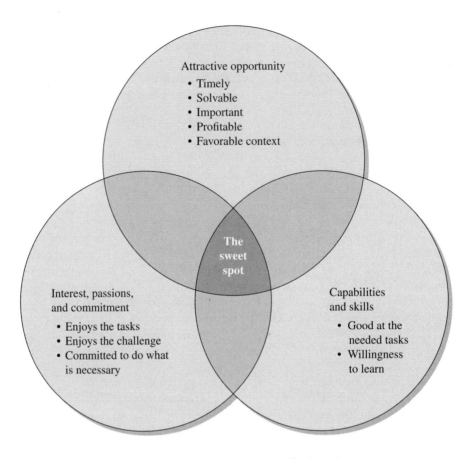

**FIGURE 1.1** Selecting the right opportunity by finding the sweet spot.

Entrepreneurship is not easy. Only about one-third of new ventures survive their first three years. As change agents, entrepreneurs must be willing to accept failure as a potential outcome of their venture. But, regardless of whether the right opportunity has emerged, a person can learn to act as an entrepreneur by trying the activity in a low-cost manner. To avoid the realm of daydreams and fantasy, a person needs to start the practice of experimenting, testing, and learning about his or her entrepreneurial self [Ibarra, 2002]. The would-be entrepreneur should, therefore, engage in this sequence: do it, then reflect on it.

The first step is to identify the hypotheses associated with an idea: what assumptions is the entrepreneur making when concluding that an identified problem is really a problem and that a proposed solution is a good and realistic solution? Then, the entrepreneur can test these hypotheses by engaging with knowledgeable individuals, such as potential customers, employees, and partners. Through these small experiments, the entrepreneur not only develops contacts and mentors

**TABLE 1.2  Four steps to starting a business.**

1. The founding team or individual has the necessary skills or acquires them.

2. The team members identify the opportunity that attracts them and matches their skills. They create a solution to match the opportunity.

3. They acquire (or possess) the financial and physical resources necessary to launch the business by locating investors and partners.

4. They complete an arrangement or contract with their partners, with investors, and within the founder team to launch the business and share the ownership and wealth created.

critical for executing upon an idea [Baer, 2012], but also learns more about the opportunity, and what changes may be necessary to make it viable. In this way, entrepreneurship is akin to the scientific method, in that entrepreneurs seek to gather data in connection with hypotheses, and they refine their ideas based upon their findings [Sarasvathy and Venkataraman, 2011]. Put simply, as Y Combinator founder Paul Graham advises, there are three key things necessary to creating a successful startup: start with good people, make something that people actually want and are willing to pay for, and spend as little money as possible while you validate the market and your product acceptance by buyers [Graham, 2005].

If team members identify an opportunity that attracts them and matches their skills, they next obtain the resources necessary to implement their solution. Finally, they launch and grow an organization, which can grow to have a massive impact, like those enterprises listed in Table 1.1. The four steps to starting a business appear in Table 1.2. Most entrepreneurs repeat these steps multiple times as they work to validate an opportunity, making continual adjustments as they learn more.

Ultimately, entrepreneurship is centrally focused on the identification and exploitation of previously unexploited opportunities. Fortunately for the reader, successful entrepreneurs do not possess a rare entrepreneurial gene. Entrepreneurship is a systematic, organized, rigorous discipline that can be learned and mastered [Drucker, 2002]. This textbook will show you how to identify true business opportunities and how to start and grow a high-impact enterprise.

## 1.2 Economics and the Firm

All entrepreneurs are workers in the world of economics and business. **Economics** is the study of the production, distribution, and consumption of goods and services. Society, operating at its best, works through entrepreneurs to effectively manage its material, environmental, and human resources to achieve widespread prosperity. An abundance of material and social goods equitably distributed is the goal of most social systems. Entrepreneurs are the people who arrange novel organizations or solutions to social and economic problems. They are the people who make our economic system thrive [Baumol et al., 2007].

According to Global Entrepreneurship Monitor (GEM) researchers, the United States maintained about a 12 percent entrepreneurial activity rate between 1999 and 2011. Thus, one in ten U.S. adults was engaged in setting up or managing a new enterprise during that period [Phinisee et al., 2008].

These entrepreneurs have had a tremendous impact on U.S. economic growth. For example, venture capital funds, which invest in companies led by entrepreneurs, accounted for just two-tenths of one percent of U.S. Gross Domestic Product in 2012. VC-backed companies, however, accounted for 11 percent of private sector jobs and 21 percent of U.S. GDP [National Venture Capital Association, 2013]. Another 2010 study found that for all but seven years between 1977 and 2005, existing firms were net job destroyers, losing a combined average of 1 million jobs per year [Kauffman Foundation, 2010]. By contrast, new firms in their first year added a combined average of 3 million jobs. Venture-backed companies such as Amazon, Netflix, Apple, Google, Intuitive Surgical, and Salesforce have accounted for tremendous new employment over the past two decades.

An economic system is a system that produces and distributes goods and services. Given the limitations of nature and the unlimited desires of humans, economic systems are schemes for (1) administering scarcities and (2) improving the system to increase the abundance of goods and services. For a nation as a whole, its wealth is its food, housing, transportation, health care, and other goods and services. A nation is wealthier when it has more of these goods and services. Nations strive to secure more prosperity by organizing to achieve a more effective and efficient economic system. It is entrepreneurs who organize and initiate that change.

Almost all variation in living standards among countries is explained by **productivity,** which is the quantity of goods and services produced from the sum of all inputs, such as hours worked and fuels used. A model of the economy is shown in Figure 1.2. The inputs to the economy are natural capital, financial capital, and intellectual capital. The outputs are the desired benefits or outcomes and the undesired waste. An appropriate goal is to maximize the beneficial outputs and minimize the undesired waste [Dorf, 2001].

**Natural capital** refers to those features of nature, such as minerals, fuels, energy, biological yield, or pollution absorption capacity, that are directly or indirectly utilized or are potentially utilizable in human social and economic systems. Because of the nature of ecologies, natural capital may be subject to

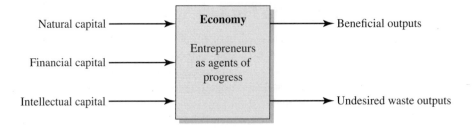

**FIGURE 1.2**  A model of the economy.

**TABLE 1.3  Three elements of the intellectual capital (IC) of an organization.**

**Human capital (HC):** The skills, capabilities, and knowledge of the firm's people

**Organizational capital (OC):** The patents, technologies, processes, databases, and networks

**Social capital (SC):** The quality of the relationships with customers, suppliers, and partners

$$IC = HC + OC + SC$$

irreversible change at certain thresholds of use or impact. For example, global climate change poses a serious threat to sources of natural capital.

**Financial capital** refers to financial assets, such as money, bonds, securities, and land, which allow entrepreneurs to purchase what they need to produce goods and services. The **intellectual capital** of an organization includes the talents, knowledge and creativity of its people, the efficacy of its management systems, and the effectiveness of its customer and supplier relations. The sources of intellectual capital are threefold: human capital, organizational capital, and social capital. **Human capital** (HC) is the combined knowledge, skill, and ability of the company's employees. **Organizational capital** (OC) is the hardware, software, databases, methods, patents, and management methods of the organization that support the human capital. **Social capital** (SC) is the quality of relationships with a firm's suppliers, allies, partners, and customers. These elements of intellectual capital appear in Table 1.3.

Intellectual capital can be thought of as the sum of the knowledge assets of an organization. This knowledge is embodied in the talent, know-how, and skills of the members of an organization. Thus, a firm needs to attract and retain the best people for its requirements in the same way that it seeks the best technologies or physical assets. Knowledge is one of the few assets that grows when shared. By organizing itself around its intellectual capital, a new firm can leverage its benefits through collaboration, development, and sharing.

The economy as portrayed in Figure 1.2 consists of the summation of all organizations, for-profit as well as nonprofit and governmental, that provide the beneficial outputs for society. These are the organizations that we study and will label as enterprises or firms*. Entrepreneurs constantly form new organizations or enterprises to meet social and economic needs.

The purpose of a firm is to establish an objective and mission and carry it out for the benefit of the customer. Thus, the purpose of Merck Corporation is to create pharmaceuticals that protect and enhance its customers' health. To carry out its purpose, each individual firm transforms inputs into desirable outputs that serve the needs of customers.

---

* Henceforth, we use firm to represent organizations, enterprises, and corporations.

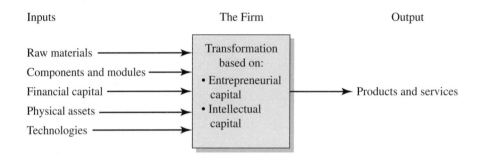

**FIGURE 1.3** The firm as transforming available inputs into desired outputs.

A firm exists as a group of people because it can operate more effectively and efficiently than a set of individuals acting separately. Furthermore, a firm creates conditions under which people can work more effectively than they could on their own. Thus, firms exist to coordinate and motivate people's economic activity [Roberts, 2004]. A firm is more effective because (1) it has lower transaction costs and (2) the necessary skills and talent are gathered together in effective, collaborative work.

A model of the firm as a transformation entity is shown in Figure 1.3. The transformation of inputs into desired outputs is based primarily on the intellectual capital and the entrepreneurial capital of the firm. As an example, consider Microsoft, a powerful software firm. It creates and purchases technologies, develops new software, and builds a client base. The transformation of its inputs into outputs is based on its formidable stock of entrepreneurial capital and intellectual capital.

**Entrepreneurial capital** (EC) can be formulated as a product of entrepreneurial competence and entrepreneurial commitment [Erikson, 2002]. **Entrepreneurial competence** is the ability to (1) recognize opportunity, and (2) gather and manage the resources necessary to capitalize upon the opportunity. **Entrepreneurial commitment** is a dedication of the time and energy necessary to bring the enterprise to initiation and fruition. The presence of competence without any commitment creates little entrepreneurial capital. The presence of commitment without competence may waste both time and resources. Both commitment and competence are required to provide significant entrepreneurial capital. Thus, we can say that

Entrepreneurial Capital = entrepreneurial competence
$$\times \text{ entrepreneurial commitment}$$

or

$$EC = Ecomp \times Ecomm \tag{1.1}$$

where Ecomp is entrepreneurial competence and Ecomm is entrepreneurial commitment. Note that the symbol $\times$ is a multiplication sign, but it should be recognized that this equation is qualitative in nature.

The accretion of knowledge and experience over time leads to increased competence as people mature. However, commitment of energy and time may decline when people become less interested in or available for the necessary entrepreneurial activities. Both commitment and competence are qualities of the leadership team, and they may be complementary qualities shared among the team members.

To transform inputs into outputs, the firm also acts to develop, attract, and retain intellectual capital. Thus, celebrated wine producer Antinori succeeds because of the human capital of its grape growers and wine makers. KFC relies on the organizational capital of its recipes and processes. A local café where the waiter recognizes you and knows your favorite latté relies on its social capital. Social capital is based on strong, positive relationships. The firm provides a place where people can collaborate, learn, and grow, thus furthering the firm's intellectual capital.

In fact, the firm's intellectual capital is critical to its mission and purpose. Figure 1.4 depicts the business theory of a firm, or how it understands its total activities, resources, and relationships. First, a firm is clear about its mission and purpose. Second, the firm must know and understand its customers, suppliers, and competitors. Third, a firm's intellectual capital is understood, renewed, and enhanced as feasible. Finally, the firm must understand its environment or context, which is set by society, the market, and the technology available to it. One hundred years ago, firms were hierarchical and bureaucratic with a theory of business that emphasized making long runs of

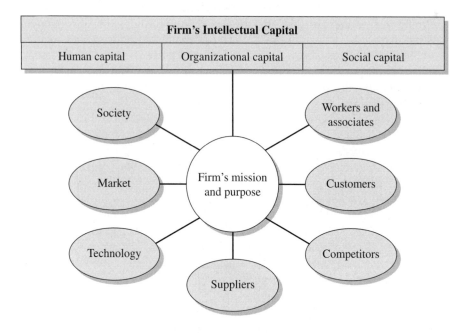

**FIGURE 1.4**  A firm's theory of business depicts how it understands and uses its total resources, activities, and relationships.

standardized products. They regularly introduced "new and improved" varieties and provided lifetime employment. Today, firms compete globally with high-value, customized products. They use flattened organizations and base their future on intellectual capital. Firms look to brands and images to cut through the clutter of messages. In the future, a firm's human capital—talent—will become even more important.

One way to look at the future of a firm is as a competition among its stakeholders. Flexibility and leanness mostly benefit the firm's shareowners. Stakeholders include not only these shareholders, but also workers, customers, people in the community, and society in general. Placing a high valuation on talent gives more power to the workers. Customers stand to gain power as competitors vie for their attention. A good reputation means the firm needs to look after its community and society. The entrepreneur in the new firm strives to build a firm that serves all its stakeholders well.

## 1.3 Creative Destruction

One view of economic activity describes a world of routine in which little changes. In this static model, all decisions have been made, and all alternatives are known and explored. But clearly, no economy is static, and change appears to be certain.

**Dynamic capitalism** is the process of wealth creation characterized by the dynamics of new, creative firms forming and growing and old, large firms declining and failing. In this model, it is disequilibrium—the disruption of existing markets by new entries—that makes capitalism lead to wealth creation [Kirchhoff, 1994]. New firms are formed by entrepreneurs to exploit and commercialize new products or services, thus creating new demand and wealth. This renewal and revitalization of industry leads to a life cycle of formation, growth, and decline of firms.

The recorded music industry provides a good example of waves of change. Music lovers listened to their favorite music recorded on vinyl discs until about 1980, when cassette tapes grew in popularity. The compact size and recordability of the cassette tape caused a massive shift from vinyl records to tape. By the late 1980s, however, compact discs (CDs) overshadowed cassettes, due to the CD's better sound quality and instant access to tracks. In turn, the CD business peaked in 1995 just as the Internet was gaining momentum in society at large. A few years later, peer-to-peer file transfer began to allow piracy of music. By 2001, Apple had introduced the iPod and iTunes and eventually gained a commanding position in the music distribution and sales business. In a dynamic economy, companies need to reinvent their business arrangements or end up becoming irrelevant [Knopper, 2009].

Joseph Schumpeter (1883–1950) described this process of new entrepreneurial firms and waves of change as **creative destruction.** Born and educated in Austria, Schumpeter taught at Harvard University from 1932 until his death in 1950. His most famous book, *Capitalism, Socialism and Democracy,* which appeared in 1942 [Schumpeter, 1984], argued that the economy is in a perpetual state of **dynamic disequilibrium.** Entrepreneurs upend the established order, unleashing a gale of creative destruction that forces incumbents to adapt or die. Schumpeter argued that the concept of perfect competition is irrelevant because it focused entirely on market (price) competition, when the focus should be on technological competition. Creative destruction incessantly revolutionizes the economic structure from within, destroying the old structure and creating a new one. The average life span of a company in the Standard and Poors 500 declined from 35 years in 1975 to less than 20 years today. Less than 4 of the top 25 technology companies 30 years ago are leaders today—perhaps only IBM and Hewlett-Packard.

In a world of change, entrepreneurs seek to embrace it. Entrepreneurs match ideas for change with opportunity. These changes include the adoption of new and better (or cheaper) sources of input supplies, the opening of new markets, and the introduction of more profitable forms of business organization.

The profit of the new firm is the key to economic growth and progress. By introducing a new and valuable product, the innovator obtains temporary monopoly power until rivals figure out how to mimic the innovation. Lower costs may give the innovative firm profits higher than those of its rivals, which must continue to sell at higher prices to cover their higher expenses. Alternatively, a superior product may permit a price above that charged by other firms. The same concept clearly fits all forms of successful change. The business system works to drive out inefficiency and forces business process renewal.

Economic progress is reflected in productivity growth, which provides for increases in people's standard of living. Over the past half-century, the U.S. workforce (including immigration) has grown at about 1.7 percent annually, and productivity per worker has risen at 2.2 percent, generating real economic growth (excluding inflation) averaging 3.9 percent. This is an excellent record, due in great part to the impact of technology entrepreneurship.

Rising output per worker comes from two sources: (1) new technology and (2) smarter ways of doing work. Both paths have been followed throughout human history, and they became faster tracks with the coming of the Industrial Revolution. The twentieth century started with new techniques of management and many new inventions. The century ended with smarter management techniques and dramatic advances in electronic technology, which helped revive productivity growth after limited gains through much of the 1970s and 1980s.

The free spirit of entrepreneurs provides the vital energy that propels this capitalist system. During the past 30 years, the forces of entrepreneurship, competition, and globalization have encouraged new technologies and business methods that raise efficiency and efficacy. In recent years, due to competition, many of the benefits of strong productivity have flowed to consumers in the form of lower prices. Together, innovation, entrepreneurship, and competition are important sources of productivity growth.

## 1.4 Innovation and Technology

Little doubt now exists that the economy is driven by firms that capitalize on change, technology, and challenge. This book is focused on helping the reader to purposefully become an agent of creative destruction by creating his or her own firm. An example of an agent of creative destruction is Craig Venter, who founded Synthetic Genomics in order to use modified or synthetically produced microorganisms to create ethanol and hydrogen. The company is attempting to capitalize on the growing interest in alternative fuels and to design and synthesize specifically engineered cells to perform particular tasks.

New technologies such as these are often a source of disequilibrium or discontinuity, and Schumpeter's theory was based on disruptive, or "radical," innovations. **Technology** includes devices, artifacts, processes, tools, methods, and materials that can be applied to industrial and commercial purposes. For example, Intel was formed to apply semiconductor technology to the design and manufacture of semiconductor circuits. Microsoft was formed to create and distribute computer software products for applications in industry and the home. Apple has reshaped itself around mobile communications and mobile media technologies.

Modern entrepreneurial firms breed a constant flow of high-impact products that create value and stimulate economic growth by bringing new methods, technologies, and ideas to the global marketplace [Schramm, 2004]. Figure 1.5 illustrates "waves" of innovation based upon different technologies throughout history. Modern entrepreneurial firms are at the forefront of the sixth wave, which places a special emphasis on sustainability.

Population growth and a worldwide rising middle class, combined with tightening energy supplies and fears of climate change, have prompted a move toward socially and environmentally responsible business. The goal is to provide housing, transportation, and energy systems that use less energy and emit less pollution and carbon dioxide. The concept is to use knowledge and innovation to create and implement sustainable energy systems and to increase resource productivity [Friedman, 2008].

A clean energy system would consist of a mixture of energy generation, transmission, and utilization in ways that best use natural resources and minimize environmental impacts. By clean and green we mean a system based on conservation, best uses of natural resources, and minimizing environmental impacts. Examples of green technology solutions include installing carbon capture systems at power plants, increasing the use of wind power systems, and developing high-efficiency biofuel systems. Improving the reliability and smart control of the electricity grid also offers a good opportunity for entrepreneurs.

As the green technology movement highlights, technology entrepreneurship is based upon intellectual capital. One hundred years ago, successful companies such as U.S. Steel were primarily managing physical assets. By contrast, today's successful firms, such as Microsoft and Genentech, manage knowledge

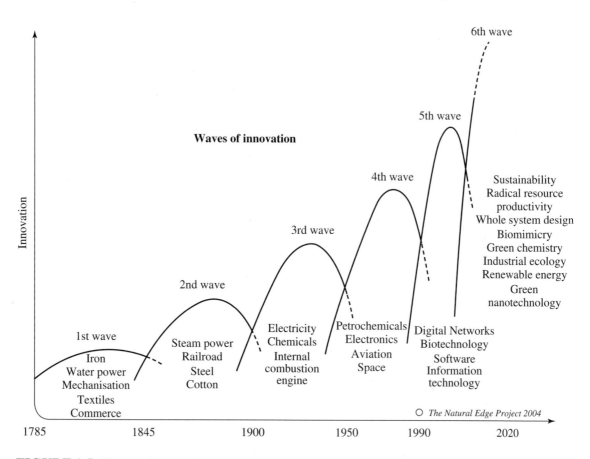

**FIGURE 1.5** Waves of innovation throughout history.

and intellectual capital. In fact, for many, if not most, firms, intellectual capital is the organization's most important asset, more valuable than its other physical and financial assets. Many firms depend on their patents, copyrights, and software, and the capabilities and relationships of their people. This intellectual capital, appropriately applied, will determine success or failure. Thus, knowledge has become the most important factor of production.

While innovation and intellectual property are critical, however, a dynamic economy ultimately rests on the actions of entrepreneurs who assume and accept the benefits and risks of an initiative. It is people acting as leaders, organizers, and motivators who are the central figures of modern economic activity.

Three factors make up entrepreneurial action: (1) a person or group who is responsible for the enterprise, (2) the purposeful enterprise, and (3) initiation and growth of the enterprise. The individuals responsible for the organization are described in Section 1.5. The purposeful enterprise may be a new firm organized for a suitable and attractive purpose or a new unit within or separated from an existing business corporation. Furthermore, the organization may be based on radical innovation, incremental changes, imitation, or rent-seeking behavior.

In the first type of enterprise, the entrepreneur engages in an innovative activity that results in novel methods, processes, and products. The second form emphasizes the founding and management of a business that builds upon and improves an existing product or service. The imitative venture is founded by an entrepreneur who is involved in the rapid dissemination of an innovative idea or process. This person or group finds a novel innovation and transfers it to another environment, region or country. The final means of entrepreneurship is called rent-seeking or profit-seeking and focuses on the use of regulation, standards, or laws to appropriate some of the value of a monopoly that is generated somewhere in the economy.

In this book, we emphasize the creation of the venture that capitalizes on technological changes and that will have a significant impact on a region, a nation, or the world. A new regulation or clever financial restructuring may afford the entrepreneur a new opportunity, but a radical or transforming innovation may provide the entrepreneur with an important opportunity to make a productive and significant contribution to the world as we know it.

## 1.5 The Technology Entrepreneur

The critical people at the center of all of these activities are entrepreneurs. The entrepreneur is a bold, imaginative deviator from established business methods and practices who constantly seeks the opportunity to commercialize new products, technologies, processes, and arrangements [Baumol, 2002]. Entrepreneurs thrive in response to challenges and look for unconventional solutions. They apply creativity, create visions, and build stories that explain their visions, and then act to be part of the solution. They forge new paths and risk failure, but persistently seek success. Entrepreneurs distinguish themselves through their ability to accumulate and manage knowledge, as well as their ability to mobilize resources to achieve a specified business or social goal [Kuemmerle, 2002].

Entrepreneurs engage in eight key activities, as described in Table 1.4. They identify and select opportunities that match their skills and interests they acquire and mobilize financial, physical and human resources; and they start and grow organizations, cognizant of their broader context.

In order to successfully pursue these activities, entrepreneurs should possess several important capabilities, as noted in Table 1.5. Entrepreneurs are opportunity driven and work to find a strategy that can reasonably be expected to bring that opportunity to fruitful success. They seek new means or methods and are willing to commit to solving a social or business problem that will result in success. Entrepreneurs work toward needing shorter time periods to decide on an appropriate strategy and seize opportunities. Entrepreneurs have a passion to build an enterprise that will solve an important problem. They seek ways to express themselves and validate their ideas. They are creative, internally motivated, and attracted to new, big ideas or opportunities.

Entrepreneurs exhibit robust confidence, sometimes bordering on overconfidence [Hmieleski and Baron, 2009]. Entrepreneurial innovators tend to exhibit

## TABLE 1.4  Eight key activities by entrepreneurs.

- Entrepreneurs initiate and operate a purposeful enterprise
- Entrepreneurs operate within the context and industrial environment at the time of initiation
- Entrepreneurs identify and screen timely opportunities
- Entrepreneurs accumulate and manage knowledge and technology
- Entrepreneurs mobilize resources—financial, physical, and human
- Entrepreneurs assess and mitigate uncertainty and risk associated with the initiation of the enterprise
- Entrepreneurs provide an innovative contribution or at least a contribution that encompasses novelty or originality
- Entrepreneurs enable and encourage a collaborative team of people who have the capabilities and knowledge necessary for success

## TABLE 1.5  Required capabilities of the entrepreneurial team.

- Has talent, knowledge, and experience within the industry where the opportunity occurs
- Seeks important opportunities with sizable challenges and valuable potential returns
- Able to select an opportunity in a short period: timely
- Creatively explores a process that results in the concept of a valuable solution for the problem or need
- Able to convert an opportunity in to a workable and marketable enterprise
- Wants to succeed: achievement-oriented
- Able to accommodate uncertainty and ambiguity
- Flexibly adapts to changing circumstances and competitors
- Seeks to evaluate and mitigate the risks of the venture
- Creates a vision of the venture to communicate the opportunity of staff and allies
- Attracts, trains, and retains talented, educated people capable of multidisciplinary insights
- Skilled at selling ideas and has a wide network of potential partners

high self-efficacy—the belief that they can organize and effectively execute actions to produce desired attainments [Markman et al., 2002]. They believe they possess the capabilities and insights required for the entrepreneurial task. Empirically, entrepreneurs with prior experience in the industry in which the new venture will be operating perform better, due not only to their related technical skill but also to their knowledge of marketing and the regulatory environment in a sector [Chatterji, 2009]. The best entrepreneurs combine both experience and talent [Eesley and Roberts, 2012].

Good entrepreneurs seek to be flexible so they can adapt to changing conditions and reduce the risks of the venture. They are resilient in the face of setbacks, able to multitask, and exercise well-developed problem-solving skills to overcome challenges. Table 1.6 lists some of the elements of this ability.

Finally, entrepreneurs create an overarching vision of the venture and use it to motivate employees, allies, and financiers. Perhaps the most important qualities or characteristics of an entrepreneur are the abilities to accomplish the necessary tasks, meet goals, and inspire others to help with these tasks. Successful

**TABLE 1.6 Elements of the ability to overcome a challenge.**

| | |
|---|---|
| ■ Able to deal with a series of tough issues | ■ Willing to work hard and not expect easy solutions |
| ■ Able to create solutions and work to perfect them | ■ Well-developed problem-solving skills |
| ■ Able to handle many tasks simultaneously | ■ Able to learn and acquire the skills needed for the tasks at hand |
| ■ Resilient in the face of setbacks | |

entrepreneurial teams attract, train, and retain intellectually brilliant and educated people capable of multidisciplinary insights [van Praag, 2006].

Members of the entrepreneurial team must, therefore, exhibit leadership qualities. **Leadership** is the ability to create change or transform organizations. Leadership within an organization enables the organization to adapt and change as circumstances require. A real measure of leadership is the ability to acquire needed new skills as the situation changes.

Entrepreneurs vary widely in their backgrounds. Recall the list of entrepreneurs in Table 1.1. The age of these people when they launched their enterprises ranges from 19 to 43. The median age of all technology-based company founders is 39 and many founders are much older [Wadha et al., 2008]. Entrepreneurship is a lifelong pursuit that is accessible to people of all ages. Entrepreneurs are also well educated. Ninety-two percent of technology entrepreneurs surveyed by the Kauffman Foundation hold a bachelor's degree, 31 percent hold a master's degree, and 10 percent hold a Ph.D. At the same time, however, institutions such as the Grameen Bank, which lends primarily to women in the third world so that they can start businesses, have opened up entrepreneurship as a possibility for a wide range of people.

In general, entrepreneurs should have most of the qualities listed in Table 1.5 in order to participate in a new venture. But, not everyone will have the same blend of capabilities. In order to strengthen, diversify, and complement an organization's skills, insights, resources, and connections, most entrepreneurs work as part of a team.

Moreover, entrepreneurship is an attitude and capability that diffuses beyond the founding team to all members of an organization. Most growing firms strive to infuse the culture of the entire company with the entrepreneurial spirit. For example, Thomas Edison created an enterprise that became General Electric; Steve Jobs and Steve Wozniak founded Apple Computer; and Azim Premji started Wipro Technologies. These entrepreneurs combined their knowledge of valuable new technologies with sound business practices to build important new enterprises that continued to maintain their entrepreneurial spirit for years after founding.

Members of an entrepreneurial team decide whether to act as entrepreneurs based on the seven factors listed in Table 1.7 [Gatewood, 2001]. Good entrepreneurs tend to seek independence, financial success, self-realization, validation of achievement, and innovation, while fulfilling leadership roles. At the same time, potential entrepreneurs evaluate the risk and work efforts associated with an

**TABLE 1.7 Factors people use to determine whether to act as entrepreneurs.**

| Positive factors or benefits | |
|---|---|
| ■ Independence: Freedom to adapt and use their own approach to work and flexibility of work, autonomy<br><br>■ Financial success: Income, financial security | ■ Self-realization: Recognition, achievement, status<br><br>■ Innovation: Creating something new<br><br>■ Roles: Fulfilling family tradition, acting as leader |
| **Negative factors** | |
| ■ Risk: Potential for loss of income and wealth | ■ Work effort and stress: Level of work effort required, long hours, constant anxiety |

opportunity and balance them with the benefits. Successful entrepreneurs are able to answer positively the five questions listed in Table 1.8 [Kuemmerle, 2002].

Context can have an important effect on whether or not someone becomes an entrepreneur [Sørenson, 2007]. For example, people whose colleagues are entrepreneurial are more likely to become entrepreneurs themselves [Stuart and Ding, 2006]. Similarly, younger and smaller organizations are more likely to spawn entrepreneurs [Dobrev and Barnett, 2005]. Work environments with an unfavorable innovation climate or a lack of technical-excellence incentives can positively influence the entrepreneurial intentions of skilled and ambitious employees through low job satisfaction [Lee et al., 2011]. Finally, environmental changes, such as an increase in the availability of venture capital financing, also affect the decision to become an entrepreneur [Hsu et al., 2007].

On an individual level, people act as self-employed entrepreneurs when that career path is felt to be better than employment by an existing firm. Consider the satisfaction (utility) derived from an employment arrangement. A utility function, U, is [Douglas and Shepherd, 1999]:

$$U = f(Y, I, W, R, O)$$

where $Y$ = income, $I$ = independence, $W$ = work effort, $R$ = risk, and $O$ = other working conditions. It may be assumed that income depends in turn

**TABLE 1.8 Five questions for the potential entrepreneur.**

| | |
|---|---|
| ■ Are you comfortable stretching the rules and questioning conventional wisdom?<br><br>■ Are you prepared to take on powerful competitors?<br><br>■ Do you have the perseverance to start small and grow slowly? | ■ Are you willing and able to shift strategies quickly?<br><br>■ Are you a good deal closer and decision maker? |

on ability. People will have an incentive to be entrepreneurs when the most satisfaction (utility) is obtained from the entrepreneurial activity. In other words, entrepreneurship pays off due to higher expected income and independence when reasonable levels of risk and work efforts are required.

For new entrepreneurial activities, the results of the venture are less known, and expected returns, independence, work effort, and risk can only be estimated. Potential entrepreneurs must be careful to do an honest assessment of their motivation and skills [Wasserman, 2012]. Regrettably, many entrepreneurs overweigh the benefits of independence and income, and underestimate the work effort required.

Based on the utility function above, we may postulate a utility index that we will call the Entrepreneurial Attractiveness (EA) index [Lévesque et al., 2002]. For each factor (Y, I, W, and R), we use a scale of 1 to 5 with $1 =$ low, $3 =$ medium, and $5 =$ high.

$$EA = (Y + I) - (W + R) \tag{1.2}$$

As a simple example, consider the straightforward alternatives for a successful marketing manager in the electronics industry. She can earn \$60,000 annually in her existing job (Y in equation 1.2). However, she values the independence of the new venture highly (I). The work effort for the new venture is estimated to be the same as for her current work (W). However, the risk is higher for the new independent venture (R). The potential entrepreneur estimates that she can obtain the same income over the next two years, although she will need a four-month period with a lower income at the start. The entrepreneur can compare the two options across these dimensions as shown in Table 1.9. In this case, over the first two years, the benefits of the new venture are $Y + I = 8$, and the costs of the venture are $W + R = 7$. The benefits of the existing job are equal to 5, and the costs are 6. Therefore, the EA for the new venture is $+1$ and the EA for the existing job is $-1$. The new opportunity looks more favorable due to this entrepreneur's desire forindependence, and it warrants in-depth analysis.

To mitigate risks, many entrepreneurs transition slowly, retaining their wage jobs while entering into self-employment [Folta et al., 2010]. Entrepreneurs must be careful, however, to dedicate the necessary effort to a new venture. Technology startups are not lifestyle businesses that entrepreneurs can pursue with limited hours [Ogle, 2012]. If an idea is not attractive enough to pursue wholeheartedly, then it may not be a good opportunity for that entrepreneur.

**TABLE 1.9  Summary of the entrepreneur's analysis of a new opportunity and the opportunity cost using a two-year period.**

| Factor | New venture | Existing job |
|---|---|---|
| Income over | \$120,000 | \$120,000 |
| two years (Y) | Y = 3 | Y = 3 |
| Independence (I) | I = 5 | I = 2 |
| Work effort (W) | W = 4 | W = 4 |
| Risk (R) | R = 3 | R = 2 |

In summary, entrepreneurs are multitalented individuals who leverage their capabilities and interests to pursue a particular opportunity, almost always with the help of a team.* The decision to pursue an entrepreneurial path and a particular opportunity is determined by weighing the benefits of independence and income against the work effort required and the risk of the venture. In Chapter 2, we learn how a potential entrepreneur can evaluate an idea to determine if it is an actual opportunity.

## 1.6 Spotlight on Facebook

Facebook is a social networking service that was founded in 2004. Harvard University student Mark Zuckerberg and several classmates started the service as an online student directory with photos and basic information. Zuckerberg was the lead founder and entrepreneur; he attracted other Harvard students to co-found and develop the website. In late 2004, Facebook expanded to other Ivy League schools, Stanford, and many other universities. That year, Facebook also moved its base of operations to California.

Zuckerberg displayed leadership skills, entrepreneurial commitment, and competence. He leveraged innovation and technology to expand his service. Zuckerberg's journey from unknown teenager to dotcom celebrity began in the summer of 2006, when he walked away from a billion-dollar Yahoo acquisition offer. He focused on his grand vision of recreating real-world relationships on the web. Facebook held an initial public offering of stock on May 17, 2012 at an initial price of $38 per share, valuing the company at $104 billion dollars.

## 1.7 Summary

Entrepreneurship is the process through which individuals and teams bring together the necessary resources to exploit opportunities and, in doing so, create wealth, social benefits, and prosperity.

The critical ideas of this chapter are:

- Entrepreneurs identify problems or needs and they execute upon solutions.
- Entrepreneurship is an engine of economic growth.
- Entrepreneurs use knowledge to create innovations and new firms.
- Positive entrepreneurship activity flows from a combination of entrepreneurial capital and intellectual capital that leads to productivity and prosperity.
- Entrepreneurs identify and develop opportunities, acquire resources, and start organizations.
- A person can learn to be an entrepreneur.

---

*Throughout this book, the word *entrepreneur* will refer to an individual or a team of individuals.

> **Principle 1**
> Entrepreneurs develop enterprises with the purpose of creating pros-
> perity and wealth for all participants—investors, customers, suppliers,
> employees, and themselves—using a combination of intellectual capital
> and entrepreneurial processes.

## Video Resources

Visit http://techventures.stanford.edu to view experts discussing content from
this chapter.

| | | |
|---|---|---|
| Entrepreneurial Skills Learned | Mark Zuckerberg | Facebook |
| Reasons Not to Be an Entrepreneur | Phil Libin | Evernote |
| You Gotta Grind | David Friedberg | The Climate Corporation |

## 1.8 Exercises

**1.1**   What is the difference between an idea and an opportunity? Why is this difference important to entrepreneurs?

**1.2**   Consider opportunities that have occurred to you over the past month and list them in a column. Then, describe your strong interests and passions, and list them in a second column. Finally, create a list of your capabilities in a third column. Is there a natural match of opportunity, interests, and capabilities? If so, does this opportunity appear to offer a good chance to build an enterprise? What would you need to do to make this opportunity an attractive chance to build an enterprise business?

**1.3**   Name an entrepreneur that you personally admire. Why do you consider this person to be an entrepreneur? What sets him or her apart from other business leaders? What path did this person take to entrepreneurship? What personal sacrifices or investments did this person make in the journey? What people were important to this person's success?

**1.4**   Name a successful entrepreneurial team you personally admire. How would you classify it in the context of the entrepreneur capabilities shown in Table 1.5? Do these elements of entrepreneurship apply to it?

**1.5**   Research the number of companies that either had an IPO (initial public offering) or have been acquired in the last five years. What industries were these companies in? Where is the number of IPOs vs. M&As (mergers and acquisitions) trend leading? What implications does this have on the number of new ventures being started?

**1.6**   Given an understanding of the waves of innovation throughout history (Figure 1.5), explore opportunities that are created in a wave after the peak. For example, how can an entrepreneur take advantage of a mature or declining market?

# VENTURE CHALLENGE

Select a high-potential opportunity that interests you and then use it for the venture challenge exercises at the end of each chapter.

1. Describe the opportunity that attracts you and why you think it is a new venture opportunity.

2. Describe the competencies and skills you and your team members possess.

3. What important stakeholders will you need to be successful?

4. Describe the passion and commitment you have for the opportunity.

5. Is this a good opportunity for you?

# CHAPTER 2

# Opportunities

*In the field of observation, chance only favors minds which are prepared.*
**Louis Pasteur**

**How can an entrepreneur identify and select a valuable opportunity?**

The identification and evaluation of opportunities is one of the entrepreneur's most important tasks. Good opportunities address important market needs. Examining social, technological, and economic trends can lead to the identification of emerging needs. Entrepreneurs seek to build new ventures and to act on a good opportunity when it matches their capabilities and interests, exists in a favorable context, exhibits the potential for sustainable long-term growth, and facilitates the acquisition of required resources. Such opportunities offer a reasonable chance of success and require the entrepreneur to make a difficult decision to act or not act. The choice of an opportunity and the decision to act is a critical juncture in the life of an entrepreneur. With the decision to act, the entrepreneur prepares a business summary for the venture that is used to test the new venture with potential investors, employees, and customers. Table 2.8 outlines a five-step process for evaluating opportunities. ■

## 2.1 Types of Opportunities

The first role of the entrepreneur—an individual or a group of people—is to identify and select an appropriate opportunity. An opportunity is a timely and favorable juncture of circumstances providing a good chance for a successful venture. Effective entrepreneurs often find that opportunity identification is a creative process that relates a need to the methods, means, or services that address or solve it—all within a reasonable period of time.

Opportunities can develop from market demand or from new technological possibilities. These opportunities are "demand pull" and "technology push" respectively [Di Stefano et al., 2012]. With demand-pull opportunities, an entrepreneur begins by assessing a need or problem that cries out for a solution. An example is the need for a pharmaceutical that can mitigate or cure the effects of AIDS. The founders of new industries capitalize on demand pull to create disruptive innovations that lead to new products that satisfy the demand.

As Vinod Khosla, the co-founder of Sun Microsystems and a well-known venture capitalist, is fond of saying, "Every problem is an opportunity." In turn, great opportunities are often disguised as difficult problems.

An effective tactic for identifying these problems is to focus on situations where a potential customer experiences significant "pain," which represents the extent of need for the solution to a problem: a customer who feels significant pain of need seeks a high-value solution. For example, in 2011, the World Health Organization found that 64 percent of anti-malarial drugs were counterfeit. False medications cause pain for the patients, who desire effective treatments, and for manufacturers, who desire to get their treatments to patients. Thus, an opportunity exists for companies to track raw materials, manufacturing processes, and product shipments in order to validate that a drug is genuine. One such venture is called TruTag [Economist, 2012b].

Successful new ventures are often initiated by people who have experienced significant painful problems. Sam Goldman, the founder of d.light, grew up in Mauritania, Pakistan, Peru, India, and Rwanda before becoming a Peace Corps volunteer in Benin. He then moved on to study biology and environmental studies in Canada before receiving his MBA from Stanford. While Goldman was living in Benin, his neighbor's son was badly burned by a kerosene lamp. This inspired him to create a new source of light, which could match kerosene lamps on price, but be safe for use around small children. d.light now creates extremely efficient LED lights that are 8 to 10 times brighter than a kerosene lamp and 50 percent more efficient than fluorescent lights.

Once an entrepreneur has identified a problem, he works to develop a solution. For example, Scott Cook, who founded Intuit, saw a problem experienced by individuals who wanted to easily and reliably keep their own home budget records, do their taxes, and pay the bills. Cook thought this problem could be solved through financial software that was so intuitive that most people could use it without resorting to the manual—thus, the name of the firm: Intuit (www.intuit.com). Entrepreneurs like Cook develop a solution by first asking, "How

**Demand Pull at ResMed**

Obstructive sleep apnea (OSA) was a widespread but underdiagnosed problem during the 1980s and early 1990s. OSA occurs when tissue at the back of the throat collapses during sleep, blocking the airway and preventing breathing. Oxygen levels drop in the bloodstream causing sharp fluctuations in heart rate and blood pressure. OSA is strongly correlated with other severe conditions—nearly half of all heart failure patients and 60 percent of type 2 diabetes patients suffer from OSA. It was estimated that 2 percent of the U.S. population suffered from OSA in some form. It was clearly a massive problem waiting for a solution.

ResMed was founded in Australia to combat this problem. The company created a novel device that pressurized the airway during sleep to prevent the airway from blocking. The device was fantastically successful and as recognition of OSA expanded during the early 1990s, ResMed took off. ResMed correctly identified a huge unsolved problem and provided a solution that fit into both the patient's life and the health insurers' plans. As a result, the company has been incredibly successful. It is now public on the New York Stock Exchange with revenues of almost $1.4 billion in 2012.

would an unconstrained person solve the problem?" Starting without constraints such as price and physical limits opens up many possibilities.

Once a good unconstrained solution appears attractive, it can often be rearranged to accommodate reasonable constraints [Nalebuff and Ayres, 2003]. Cook solved a big problem with an easy-to-use solution.

Technology-push opportunities might be said to begin with a solution, such as the one developed by Cook. The discovery of a new technology, such as HDTV in consumer electronics or stem cells in biotechnology, leads entrepreneurs to search for ways to apply it. For example, Sandra Lerner and Leonard Bosack formed Cisco Systems in 1984 to exploit the capabilities of a router to transmit and translate data to and from disparate computers [Bunnell, 2000]. By 2012, Cisco had revenues of over $46 billion.

The power of *serendipity*—making useful discoveries by accident—can also lead to good opportunities. Working in a microwave lab, Percy Spencer observed a chocolate bar melting by microwave power—thus, leading to the microwave oven. Clarence Birdseye was a fur trader in Canada when he noticed a phenomenon while ice fishing. At 50 degrees below zero, fish froze rock-hard almost instantly, yet when thawed, they were fresh and tender. After some experimentation, he learned that the key was the speed at which foods were frozen. That observation led to the flash freezing process that created a multibillion-dollar industry and made Birdseye a success.

Would-be entrepreneurs need to be careful, however, not to mistake a new technology for a solution in itself. Ultimately, customers want a need filled or a

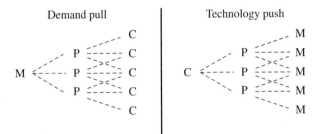

$M$ = Market need;   $P$ = Product (or service);   $C$ = Technological capability

**FIGURE 2.1** Demand Pull and Technology Push Opportunities
*Source:* Kingon, 2014.

problem solved. Usually, they do not care what technology is employed towards this goal. Thus, entrepreneurship is not about having a great technological idea, but rather about creating a new business that solves a problem.

Figure 2.1 summarizes and illustrates demand-pull and technology-push opportunities. In the demand-pull case, the entrepreneur begins with a market need. That need may be filled with several potential products, which may (or may not) have special technological capabilities. In the technology-push case, the entrepreneur begins with a technological capability, which is often a new technology or a new application of an existing technology. This capability needs to be combined with other capabilities to make a cohesive product (and, often, the capability can be applied to multiple products). In turn, the selected product may be applied to various market and customer needs. In both cases, however, the ultimate task of the entrepreneur is to match an important need with a good solution.

Table 2.1 further dissects different opportunities into nine categories, which we use to describe additional ways of identifying opportunities. The first category, and perhaps most common, is to increase the value of the product or service. This increased value can include improved performance, better quality or

**TABLE 2.1  Nine categories of opportunity.**

1.  Increasing the value of a product or service

2.  New applications of existing means or technologies

3.  Creating mass markets

4.  Customization for individuals

5.  Increasing reach

6.  Managing the supply chain

7.  Convergence of industries

8.  Process innovation

9.  Increasing the scale of the firm

experience, and improved accessibility or other values unique to the product. For example, Shokay is a for-profit social enterprise based in Tibet that manufacturers and distributes 100 percent yak down products including scarves and throws. The products are sourced from impoverished Tibetan yak herders with the broader mission of fostering economic development to remote areas of western China. Their products are luxurious, soft, and functional, and have wide appeal.

The second category seeks new applications of existing means or technologies. Credit cards with magnetic stripes were available in the 1960s, but a thoughtful innovator recognized the application of this technology to hotel door cards and created a wholly new application and industry.

The third category concentrates on creating mass markets for existing products. A good example is VMWare's development. Initially, VMWare focused on a product that enabled software testers to check the quality of code using virtual environments. Thus, VMWare enabled a single machine to operate like multiple machines. VMWare recognized that with modest enhancements, the company could expand its product offering to run critical systems called operation workloads. The market for these operational application environments is at least an order of magnitude larger than that for quality assurance and test environments. As a result, VMWare expanded its product to a $10 billion market from an initial market of several hundred million dollars.

Customization of products for individuals, category 4, affords a new opportunity for an existing product or technology. Examples of customization can be found in the streaming music business. Spotify lets users play songs from their own libraries as well as from Spotify's collection of millions of tracks. Users can access music via a desktop app, tablet, or smart phone. Spotify also integrates with social network sites, including Facebook.

Expanding geographic reach or online reach, category 5, allows a new venture to increase its number of customers. Founded in Scotland, Optos developed a novel eye exam technology and innovative pay-per-use business model. Backed by angel investors for many years, it carefully expanded its operations into the United States and Germany. It is now a viable public company listed on the London stock exchange.

Managing the supply chain, category 6, is a powerful force for improvement. Wal-Mart integrated inventory-information systems in each store with its broader distribution system to reap economic benefits from improved inventory management.

Convergence of industries, category 7, affords new opportunities by creating novel combinations of markets and technologies. For example, genetic engineering is the convergence of electron microscopy, micromanipulation, and supercomputing.

Business and manufacturing process innovations, category 8, are another source of opportunity. For example, FedEx and other airborne shipping systems changed the way in which individuals and organizations ship goods to one another.

Finally, the ninth category of opportunity is the increasing scale or consolidation of industry. Historically, the railroad industry provides a powerful example of consolidation in the United States. Consolidation of the railroads began by the turn of the twentieth century. Today, there are five major railroad companies, down from the thousands of companies in the late 1890s. More recent consolidation examples include the automobile manufacturing industry, cable and satellite TV broadcasters, and telecommunications carriers. Through mergers and acquisitions, an industry can be consolidated with attendant cost savings and value for the customer.

Any specific business opportunity may be portrayed in the three-dimensional cube of Figure 2.2. The entrepreneur identifies the customer, the required technology, and the application of this technology to create a solution. Different opportunities typically leverage different combinations of technologies, applications, and customers.

The entrepreneur's personal background is key to the process of opportunity identification. In fact, since prior knowledge, experience and motivation play a major role in opportunity identification, different people may not recognize or perceive the same opportunity [Gregoire and Shepherd, 2012]. The people whom an entrepreneur knows, the activities she pursues, and even the books and magazines she reads can shape her awareness of opportunities [Ozgen and Baron, 2007].

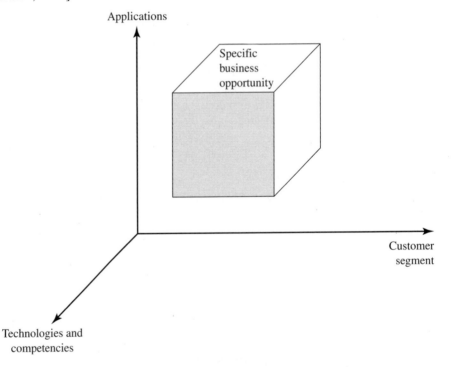

**FIGURE 2.2** Finding a specific business opportunity with a combination of customer segment, technology and competencies, and applications.

Good opportunities can also vary from person-to-person since it is the entrepreneur who adds value to the opportunity by creating a response. The opportunity, and even a general response to it, is not unique—many people may recognize it. Few of them, however, will possess the relevant passion and capabilities to solve the problem. For example, many people propose to exploit the new science of nanotechnology to solve various problems, but few of them will act. The true entrepreneur finds the best opportunity that matches his or her interests, skills, and knowledge—and acts to get it done. Thus, it is really the passion and capabilities that distinguish an entrepreneurial team.

Frank Slootman is a good example of an entrepreneur who found an opportunity that matched his interests, capabilities and passion. Slootman joined Data Domain, a company with 20 employees, no customers, and no revenue. He felt he could make a meaningful difference to an organization that was pursuing a large market if successful, under his guidance, it could become a highly valuable company. Slootman said that the problem Data Domain was solving – enabling disk economies for backup/recovery storage "spoke to him." He also felt going in that he could provide leadership to a group of very talented technical engineers. During the next six years he led the company to become the undisputed leader in its field, taking it public and through an eventual acquisition by EMC for $2.4 billion in 2009.

## 2.2 Market Engagement and Design Thinking

Entrepreneurs must engage with the market on a continual basis in order to identify and validate opportunities. Market research is the process of gathering information that can be used to refine an opportunity and to plot and execute upon a strategy. This information is critical to the new venture team; without it, a new venture may launch a product only to find out that the customer does not value it.

Customer development is the process of the discovery, validation, and creation of customers leading to company building, as shown in Figure 2.3 [Blank, 2012]. In the first step, the entrepreneur works to identify the customers, to determine whether the problem that the entrepreneur is solving is important to them, and to assess whether they value the proposed solution. In the second step,

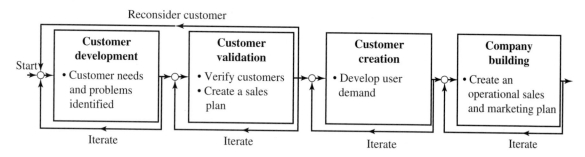

**FIGURE 2.3** The customer development process.
*Source:* Blank, 2013

the entrepreneur builds a repeatable sales roadmap based upon actual customer commitments. Iteration – the reconsideration and adjustment of the problem, product, and customer hypotheses – is especially common in these first two steps as the entrepreneur works to refine an idea. In the third step, customer creation, the entrepreneur works to build upon initial sales by creating end-user demand. Finally, the startup transitions from an informal learning and discovery organization into one with formal sales, marketing and business-development departments that build on the startup's success [Blank, 2013].

Primary data – that is, data collected for your specific proposed venture – are particularly important to the customer development process; there is no substitute for engaging with actual customers. One popular form of primary research uses the focus group, which is a small group of people from the target market. These people are brought together in a room to have a discussion about the problem or proposed product. This discussion can be led by someone from the venture team or by a professional moderator who may elicit more honest feedback due to her detachment from the venture. Table 2.2 outlines the general process for convening and running a focus group. The overall goal is to obtain honest feedback from a group that is representative of your intended market.

Entrepreneurs can also collect data from surveys and from customer interviews. One advantage of interviews over focus groups is that individual interviewees can avoid "groupthink" that leads participants to converge towards the same responses. Typically, an entrepreneur also can entice individuals to a one-on-one interview who may not participate in a broad focus group.

Table 2.3 offers several tips for conducting customer interviews. As with focus groups, effective interviews employ open-ended questions and encourage honest feedback. It is important for entrepreneurs to use customer engagement to elicit new insights and to challenge preconceptions – not simply to seek information that confirms current assumptions [Sanchez Garcia et al., 2011].

When using customer interviews or focus groups to obtain feedback on a proposed solution, it also can be effective to use the "a day in your life" format. This format takes customers through their day before and after the new product launch to expose the expected benefits of the new product. Another widely used approach

## TABLE 2.2  Steps to conduct a focus group.

1. Determine the purpose and goal of the focus group

2. Identify and attract the appropriate people to participate, based on the purpose, goal, and target market

3. Prepare opening questions to get the conversation started and key questions that are central to the purpose and goal of the meeting

4. Use simple, open-ended questions

5. Encourage critical feedback

6. Analyze the results by determining general patterns and looking for interesting outliers

*Source:* Krueger and Carey, 2008.

**TABLE 2.3  Tips for customer interviews.**

1. Know your goals and questions ahead of time

2. Identify and attract the appropriate people to participate, based on the purpose, goal, and target market

3. Ask open-ended questions

4. Encourage customers to be brutally honest and be prepared for critical feedback

5. Focus on listening, not talking

6. Do not attempt to steer or influence the interviewee

7. Ask follow-up questions

8. Confirm what you heard by paraphrasing it back

9. Ask for introductions to other customers

10. Write up your notes as quickly as possible

*Source:* Constable, 2011.

is to ask customers to compare two alternatives and to reflect on what they value about each of them. These comparisons can be quantified through conjoint analysis, where the researcher asks a respondent to make trade-off adjustments and decisions. Conjoint analysis provides a quantitative measure of the relative importance of one attribute as opposed to another. Although this method requires an investment of time and money in the research process, it may be worth it to avoid misreading customers' preferences [Aaker et al., 2001].

Focus groups, surveys, and customer interviews all face a challenge in that they have a limited ability to identify new problems and solutions. As Henry Ford, the automotive pioneer, once remarked, "If I'd asked my customers what they wanted, they'd have said a faster horse." Similarly, no potential customer ever asked for ATMs, personal computers, the Internet, or highly portable music players. Akio Marita, the founder of Sony, seized on this last opportunity. Marita directed his engineers to design a small portable radio and cassette player that would provide good audio quality and be attached to a person's head. Although no customer was asking for this product, the Sony Walkman became one of the twentieth century's most successful applications of miniaturized electronics and it fueled a market later addressed by the Apple iPod and other innovations.

Since customers cannot always verbalize their problems and needs, direct observations are another important means of gathering customer information. As human factors expert Leon Segal says, "Innovation begins with an eye" [Kelley, 2001]. By studying potential customers engaged in their everyday activities, entrepreneurs can see first-hand the behaviors, frustrations, bottlenecks and other cues that may indicate a need or problem – and, therefore, a potential opportunity. For example, the healthcare provider Kaiser Permanente had a project team conduct observations at four hospitals in order to improve both patient and medical-practitioner experiences. The project team quickly focused on nursing-staff shift changes. They found that nurses spent a great deal of time trying to convey information to the new shift and that much of the most important information was still missed or miscommunicated since there was no easy or standardized means of sharing. These observation-based insights enabled Kaiser Permanente to introduce a new system that improved patient care and nurses' job satisfaction [Brown, 2008].

There are several ways that entrepreneurs can act as effective observers. First, rather than rely on expertise to guide observations, entrepreneurs may find that naivety is a gift. For example, Jane Goodall attributes her groundbreaking research on chimpanzees to the fact that she had little prior knowledge about chimpanzee behavior. She, therefore, was not blinded to novel findings by any preconceptions [Sutton, 2002]. Second, when selecting people to observe, it can be particularly effective to study "extreme" users – such as children – and "rule breakers" who face novel constraints or who apply products in new ways. Finally, observations are most effective when the observers themselves come from multiple backgrounds and perspectives. Diversity enables a team of entrepreneurs to imagine the world from multiple perspectives in order to identify meaningful problems and novel solutions [Kelley, 2001].

The customer development process and the product development process proceed in parallel. Thus, as entrepreneurs gather data from the market, they should also work to create and refine product mockups and prototypes. Eric Ries and Steve Blank are the leading supporters of the idea of a "**lean startup.**" Ries coined the term "lean startup," and Blank created a course called "The Lean Launchpad" that teaches entrepreneurs many of the concepts initially discussed by Ries [Ries, 2011]. When building lean startups, entrepreneurs focus on learning what customers may be interested in, and proceed to build a "minimum viable product" in order to see if there is demand for their initial product offering. Thus, rather than waiting until the product is "perfect," entrepreneurs are advised to start selling early [Onyemah, 2013]. Additional product features can then be added over time. Startups that utilize the lean approach are typically funded with smaller amounts of initial capital and seek incremental capital once the founders gain confidence that there is a market and have proven that they can build viable products. These startups also make heavy use of crowdsourcing, outsourcing, cloud computing, and Software as a Service (SaaS) in order to limit their cash outlays [McQuivey, 2013].

Sahl Khan followed this approach when launching the Khan Academy, a virtual school providing software and video tutorials in math, science and other subjects. Kahn created almost all of the company's initial videos himself, often creating a video in just one or two days. His approach was to get videos out on the web quickly to see if they gained traction, thus testing the market and gathering feedback. Khan also runs his service on outsourced hardware. By following the lean approach, Khan proved his business model on very limited initial capital. By 2013, Khan Academy had over five million unique visitors each month and over one million page views per day. The company is well on its way to revolutionizing the way education is taught and consumed.

# 2.3 Types and Sources of Innovation

Opportunities often reside at the nexus of a market need and a technological capability. Thus, entrepreneurs should simultaneously engage with the market and with the process of technological innovation. **Innovation** is based on teamwork and creativity, and is defined as invention that has produced economic value in the marketplace. Innovations can be new products, new processes, new services, and new ways of doing business.

There are several different types of innovation, as illustrated in Figure 2.4. **Incremental innovation** is characterized by faster, better, and/or cheaper versions of existing products. Thus, entrepreneurs engaged in incremental innovation take an existing idea and creatively expand on it. To be successful, the incremental innovator must understand specific customer needs that are unmet by current offerings. For example, portable, battery-driven radios have been used since the 1950s. But, Trevor Bayles saw an opportunity to bring information to remote Africa by creating a wind-up spring- and dynamo-powered radio. Twenty-five seconds of winding gives the user one hour of listening. Bay Gen in Cape Town, South Africa, now manufactures more than 60,000 of these radios a month [Handy, 1999].

Like incremental innovation, architectural innovation leaves core design concepts untouched. But, **architectural innovation** changes the way in which components of a product are linked together. Thus, the components remain unchanged, but the architecture of module connection is the innovation. (The overall architecture of the product describes how the components will work together.) The essence of an architectural innovation is the reconfiguration of an established system to link together existing components in a new way [Henderson and Clark, 1990]. By contrast, **modular innovation** is focused on the innovation of new components and modules. But, it does not disrupt the linkages between modules.

|  | **Basic design concepts** | |
| --- | --- | --- |
|  | Reinforced | Overturned |
| Unchanged | Incremental innovation ("faster, better, cheaper") | Component or modular innovation |
| **Linkages between modules** Changed | Architectural innovation | Radical or disruptive innovation ("brave new world") |

**FIGURE 2.4** Four types of innovation.

Finally, **radical innovation** or **disruptive innovation** uses new modules and new architecture to create new products. The Internet is an example of a network system with new modules and new architecture – a radical or disruptive innovation. Disruptive innovation transforms the relationship between customer and supplier, restructures markets, displaces current products, and often creates new product categories [Leifer et al., 2000]. Disruptive products also introduce a new value proposition [Christensen et al., 2004]. For example, e-mail is an application on the Internet that is a disruptive application (often called a "killer app") and is one reason that the Internet so widely used.

Salesforce.com introduced a disruptive application for sales force activity tracking. Salesforce.com started selling sales force automation on a Software as a Service (SaaS) basis, in which customers pay a monthly licensing fee. This arrangement contrasts dramatically with the way that the incumbent leaders in sales force automation, Oracle and SAP, were selling software: Oracle and SAP would charge a one-time license fee that could run into hundreds of thousands of dollars, and would additionally charge yearly maintenance fees. Furthermore, they made customers responsible for running and updating their own software. By contrast, Salesforce.com would run and maintain the customer's software. New customers implementing sales force automation were much more likely to try to implement their solution with lower up-front costs.

An emerging disruptive innovation is three-dimensional (3D) printing. 3D printers enable the making of parts and products using a computer-driven, additive process, one layer at a time. 3D printing builds plastic and metal parts directly from CAD drawings that have been cross-sectioned into thousands of layers. It provides a faster and less costly alternative to machining (cutting, turning, grinding and drilling solid materials). Used for making prototypes as well as final products, 3D printing evolved from the "rapid prototyping" industry, pioneered by Chuck Hull of 3D Systems in the mid-1980s. Time will tell the extent of 3D printing's impact, but, for now, dozens to hundreds of startups and many bigger companies are betting their time and money that it will be the next big thing [*Economist*, 2012a].

Other good examples of disruptive innovations include the invention of the "800" number toll-free telephone call by AT&T and the development of the CT scanner for medical imaging, which combined X-ray technology with computer technology [Adner and Levinthal, 2002]. These types of disruptive applications bring significant value to a product and cause an industry to grow exponentially. They also are associated with improved odds of success for an entrepreneur. New firms that attempt to make an incrementally-better product and sell it to the same customers as existing firms experience only a 6-percent success rate; new firms that introduce disruptive innovations experience a 33-percent success rate [Christensen, 2002].

**Sources of innovation** include existing companies, research laboratories and universities, independent inventors and "user innovators," and open technical communities. Many companies are formed when employees in an existing enterprise have ideas for a new product and they decide that the best way to pursue it is as an independent venture. Of course, employees need to be careful about legal restrictions, which may restrain competition or require a license. Often, however, employees can take existing market insights and even technical knowledge and apply it to a new venture.

Universities are another important source of innovation. Professors, graduate students and other university researchers conduct an enormous amount of cutting-edge research. Since academic research typically is not driven by direct market needs, however, a major challenge in the commercialization of university breakthroughs lies in the need to move this research from lab prototypes and concepts into complete working products that can be manufactured reliably at a reasonable cost [Jensen and Thursby, 2001]. To get the most benefit from a relationship with a university, a new venture should take a long-term view and imagine a partnership focused on both technical and strategic issues. When companies take a transactional approach to the relationship, attempting to pick technologies, sign a contract, and quickly commercialize them, they are likely to fail [Wright, 2008]. By contrast, when an inventor stays involved with product development as it moves from the university to a start-up, the chances of success increase dramatically [Thursby and Thursby, 2004]. University-sourced innovations also present special legal challenges, which we will discuss in Chapter 11.

Many innovations come from end-users themselves [von Hippel, 2006]. For example, snowboarding, skateboarding, and windsurfing trace their origins to users who experimented with new combinations of equipment [Shah, 2003]. Similarly, end-users have developed innovations ranging from library information systems to baby strollers to chemical processing techniques [Shah and Tripsas, 2007]. Because users may understand their own needs and the limitations of existing products, they are uniquely positioned to innovate. Entrepreneurs are wise to engage with user-innovators to understand emerging technologies [Chatterji and Fabrizio, 2012].

Many user-innovators also participate in open technical communities [Baldwin and von Hippel, 2011]. These communities may be autonomous or sponsored and their members may include both individuals and organizations. Open technical communities play a crucial role in helping new enterprises to develop and deploy innovations [West and O'Mahony, 2008]. Firms benefit from participating in open communities by gathering information on potential alliances, identifying new opportunities, and sharing work and risk. Furthermore, such communities offer the benefit of transparency of developments and ready accessibility to shared knowledge.

An **open source innovation** community may be defined as a collection of many firms and individuals collaborating to develop and deploy an innovation. These communities share a common goal and an agreed-to governance system. Open source software communities are important examples of open technical communities. Individual members of these communities share common goals, but not a common employer. Examples are the Mozilla and Hadoop communities.

Effective open source organizations enable new ventures to build on the ideas of others to create new innovations [Murray and O'Mahony, 2008]. Cumulative innovation is enabled by ready access, disclosure, and incentives associated with community activities. The sharing process enables knowledge to be reused, recombined, and accumulated, often in a modular format that all members can readily access and use.

One example of a good open source effort is the Collaborative Software Initiative (CSI), based in Portland, Oregon. Their enterprise software solutions are based on the efforts of contributors worldwide and are potentially more effective than individual or independent development. Other examples include Wikipedia, a free online encyclopedia that anyone can edit, and YouTube, which contains an ever-expanding video library based upon the contributions of users [Tapscott and Williams, 2008].

## 2.4 Trends and Convergence

Both market needs and technology developments tend to align with broader trends. Moreover, the convergence of different markets and technologies can point to new opportunities. For example, just 30 years ago both online shopping and the use of mobile phones to maintain social networks were far-off dreams. Today, hundreds of millions of people shop online and mobile phones are widespread on every continent.

One of the most important trends lies in the globalization of business. Specifically, the Internet and overnight shipping have negated time and distance. This trend enables entrepreneurs to reach vastly larger markets and to build businesses that bring together a far more diverse set of resources. At the same time, of course, entrepreneurs face new competitors around the globe.

Growing environmental consciousness and associated challenges with climate change and pollution are other trends that shape the opportunity landscape. Consumers increasingly demand products that minimize negative environmental impacts. In turn, the redesign of products for sustainability presents enormous opportunities [Russo, 2010].

Social and cultural trends offer many examples of opportunity. Table 2.4 lists several social and cultural trends. One of the biggest current trends in America is the aging of the baby-boom generation—those born between 1946 and 1973. During those years, 107.5 million Americans were born, making up

**TABLE 2.4  Social and cultural trends that will create opportunities.**

| | |
|---|---|
| ■ Aging of the baby-boom generation | ■ Changing role of religious organizations |
| ■ Increasing diversity of the people of the United States  (e.g., Latino population) | ■ Changing role of women in society |
| ■ Two-working-parent families | ■ Pervasive influence of media and constant connectivity |
| ■ Rising middle class of developing nations | ■ Broadening access to education |

50 percent of everyone alive in 1973 [Hoover, 2001]. Those born in the peak-birth year of 1961 will be 55 in 2016 and acting as consumers of goods and services such as new homes, travel, healthcare and retirement plans. Other critical trends include the changing role of women in many societies and broadening access to education, including online education.

Entrepreneurs can sometimes study a trend to see how it might apply in other settings. Peer-to-peer services are one example. Airbnb is best known for enabling one person to work with another person to rent a room rather than staying in a more expensive hotel. That same model can be applied to many other services, such as car rentals and money transfers [de Boer, 2012]. For example, Lending Club and Prosper are establishing person-to-person lending, eliminating the bank as a middleman [Landes, 2012].

Opportunities often lie at the intersection of social and technological change. For example, new online education businesses are positioned at the intersection of growing interest in education and lifelong learning, on one hand, and increasing penetration of high-speed Internet and mobile Internet devices on the other hand. As another example, new business are forming at the intersection of Unmanned Aerial Vehicles (UAVs) and regulatory changes that are permitting these UAVs to access U.S. airspace.

One of the most promising areas of science and engineering is based on several breakthroughs that enable the manipulation of matter at the molecular level. Mass production of products with these molecular adjustments now offers a world of possibilities. Nanotechnology will make materials lighter, more durable, and more stain resistant, and will enable the creation of products with wholly new capabilities. (One nanometer is one-billionth of a meter.)

With the need for security and safety of personal information, the emergence of personal identification cards, or smart cards, may be the next trend in America. A smart card is a plastic card incorporating an integrated circuit chip and memory that stores and transfers data such as personal data and identification information for example, finger or palm print or facial scans. These cards have been adopted in several European and Asian countries and could spread worldwide. One form of smart card, the Octopus Card, is used in Hong Kong to pay for everything from subway rides to groceries. Another important use of smart cards would be a common approach for driver's licenses and personal information. Table 2.5 lists several other technology-related trends and opportunities.

Opportunities also emerge as the boundaries between many once-distinct businesses, from agribusiness and chemicals to energy and computing, continue to blur. The **convergence** of technologies or industries is the coming together or merging of several technologies or industries thought to be different or separate. Often, opportunities can be found in creative combinations that build on complementary technologies.

One example of industry convergence is that of computing and communications, which merged into the field of networks. Another example is convergence of a handheld computer and a mobile phone, now evident in Apple's iPhone and Samsung's Galaxy.

## TABLE 2.5 Trends and opportunities.

- Life science: Genetic engineering, genomics, biometrics
- Information technology: Internet, wireless device, cloud computing
- Food preservation: Improved distribution of food
- Video gaming: Learning, entertainment
- Speech recognition: Interface between computers and people
- Security devices and systems: Identification devices, baggage checkers, protective clothes
- Nanotechnology: Devices 100 nanometers or less for drug delivery, biosensors
- Renewable energy: Solar cells and wind turbines
- Fuel cells: Electrochemical conversion of hydrogen or hydrocarbon fuels into electric current
- Superconductivity: Energy savings on utility power lines
- Designer enzymes: Protein catalysts that accelerate chemical reactions in living cells for consumers and health products
- Cell phones: Communications and computing
- Software security: Blocking unsolicited e-mail (spam), preventing "phishing"
- Robots: Teams of small coordinated robots for monitoring and safety

Satellite imaging and data and the handheld computer converged into global positioning system (GPS) devices, which are widely used, inexpensive devices that address the need for accurate locational data. Scanners, computers, and security systems converged into self-checkout systems for shoppers. Gene chips use semiconductor technology to speed up biochemistry lab analysis. Finally, medical and robotic technologies have converged to facilitate highly-precise minimally-invasive robotic surgery. Intuitive Surgical's da Vinci medical robot has four arms and flexible wrists mounted with tools and cameras that can be controlled by a surgeon. The robots are being used for prostate surgeries, hysterectomies, and even more complicated surgeries like heart valve repair. As these examples highlight, effective entrepreneurs carefully examine social and technological trends. They strive to identify novel combinations that may point to significant opportunities.

## 2.5 Opportunity Evaluation

Alongside the identification of opportunities, entrepreneurs should work to evaluate these opportunities. In fact, a critical task of the entrepreneur is to distinguish between an idea and a true opportunity. Good opportunities have the potential to solve important and timely problems. As shown in Table 2.6, attractive opportunities also have the potential to be profitable and can be pursued under a favorable regulatory and industry context.

Choosing the right opportunity is both difficult and important. This choice is analogous to the selection of an equity investment in a company: Entrepreneurs will invest time, effort, and money in the venture they choose in a manner similar to how people invest in the stock of a company. Some sound investment principles that can be used for selecting opportunities are listed in Table 2.7.

## TABLE 2.6  Five characteristics of an attractive opportunity.

- Timely—a current need or problem
- Solvable—a problem that can be solved in the near future with accessible resource
- Important—the customer deems the problem or need important
- Profitable—the customer will pay for the solution and allow the enterprise to profit
- Context—a favorable regulatory and industry situation

## TABLE 2.7  Guiding principles for selecting good opportunities.

- Only one or two very good opportunities are needed in a lifetime.
- Invest less time, money, and effort in the venture than it will be worth in one or two years. Calculate the probability of a large return in four years.
- Do not count on making a high-priced sale of your firm to the public or another company.
- Carry out a solid analysis of the current and expected conditions of the industry where the opportunity resides.
- If the opportunity is selected and turns out unfavorably, can you exit with minor losses?
- Does this opportunity provide a potential for a long-term success, or is it a fad? Go to where the potential future gains are significant.
- Can the management team execute the strategy selected for this opportunity?
- Will the customer enable your firm to profit from this venture?

The entrepreneur finds and thoroughly analyzes the best opportunities, since for many people, only one or two opportunities are needed to make a good life of entrepreneurial activity. One goal is to invest in a firm for which you pay less than it is worth; this provides some cushion for unforeseen challenges. Also, entrepreneurs try to find an opportunity with solid long-term potential in an industry they understand. They put together a good management team that can execute the strategy for this opportunity. And, they ensure that the customer will allow their firm to profit from the venture. Thus, they avoid industries that sell commodities where price is the only differentiation unless they have a new, innovative business process that enables their firm to be the low-cost provider.

Tom Stemberg, the founder of Staples, conceived the idea of a supermarket store for office supplies in the mid-1980s. He didn't like the politics of big companies and sought independence. He started with a single store in Brighton, Massachusetts, and built 1,500 outlets. He carried out a complete analysis of the opportunity and determined it was a $100 billion market growing at 15 percent per year with large profit margins.

The review of opportunities will always include the evaluation of alternatives. The **opportunity cost** of an action is the value (cost) of the forgone alternative action. Selecting one opportunity will involve rejecting others. Chapter 1 discussed some of the considerations that people should use when deciding whether to become an entrepreneur by pursuing a specific opportunity. A critical part of this decision hinges on the quality of the opportunity in terms of a market assessment, feasibility of implementation, and differentiation of the product.

## TABLE 2.8  Basic five-step process of evaluating an opportunity.

1. **Capabilities:** Is the venture opportunity consistent with the capabilities, knowledge, and experience of the team members?

2. **Novelty:** Does the product or service have significant novel, proprietary, or differentiating qualities? Does it create significant value for the customer—enough so that the customer wants the product and will pay a premium for it?

3. **Resources:** Can the venture team attract the necessary financial, physical, and human resources consistent with the magnitude of the venture?

4. **Return:** Can the product be produced at a cost so that a profit can be obtained? Is the expected return of the venture consistent with the risk of the venture?

5. **Commitment:** Do the entrepreneurial team members feel compelled to commit to this venture? Are they passionate about the venture?

There is an inherent tension in entrepreneurship in that much of this analysis requires additional information, yet a comprehensive analytical approach to evaluation of the opportunity does not suit most start-ups; entrepreneurs often lack the time and money to fully engage with all potential customers, analyze substitutes, reconstruct competitors' cost structures, and project alternative planning scenarios.

Most entrepreneurial teams instead follow a basic five-step process, as outlined in Table 2.8. The goal is to assess these five characteristics – capabilities, novelty, resources, return, and commitment – to quickly weed out unpromising ventures and thus conserve energy and time for the promising ones. The appropriate analytical effort and the issues that are most worthy of research and analysis depend on the characteristics of each venture. For example, the exploration process should be short for potential ventures with low degrees of novelty [Choi et al., 2008]. In general, however, an entrepreneur works through the five steps and eliminates the opportunities that do not pass muster. Those that do pass a quick review are worth looking into further.

### The iPod Opportunity

In the late 1990s, many people had been listening to music stored on their computers that was obtained through the Napster file-sharing service. The challenge was to design and sell a portable music storage and player device. Tony Fadell, who had worked at General Magic, started his own company, Fuse, to design consumer electronic products. He tried to secure financing for the design of a portable music player. Without financing, he went to Apple in February 2001 as a contractor and then in April 2002 joined Apple as an employee to lead the iPod project. Fadell and Apple together recognized that the iPod opportunity possessed all the characteristics of Table 2.7. The capabilities, resources, and commitment of Apple and Fadell enabled them to build a device, the iPod, that was truly novel and that could offer a significant return.

The entrepreneur has to live with critical uncertainties, such as the relative competence of rivals or the preferences of customers, which are not easy to analyze. Who could have forecast, for example, that IBM would turn to Microsoft for an operating system for its personal computer, allow Microsoft to retain the rights to this operating system, and thus enable Microsoft to gain monopolistic dominance of the operating system marketplace? Entering a race requires faith in one's ability to finish ahead of whoever else might participate.

When evaluating an opportunity, the entrepreneur considers whether it fits or matches the contextual conditions, the team's capabilities and characteristics, and the team's ability to secure the necessary resources to initiate a new venture based on the opportunity. Figure 2.5 shows a diagram of fit or congruence that can be used to review an opportunity. A big diamond with high grades of fit are best.

Consider an opportunity that has existed for over 100 years—the electric automobile. We will assume that a capable set of engineers is available and the entrepreneurial team has the attitudes and capabilities required. However, the team is insecure about the risky nature of the venture, given the numerous failures over the past century. We will rate the entrepreneurial team at 75 percent on the team scale. The characteristics of the context are very mixed since regulations and support for electric cars are continually changing as potential customers and government organizations adjust their assessment of the benefits and costs of these vehicles. We will rate this opportunity as only 50 percent on the context scale. Next we turn to the opportunity, which is challenged by costs, limited life batteries, and short ranges before a recharge is required. The characteristics of the opportunity call for a rating of 75 percent on the opportunity scale. Given these ratings, most teams would be severely limited in their ability to secure the tens of millions of dollars required to launch this venture. Thus, we rate it only 50 percent on the resource scale. Clearly, this opportunity is a challenging one. At the same time, however, Tesla and other electric car companies have demonstrated that the right team backed by significant resources can make the electric automobile opportunity much more attractive.

Another way of envisioning the concept of a fit with an opportunity is shown graphically in Figure 2.6. Both markets and industries must be examined on the macrolevel and the microlevel. Moreover, the team must be evaluated across

---

**A Big Opportunity in Television**

What is the next big opportunity in television? Flat-panel television has been one of the compelling trends in consumer technology. A switch from the bulky cathode-ray tube to flat panel displays is under way in the multi billion unit global market. Opportunities exist in the market for glass and the devices and chips for digital light processing. Another opportunity is making the sets and selling them. As nations switch to high-definition TV (HDTV), the market for flat panel displays has grown significantly. Which of these opportunities pass the evaluation process of Table 2.8?

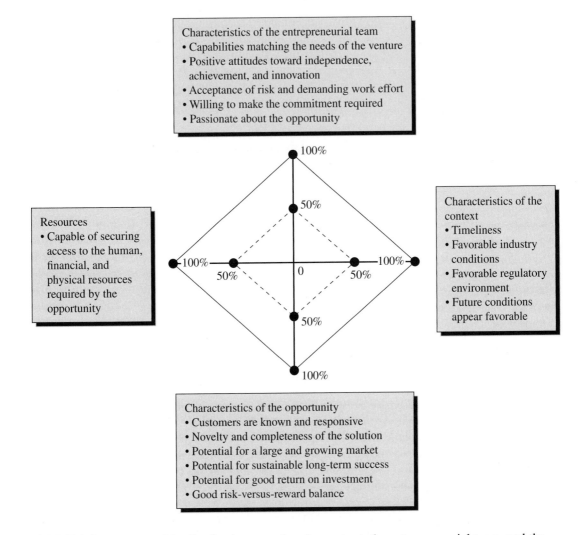

**FIGURE 2.5** Diagram of the fit of an opportunity, the context, the entrepreneurial team, and the resources required. Rate each factor on a scale of 0 to 100 percent.

multiple dimensions. An ideal opportunity lies where the market and industry are attractive, customer benefits are compelling, the start-up's advantage is sustainable, and the team can deliver the results it seeks [Mullins, 2006].

After evaluating an opportunity by using the factors in Table 2.8, the entrepreneurs should decide whether to act. With the knowledge generated by using the five-step process in Table 2.8, the entrepreneurs will tend to act on their estimate of the potential benefits and gains, B, while accounting for the total costs of the venture, C. Within the total cost accounting, there will need to be a recognition of their security needs and loss aversion. An individual will tend to act if the

**FIGURE 2.6** The seven domains of attractive opportunities.
*Source:* Mullins, 2006.

ratio B/C is greater than 1. The lucrative opportunity (high benefits and low losses) will tend to cause higher intention to act [McMullen and Shepherd, 2002]. If one acts and it is a false choice, the cost of that choice is important. Opportunities that can be attempted with low initial financial and time commitment costs may offer the chance for lucrative returns at a low initial cost.

The matrix in Figure 2.7 shows the decision to act or not act. Then, the actual resulting quality of the opportunity is shown (this can be determined only after the decision). Life is about choices, and the best case is when we choose to act and it turns out we are right!

**FIGURE 2.7** Decision matrix.

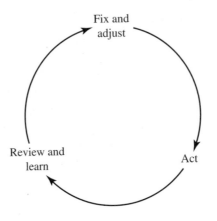

**FIGURE 2.8** Act-learn-fix cycle
of building a new venture.

The entrepreneur attempts to make a rational decision based on (1) his or
her current psychological and financial assets, and (2) the possible consequences
of the choice [Hastie and Dawes, 2001]. The decision challenge is the task of
turning incomplete knowledge of an opportunity into an action consistent with
that knowledge. Competitive advantage comes from actually doing something
that others cannot do; analyses and reports cannot substitute for action, and
reworking a plan is no substitute for acting to get things done. In the end, an
opportunity can be evaluated only so much [Pfeffer and Sutton, 2000]. Ambiguity remains, and the entrepreneur needs to act on or reject the opportunity. Fear
of failure may overwhelm all but the best opportunities.

Ultimately, identifying a good opportunity is less about making an initial
selection and more about gathering data and refining the idea accordingly.
Thus, while it is important to estimate fit, as in Figure 2.5, it is critical to
act on the opportunity, trying it out in the marketplace of ideas and investors. Data gathered and lessons learned from these initial attempts can lead to
refinement of the opportunity. Tom Peters and Robert Waterman [1982] called
this approach "ready, fire, aim." Banishing fear of failure and learning from
a series of small failures can lead to a good new venture. The act-review-fix
cycle, as shown in Figure 2.8, summarizes the critical ability to act, review,
and learn from the results, and then, fix and adjust the business scheme as
required. As John Stuart Mill stated: "There are many truths of which the full
meaning cannot be realized until personal experience has brought it home."

## 2.6 Spotlight on Solazyme

Solazyme Inc. is a publicly held alternative energy company specializing in the
production of algae-based fuel. Founded in March 2003, Solazyme developed a
means of converting low-cost plant-based sugars into high-value oils. The South

San Francisco firm's initial focus was the manufacture of transportation oil. By 2010, Solazyme expanded to making oil-based nutrition and cosmetics products as well.

Solazyme combines features of technology-push and demand-pull opportunities. The company has taken advantage of the convergence of new production technologies, extreme fluctuations in the supply and price of crude oil, and trends in energy and the environment. They emphasize a customer-driven approach to technology and product development. In May 2011, Solazyme made an initial public offering that raised $198 million. Solazyme is leveraging a strong team and a novel product, in a favorable context with significant resources. Thus, the opportunity appears promising.

## 2.7 Summary

The entrepreneur identifies numerous problems and needs that may point to good opportunities that can be made into great companies. However, he or she searches for the one that best fits the capabilities of the team, the characteristics of the business context, the characteristics of the opportunity, and the team's capability to secure the necessary resources. Then, the entrepreneur decides whether to act or not act on that best-fit opportunity.

The important ideas of the chapter are:

■ Great opportunities are often disguised as problems that are difficult to describe.
■ Opportunities can be tied to market demands and new technological capabilities.
■ New technologies can arise from existing companies, universities, users, and open source communities.
■ Social and technological trends shape opportunities.
■ Entrepreneurs must engage with the market on a continual basis in order to identify and validate opportunities.
■ Attractive opportunities are timely, solvable, important, profitable, and they exist in a favorable context.
■ The entrepreneurial team should cumulatively possess all the necessary capabilities.
■ Entrepreneurs should, if possible, act on favorable opportunities in a timely manner.

---

**Principle 2**
The capable entrepreneur knows how to identify, select, describe, and communicate an opportunity that has good potential to become a successful venture.

## Video Resources

Visit http://techventures.stanford.edu to view experts discussing content from this chapter.

| | | |
|---|---|---|
| Disruptive Technologies | John Doerr | KPCB |
| Pressure Points Around Opportunity | Brad Feld | Foundry Group |
| Any Big Problem is an Opportunity | Vinod Khosla | Khosla Ventures |

## 2.8 Exercises

**2.1**  One approach to classifying market entry is by (a) creating a new market, (b) attacking an existing market, or (c) resegmenting an existing market. Using Table 2.1, indicate how each of these categories of opportunities would be applicable to these market-entry approaches.

**2.2**  What were some of the key customer, technology, and market trends that drove entrepreneurship during the last decade? What factors do you predict will drive entrepreneurial challenges in the next decade?

**2.3**  The next big wave of innovation may be the convergence of bio-, info-, and nanotechnologies. Each holds promise in its own right, but together in combination, they could give rise to many important products. Describe one opportunity motivated by the convergence of these new areas, and develop a story about the opportunity.

**2.4**  Some imagine that within a few years it will be possible, through the use of stem cells, to create new cells and eventually new organs to replace those that fail. Summarize the potential opportunity for stem cell enterprises. How would you begin to estimate the size of this opportunity? Develop a story depicting the opportunity.

**2.5**  As energy costs rise and the impact on the environment becomes clearer, clean tech has become an area of significant new investment. Quantify the trends driving this renewed investment interest. How would you evaluate and market size the clean tech opportunity?

**2.6**  Consider a software application you use regularly. What task(s) does it improve or enable? Suggest three ways the application could be improved. Would any of these improvements be considered an opportunity for a new venture? Why or why not?

**2.7**  Global sales of radio frequency identification tags (RFID) and related equipment have been forecasted to explode multiple times in the last decade. Describe the problems solved by RFID and the opportunities presented. What have been the barriers to commercialization of this technology? What types of opportunities will be created when RFID tags are widely adopted in products?

**2.8**    The trend of performance of two electronic technologies is given in Figure 2.9. Determine the performance trend of another technology. Prepare a chart of its performance over time.

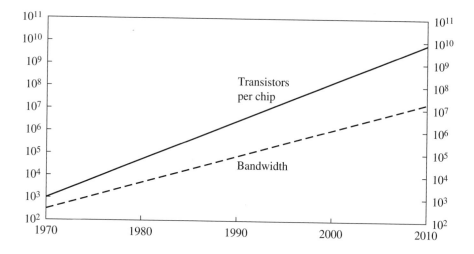

**FIGURE 2.9** Technology trends: (a) transistors per chip; (b) bandwidth per household (bits/second).

*Source:* Dorf, 2004.

## VENTURE CHALLENGE

Consider the opportunity that you identified at the end of Chapter 1.

1.    Using Table 2.1, attempt to categorize the type of opportunity you have selected.

2.    How could you develop your customer at the same time as developing your product or service as shown in Figure 2.3?

3.    Evaluate your opportunity using Table 2.7's principles and Table 2.8's process. Sketch a diagram using Figure 2.5 to illustrate your results.

# CHAPTER 3

# Vision and the Business Model

*Success in any enterprise requires the right product, methods, and workers, and each must complement the others.*
**Joseph Burger**

**How do successful entrepreneurs create a compelling business design for their new ventures?**

A new business should develop a clear vision or long-term destiny and a mission statement that captures its goals. Any new business is defined by the wants or needs that customers satisfy when they buy a product or service. To create a theory of a new business, the entrepreneur must cogently and clearly describe the customers and their needs and how the new venture will satisfy those needs. A business model describes the relationship between a venture's customers, value proposition, differentiation, scope, organization design, and profit model. Figure 3.1 outlines the steps towards the creation of a business model. ∎

## 3.1 The Vision

Once the entrepreneur identifies a good opportunity and decides to pursue it, the next step is to formulate a vision. A **vision** is an informed and forward-looking statement of purpose that defines the long-term destiny of the firm. Thus, if the entrepreneur recognizes a good opportunity to meet a real customer need, he or she describes a vision of a future venture that will respond effectively to that opportunity. The vision is a statement of insight, intention, ambition, and purpose. It reflects clearly the novelty of the solution and uniqueness of the entrepreneur's commitment. Successful entrepreneurs are able to communicate their vision and their enthusiasm about that vision to others. For example, Google's vision, conveyed by founders Larry Page and Sergey Brin, is: online search that reliably provides fast and relevant results.

A good vision is clear, consistent, unique, and purposeful, as summarized in Table 3.1 [Hoover, 2001]. Clarity means that the vision is easily understood and focused. Consistency means that the vision does not change in response to daily challenges and fads. Uniqueness means that the vision is tailored to the specific venture. Finally, the purpose of the firm defines the enduring character of the organization, consistently held to and understood through the life of the firm. The purpose, or core ideology, of Hewlett-Packard has been a respect for the individual, a dedication to innovation, and a commitment to service to society. The purpose of Merck is to gain victory against disease and help mankind. This core ideology provides the glue that holds an organization together [Collins and Porras, 1996]. The vision provides a clear picture of the future for all concerned. The core ideology is based on the core values of the organization, such as respect for the individual.

A vision describes a specific desired outcome and promotes action and change. It serves as a picture of the organization's destiny as it moves through challenge and change. It also provides the basis for a strategy. A vision is an imaginable picture of the future. It is like a rudder on a boat in a turbulent sea. An example of a simple, clear vision is given in Table 3.2. This vision statement provides the reader with a clear mental model of where the firm is going and how it will get there. Notice that the vision statement of Table 3.2 states the values and goals of the firm, and it inspires and motivates people.

Entrepreneurs need to create a shared vision or meaning for their venture. A dialogue of meaning and commitment will help bring a shared sense of urgency

**FIGURE 3.1**

Creating a business model for a new venture.

**TABLE 3.1  Elements of a vision.**

- **Clarity:** Easily understood and focused
- **Consistency:** Holds constant over a time period, but is adjustable as conditions warrant
- **Uniqueness:** Special to this enterprise
- **Purposeful:** Provides reason for being and others to care

**TABLE 3.2  Example of a vision for an innovative firm.**

We strive to preserve and improve people's lives through the innovation of biomedical devices while supporting, training, and inspiring our employees so that individual ability and creativity are released and rewarded. Our goal is to be the leader in our industry by 2017 and be widely known throughout the world for devices that save and extend lives.

and importance for the venture. Thus, the vision is used as a part of the business plan and described often to potential team members and investors.

The vision can be verbally expressed as a story. A story is a narrative version of the vision, told in an engaging way. Stories play an important role in the processes that enable new businesses to emerge. Stories help to make the unfamiliar new enterprise more familiar, understandable, acceptable, and thus more legitimate to key constituencies [Lounsbury and Glynn, 2001]. By clarifying the core idea behind an enterprise, a story can also help an enterprise raise money and gather other resources [Martens et al., 2007].

Jim Clark started three companies: Silicon Graphics, Netscape, and Healtheon (now WebMD). As recounted in *The New New Thing* [Lewis, 2000],

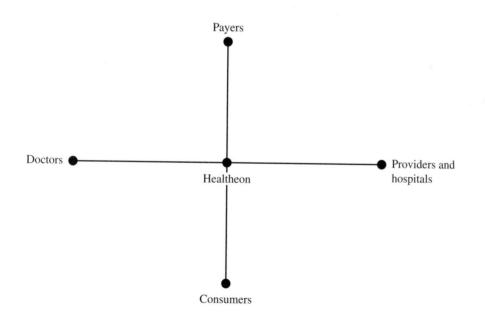

**FIGURE 3.2**  Vision for Healtheon (now WebMD). WebMD's three main businesses are providing electronic transaction services to doctors and hospitals, marketing software to help doctors run their practices, and providing online health information to doctors and consumers (see www.webMD.com).

Clark stated, "The only thing I can do is start 'em." As Lewis [2000] described Clark's realization:

> His role in the Valley was suddenly clear: he was the author of the story. He was the man with the nerve to invent the tale in which all the characters— the engineers, the VCs, the managers, the bankers—agreed to play the role he assigned them. And if he was going to retain the privilege of telling the stories, he had to make sure the stories had happy endings.

Clark had a vision for eliminating waste in the $300 billion costs of the U.S. health system by using the Internet to enable all the parties of any health transaction to connect via an online network—no paper forms, no hassle. Clark sketched a diamond depicting the players, as shown in Figure 3.2, and placed his proposed company, Healtheon, in the middle as intermediary. This was the way Clark told his ambitious stories—graphically, using sketches. Tales told by the entrepreneur aim to show plausibility and build confidence that the enterprise can succeed. To construct an identity that legitimates a new venture, entrepreneurial stories must have narrative clarity and resonate with the expectations, interests, and agendas of potential stakeholders [Lounsbury and Glynn, 2001].

Entrepreneurs need to learn how to tell their story about their team and venture, and to explain how their products will solve a problem. Their vision of the future can capture the interest of investors and team members.

## 3.2 The Mission Statement

The mission statement for a new venture more completely describes the organization's goals and customers, while incorporating the basic tenets of the vision statement. A vision is an imaginative picture of the future, while a mission is a description of the course of action to implement the vision. The mission of an organization is lofty and audacious. It provides for a theory of change.

The potential elements of a mission statement are shown in Table 3.3. Most mission statements include only some of these elements. For example, the mission statement of eBay is given in Table 3.4.

Most mission statements are short—fewer than 100 words. The mission statement should be a concise, clear explanation of the purpose, values, product, and customer. The eBay statement, for example, clearly describes that venture's mission. An example of a concise yet clear mission statement for an electronics firm might be: "Our mission is to design and manufacture electronic devices that serve the needs of the aerospace industry on a timely basis and at reasonable prices."

**TABLE 3.3  Possible elements of a mission statement.**

| | |
|---|---|
| ■  Core values | ■  Competitive advantage |
| ■  Customers and/or stakeholders | ■  Values provided to customer |
| ■  Products | ■  Markets or industry |

**TABLE 3.4  Mission statement of eBay.**

We help people trade practically anything on earth. eBay was founded with the belief that people are basically good. We believe that each of our customers, whether a buyer or a seller, is an individual who deserves to be treated with respect.

We will continue to enhance the online trading experiences of all—collectors, hobbyists, dealers, small business, unique item seekers, bargain hunters, opportunistic sellers, and browsers. The growth of the eBay community comes from meeting and exceeding the expectations of these special people.

**TABLE 3.5  Mission statement of Genentech.**

Our mission is to be the leading biotechnology company, using human genetic information to discover, develop, manufacture, and commercialize biotherapeutics that address significant unmet medical needs. We commit ourselves to high standards of integrity in contributing to the best interests of patients, the medical profession, our employees, and our communities, and to seeking significant returns to our stockholders, based on the continual pursuit of scientific and operational excellence.

A good mission statement can help align all the stakeholders and provide a rationale for allocating resources. If possible, the mission statement should be developed by the entrepreneurial team with other employees. Table 3.5 shows the mission statement for Genentech. This statement is very complete and describes its commitment to all its stakeholders—customers, employees, and community.

## 3.3 The Value Proposition

Value delivered to the customer results in a satisfied customer who will pay a reasonable price in return for the product or service. **Value** is the worth, importance, or usefulness to the customer. In business terms, value is the worth in monetary terms of the social and economic benefits a customer receives from paying for a product or service. To be successful, firms must offer products that meet the needs and values of the customer. The needs of the customer often include ease of locating or accessing the product as well as its qualities and features.

The five key values held by a customer can be summarized as product, price, access, service, and experience. Table 3.6 lists these five values, along with specific descriptors for each value. Price, for example, can have high value to the customer when it is fair, visible, and consistent. A product may have value if it

**TABLE 3.6  Five values offered to a customer.**

| | | |
|---|---|---|
| 1. | **Product:** | Performance, quality, features, brand, selection, search, easy to use, safe |
| 2. | **Price:** | Fair, visible, consistent, reasonable |
| 3. | **Access:** | Convenient, location, nearby, at-hand, easy to find, in a reasonable time |
| 4. | **Service:** | Ordering, delivery, return, check-out |
| 5. | **Experience:** | Emotional, respect, ambiance, fun, intimacy, relationships, community |

has high performance and quality, and is easy to find and use. Most technology-based products are initially focused on performance and functionality [Markides and Geroski, 2005].

The value proposition defines the company to the customer. Most value propositions can be described using the five key values. Crawford and Matthews have shown that one value is selected to dominate the value proposition offered to the customer. A second value differentiates the offering, and the remaining three values must meet the industry norm [Crawford and Matthews, 2001]. Consider a performance rating on a 1-to-5 scale where 5 is world-class, 1 is unacceptable, and 3 is industry par. Crawford states that a venture should plan a good product offering to have a value score of 5, 4, 3, 3, 3 for its five value proposition attributes in the following order: dominant, differentiating, norm, norm, norm.

Consider Wal-Mart, where price is the dominant value of its offering. Wal-Mart differentiates itself on product in terms of selection and quality. By contrast, the values offered by Target are dominated by product and differentiated by price. Many firms focus on good service, which is about human interaction. For example, Honda has great service as its dominant value, and its secondary, differentiating value is product.

Access can be described by ease of locating, connecting to, and then navigating the physical or virtual facility of a business. Very good accessibility is offered by Amazon.com, and its website is relatively easy to navigate. Accessibility can also be described as convenience or expedience. For a customer with a high demand for time, convenience is very important. A readily accessible website can be very valuable to a time-starved customer.

Zappos.com sells shoes and other clothing and accessories through its website. The company is known for providing an excellent customer experience. According to Zappos, "Customer service isn't just a department, it is the entire company." As a result, the company has a 75 percent repeat business rate, and enjoys a very good reputation through word-of-mouth referrals.

Apple realized that by opening retail stores, it could make buying its products more of a recreational experience. Apple Stores provide a place to gather casually and learn how to do interesting things with Apple products. Customers can do everything from buying a computer or phone to learning how to record their own music to interacting with other Apple aficionados. As important, these extra services that Apple Stores provide are complimentary to all customers.

Most customers seek a provider of a product or service who saves them time, charges a reasonable price, makes it easy to find exactly what they want, delivers where they ask, pays attention to them, lets them shop when they want to, and makes it a pleasurable experience. Any firm that fashions a value proposition to that set of customer values and actually delivers on that promise should do well.

The product value is described by its performance, range of selection, ability to search for it, and quality. Volvo built its business on the idea of product safety. Volvo became the first car company to offer three-point, lap and shoulder seat belts. Home Depot focuses on providing a very wide selection of quality

**TABLE 3.7  Primary and secondary values for leading firms.**

| | | Primary Value | | | | |
|---|---|---|---|---|---|---|
| | | **Product** | **Price** | **Access** | **Service** | **Experience** |
| **Secondary Value** | **Product** | —— | Wal-Mart | Amazon.com | Honda | Harley-Davidson Disney World |
| | **Price** | Target | —— | Holiday Inn | Wal-Mart | Olive Garden |
| | **Access** | Google Barnes & Noble | Priceline Visa | —— | Dell Computer | Starbucks |
| | **Service** | Toyota Home Depot Intel | Southwest Airlines | McDonald's | —— | Carnival Cruise Line |
| | **Experience** | Mercedes | Virgin Atlantic Best Buy | AT&T | Nordstrom | —— |

products. The differentiating (secondary) value for Home Depot is its service. The primary and secondary values for selected leading firms are shown in Table 3.7.

One value of product is range of selection or choice. Often to appeal to many different customers, a firm offers many versions of a product. However, too much choice is often debilitating [Schwartz, 2004]. If a firm offers extensive choice, it should help the customer search and select the right version. Amazon and TiVo offer such help to their customers.

Remember, a firm must meet at par the three remaining variables. Consider the plight of today's department stores. Their primary value is product selection. However, they are struggling to be accessible to today's shopper and just be at par on service, price, and experience.

---

**Google's Value Proposition**

What are the primary and secondary values for Google? It offers product as its primary value with fast, relevant results for the most ill-described inquiry. Its secondary value is access, which is embodied in the easy online connection right to the search page without annoying pages or advertisements obscuring the search box.

---

In general, the more value, V, customers place on a firm's products, the higher price, P, the company can charge for these products. The cost of producing the products is C, and the profit margin is P − C. The company profits as long as P > C. The value created is V − C, and the net value to the customer is V − P.

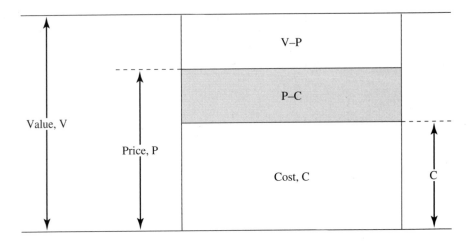

**FIGURE 3.3**  Value and return to the customer and the firm.

These relationships can be portrayed as shown in Figure 3.3. The most powerful new venture provides a great sense of value at a reasonable price, resulting in a high ration of value to price for the customer.

American Express invented the traveler's check as a means of getting money while traveling abroad. The value to the customer was high. The cost to issue the checks was low, and $(V - P)$ remained high to the customer. All parties have been pleased with this business model for more than 100 years [Magretta, 2002].

The **value proposition** states who the customer is and describes the values offered to this customer. The value proposition for Amazon.com could be described as:

> An easily accessible Internet site that is always convenient and provides a wide selection of products at a fair price to the busy, computer-literate customer.

The value proposition for Starbucks could be described as:

> A friendly, comfortable, well-located café offering a wide range of fresh, customized quality coffees, teas, and other beverages for the person who enjoys a good experience and a good beverage.

Home Depot and Lowe's stores are the two large home-improvement chains in the United States. Home Depot's dominant value is product selection, and its secondary value is service. Lowe's has a dominant value of accessibility and a secondary value of product selection. These value differences lead to clearly separate value propositions offered to the customers of these two competitors.

The **unique selling proposition** (USP) is a short version of a firm's value proposition and is often used as a slogan or summary phrase to explain the key benefits of the firm's offering versus that of a key competitor. For example, Hewlett-Packard uses a USP as follows:

> Excellent technical products with reliable service at a fair price.

The clear, simple USP for FedEx is:

Positively, absolutely overnight.

USPs are useful for succinctly describing a new venture to would-be investors, customers, or team members.

New ventures can use their value proposition and unique selling proposition to clarify the business values offered to the customer. This will help all stakeholders understand the purpose of the firm's business concept.

## 3.4 The Business Model

The design of a business is the means for delivering value to customers and earning a profit from that activity. The **business design** incorporates the selection of customers, its offerings, the tasks it will do itself and those it will outsource, and how it will capture profits. A successful business design represents an improvement over existing alternatives. It also functions like a story, which can be used to attract investors, customers, and team members.

A good business design addresses what the venture will and will not do and how the venture will create a sound value proposition. The business design answers three key questions: who is the customer, how are the needs of the customer satisfied, and how are the profits captured and profitability protected. The resulting outcome of the business design process is the **business model,** which is an integrated description of a new venture's unique customer value proposition and of how the venture will configure its resources and activities to deliver that value and to earn sustainable profits [Eisenmann, 2012]. A business model is the framework that connects a technology-based solution to economic profits.

Table 3.8 depicts the elements of a business model. A description of a business model for Apple is given in Table 3.9. The first critical element of a business model is the selection of the customer. The business design aims to specify the customers with unmet or latent needs, which will then define the target market. Apple succeeds by focusing on customers who are less cost-conscious and willing to pay a premium for Apple products.

**TABLE 3.8  Elements of a business model.**

| | |
|---|---|
| ■ **Customer selection:** | Who is the customer? Is our offering relevant to this customer? |
| ■ **Value proposition:** | What are the unique benefits? |
| ■ **Differentiation and control:** | How do we protect our cash flow and relationships? Do we have a sustainable competitive advantage? |
| ■ **Scope of product and activities:** | What is the scope of our product activities? What activities do we do, and what do we outsource? |
| ■ **Organizational design:** | What is the organizational architecture of the firm? |
| ■ **Value capture for profit:** | How does the firm capture some of the total value for profit? How does the firm protect this profitability? |
| ■ **Value for talent:** | Why will good people choose to work here? How will we leverage their talent? |

**TABLE 3.9  Apple business model.**

| | |
|---|---|
| ■ Customer selection: High relevance | Target customer is the consumer who values breakthrough technology and world-class user interface over open standards. |
| ■ Value proposition: Unique benefits | Seamless integration of products and services, intuitive interfaces, elegant designs. |
| ■ Differentiation and control: Sustainable competitive advantage | Control over both hardware and software, critical mass in App Store, constant innovation, and strong intellectual property protection. |
| ■ Scope of product and activities | Desktops, laptops, smartphones, tablets, cloud based services, customized devices for experiencing media, and an applications store with many applications developed by third parties. |
| ■ Organizational design | Engineering and design are valued above all other disciplines. Marketing is used to create a culture around what it means to be an Apple user. |
| ■ Value capture for profit | Premium price charged for a premium product. Tight control over manufacturing leads to greater product margins and differentiated product. |
| ■ Value for talent: | Opportunity to work with a world-chanigng company and with some of the best and brightest engineers. |

The second element is a clear and unique value proposition that addresses the identified market segment. The business model also should describe the venture's differentiation, which is the third element. For example, Apple sells premium product with leading-edge technologies that are easy and fun to use.

The fourth and fifth elements explain the scope of product and activities, and the organizational design that will enable you to implement the value proposition. Finally, the sixth and seventh elements explain how you will capture value for profit and how you will attract and leverage talent. A clear path to profitability is critical. You should determine your company's actual and projected revenues and expenses, identifying the key factors that influence total revenues and costs. Then, plot cash flow versus time to determine your financing needs [Hamermesh et al., 2002]. More on this process is detailed in Chapter 17.

Furthermore, it is important that you can retain good profit margins so that you can invest for the future. For example, Apple controls both the hardware and software components of its products, which enables not only superior integration but also an increased ability to demand significant price concessions from its suppliers. By contrast, it is best to avoid competing solely on price and making low price the dominant value of the value proposition. Those companies that do make price the dominant value, such as Wal-Mart, Costco, Dollar General, and Family Dollar, are careful to differentiate themselves along other dimensions, too. Dollar General and Family Dollar use smaller stores in well-located strip malls so that accessibility is their differentiating value. Wal-Mart and Costco compete on the product quality and selection as their secondary value. The business model of Wal-Mart is successful because of its use of technology to achieve strong supply-chain management and store inventory control.

Southwest Airlines is another example of a business with price as the dominant value in its value proposition. Its secondary value is service: on-time arrival and departure, online ticket ordering, and a customer-friendly attitude. It captures profit from the valued service by controlling costs. It uses one type of aircraft, which keeps its costs of maintenance and training lower than its competitors'. It also heavily promotes the online sales of tickets. As a result, Southwest has been profitable every year since 1973 [Freiberg and Freiberg, 1997]. The business model of Southwest Airlines is compared with the business model of American Airlines in Table 3.10.

Alex Osterwalder has proposed a compelling visualization of the business model, "the business model canvas," as depicted in Figure 3.4 [Blank, 2013]. In Osterwalder's canvas, an organization's business model is depicted using nine building blocks: (1) customer segments, (2) the value proposition for each segment, (3) channels to reach customers, (4) customer relationships, (5) revenue streams, (6) key resources required to create value, (7) key activities required to create value, (8) key partnerships, and (9) the cost structure. A key feature of the business model canvas is that it highlights the relationships between these elements. Thus, it is not enough to simply describe each element independently; instead, entrepreneurs also should focus on the ways in which the elements reinforce one another to form a coherent and robust business model.

Successful entrepreneurs treat the business model itself as an opportunity to be innovative [Zott and Amit, 2007]. For example, the success of Apple's iPod

**TABLE 3.10  Business model of two airlines.**

|  | **American Airlines** | **Southwest Airlines** |
|---|---|---|
| Customer | Traveler who needs to fly many places throughout the world | Traveler who desires to fly routes served point to point in the U.S. |
| Value proposition: Dominant value Differentiating value | Product Accessibility | Price Service |
| Differentiation | Wide scope of product: goes almost anywhere | Limited point-to-point flights Easy maintenance and training for low cost |
| Scope of products and activities | Very broad: connects everywhere | Narrow: only flies to selected cities (point to point) |
| Organizational design and implementation | Hub-and-spoke High fixed cost | Point-to-point Lower, flexible costs Control costs |
| Value capture for profit | Dominate hub city Requires high occupancy | Requires high occupancy |
| Value for talent | High pilot salaries Good career | Participation in stock options and camaraderie |

*TEAM NAME HERE*                    *FILL OUT ALL 9 BOXES OF THE CANVAS IN ORDER 1 THRU 9*

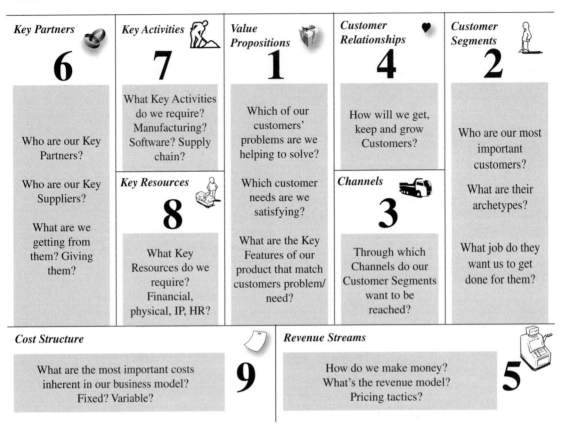

FIGURE 3.4  Business model canvas.

depended in large part upon Apple's use of a novel business model. By combining hardware, software, and service, Apple provided customers with significant value—and realized record-breaking profits [Johnson et al., 2008].

Almost every aspiring entrepreneur assumes that his or her initial business model (Plan A) will be successful. Empirically, however, most Plan A's fail [Mullins and Komisar, 2009]. PayPal, for example, morphed from infrared beaming of cash between PDAs to its current model based on online transactions tied to sites like eBay. Good entrepreneurs, therefore, test the hypotheses and assumptions embedded in their business model. Based upon data, they then adjust their business model [Ries, 2011]. As Figure 3.5 illustrates, the business design process, therefore, is iterative and is responsive to changing conditions and to new data.

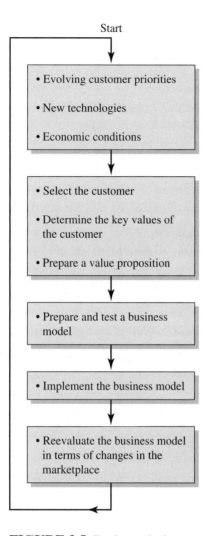

Start

- Evolving customer priorities
- New technologies
- Economic conditions

- Select the customer
- Determine the key values of the customer
- Prepare a value proposition

- Prepare and test a business model

- Implement the business model

- Reevaluate the business model in terms of changes in the marketplace

**FIGURE 3.5**  Business design process.

## 3.5 Business Model Innovation in Challenging Markets

Changing market conditions can signal that a firm's business model should also change. For example, a firm may have an opportunity to address the needs of large groups who find existing solutions too expensive or complicated, to capitalize on new technology, or to leverage existing technology in a new market. Alternatively, a firm may find that it needs to fend off low-end disrupters or to respond to shifts in competition. In all of these cases, a business model change may be needed [Johnson et al., 2008].

Business model innovation is the capacity to reconceive existing business models in new ways that create new value for customers [Hamel, 2000].

Monster.com, Indeed, and Simply Hired have introduced new ways of searching for jobs, with hundreds of thousands of listings across the country. Salesforce.com pioneered the Software as a Service (SaaS) model that enables customers to pay for their customer relationship management software on a monthly subscription basis and to run it on Salesforce.com's own systems. This approach has become the preferred method by which many companies now buy their software. As of January 2013, Salesforce.com had a market capitalization of $20 billion. Many other companies have copied the SaaS model for other applications. For example, Workday, which went public in 2012 and has a market capitalization of $8 billion, sells financial and human resources software as a service.

Existing ventures should also engage in business model innovation as markets change. For example, Hewlett-Packard, which was founded in 1938, changed its model to become America's premier brand for computer printers. Amazon.com started as a bookseller, but it now looks like the Wal-Mart of the Internet. Gas stations have evolved into convenience stores selling beverages, food, newspapers, and fuel.

Netflix is a prime example of a company that creatively destroyed video rental retail stores by offering to mail videos directly to the customer's home. Netflix was thus able to offer greater convenience and hundreds of thousands of titles. Retail video rental stores like Blockbuster, which went out of business, could not match Netflix for convenience or selection. Then as higher bandwidth speeds made it possible to download movies over the web, Netflix in 2012 found itself competing with a new set of players like Amazon and Apple, who were enabling customers to download movies to tablets and e-readers. Soon, all of the players in this industry will need to figure out a strategy for delivering movies to web-connected televisions. The dynamic marketplace for movie rentals will continue to offer significant promise as well as significant risk.

### Kiva Systems Uses Robots to Forever Change How Warehouses Operate

Kiva Systems uses robots to pick, pack, and ship products from warehouses. These large warehouses used to be dependent upon manual labor to store, move, and sort product inventories. With Kiva Systems, it has become economical to manage larger inventories and inventories with many more product lines. In addition to the labor savings, companies save money because Kiva Systems uses analytics to optimize product placement within warehouses. Thus, companies can ship up to four times as many packages as they could with traditional labor. Companies like Office Depot, Toys R Us, Crate & Barrel, and Timberland are able to operate more efficiently and at lower costs due to Kiva's innovations.

## 3.6 Spotlight on Stratasys

S. Scott Crump and his wife Lisa Crump founded Stratasys Inc. in 1989. Stratasys manufactures additive fabrication (3D printing) machines that deposit layers of plastic, which build up to become prototype models. Stratasys devices enable engineers and designers to create physical models and prototypes directly from their computer-aided-design workstations. Stratasys merged with Israel-based Objet Ltd. in 2012 in order to increase entry to potential markets.

Stratasys' value proposition is that it enables faster and cheaper manufacturing of prototypes and low-volume products. Stratasys has a powerful business model based on attractive costs, a wide range of users, and few competitors. Stratasys offers lease or rent systems and maintenance services. Their customers include aerospace, medical, automotive, electronics, and consumer products firms. Stratasys has utilized a merger and acquisition strategy to grow quickly and to become a leader in an emerging industry.

## 3.7 Summary

The theory of a business is a description of the elements required for the entrepreneur to act to build a business that satisfies the customers' needs. Coupled with the firm's core competencies and resources, the firm uses the elements of its business design to build a sustainable competitive advantage. The elements of a firm's theory of its business include: vision, mission, value proposition, and business model.

- Great vision is a statement of purpose (or story) in response to an opportunity.
- The mission describes the firm's goals, products, and customers, providing a theory of change for all to see.
- The value proposition describes customer needs that will be satisfied.
- The business model describes the economics and activities of the new enterprise.

---

**Principle 3**
The vision, mission, value proposition, and business model embodied within the business design of a firm can lead to compelling results.

---

## Video Resources

Visit http://techventures.stanford.edu to view experts discussing content from this chapter.

| Don't Write a Mission Statement, Write a Mantra | Guy Kawasaki | Garage |
| Innovate in Technology and Business: The Founding of Google | Larry Page, Eric Schmidt | Google |
| Business Models Matter | Ann Miura Ko | Floodgate Partners |

## 3.10 Exercises

**3.1**    How would you define Google's vision? Construct a mission statement for Google. After completing both of these tasks, go to Google's website and compare its actual corporate mission statement to your impression.

**3.2**    Compare the business models for Yahoo and Google using Table 3.8. Make sure to identify how they are different. How do you see their business models evolving over the next five years?

**3.3**    Purchasing a used car is one of the least desirable experiences for most people. eBay Motors offers fraud protection, a warranty, and a title history (www.ebaymotors.com). What is the value proposition for eBay Motors? Would you buy a car using eBay?

**3.4**    Kik's mobile messaging app has seen explosive user growth. However, a business model has yet to materialize. Describe three business models Kik could pursue to become a profitable business.

**3.5**    Apple has been successful in expanding its product and service portfolio from computers to MP3 players to mobile phones. What are Apple's business models? Describe the core competencies that have allowed Apple to make the moves from Mac to iPod, and from iPod to iPhone.

**3.6**    Woot.com is an online seller of mainly closeout products at a cheap price. This site provides low-priced sales and an online community to talk about the product of the day. Visit Woot.com and determine the business model of the firm. How does Woot generate a profit for this service?

## VENTURE CHALLENGE

**1.**    Create a brief vision statement for your venture.

**2.**    State the value proposition for the venture.

**3.**    Create a draft business model for the venture using the canvas elements in Figure 3.4.

# Competitive Strategy

*Praise competitors. Learn from them. There are times when you can cooperate*
*with them to their advantage and to yours.*
**George Mathew Adams**

**How can a venture create a strategy to fit the new business opportunity?**

Every new venture has a strategy or approach to achieve its goals. This strategy is in response to its plan to implement a solution to an important problem or opportunity. The process for creating a strategy for a new firm is shown in Table 4.1. Step 1 was described in Chapter 3. With sound vision and mission statements and an initial business model, the entrepreneur identifies the venture's core competencies and examines the political and economic context of the industry (steps 2 and 3). Once the industry is understood, steps 4 and 5 are used to describe the firm's strengths and weaknesses and its opportunities and threats (SWOT). In step 6, the entrepreneur integrates his or her knowledge of the industry and competitors with his or her own SWOT to identify key success factors. Based on the information gathered in the preceding steps, the entrepreneur refines his or her vision, mission, and business model and creates a strategy to achieve a sustainable competitive advantage. The formation of cooperative alliances with other enterprises can be an important way for a new venture to position itself within an industry. Long-term success depends upon addressing the needs of all stakeholders and acting in a responsible manner. ■

## 4.1 Venture Strategy

A **strategy** is a plan or road map of the actions that a firm or organization will take to achieve its mission and goals. To be useful, the plan must be action-oriented and based on the firm's opportunities, strengths, and competencies. The strategy also should be simple and clear, so that everyone in the organization can follow a commonly understood plan. The desired outcome of a strategy is a sustainable competitive performance.

A strategy is a response to opportunity. The word *opportunity* is derived from the Latin expression "toward the port." The builder of value is like a merchant sea captain who secures the right payloads from the best customers, manages his crew, and adjusts his mix of established ports and new ports with high potential [McGrath et al., 2001]. Thus, the formulation of a sound strategy is based on deep knowledge of the opportunity, the industry, and its context.

In describing the opportunity as a vision, a sense of drama and vitality emerges. With this vitality, the entrepreneur motivates the team and the investors to share the vision, embrace the strategy, and act on it.

Strategies help to set a firm on a course and then focus their efforts on it. In fact, for some, the essence of strategy is choosing what not to do [Magretta, 2002]. Often, a strategy emerges as actions are taken and tested, eventually converging toward a pattern [Mintzberg et al., 1998].

Because of the dynamic nature of the competitive business world, long-term planning is very difficult and the strategy also should be dynamic. Industries are not in equilibrium, and industry analysis is difficult. For example, it is hard to define where an industry begins and ends, and it is difficult to distinguish competitors from collaborators from suppliers. Thus, all strategies are subject to change and reemergence as conditions, alliances, and competition change.

Table 4.1 summarizes the process for developing a strategy. Entrepreneurs start in the center of Figure 4.1 by building and aligning their capabilities, resources, and products. They then act on their initial strategy or business plan. Entry into the competitive marketplace will force a reassessment of the marketplace and industry as well as their competitor analysis. This leads strategic

**TABLE 4.1  Management process for developing a strategy.**

1. Develop the vision and mission statements, and the business model.
2. Describe the firm's core competencies.
3. Describe the industry and context for the firm and its competitors.
4. Determine the firm's strengths and weaknesses in the context of the industry and environment.
5. Describe the opportunities and threats for the venture.
6. Identify the key factors for success.
7. Formulate strategic options and select the appropriate strategy.
8. Translate the strategy into action plans with suitable measures and controls.

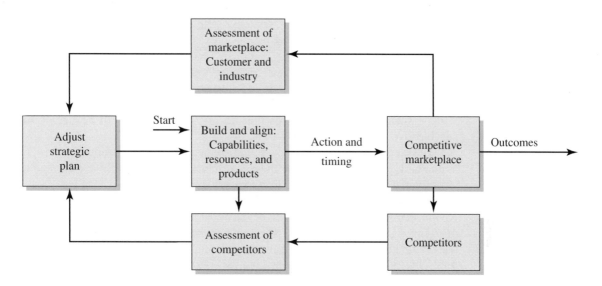

**FIGURE 4.1** Framework for a firm operating in a dynamic marketplace.

managers to redeploy and adjust the capabilities, resources, products, and actions to effectively compete in the dynamic market. These managers strive to attain a competitive advantage by securing and managing the assets of the firm. How the internal management responds to a changing customer, industry, and competition is crucial in the reestablishment of the strategic plan and the firm's assets to act competitively. Venture leaders strive to identify the fundamental forces for creating and capturing customer value. Those ventures that focus on continuously adjusting and aligning their strategy and capabilities will constantly evolve from one strategic position to the next strategic position in response to changing conditions.

GE Aircraft Engines (GEAE) provides an example of an adjustment in a strategic plan as a result of changes in the market. GEAE had a product strategy to develop engines with more power, efficiency, and better reliability. Because of relentless competition and shorter product cycles, sustaining profitability was difficult. GEAE shifted to operating as an engine production and services provider, generating significant profits in the after-market services business [Demos et al., 2002]. Faced with a dynamic marketplace, the strategic leader develops a strategic response and adapts to the changes in the market.

To summarize Figure 4.1, the first step is to determine the basic driving forces in the industry: the economic, demographic, technological, or competitive factors that either constitute threats or create opportunities. The second step is to formulate a strategy that addresses the driving forces identified in step 1. The third step is to create a plan to implement the new strategy. Finally, the new strategy is implemented by building and realigning the firm's capabilities, resources, and products.

Entrepreneurs define their strategy within their perception of opportunity. They are not constrained by the present resources or capabilities but seek to acquire the necessary resources and capabilities. The theory of resource dependence states that a company's freedom of action is limited to satisfying the needs of customers and investors that give it the resources to survive [Christensen, 1999]. Investors and customers dictate how money will be spent because companies that do not satisfy them will be unable to survive.

The six questions for creating a sound, dynamic strategy appear in Figure 4.2. With solid, effective answers to these six questions, a firm will have formed a strategy that has the potential to lead to profitability.

The development of a strategy often uses reasoning by analogy [Gavetti and Rivkin, 2005]. For example, Staples began by asking: "Could we be the Toys-R-Us of office supplies?" Analogical reasoning makes efficient use of information, but can be misleading if built on superficial similarities and inaccurate information. Thus, it is important to understand the source of the analogy and check the similarities.

Ultimately, a strategy can be viewed as a plan that integrates a firm's goals and actions into a cohesive whole that draws effectively on its resources and capabilities. The essence of strategy is choosing the priorities and deciding what to do and what not to do. The strategic priorities determine how a business is positioned relative to the alternatives. As the competitive conditions change, the new venture adjusts its strategy to meet the new conditions.

| Profitability | | |
| --- | --- | --- |
| Why are we pursuing this objective?<br><br>• Vision<br>• Mission | Where will we be active?<br><br>• Customer<br>• Market | How will we achieve our objective?<br><br>• Innovation<br>• Acquisitions |
| When will we act and at what speed?<br><br>• Timing<br>• Execution | What will differentiate our product?<br><br>• Positioning<br>• Competitor response | With whom will we compete and cooperate?<br><br>• Competition<br>• Alliances |

**FIGURE 4.2** The six questions for creating a dynamic strategy. Profitability rests on six solid answers to these questions.

## 4.2 Core Competencies

The core competencies of a firm are its unique capabilities and resources that enable it to implement its business model and thus deliver a valuable product or service to its customers. A capability is the capacity of the firm, or a team within the firm, to perform some task or activity. Thus, the capabilities of a firm include the collective learning in the organization, the skills of its people, and its capabilities to coordinate and integrate know-how and proprietary knowledge. Resources are human, physical, financial, and organizational, and include patents, brand, know-how, and capital. It is the usefulness of both capabilities and resources in a coordinated way that leads to distinctive core competencies – those activities that a firm does better than its competitors.

To be competitive, a firm must have: (1) unique and valuable resources and the capability to exploit these resources, or (2) a unique capability to manage common resources. Intel possesses unique patent and know-how resources and the capabilities to exploit that knowledge and intellectual property. Ryanair and Southwest possess common resources – aircraft and aircraft equipment – but have unique capabilities to manage these resources. Disney has unique resources in its film library, brand, and theme parks but a mixed record of managing them well.

Core competencies are the critical asset of a technology venture. Unlike physical assets, core competencies do not deteriorate as they are applied and shared. In fact, they can grow as a firm learns to build its competencies. For example, the core competency of Intel is the ability to design and manufacture integrated circuits for computers and communication systems. The core competencies of 3M are in designing and manufacturing materials, coatings, and adhesives, and devising various ways of combining them for new, valuable products. Ideally, a firm's core competencies will be rare, difficult for others to imitate, and difficult to substitute.

Core competencies provide potential access to a wide variety of markets. For example, Honda's core competencies in engines and power trains have enabled it to provide distinctive products for lawnmowers, motorcycles, automobiles, and electric generators. Core competencies are the wellspring of new business ventures.

It is very important that the core competencies of a firm match the requirements of its business. Firms with core competencies that match those necessary to effectively implement their business model have the best chance to succeed. For example, the core competency of Google is the design and operation of massively scaled Web services. It is the dominant online search engine. After starting as a search tool for finding information on diverse subjects, it also has become the leader in the Internet advertising industry.

We care about competencies since they are the roots of competitive advantage, as Section 4.6 will explain. Competitive advantage also depends, however, upon the firm's industry and context, as we discuss in the next section.

## 4.3 The Industry and Context for a Firm

A full description of the customer and the industry will help the entrepreneur build a sound strategic plan. The main elements of an industry analysis are given in Table 4.2. The first step is to accurately name and describe the industry in which the firm is or will be operating. The definition should be narrow and focused. An **industry** is a group of firms producing products that are close substitutes for each other and serve the same customers. Thus, selecting the telecommunications industry may be too broad. The definition of the industry should be more focused, such as "the Internet service provider industry serving homes and businesses in Ohio and Indiana." If data are not available for the targeted area of the market, the closest proxy should be used. For example, if statistics are not available for Ohio and Indiana, they may be available for the Midwest or the United States. Then, define this market and describe the customer. The second step is to describe the regulatory and legal issues within the industry. Both national, as well as state and local regulations, should be considered. Also, changes in regulations can influence both industry-funding trends and particular types of companies within an industry [Sine et al., 2005].

The third step of Table 4.2 suggests describing the growth rate and state of evolution of the industry. Most industries tend to emerge through an initial period of slow growth with limited sales and few competitors. Then, they expand through a period of rapid growth as sales take off and many firms enter the industry. This is followed by a third period of maturation marked by slower growth and stability. Eventually, the number of firms in the industry declines [Low and Abrahamson, 1997]. We depict in Table 4.3 these four stages as (1) emergence, (2) growth, (3) maturation, and (4) decline. It is important to know where your industry is in the evolution cycle. In the emerging phase, significant product and market uncertainty exists. Producers are unsure of what features are required for the product. Customers may be unsure of the elements of the product they need. Many technology ventures begin in the emerging phase of an industry. For a technology venture, an emerging industry will not yet have a dominant design and will respond well to new firms with a wealth of knowledge that can be used to build a powerful new venture [Shane, 2005].

The growth stage emerges when the necessary features and performance become clear and a dominant design emerges. A **dominant design** is one whose major components and core concepts do not substantially vary from one

### TABLE 4.2  Five elements of an industry analysis.

1. Name and describe the industry.
2. Describe the regulatory, political, and legal issues in this industry.
3. Describe the growth rate of the industry and the state of the evolution of the industry.
4. Describe the profit potential and the typical return on capital in the industry.
5. Describe the competitors in the industry and the rivalry among them.

**TABLE 4.3  Four stages of an industry life cycle.**

| Stage | Examples |
|---|---|
| 1.  Emergence | Artificial organs |
|  | Nanotechnology |
|  | Genomics |
| 2.  Growth | Medical technology |
|  | Software |
|  | Smart phones |
| 3.  Maturation | Electric appliances |
|  | Automobiles |
|  | Personal computers |
| 4.  Decline | Steel |
|  | Fax machines |
|  | Car phones |

product offering to another. With the emergence of a dominant design, the number of competitors stabilizes.

Eventually, an industry enters its mature phase as the number of competitors stabilizes and profit margins slowly decline as price becomes the primary competitive weapon. Finally, an industry enters a declining phase as the number of firms decline and profit margins erode.

The personal computer market began in 1978, with a number of small, emerging firms such as Apple Computer. IBM entered the personal computer market in 1982, and its PC quickly emerged as the dominant design. Many other firms entered after IBM made the design open to all, and the PC industry experienced a growth phase between 1984 and 1998. Eventually, the market reached a period of maturity, with only a few dominant firms having standardized or slightly differentiated products and relatively stable sales and market shares.

Table 4.2 shows that the next step in the industry analysis is a statement of the profit potential and the typical return on investment capital in the industry. The **return on capital** is defined as the ratio of profit to the total invested capital of a firm. The average return on capital in the computer software industry is about 16 percent, while the return on capital in the steel industry is about 6 percent. The steel industry is less attractive, while the computer software industry is attractive. One of the most effective ways to identify realistic profit opportunities for a new venture is to look at the Securities Exchange Commission filings of a young representative firm in the industry (www.sec.gov).

The **six forces model,** shown in Figure 4.3, is one popular method for evaluating the competitive forces in an industry. The six forces are: (1) firm rivalry, (2) threat of entry by new competitors, (3) threat of substitute products,

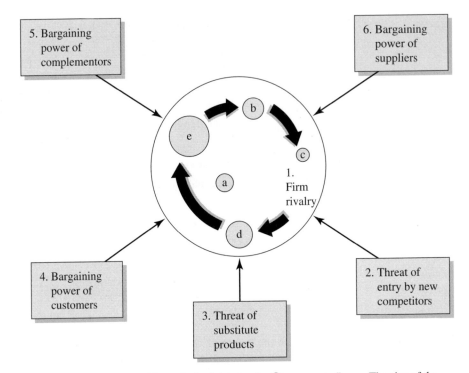

Note: Firms are represented by a circle; for example, ⓐ represents firm a. The size of the circle indicates the size of revenues of the firm. The six forces are numbered for clarity. The rivalry of the firms is shown as a vortex of competition illustrated by the solid arrows.

**FIGURE 4.3**  Six forces model.

(4) bargaining power of customers, (5) bargaining power of complementors, and (6) bargaining power of suppliers. This framework is an extension of the five forces model [Porter, 1998]. The six forces model enables the analyst to consider all the issues facing a new entrant by describing the key industry factors. The rivalry among the industry competitors may be intense or modest. In some industries, the bargaining power of the customer may be modest.

Consider the automobile industry, which has about 10 competitors. The rivalry is extremely intense. The bargaining power of customers regarding a new vehicle is very high since they have access to broad information on the relative performance and price of the products of the competitive companies and their dealers. The bargaining power of the suppliers in the industry is modest. Furthermore, the threat of a substitute product is small. The threat of new entrants is very small, due to the costs of developing a new product and dealer network. Thus, the auto industry experiences intense competition with the buyer wielding significant power.

Consider the online bookselling industry: Amazon.com and Barnesand-Noble.com are the two large online booksellers in the United States, but there are many regional competitors such as Powells.com. The rivalry among these competitors is high. Their suppliers have low bargaining power, and the barriers to entry are moderate. The bargaining power of the customer is large, resulting in low prices, and profitability is modest. The threat of substitute products is low. However, e-books could undermine the printed book industry as more attractive devices emerge such as Amazon's Kindle.

By contrast, many new firms enter the computer software industry each year. The bargaining power of customers is moderate, and the threat of substitute products is low. As a result, profitability in the industry is high. However, the rivalry of the firms is intense.

A competitive analysis explains how you will do better than your rivals. And doing better, by definition, means being different. Organizations achieve superior performance when they are unique, when they do something no other business does in ways that no other business can duplicate. In military competition, strategy refers to the large-scale plan for how the generals intend to fight and win a war. The word *tactics,* in contrast, refers to small-scale operations, such as the conduct of a single battle [Clemons and Santamaria, 2002]. Very few strategic plans survive the first contact with competitors. Competitors respond and change the situation.

**Complementors** are companies that sell complements to the enterprise's own product offerings. A **complement** is a product that improves or perfects another product. For example, the complementors to Sony's PlayStation 4 and Nintendo's Wii are the companies that produce the video games that run on these consoles. Without an adequate supply of complementary products, demand for the player product would be modest. Complements to the automobile include the interstate road system that enables automobiles to safely and rapidly travel long distances and the large number of dispersed filling stations that enable convenient refueling. By contrast, without suitable, widely located electric recharge stations, the future of electric vehicles may be very limited.

The bargaining power of the suppliers depends, in part, on the number of suppliers and buyers. When the supplier industry is composed of many small companies and the buyers are few and large, the buyers tend to dominate the supply companies. An example is the automotive component supply industry, in which the buyers are few and large and dominate the many small suppliers.

The entrepreneurial firm is likely to be a new entrant to the industry. Thus, the new venture should describe the barriers to entry, the threat of substitutes, and the bargaining power of the suppliers, customers, and complementors. One of the main factors that drives traditional analyses of the determinants of market structure involves comparing the size that a firm must be to compete efficiently to the overall size of the market in which it competes. If the industry has few firms, a new firm may be able to readily enter and gain market share. Using the six forces model, a new technology venture is likely to perform better when it operates in an industry with high barriers to entry, low rivalry, low threat of substitutes, low buyer power, low supplier power, and low bargaining power of complementors.

To complete the industry analysis, it will be necessary to name the competitors and describe the profitability of the industry. One method is to use *Standard and Poors Reports* or the *Value Line Investment Survey*. For example, if the new firm is entering the biomedical devices industry, the leading competitors are Medtronic and Boston Scientific. Using Value Line, we note that the average return on total invested capital for these companies is 15 percent. Value Line projects a 13 percent future growth rate of sales for this industry. With these attractive measures, the industry appears to be very attractive to new entrants with well-differentiated, fairly priced products.

## 4.4 SWOT Analysis

Steps 4 and 5 of the management process for developing a strategic plan (Table 4.1) suggest that a strategy is based on the firm's strengths and opportunities, while avoiding or mitigating its weaknesses and managing threats. As discussed in chapters 2 and 3, a new firm is focused on securing the capabilities and resources necessary to succeed in its industry. Furthermore, the new firm concentrates on an attractive opportunity that was selected using Table 2.6. Thus, a strategy addresses the four aspects of the setting in which a firm operates: (1) a firm's strengths, (2) its weaknesses, (3) the opportunities, and (4) the threats in its competitive environment. This analysis is often called a SWOT analysis, which allows a firm to match its strengths and weaknesses with opportunities and threats and find the purpose for which it is best suited.

A firm's strengths are its resources and capabilities. Its weaknesses are its limitations of organization or lack of capabilities or resources. A firm's opportunities are its chances for success in a new entry or product in its industry. The threats are actions or events outside its control in the competitive environment.

A basic SWOT analysis for Amgen, a leading biopharmaceutical company, is given in Table 4.4. The SWOT analysis provides the questions for a strategic

**TABLE 4.4  SWOT analysis for Amgen.**

| Organizational (internal) | Environmental (external) |
|---|---|
| 1. Strengths:<br>• Expertise in development and manufacturing of biologic drugs (e.g., proteins and antibodies)<br>• High-margin products and limited competition | 1. Opportunities:<br>• Expansion of marketed products for new geographies, indications, and formulations<br>• Allocation of resources to discover novel therapeutics to sustain growth |
| 2. Weaknesses:<br>• Inability to discover novel therapeutics to avoid declines in revenue | 2. Threats:<br>• Pharmaceutical companies entering the biologics arena<br>• Competition from follow-on biologics and pricing pressures |

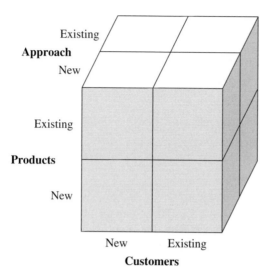

**FIGURE 4.4**  Three dimensions for examining opportunities.

response and helps a firm exploit its strengths, avoid or fix its weaknesses, seize its good opportunities, and mitigate its threats. Examples of threats are market shifts, regulatory changes, and delays in product development. Positive opportunities include increasing demand, repeated use, and willingness to pay.

We can examine opportunities in three dimensions, as shown in Figure 4.4. Perhaps the safest strategy is to take new products to existing customers via existing distribution channels using existing approaches. We can call the three dimensions: products, customers, and approach [Black and Gregersen, 2002]. Approach is the method or means of taking the product to the customer. The most risky strategy would be a new product taken to new customers via a new approach. Amazon.com started selling books (existing products) to book buyers (existing customers) via a new approach—online.

## 4.5 Barriers to Entry

**Barriers to entry** are factors that make it costly for companies to enter an industry. The greater the costs that potential competitors must bear to enter an industry, the greater are the barriers to entry. The six potential barriers to entry are in Table 4.5. Economies of scale can be a barrier in industries where the costs of production are low for a narrow range of volume or occur only for higher volumes. An example is the aircraft design and production industry. It is difficult to enter that industry since a low volume of production of aircraft is most likely uneconomic for the new entrant [Barney, 2002].

Cost advantages independent of scale may be held by existing companies and will deter a new company from entering. For example, incumbent firms

### TABLE 4.5  Potential barriers to entry into an industry.

- Economies of scale
- Cost advantages independent of scale
- Product differentiation
- Contrived deterrence
- Government regulation
- Switching costs

may have proprietary technology, know-how, favorable geographic locations, and learning-curve advantages. These can all be barriers to a new entrant.

Product differentiation means that incumbent firms possess brand identification and customer loyalty that serve as barriers to new entrants. For example, Dell, Hewlett-Packard, and Apple have brand and customer loyalty, making it difficult for a new personal computer company to enter the industry on a large scale. Of course, this barrier may be less important to a specialty manufacturer that seeks a small niche in the personal computer market. A formidable barrier to entry is the reputation or brand equity of the incumbents. Providing ratings for bonds is an attractive industry since it is not asset-intensive and the profit margins are very good. If a new firm tries to enter this market, however, it will have to compete with Moody's and Standard and Poor's, both competitors with strong reputations.

Contrived deterrence as a barrier occurs when incumbent firms strive to throw up unnatural barriers at a cost to them. They can use lower prices, newer products, or brand building to send a signal to potential entrants that intense responses will result if they try to enter.

Government regulation can also act as a barrier to entry. For example, a potential entrant to television broadcasting is deterred by government allocation of regular broadcast channels. A response to this limitation is for the new entrant to choose another means such as cable as the distribution channel—for example, the Fox Channel.

Two kinds of economic markets exist: substitutable and nonsubstitutable. Substitutable products are commodities such as groceries, cola drinks, and gasoline. In a nonsubstitutable market such as semiconductor manufacturing equipment, the required associated infrastructure means that once purchasers choose a system, they are not inclined to switch due to high switching costs.

**Switching costs** are the costs to the customer to switch from the product of an incumbent company to the product of the new entrant. When these costs are high, customers can be locked into the product of the incumbents even if new entrants offer a better product. An example is the cost of switching from Microsoft to the Apple computer operating system. Users would need to purchase a new set of software to use on the Apple computer as well as train their employees to use the new software.

**Low Barriers to Entry in Web 2.0**

Web 2.0 start-ups have been attractive to some entrepreneurs because the market seems relatively easy to enter. To them, it is a growth industry with low barriers to entry. The cost to set up a website is relatively low. Creating a website requires technical and programming knowledge, but modest capital investment. This industry offers an opportunity for entrepreneurs with little financial backing to create a product for a huge market quickly. Website services can easily be made accessible worldwide, without the need for physical distribution channels.

## 4.6 Achieving a Sustainable Competitive Advantage

Ultimately, a firm strives to maintain a competitive advantage - those distinctive factors that give it a superior or favorable position in relation to its competitors. Competitive advantage brings together the firm's core competencies and its strategic relationship within its industry.

As shown in Figure 4.5, a firm uses its distinctive competencies to manage its innovation, efficiency, product quality, customer relations, and supplier relations in order to differentiate its product and manage its costs. A technology venture works to design and produce at low cost the highest-quality product that possesses unique differentiating factors. Four common ways in which a firm will distinguish itself from its competitors are niche, cost, differentiation, and combined differentiation and cost, as summarized in Table 4.6.

The niche strategy is directed toward one or two smaller segments of a larger market. This niche can be geographic or a product or price segment. The low-cost strategy is based on unique competencies that enable the efficient management of processes. The goal of a differentiation strategy is to create a unique product based on a firm's unique competencies. Many firms can achieve a combined differentiation–low-cost strategy that blends the best of low cost and differentiation.

**FIGURE 4.5** Distinctive competencies lead to a competitive advantage.

**TABLE 4.6 Four common types of strategies and their characteristics.**

| | Type of strategy | | | |
|---|---|---|---|---|
| **Factor** | **Niche** | **Low cost** | **Differentiation** | **Differentiation-cost** |
| Distinctive competencies | Relationships | Processes, logistics | Innovation and relationships | Innovation and processes |
| Product differentiation | Medium | Low | High | Medium |
| Market segmentation | One or two segments | Mass market | Many segments | Many segments |
| Examples | Getty Images | RadioShack | Intel | Dell |
| | Incyte | Wal-Mart | Microsoft | Southwest Airlines |

---

**Intel's Competitive Advantage**

Since the founding of Intel, its strategy was focused on technology leadership, first-mover advantage, and the dominance of important new markets. Intel emerged as the dominant supplier of microprocessors, which are used in 90 percent of personal computers. Intel is also a leading manufacturer of flash memory, embedded control chips, and communication chips. A unique competency is Intel's ability to build, manage, and exploit the world's best semiconductor manufacturing facilities. As an example of its technology leadership strategy, Intel announced a new material that will replace silicon, enabling Intel to build more density (transistors per area) while reducing heating and current leakage. For decades, Intel has had a successful differentiation strategy.

---

Niche ventures often require less capital and achieve financial success rather quickly. Typically, a niche business is too small for the mass-market supplier, and thus, competition is low. A niche can be geographic or a product or price segment. Niche businesses typically are started in one market segment and based on a focused core competency and good customer and supplier relationships.

Fastenal Company of Winona, Minnesota, is the largest distributor of nuts and bolts in the United States. It uses a low-cost strategy for its manufacturing and distribution business with about 2,200 warehouse stores achieving total sales exceeding $2 billion. Each store has at least one delivery truck. Customers talk to the local store and receive personal service. Fastenal sees itself as an inventory and delivery manager offering excellent customer service. It currently has a fleet of 4,100 pickup trucks that respond to customer orders on an expedited basis (www.fastenal.com).

Southwest Airlines is an example of an airline that started as a low-cost, niche low-cost business operating only in Texas. It served three cities—Dallas, Houston, and El Paso—and operated using standardized Boeing 737 aircraft. It used highly productive crews, frequent, reliable departures, and a no-frills (low-cost), short-haul, point-to-point system. Eventually, Southwest moved to other western states and many locations across the nation. Thus, its strategy evolved to a differentiation-cost strategy.

Many profitable firms are built on differentiation: offering customers something they value that competitors don't have. This unique offering can be in the product, service, or sales, delivery, or installation of the product. While the basic product may be a commodity, the differentiation can be obtained somewhere in the various interactions or services for the customer. Some firms selling personal computers attempt to differentiate themselves by offering high-quality service. Harley-Davidson differentiates itself from other motorcycle manufacturers, in part, by lending money to people to buy its motorcycles.

Miox is a New Mexico-based venture that produces water purification systems. Traditionally, these systems used volatile and hazardous chlorine gas. Miox developed a technology that allows their products to function using only salt and water (www.miox.com).

Paychex is an example of a company with a differentiation-cost strategy. It provides payroll-processing services and began by targeting small- and medium-sized businesses that needed this service. The company offers customer service and payroll accuracy at a reasonable price, leading to wide acceptance. Once it has a satisfied customer, the switching costs for this customer are sizable. Paychex's revenues have grown to over $2 billion, and it serves more than 572,000 businesses. Paychex's annual revenue growth rate has been greater than 18 percent for over 20 years.

With good hardware and friendly software, Apple's portable devices made a profitable business out of digital music and applications—a business that eluded Sony, Microsoft, and Napster. What was the strategy that Apple adopted that led to success? The iPod was introduced in 2001, but it was the iTunes online-music store, introduced in 2003, that caused the iPod to take off. Apple sold more than 35 million iPods in 2012. Furthermore, iTunes sold 400 million songs in the first year and 16 billion songs through 2011. Apple introduced the iPhone in 2007 and the App Store in 2008. With rapid growth in the number of applications, Apple was able to differentiate itself in the phone market. Following the success of the iPod and iPhone, Apple then made another strong move with the introduction of the iPad in 2010. The iPod, iPhone, and iPad are easy-to-use, readily portable, and able to synchronize automatically with other Apple products including iTunes and iCloud. Apple used the differentiation-cost strategy of Table 4.6 to achieve rapid success.

IKEA provides furniture to customers who are young, not wealthy, likely to have children, and work for a living. These customers are willing to forgo service to obtain low-cost furniture. IKEA designs its own low-cost, modular, and

**TABLE 4.7   Ten types of sustainable competitive advantage.**

| Type | Example |
| --- | --- |
| ■ High quality | Hewlett-Packard |
| ■ Network size | eBay |
| ■ Low-cost production or operation | Wal-Mart |
| ■ Product design and functionality | Google |
| ■ Market segmentation | Facebook |
| ■ Product-line breadth | Amazon.com |
| ■ Product innovation | Medtronic |
| ■ Effective sales methods | Pfizer |
| ■ Product selection | Oracle |
| ■ Intellectual property | Genentech |

ready-to-assemble furniture. In large stores, it displays a wide range of products. While IKEA is a low-cost provider, it also offers several differentiated factors, such as extended hours and in-store childcare. Its competitive advantage is based upon a differentiation-cost strategy.

A competitive advantage is a significant difference in a product or service that meets a customer's key buying criteria. Thus, competitive advantage can be based on lower costs or differentiation of product or both. Most firms try to improve the efficiency of their operations to lower costs. They also strive to innovate or provide superior quality to outdo their competitors.

All firms seek to erode competitors' advantages by acting to imitate their product or service attributes or innovation. A sustainable competitive advantage is a competitive advantage that can be maintained over a period of time—hopefully, measured in years. The duration, D, of a competitive advantage, CA, leads to the estimate of the market value, MV, of a firm as

$$MV = CA * D \tag{3.1}$$

That is, the market value of the firm is proportional to the size or magnitude of the competitive advantage and dependent on the expected duration of that advantage. A pharmaceutical firm with a 20-year patent and a strong competitive advantage will be highly valued indeed!

The sustainability or duration of a firm's competitive advantage, D, is a function of the competitors' difficulty in imitating or innovating around the incumbent's unique product or service attributes. This difficulty is present when unique skills and assets are required and are hard for a competitor to replicate or obtain. A firm such as General Electric is said to have a sustainable competitive advantage in the electric power industry. It has higher profit margins than all its competitors in this field.

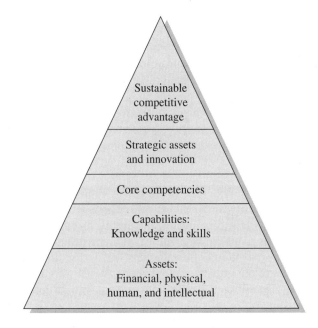

**FIGURE 4.6** Pyramid of value creation.

The pyramid of value creation is in Figure 4.6. From a solid base of assets, a firm builds its capabilities, which lead to its core competencies. With its core competencies and knowledge, it develops new products, processes, and other activities to build a competitive advantage. The sustainability of a firm's competitive advantage depends on its ability to continually innovate. Ultimately, the winning companies in the long term will be those that excel at learning how to do new things – not those who learn only how to improve existing practices [Reeves and Deimler, 2011].

In 1876, Sir Joseph Lister, a longtime advocate of improving sanitation conditions, was invited to speak at a medical conference in Philadelphia. In attendance was Robert Wood Johnson, who became inspired by Lister's speech. In 1886, Johnson, joined by two of his brothers, started Johnson and Johnson in order to manufacture a line of sterile surgical dressings. After experiencing some success, the company diversified into other segments of the medical industry. The company now manufactures pharmaceuticals and medical devices for physicians and consumers alike. Their products, including Tylenol, Band-Aids, and Listerine, are well known and trusted in the United States. The core competency of Johnson and Johnson is the ability to select and market trusted, useful products. Its consistency in doing so has allowed the company to retain a sustainable competitive advantage over the last century.

See examples of ten types of sustainable competitive advantage in Table 4.7.

## 4.7 Alliances

Many businesses use competitive strategies to shape their business strategies but often ignore *cooperative* strategies. Business is a complex mix of both competition and cooperation – sometimes called "coopetition." A new venture possesses valuable novelty and innovation that will attract the attention of suppliers, customers, competitors, and complementors, acting as a value network, as shown in Figure 4.7. All the participants are connected and participate in this network of activity. Consider the value network for a university, shown in Figure 4.8 [Brandenburger and Nalebuff, 1997]. The complementors to a university include kindergarten through grade-12 schools, local housing, community activities, and computing systems. All the members of the value network are connected together in the higher education value network. The university, to succeed, must cooperate with its suppliers, customers, competitors, and complementors. Competitors can be seen as rivals but also will be, in many instances, collaborators.

Many technology ventures offer products or services that require distinctive strategies because the products are parts of systems with complements provided by others. If a platform leader emerges and works with complementors, an ecosystem of innovation is formed [Gawer and Cusumano, 2008]. A platform strategy requires a compelling vision and strong leadership. A platform product or technology should provide a core function and be easy to connect for complementors. Examples of platform leaders are Google and Microsoft.

The value network is important to entrepreneurial ventures as they strive to accumulate the resources and capabilities required for success. The value of exploiting complementary resources can be significant [Hitt et al., 2001]. For example, a smaller, new biotechnology firm and a large pharmaceutical firm

**FIGURE 4.7**  Value network.

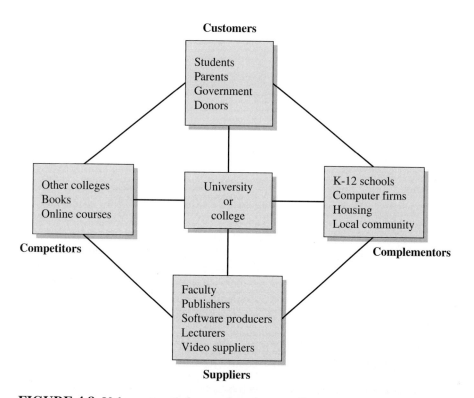

**FIGURE 4.8** Value network for a university or college.

can both benefit from an alliance. The biotech firm provides new technologies and innovation, while the pharmaceutical firm provides the distribution networks and marketing capabilities to successfully commercialize the new products. The larger established pharmaceutical firm also gains value through access to its partner's innovation. Thus, firms usually search for partners with complementary assets or capabilities. This is a partnership that enhances the value of each participant.

Few start-up firms will have all the necessary capabilities and resources. Alliance networks can enable them to move forward effectively. A **partnership** or **alliance** is an association of two or more firms that agree to cooperate with one another to achieve mutually compatible goals that would be difficult for each to accomplish alone [Spekman and Isabella, 2000]. Proactive firms take the initiative rather than react to events. Proactive formation of strategic alliances is an important dimension of entrepreneurial activity that enables a new firm to acquire access to unowned but required strategic assets. All alliances are based on some exchange of knowledge in addition to a flow of products, capital, or technology. Alliances function best when mutual benefits and commitment are clear to all parties [Lee et al., 2005].

The benefits of alliances can be significant. Both firms learn and acquire new capabilities. Furthermore, they have access to complementary resources that they cannot easily duplicate. Thus, a firm's decision to enter an alliance can be motivated by a desire to exploit an existing capability or technology, or to explore for new opportunities and new technologies [Rothaermel and Deeds, 2004]. For example, firms with diverse alliance partners increase their ability to innovate [Phelps, 2010].

To select an alliance partner, a firm must be clear which missing capabilities or resources are required. Then, it must determine which firms possess those assets and look at their characteristics. For example, the alliance that Alibaba.com and Yahoo! made in 2005 allows Yahoo! access to the Chinese market while Alibaba gains significant capital and expertise in scaling from Yahoo!. Startups with little track record may need to rely on the existing connections of their founders to establish these alliances [Hallen and Eisenhardt, 2012].

Entrepreneurs are well advised to visualize their alliance portfolios in the context of an entire industry, rather than as a series of single relationships. High-performing alliance portfolios emerge when entrepreneurs simultaneously form ties with multiple partners [Ozcan and Eisenhardt, 2009]. At the same time, alliances also place significant demands on an organization's management capability. In fact, too many alliances can actually harm an organization's performance. Thus, it is critical to examine each potential alliance for both the benefits it brings and the time, resources, and attention it will require [Rothaermel and Deeds, 2006].

---

### Cell Phones and Gaming

As mobile phones became more and more popular, a significant opportunity arose for the development of software for these devices. In particular, significant potential was seen for sales of games for mobile phones. Jamdat developed a number of these games, including the popular Bejeweled. In order to distribute these games, it had to form an alliance with mobile phone service providers like Verizon Wireless and Cingular (now AT&T). The service providers controlled the only distribution channel for these games. Because of the partnerships it was able to establish, the company also served as a link between other game developers and service providers. Acting as a gaming publisher in this way is now one of the company's main functions. Jamdat was purchased by Electronic Arts in 2006 and is now called EA Mobile.

---

Keep in mind that complementor firms may also be potential competitors. Many a well-conceived alliance has fallen apart due to the tension between cooperative and competitive forces. Such tensions can arise from culture clashes, poor conflict management, and lack of effective coordination mechanisms. Furthermore, the entrepreneurial firm may be seeking access to needed assets but may, as a result, be exposed to the risk of losing its own vital internal knowledge. An example of this occurred during the development of the Apple Macintosh. Apple partnered with

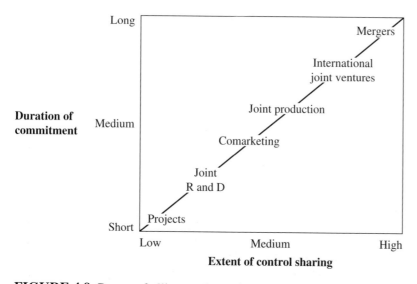

**FIGURE 4.9** Range of alliances dependent on commitment and control sharing.

Microsoft to develop spreadsheet, database, and graphical applications for the Mac. As a result, Microsoft acquired critical knowledge about Apple's graphical user interface products, which eventually enabled its engineers to develop the Windows operating system [Norman, 2001]. At the same time, knowledge sharing is important to maximize the success of a partnership. For example, when Starbucks places its coffee shops in Safeway stores, it makes sure that the Safeway employees at its counters are well-versed in the Starbucks way of doing things, so that the Starbucks brand and experience are maintained.

The type of alliance can range from a joint short-term project to a merger, as shown in Figure 4.9.

One approach to successful collaboration is to rotate leadership roles over the course of a project. Alliances continually dominated by a single organization or by the need to have consensus around every decision, by contrast, are associated with less innovation [Davis and Eisenhardt, 2011]. Table 4.8 outlines five other, simple rules for effectively managing alliances [Hughes and Weiss, 2007].

**TABLE 4.8  Five simple rules for making alliances work.**

1. **Develop the right working relationship** by specifying how you will work together
2. **Peg metrics to alliance progress**, not just progress toward alliance goals
3. **Leverage differences**, rather than trying to eliminate them
4. **Encourage collaboration** by moving beyond formal structures
5. **Manage internal stakeholders** to ensure that all involved players are committed to the success of the alliance

*Source:* Hughes and Weiss, 2007.

## 4.8 Matching Tactics to Markets

A company can be said to be successful if it outperforms its competitors over time. Another view of how to formulate the best strategy for a venture is to match the firm's approach to the pace of the market. Table 4.9 summarizes three competitive approaches [Eisenhardt and Sull, 2001]. The first approach is based on establishing a *position* in an industry and defending it. The goal is to position the company so that its capabilities provide the best defense against the competitive forces of Figure 4.3 [Porter, 1998]. Furthermore, the positioning approach can be defended by anticipating shifts in the six forces of Figure 4.3 and responding to them.

The second method focuses on *resources,* such as patents and brand, and attempts to leverage those resources against the resources of the competitors. For example, the powerful brand of Southwest Airlines has enabled the firm to issue its own Visa card to many of its customers.

The third approach may be called *emergent* and is based on flexible and simple rules [Eisenhardt and Sull, 2001]. Firms using this method to develop a strategy select a few significant strategic processes and build simple rules to guide them through the ever-changing marketplace. The strategic processes could be innovation, alliances, or customer relationships. Apple for example, has chosen its user interface design, customer relationships and customized proprietary products as its basic strategy. It then adjusts this strategy as conditions require.

Cisco Systems used an innovation strategy to guide it through emergent opportunities for its first years of operation. Later, it changed to a basic strategy of acquisitions to respond to rapidly changing markets. These basic tenets for guidance in emerging markets may be called simple rules and are summarized in Table 4.10 [Eisenhardt and Sull, 2001]. These rules allow a firm to compete in

**TABLE 4.9  Three types of competitive tactics.**

|  | Position | Resources | Emergent |
|---|---|---|---|
| Approach | Establish a position and defend it | Leverage resources such as brands, patents, or assets | Pursue emerging opportunities |
| Firm's basic question | Where should we be? | What should we be? | How should we take our next step? |
| Basic steps | Identify an attractive market Locate a defensible position, and fortify and defend it | Acquire unique, valuable resources | Choose one or two core strategic processes and use them to guide to the next step |
| Works best in | Slowly changing, well-understood markets | Moderately changing, well-understood market | Rapidly changing, uncharted markets |
| Duration of competitive advantage | Relatively long (3–6 years) | Relatively long (3–6 years) | Short period (1–3 years) |
| Risk or difficulty | Difficult to change position | Difficult to build new resources, if needed | Difficult to choose best opportunities |
| Performance goal | Profitability | Long-term dominance | Growth and profitability |

**TABLE 4.10  Simple rules for emergent markets.**

| Rules | Purpose | Example |
|-------|---------|---------|
| Boundary | State which opportunities can be pursued | Cisco acquisition rule: No more than 75 employees in an acquired company |
| Priority | Rank the possible opportunities | Expected return on investment |
| Timing | Synchronize the selection of opportunities and the conditions of the firm | When product must be delivered |
| Exit | Know when to pull out of opportunities | Key team member leaves |

a fast-moving marketplace such as the emergent markets that many technology ventures start in.

A good way to understand strategic planning in emerging industries is to imagine an American football team trailing by a touchdown with only two minutes left to play, and it has the ball. The team refuses to panic. It has well-established rules of play for this situation. It switches to the "no-huddle" offense, with the quarterback calling the plays at the line of scrimmage as he surveys the defense.

Uncertainty is endemic in strategy formulation. Thus, the quality of a strategy cannot be fully assessed until it is tried. Strategy making can be thought of as an organizational capability, where different approaches are generated and considered, and where past successful approaches are just options for the future among many.

Sam Walton started with a strategy based on low-cost retail discount stores. He gained differentiation by locating many of these stores in relatively rural cities that were only large enough to support one large discount retail operation. His second differentiating factor was his organizational culture, which inspired his employees. As competition emerged, he developed one of the most cost-efficient distribution networks based on information technology systems. Walton's simple rules of strategy and operation were part of Wal-Mart's success.

In addition to matching the approach to the pace of the market, Table 4.11 highlights two key factors for determining a successful strategy: (1) specific industry-related competence, and (2) the existing level of competitive rivalry in the industry [Shepherd et al., 2000]. The venture capitalists who participated in Shepherd's study stated in summary: The most attractive strategy is led by a team that has strong competence in an industry that has not yet built up intense rivalries. The timing of entry may be favorable in these circumstances.

The success of a new venture arises, in part, from a fit between the distinctive competencies of the venture team and the major success factor requirements of the industry. The better the fit, the greater is the competitive advantage. A competitive advantage is sustainable if the competencies of the venture quickly track and match the changing requirements of the industry. In addition, it is important to build alliances with critical stakeholders, such as suppliers or distributors, thus erecting barriers to new entrants.

**TABLE 4.11  Factors for determining a successful strategy, in priority order.**

1.  Industry-related competencies: Distinctive competencies

2.  Competitive rivalry: Low rivalry in the industry

3.  Time of entry: Enters industry early at appropriate time

4.  Educational capability: Able to obtain the skills, knowledge, and resources required to overcome market ignorance

5.  Lead time: Significant time between the pioneer's entry and the appearance of the first follower

Every business strategy is unique since it is a unique mix of resources, context, goals, competencies, and organizational values. The potential for differentiation of a firm's strategy can also occur along a selected part of the consumption sequence shown in Figure 4.10 [McGrath et al., 2001]. Unique methods, tools, or arrangements can be used at each step in the sequence. Every new technology venture should look at the consumption sequence and decide where it can differentiate its product or service.

The power of computer maker Dell is its direct sales model offered to three different customer segments. The Dell direct sales model incorporates all 15 steps of the consumption sequence. On the other hand, CDW (www.cdw.com) acts as a middleman reseller for Hewlett-Packard and offers excellent customer service for the purchaser who needs help in choosing a computer. Its large sales force helps customers choose a total system that fits them, and a single salesperson is assigned to each customer for follow-up and later purchases. The CDW sales model incorporates steps 3 through 10 of the consumption sequence.

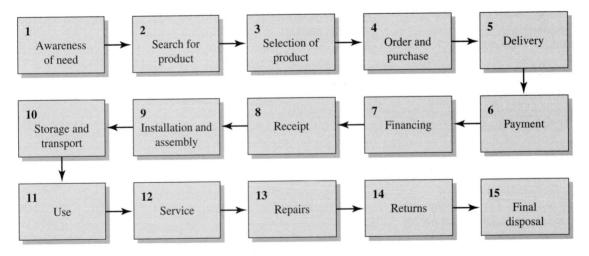

**FIGURE 4.10**  Consumption sequence.

# 4.9 The Socially Responsible Firm

Any strategy adopted by a new venture firm inevitably affects the welfare of its stakeholders: customers, suppliers, stockholders, and the community. While a specific strategy may enhance the welfare of some stakeholders, it may harm others. The leaders of new ventures are challenged to build a strategy that attempts to meet the economic and social needs of stakeholders while protecting the social and environmental needs of its region. An explicit statement of a new firm's strategy for acting responsibly and ethically may be an appropriate part of a business plan [McCoy, 2007].

The quality of life on our planet depends on three factors, as illustrated in Figure 4.11. The quality of life in a society depends on equity of liberty, opportunity, and health, and the maintenance of community and households, which can be called **social capital,** or social assets. The growth of the economy and the standard of living are critical needs for all people; we call this **economic capital,** or economic assets. Finally, the environmental quality of a region or the world can be called **natural capital.** The interrelationship between these three factors adds up to the total quality of life. Quality of life includes such basic necessities

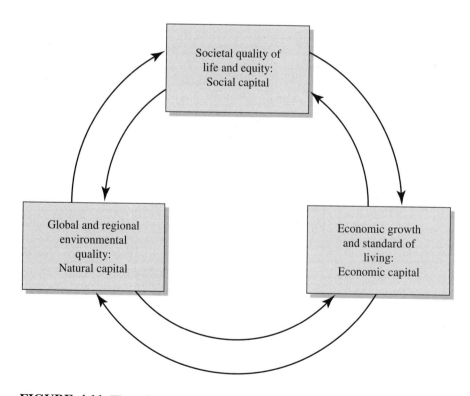

**FIGURE 4.11** Three interrelated factors that determine the quality of life on our planet.

as clothing, shelter, food, water, and safe sewage disposal. Beyond that, quality of life includes access to opportunity, liberty, and reasonable material and cultural well-being [Dorf, 2001].

Business, government, and environmental leaders need to build up capabilities for measuring and integrating these three factors and using them for decision making. We define the sum of these factors as the **triple bottom line.**

As they strive to treat nature and society respectfully while enhancing people's quality of life, corporations need to use nature only for what is necessary and in balance with what can be recycled and replenished.

Recognizing the interconnectedness and interdependence of all living things, corporate leaders can seek a balance using the triple bottom line concept. Economics, ecology, and society can be portrayed as a whole that depends on the person, the corporation, cultural values, and the community. Decisions made by corporations or society need to account for all the three factors of the triple bottom line.

For many people, there is a presumption that a company exists to enhance the welfare of society at large. For others, the only goal is the maximization of profits. We assert that the public welfare can be in the best interest of the corporation itself [Wang and Bansal, 2012]. One of the purposes of a firm is to make a profit—but service of society is also an implied expectation. In many ways, socially responsible behavior—remembering its obligations to its employees, its communities, and the environment, even as it pursues profits for shareholders—is in a firm's self-interest. A growing number of companies make corporate responsibility part of their value proposition. For example, Henry Ford believed he should pay his workers enough to afford to buy the cars they produced. His decision ultimately benefited Ford Motor Company by making it an attractive employer and stimulating demand for its products. Moreover, social responsibility can serve as an important differentiator for a firm [Russo, 2010].

Some of the best companies in history have tended to pursue a mixture of objectives, of which making money is just one—and not necessarily the primary one. For Merck, a top priority is patient welfare. For Boeing, it is advancement of aviation technology. Profitability is a necessary condition for existence, but it is not the end in itself for many visionary companies. Consider Johnson & Johnson, whose credo, published in the early 1940s, was the basis for its response to the 1982 Tylenol crisis, when a cyanide tampering incident caused the deaths of seven people in the Chicago area. The company quickly removed all Tylenol capsules from the entire U.S. market at a cost of $100 million, though the deaths occurred only in Chicago.

The social virtue matrix of Figure 4.12 illustrates the four possible responses to social responsibility challenges. The response of the lower-left quadrant (box 3) is conduct that corporations engage in by choice, in accordance with norms and customs. The lower-right quadrant (box 4) represents compliance—responsible conduct mandated by law or regulation [Martin, 2002]. These two lower quadrants represent the basic commitment of companies to

> **Solazyme Producing Fuel From Microalgae**
>
> Solazyme is a start-up that combined a profit mission with a social mission. Its founders, Jonathan Wolfson and Harrison Dillon, met while attending Emory University. They set about trying to generate fuel from low-cost plant sugars. The goal was to create a renewable source of energy from biomass sources, including waste products from sugar and corn crops. Today, Solazyme is a publicly held company that specializes in the production of renewable oil for use in ground and air transportation, nutrition, and skin and personal care products.

society's values and laws. Actions in the two lower quadrants (boxes 3 and 4) of Figure 4.12 generate little credit since the public expects actions to be in compliance with its laws and norms. The most significant impediment to the growth of corporate virtue is limited vision for actions beyond compliance and allegiance to society's norms.

The two upper quadrants encompass activities that have high social virtue. The strategic benefits quadrant (box 1) includes activities that may add to shareholder value by generating positive reactions from customers, employees, or legal authorities. These actions may ultimately benefit the firm by accruing customer goodwill and community support [Russo, 2010]. The upper-right quadrant (box 2) encompasses activities that clearly benefit society or the environment, but at a cost to the corporation.

An example of a firm active in the upper-left quadrant (box 1) is Patagonia, founded by Yvon Chouinard in 1979 as a designer, marketer, and distributor of high-performance outdoor wear with a commitment to protect the natural environment. The firm sought the use of low-impact fibers and drifted to organic cotton by 2000. Patagonia considered three criteria during the design and development of a project: its quality, its impact on the environment, and its

**FIGURE 4.12** Social virtue matrix.

aesthetics. In support of their commitment to sustainability of the environment, the firm featured a speaker series of prominent environmentalists. Furthermore, it often shared information with other clothing firms regarding the use of environmentally favorable fabrics.

One great opportunity to enter business in the top-left quadrant (by offering strategic corporate and social benefits) is to stimulate commerce at the bottom of the economic pyramid. For example, an entrepreneur can help the world's poor by partnering with them to innovate new products and services that are valuable and profitable [Prahalad, 2005]. While individual incomes may be low, the aggregate buying power of poor communities is actually quite large, representing a substantial market in many countries. In these markets, entrepreneurs need to reconsider their focus on high gross margins and shift toward securing good returns on invested capital while delivering social and environmental benefits [Prahalad and Hammond, 2002].

Environmental challenges represent another opportunity to enter business in the top-left quadrant. Businesses that combat environmental degradation can be both profitable and socially beneficial [Dean and McMullen, 2007]. New technologies and business models can help build a sustainable world incorporating clean energy, drought-resistant crops, sound fish farming, biodiversity, and much more [York and Venkataraman, 2010]. Non-fossil fuel sources of energy such as wind, solar, hydro, geothermal, and biofuels will be developed over the next 10 years as we shift to lower-impact fuel systems. New technology ventures will emerge as entrepreneurs find new means for big opportunities [Sachs, 2008].

Actions in the upper-right quadrant (box 2) may ultimately engender benefit for shareholders. However, actions that provide benefits to society at a cost to a firm are difficult to defend to shareholders. For example, if only one automaker had decided to add air bags, it would lose some profits. When such an addition is mandated, all automakers can provide added social benefits at a competitive cost. Corporate coalitions, in which firms agree to provide benefits despite the costs, can also help firms take action in the upper-right quadrant.

The public wants information about a company's record on social and environmental responsibility to help decide which companies to buy from, invest in, and work for. As an example, see Starbucks website at http://www.starbucks.com/responsibility. Starbucks estimates that it saved more than $40 million because the company's socially responsible actions increased employee loyalty and reduced turnover. The Mexican cement company Cemex helps low-income

**TABLE 4.12  Sampling of highly ranked, socially responsible companies.**

| | | |
|---|---|---|
| ■ BMW | ■ General Electric | ■ Procter & Gamble |
| ■ Cisco | ■ Google | ■ Steelcase |
| ■ Disney | ■ Johnson and Johnson | ■ Toyota |
| ■ FedEx | ■ Microsoft | ■ UPS |
| ■ Hewlett-Packard | ■ Nordstrom | |

families construct concrete homes, helping to tackle housing problems while connecting the company with a large and untapped market [Austin et al., 2007]. Good deeds can redound to a company's credit. But, they can also backfire if the company fails to live up to the good-neighbor image it tries to project. Fifteen highly ranked, socially responsible companies are listed in Table 4.12.

## 4.10 Spotlight on Google

Larry Page and Sergey Brin founded Google in 1998 while they were graduate students at Stanford University. The company initially focused on Internet search, competing against Yahoo!, Excite, and other search services. Google built a competitive advantage through continual innovation. It exploited this differentiation strategy (Table 4.6) over time as it added advertising technologies, software, mobile applications, and cloud computing. This array of innovation led to entry into many markets.

Google offers many low-cost or free services, thus leveraging the "power of free." It also has a well-developed infrastructure of bandwidth, servers, storage, and other hardware that powers its capabilities. Its business model, infrastructure, and continual innovation are difficult for competitors to match. Furthermore, Google leverages strategic alliances to expand and defend its business interests, as with the many partners who use and promote its Android operating system. In recognition of its commitment to the environment, education, and poverty alleviation, Google ranked number two in the Corporate Social Responsibility Index conducted by the Boston College Center for Corporate Citizenship and the Reputation Institute.

## 4.11 Summary

The strategy of a new business venture is its plan to act to achieve its goals. Given the challenge of an important problem (opportunity), the strategy provides a road map for the new firm to act to achieve a profitable solution to the problem. The strategy is designed to solve the problem by creating a unique and sustainable way of acting that, it is hoped, will lead to a profitable and valuable outcome for the customer and the firm. A solid strategy is based on:

- Recognition of the firm's core competencies.
- Sound knowledge of the industry and the context for the venture.
- A deep understanding of the firm's strengths and weaknesses as well as its opportunities and threats.
- A solid competitor analysis and review of the six forces encountered by firms in a rival market.
- A strategic design that can lead to a sustainable competitive advantage.
- A choice of a differentiation, low cost, differentiation and low cost, or niche strategy that provides unique value to the customer.
- Formation of productive alliances with others and always acting in a socially responsible manner.

> **Principle 4**
> A clear road map or strategy for a new venture states how it will act to achieve its goals and attain a sustainable competitive advantage in a socially responsible manner.

## Video Resources

Visit http://techventures.stanford.edu to view experts discussing content from this chapter.

| | | |
|---|---|---|
| Neutralizing Competition is a Speed Game | Geoffrey Moore | MDV |
| Social Responsibility from the Ground Up | Mitch Kapor | Foxmarks |
| Facing Competition: Creating a Barrier to Entry | Tien Tzuo | Salesforce.com |

## 4.12 Exercises

**4.1**   Zipcar offers a sophisticated form of car sharing (www.zipcar.com). The firm opened for business in Boston in late 2000. Describe the strategy of Zipcar using the six questions of Figure 4.2. Is the Zipcar strategy sustainable, and will it lead to profitability?

**4.2**   Podcasting, blogging, online photo sharing, online video, and tweeting are five technologies that are enabling a much broader set of content publishers and content consumers. Describe the nature of these industries and analyze the competitive situation using all six forces in Figure 4.3.

**4.3**   Nektar is an innovative drug delivery company creating differentiated products to allow for the inhalation of a number of medicines. Examine Nektar's website and publicly available information. Describe Nektar's strategy using Tables 4.6 and 4.9.

**4.4**   During the 1990s, DVD players became widely available and the rental DVD market took off. NetFlix (www.netflix.com) initiated an online DVD rental service creating a new market. Examine the Netflix website and determine the firm's basic strategy. What are the challenges to its strategy? Consider the timing of the initiation of NetFlix: was it too early or right on time? How have Wal-Mart, Amazon, and Apple attempted to differentiate their online services from NetFlix?

4.5    eBay has modeled worldofgood.com after early green marketplaces, positioning its activities in an environmentally friendly niche. Visit the website, describe its social mission, and describe how this fits into eBay's broader corporate mission.

4.6    Identify a technology company that incorporated more than 100 years ago. Describe the industry and context for the firm today. Describe a significant industry and context shift for the firm in its history. Has the firm maintained a sustainable, competitive advantage in the markets it competes in? If so, how?

4.7    Many online search competitors are moving to compete in the mobile local search market. Providing location tailored information to mobile phones is expected to be a large opportunity for both wireless carriers and local advertisers. Select one of these mobile local search companies and create a value network for this company (e.g., Figure 4.7).

## VENTURE CHALLENGE

1.    Develop a SWOT analysis using the format of Table 4.4.

2.    Select your strategic approach from Table 4.6.

3.    Create a partnership strategy as described in Section 4.7.

4.    Describe your strategy in one or two sentences that could be circulated to your employees and allies.

5.    Why and how will your venture be socially responsible?

# CHAPTER 5

# Innovation Strategies

*There's a better way to do it. Find it!*
**Thomas Edison**

**How can an entrepreneur build an effective strategy based on innovation that will lead to a sound technology venture?**

The decision to be the first mover needs to be addressed by all entrepreneurs. Using an idealized model of window of opportunity, the entrepreneur can decide when and how to act. The entrepreneur needs to maintain a sense of urgency but avoid being too early or too late to market. Entrepreneurs also seek to build an innovation strategy that involves new technologies, ideas, and creativity, which lead to invention and ultimately commercialization. Highly-successful new ventures may reshape entire industries through the creation of new products or services. ■

## 5.1 First Movers versus Followers

Many entrepreneurs believe that the quick survive while the slow struggle. The firm that leads the way with a new product or into a new market expects to lock in a competitive advantage that ensures superior profits over the long run. In this section, we consider the circumstances in which a pioneer may benefit from being a first mover. A **first-mover advantage** is the gain that a firm attains when it is first to market a new product or enter a new market.

We will describe the industries that a new venture enters as mature, growing, or emergent, as noted in Table 5.1. **Mature industries** have slow revenue growth, high stability, and intense competitiveness. **Growing industries** exhibit moderate revenue growth and have moderate stability and uncertainty. **Emergent industries** are newly created or newly recreated industries formed by product, customer, or context changes [Barney, 2002]. New technology ventures often start in uncertain, emergent industries.

The pioneering, first-mover firm has to bear the costs of promoting and establishing a product, including the potentially high costs of educating customers and suppliers. Furthermore, due to the high uncertainty of emergent markets, it is subject to potential mistakes in product, strategy, and execution. The follower firm can learn from the pioneer's mistakes and exploit the market potential created by the pioneer. Some firms successfully exploit a **follower strategy.**

Early entrants (second or third movers) into an emergent industry can also benefit from the additional time to develop, commercialize, and exploit new products if they possess the resources to wait for the opportunity to materialize [Agarwal et al., 2002]. Many examples exist of new start-ups that arrived early but didn't stay long. Pets.com, Helio, and Amp'd Mobile all burned through their investment capital before attracting enough customers to sustain a business. For most start-ups, the process is more like a marathon, where how fast you get out of the starting block is irrelevant.

In many cases, pioneer entrants tend to make a large and lasting impression on customers, obtaining strong brand recognition, and buyers often face high switching costs in moving their business to a later entrant. The simplest reason in favor of a first-mover strategy is the ease of recalling the first brand name

**TABLE 5.1  Three types of industries and their characteristics.**

| | Type of industry | | |
|---|---|---|---|
| **Characteristics** | **Mature** | **Growing** | **Emergent** |
| Revenue growth | Slow | Moderate | Potentially fast |
| Stability | High | Moderate | Low |
| Uncertainty | Low | Moderate | High |
| Industry rules | Fixed | Fluid | Unestablished |
| Competitiveness | High | Moderate | Low or none |

in a category. However, one study found that pioneers gained significant sales advantages but incurred large cost disadvantages relative to a fast follower entrant [Boulding and Christen, 2001]. The return on investment for pioneers was less than that for followers.

Of course, many conditions exist in which a first-mover advantage may be clear and compelling. Consider a mature industry such as restaurants or grocery stores. The attainment of a strategic resource such as a superior location may warrant acting as a first mover in a geographic market segment. Starbucks, for example, wants a store on the busiest corner in a city and acts when it finds an available site. First movers with the right set of competencies and organizational practices can reap the returns from being in the right place at the right time.

If a market is insufficiently ordered or unstable, the first entry may be too early. A market is said to be stable if the requirements necessary for success will not change substantially during the period of industry development. Amazon.com entered the online bookstore market and created intellectual property and the standard for this market. However, it incurred high development costs and had its advantages challenged by a later entry, BarnesandNoble.com. Nevertheless, Amazon.com became the leader in the race by continuous innovation.

Pioneers are often said to gain a low-cost advantage from having a head start down the experience curve, which describes improvements in productivity as workers gain experience. Often these lower costs are an advantage over later entrants [Shepherd and Shanley, 1998]. New technology ventures often act as pioneers in a new or emerging industry to gain brand, cost, and switching cost advantages. The potential advantages and disadvantages of first-mover action are summarized in Table 5.2.

When both technological innovation and consumer acceptance advance rapidly, first movers may be left behind [Suarez and Lanzolla, 2005]. However, first movers may gain advantage in an evolving market if they involve customers and

**TABLE 5.2  First-mover potential advantages and disadvantages.**

| Possible advantages | Possible disadvantages |
|---|---|
| ■ Create the standard and the rules | ■ Short-lived advantages disappear with competition |
| ■ Low-cost position | |
| ■ Create and protect intellectual property | ■ Higher development costs |
| ■ Tie up strategic resources | ■ Established firms circumvent or violate patents and intellectual property rights |
| ■ Increase switching costs for the producer | ■ Cost of attaining the resources |
| ■ Increase switching costs for the customer | ■ High uncertainty of designing the right product. If vision is wrong, then costs to switch are large |
| | ■ Customer is reluctant to buy when a large cost to switch may be incurred |

suppliers in the innovation process [Langerak and Haltink, 2005]. New technology ventures can exploit their nimbleness and competencies to build a competitive advantage. Amazon.com built a large business in a new market (e-commerce) in spite of the existence of large retailers such as Wal-Mart and Target.

Numerous examples exist of later entrants overtaking first movers and eventually bypassing them in profitability. Superior performance comes from distinctive competencies combined with an appropriate strategy leading to a competitive advantage (see Figure 4.5). Unfortunately, the first mover can develop a strategy based on uncertain or inaccurate assumptions about the six forces (see Figure 4.3). A follower who learns from the first mover's mistakes can move quickly to catch up or pass the first mover. The first mover also suffers from uncertainty about the customer, the organizational capabilities needed, and the industry context.

However, pioneering ventures can use their lead time to build relationships among suppliers, customers, and even competitors. These relationships can build trust and brand that a follower may not easily reproduce. A first-mover advantage can usually be attained under conditions of low market and internal firm uncertainty. Regrettably, most new ventures encounter large measures of uncertainty and must weigh carefully when to enter the market [Kessler and Bierly, 2002]. Entrepreneurs should emphasize speed to market in predictable markets. In an uncertain market, the new venture can probe or test the market by trying product tests, focus groups, and other means of market probes.

A commitment is an action taken in the present that binds an organization to a future course of action. A decision to act as a first mover is usually binding and should not be taken lightly. Preemptive actions can deter potential rivals from entering but may also result in heavy, irreversible investments [Sull, 2005].

The entrepreneur considers entering a market during an estimated period of opportunity often called a window of opportunity. The first mover envisions a greater cash flow as a result of early entry, as shown in Figure 5.1. Uncertainty about the period of opportunity can erode the actual results. If the first mover misestimates the timing of the window, a less attractive cash flow curve will result.

An entrepreneur's objective is to decide when to stop searching for additional information and enter the new market so as to maximize the expected profit. With insufficient information, a firm can enter too early and incur a large cost. However, if it takes too long to gather sufficient information, the firm may lose the first-mover advantage. Entrepreneurs should stop searching for information and enter a market when they estimate that the marginal benefit of additional knowledge is less than the payoff of entry [Lévesque et al., 2009].

Being a first mover means recognizing what direction existing technologies and industries are heading before competitors do. Today most personal computers (PCs) are built to support intense multimedia applications. In the early 1980s, most people did not recognize the role that gaming would play in the development of the PC. Chong-Moon Lee founded Diamond Multimedia Systems in 1982 to supply PC products such as color graphic and acceleration add-on boards. They significantly enhanced consumers' ability to play high-quality

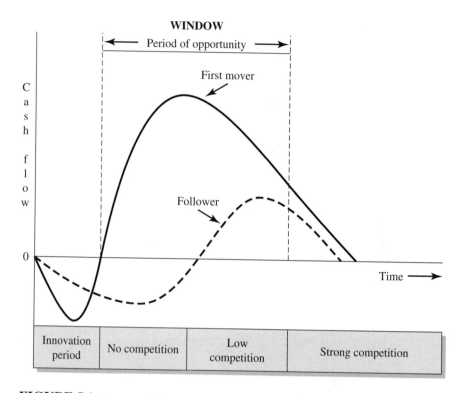

**WINDOW**

FIGURE 5.1 Expected first mover advantage and the concept of a window of opportunity.

games on their computers. Eventually, Diamond's first-mover strategy and superior products captured the attention of IBM and Tandy (then the number 1 and number 2 PC makers in the world, respectively). As these companies increased the standard multimedia features on their PCs, they continued to buy Diamond products to provide a better customer experience. In 1995, Diamond went public and raised $126 million by selling 30 percent of the company. Lee had successfully identified the direction that personal computing would be going. His first-mover position and perseverance over time allowed him to leave a lasting impression on both the computer and gaming markets.

History is replete with companies that were first movers that did not succeed. The CPM operating system preceded Apple, which preceded DOS, which eventually became the early dominant operating system for the PC. Safety razors were introduced a decade before Gillette introduced its successful safety razor. The product must have the right mix of attributes and features, and must be understood as well as demanded by the customer. Early movers don't always have all the requisite characteristics. Prodigy was the first commercial e-mail system, but it received poor acceptance. The second entrant, CompuServe, was equally unsuccessful. Only later did AOL and MSN put together the right mix of attributes to succeed.

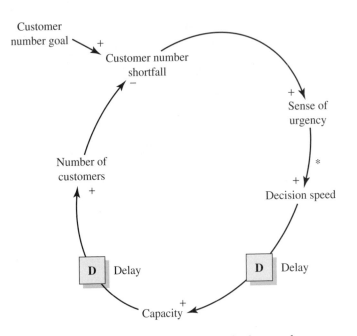

**FIGURE 5.2** The sense-of-urgency cycle that can be experienced by new enterprises.

Many new ventures set a fast pace as they and their competitors enter a window of opportunity. Many start-ups exhibit a torrid pace due to a high sense of urgency, as illustrated by the causal diagram in Figure 5.2. A causal diagram can help protray causal links in a system. Variables are related by causal links, shown by arrows. For example, the link denoted with an * implies that if "sense of urgency" increases, then "decision speed" increases. As the firm experiences a sense of urgency due to a shortfall of customers, it acts to build capacity to design, build, and sell its products. However, inevitable delays, D, slow down the buildup of capacity. As capacity increases, the firm expects customers to buy, but again it may experience delay as customers consider the purchase carefully. A slowdown in the growth of customer buildup results in a sales shortfall and an increasing sense of urgency [Perlow et al., 2002]. One way to decrease this unfortunate urgency cycle is to reduce the delay in capacity building and the time delay to customer purchase.

An encouraging case of good timing and entry into a marketplace is that of Google. Google entered the Internet search engine market in 1998, well after other search engines were firmly established. Larry Page and Sergey Brin, the founders of Google, met in 1995 as Ph.D. candidates at Stanford University. Over the next 18 months they collaborated to build a new search engine that ranked search query page results based not only on keywords, but also on popularity. Popularity was measured, in part, by the number of sites linked to each

Web page. The vision of the firm was "to organize the world's information." They allowed limited, less-intrusive advertising on the their site.

During late 1998, they wrote a business plan and raised $1 million in funding from family, friends, and angel investors. Working out of a garage in 1998, Google was answering 10,000 queries a day. In late 1999, it was answering three million queries each day. Google received $25 million in venture capital funding. In August 2004, Google sold about 20 million shares in an initial public offering at $85 per share. By 2009, the company had a market capitalization of $150 billion. Google's competitive advantages include its search technologies and its technical competencies. There is always a place for a new entry in a rapidly growing marketplace with a great technological innovation.

An example of identifying a window of opportunity both in a geographic market and for a technology is the founding of Baidu by Robin Li and Eric Xu. After spending several years working in the search industry, Li recognized that there was a need for a Chinese language Internet search engine in China. After Baidu was founded in 1999, the next four years were spent developing the best technology for China. With over a billion Chinese citizens, Li recognized that it was important to develop a search technology that would best serve everyone in China. In 2004, once Li felt Baidu had created the best search engine for the Chinese market, the firm shifted its focus to increasing brand awareness in China. 2005 saw an increased focus on revenue generation. Baidu's commitment to both search technology and the Chinese market helped it to become the second largest independent search engine in the world.

Machiavelli wrote in The Prince (XVII): "The prince ought to be slow to believe and to act, nor should he himself show fear, but proceed in a temperate manner with prudence and humanity, so that too much confidence may not make him incautious." An entrepreneur will be temperate and patient to move. On the other hand, an entrepreneur has a propensity to act. If a window of opportunity appears to be in the distant future, the entrepreneur may be wise to abandon the distant opportunity and seek one that is available and active now. If a window is about to "open," action may be prudent.

---

### Silicon Valley Bank (SVB): Founding at Optimal Time

In the early 1980s, the deregulation of the banking industry led to a need for new innovative banks. In the same period, the Bank of America, which served the San Francisco area, was discontinuing its lending to high-tech companies. At that time, one of the founders of SVB had a series of meetings with bankers who were interested in participating in this opportunity. These factors converged, and the lead founder acted on his intuition, all of which led to the formation of Silicon Valley Bank in 1983 (see www.svb.com). Since its founding, SVB has played an important role in the early days of such successful ventures as Cisco Systems, Electronic Arts, Intuit, JDS Uniphase, KLA Tencor, and Veritas. The launch of Silicon Valley Bank is a sound example of good timing.

## 5.2 Imitation

Imitation is said to be the greatest form of flattery. Many important new ventures have been based on the replication or modification of an existing business that the entrepreneur encounters through previous employment or by chance [Bhidé, 2000]. Entrepreneurs start an imitating business because they believe they can manage the business as well or better than the example they are copying. Sam Walton opened his first Wal-Mart in Rogers, Arkansas, after making numerous trips to study discount retailers in other regions of the United States. Walton once said: "Most everything I've done, I've copied from someone else." Technologists attend trade shows and conferences and often notice competitors' new products that their firm may readily produce.

Unfortunately, most attempts to replicate excellent businesses fail [Szulanski and Winter, 2002]. The difficulty of imitation springs from the lack of deep understanding of the excellent business example. Furthermore, the transfer of the best business practices from one setting to another can be fraught with unforeseen uncertainty. Imitation by independent entrepreneurs can be difficult because when they look at the existing business, they cannot fully understand what makes it work. Thus, the best approach is to copy it in detail but recognize that quick response to customer feedback will be necessary.

In 1986, Howard Schultz started his first independent effort as Il Giornale, a store modeled on his experience of Italian espresso bars. He played Italian opera in the Seattle store, and servers wore bowties. Il Giornale was set up as a stand-up bar, as is common in Italy, and it did not offer nonfat milk. Schultz had transferred the Italian coffee bar to Seattle with mixed success. People wanted chairs, and servers did not want ties. Nonfat milk quickly found its way onto the menu [Schultz, 1997].

Close copying may be the best method of imitation. It is important, however, to recognize the value of management and leadership, which is difficult to clone. A talented leader of an excellent business possesses some skills and capabilities that may be difficult to readily understand or copy.

Once the new business is up and running, customer comments can be used to adjust the business procedures to local conditions. Schultz loved the experience of Italian coffee bars but eventually adjusted his coffee café to fit Seattle's desires. Schultz was successful in adapting his Il Giornale store and ultimately created a successful system. He opened a second Seattle store after six months and a third store in Vancouver in 1987. By August 1987, Schultz arranged for his investor group to purchase all the Starbucks stores and its coffee roasting facility. He then merged his Il Giornale and Starbucks under the Starbucks name. By 2013, Starbucks had over 20,000 stores worldwide and revenues of over $13 billion.

JetBlue Airways is a good imitation of Southwest Airlines. Based on a low-cost, all-coach, point-to-point business model, JetBlue started in February 2000 with two aircraft serving New York City and Fort Lauderdale, Florida. JetBlue's initial public offering in 2002 raised almost $150 million for expansion. JetBlue was an excellent example of sound imitation.

## 5.3 Technology and Innovation Strategy

As discussed in Chapter 2, there are many different types of innovation—incremental, architectural, modular, and disruptive—and many sources of inventions, including existing companies, research labs and universities, and open technical communities. Most inventions, however, are never actually commercialized and never become true innovations. Whereas invention is the development of a product or process for the first time, innovation is the application of that product or process and its adoption by the market [Hall and Rosenberg, 2010]. Only about 6 percent of inventions developed by independent inventors actually reach a market [Astebro, 1998]. In established firms, the success rate is about four times as high, which still means that three-quarters of these inventions are never commercialized. Given this low success rate, it is critical for entrepreneurs to have a sound innovation strategy.

It is often a long road from invention to commercialization. Chester Carlson developed the photocopier process—converting an image into a pattern of electrostatic charges that attract a powdered ink—in his kitchen. He patented the process in 1942. After years of little interest from established companies, he obtained help from the Battelle Institute. Then the Haloid Company purchased a license in 1946. Haloid, which became Xerox, successfully demonstrated a working product in 1949. In 1960, Haloid Xerox finally introduced the first successful office copier, the Xerox 914.

In 1968, John Chowning invented the FM synthesis technique for creating digital sounds. The first product to incorporate FM synthesis, a musical keyboard, was not released until 1982 and it wasn't until the 1990s that the technique found application in mobile phone ring tones—three decades after its invention [Nelson, 2005].

Schumpeter asserted that the process by which independent entrepreneurs created inventions to produce new goods, services, raw materials, and organizing methods is central to understanding business organization, the process of technical change, and economic growth. An innovation strategy rests on the competencies and knowledge of the new firm. Continual product and process innovation can enable the firm to maintain a strategic advantage. Figure 5.3 illustrates the innovation and competition cycle between firms. Competitors create innovations and offer new value to customers, fueling demand and sales and increasing the innovator's market share. The struggle is for each competitor to keep up in this innovation cycle.

While inventors can license or sell technological opportunities to others, the creation of new firms is an important mechanism through which entrepreneurs use technology to bring new products, processes, and ways of organizing into existence. Three factors influence the decision to exploit an independent invention

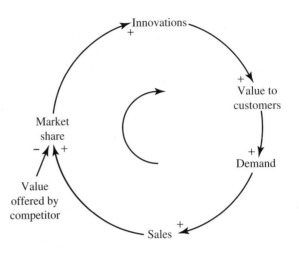

**FIGURE 5.3**  Innovation and competition cycle for market share.

through firm creation: the interests of the entrepreneurial team, the characteristics of the industry in which the invention would be exploited, and the characteristics of the invention itself. We have discussed the first two of these factors, especially, in Chapters 1 through 4. Of course, the entrepreneurial team must be interested in the opportunity to be solved by the invention, and it must be satisfied that the industry will welcome and support the commercialized invention. In this section, we will further consider the characteristics of the invention itself.

Three dimensions of technological inventions impact the probability that they will be commercialized through a new firm formation: importance, radicalness, and patent scope [Shane, 2001]. *Importance* reflects the magnitude of the economic value of an invention. The importance of an invention should increase the likelihood that a new firm will be founded to commercialize it because more important inventions have higher economic value and thus payoff to the entrepreneurs. Many inventions have limited commercial value and thus are not attractive to the entrepreneur. A critical determinant of an invention's importance is whether or not it addresses a real need. For example, if an invention makes it easier to do something that customers were not trying to do in the first place, it will fail [Christensen, 2002]. Thus, a "build it and they will come" innovation strategy will most likely fail.

*Radicalness* measures the degree to which an invention, regardless of economic value, differs from previous inventions in the field. Radical inventions have the potential, therefore, to be disruptive innovations. The radicalness of an invention is a reflection of the potential market effect of the commercialized invention. Radical technologies destroy the capabilities of existing firms because they depend on new capabilities and resources. Finally, *patent scope* describes the breadth of intellectual property protection for the invention. These three dimensions of likelihood of commercialization are listed in Table 5.3.

**TABLE 5.3  Factors that influence the entrepreneur to exploit an independent invention.**

1.  Business interests, capabilities, and experiences of the entrepreneurial team

2.  Characteristics of the industry in which the invention will be exploited

3.  Characteristics of the invention:

    a.  Importance of the invention: Economic value and potential payoff

    b.  Radicalness of the invention: Differentiation of the invention from its predecessors

    c.  Breadth of patent protection of the intellectual property

Dean Kamen, the holder of more than 440 patents, is one of the well-known inventors of the past three decades. He invented devices for infant care, for insulin delivery to diabetics, and for replacing the wheelchair [Brown, 2002]. Kamen invented the Segway Human Transporter in 2001. It is an electric scooterlike device. Gyroscopes inside the base platform make the scooter highly stable and self-balancing. There is no brake handle, engine, throttle, or gearshift. Users lean forward to go forward and backward to reverse direction. The inventor says the Segway can traverse ice, snow, or even large rocks. Although the Segway is a radical invention and is covered by patents, its economic value remains a question and it has not enjoyed the widespread adoption that many people predicted.

We can portray the new business formation process for an invention as shown in Figure 5.4. Using this process to review the potential of the Segway Human Transporter, one can obtain different conclusions for the various proposed uses: postal service, warehouse workers, or urban dwellers. Perhaps the best application for this device is not yet named.

The difficulty with deciding whether to proceed to commercialize an invention can depend, especially, on the radicalness of the invention. Disruptive or radical innovations introduce a set of attributes to a marketplace different from the ones that mainstream customers historically have valued, and the products often initially perform unfavorably along one or two dimensions of performance that are particularly important to those customers. As a result, mainstream customers are unwilling or unable to use disruptive products in applications they know and understand. At first, therefore, disruptive innovations tend to be used and valued only in new uncertain markets or applications.

Often the disruptive technology will not immediately serve a mainstream market, as shown in Figure 5.5. It will initially serve a niche market but will eventually enter the low end of the range of the mainstream market, as shown in Figure 5.5. For example, consider voice recognition software. The current performance of computer software for voice recognition is not always adequate for high-accuracy speaking (dictating) of documents for typing them; this might require 95 percent accuracy. Undoubtedly, there are many less-demanding uses for voice recognition software, such as voice-generated e-mails, customer

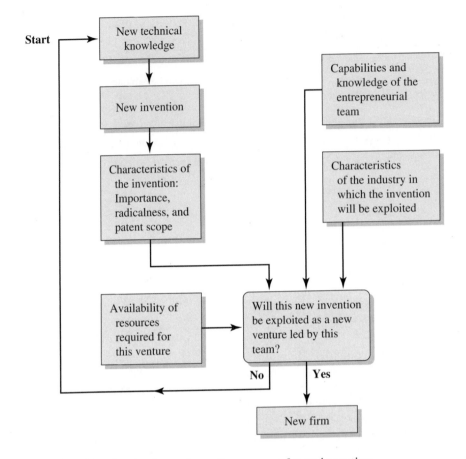

**FIGURE 5.4**  New business formation process for an invention.

service by telephone, or Siri voice-control software on iPhones. Thus, this innovation has entered the low end of the range of required performance and is progressing upward toward wider application.

Consider the disruptive innovation of U.S. discount stores in the 1960s. The increased mobility of shoppers enabled discount stores such as Kmart to select locations at the edge of town, reducing department stores' competitive advantage of prime city-center locations. The discount store had a new innovative business model: low-cost, high-unit volume and turnover provided at convenient suburban locations. It executed a trajectory from low-cost hard goods to low-cost hard and soft goods, and entered the mass markets in the 1970s and 1980s. Today, Target and Wal-Mart are in the center of the mass market. A recent disruptive challenge in the retail industry was Amazon.com, which appeared originally as an online bookstore but rapidly migrated toward becoming an online department store. Today, many people cite three-dimension (3D) printing as a disruptive

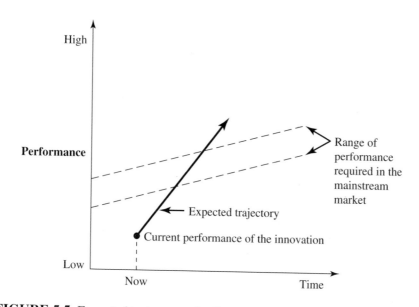

**FIGURE 5.5** Expected trajectory of a disruptive innovation.

technology, predicting that it will move from highly specialized applications to the mainstream. For example, MakerBot Industries is one of the leading start-ups in 3D printing. MakerBot has sold more than 6,000 printers to emerging technology teams within manufacturing companies and universities, and the start-up predicts rapidly growing sales. Jeff Bezos, the founder of Amazon, has invested in the start-up. Disruptive innovations, therefore, begin by addressing niche markets. With the right resources and capabilities, a new firm can satisfy initial needs in these markets, leveraging this early success toward mainstream dominance.

## 5.4 New Technology Ventures

Often, an entrepreneur encounters a new technology, but a compelling applica-tion of the technology is not obvious. Thus, a new technology sometimes can be characterized as a solution looking for a problem. Neither the first companies to use the technology nor the companies with the best technology necessarily win. Instead, the firms that find the right application for the technology succeed [Balachandra et al., 2004].

The elements of an attractive innovation strategy are provided in Table 5.4. Any new venture should have a defined customer, one or two key benefits, a short period to payback, and a proprietary advantage. Finally, the new venture team must possess the necessary core competencies to exploit the new technology.

One way of describing the potential applications is to use the model shown in Table 5.5. The new technology is described briefly, and the key assumptions

## TABLE 5.4  Elements of an attractive innovation strategy.

| | |
|---|---|
| ■ Well-defined customer | ■ Proprietary advantage that can be maintained or defended |
| ■ Key customer benefit that is measurable in dollars | ■ Core competencies required to exploit the new technology present or available to the new venture |
| ■ Short period until economic payback and positive cash flow | ■ Access to the necessary resources |
| ■ High benefit-to-price ratio for the customer | |

are listed. Then, the core competencies required for the venture are described. Finally, the possible applications market challenges are noted. Table 5.5 shows the summary of two fictional ventures. The Rotary Engine Inc. example illustrates an attractive new technology venture for vehicle engines, marine engines, appliances, and recreational vehicles. The market challenges are listed, and an attempt is made to fund the best application that will satisfy the required elements of an attractive innovation strategy.

A second example of a new technology is Fuel Cell Inc. Fuel cell technologies have been of great interest over the past decade. However, an economic application is not yet proven. With the lack of supporting infrastructure, fuel cells have limited automobile applications. On the other hand, fuel cells as energy storage devices serving as battery replacements may be viable soon.

## TABLE 5.5  Two potential new technology ventures.

| Potential venture | Rotary Engine Inc. | Fuel Cell Inc. |
|---|---|---|
| Technology | Advanced rotary gasoline engine technology | Hydrogen fuel cell technology |
| Key assumptions and benefits | Improved engine efficiency and reduced pollution | Pollution reduced to near zero |
| Core competencies required | Engine design and manufacture | Fuel cell design and manufacture |
| Potential applications | 1. Automobiles<br>2. Marine (ships)<br>3. Small appliances such as lawn mowers<br>4. Snowmobiles and off-road vehicles | 1. Automobiles<br>2. Small, local electric generators<br>3. Battery replacements<br>4. Marine (ships) |
| Market challenges | 1. Limited acceptance of rotary engines by customer<br>2. Lack of service knowledge for rotary engines<br>3. Benefits may be unclear to the customer | 1. Limited infrastructure for hydrogen fuel cells<br>2. Benefits unclear to customer<br>3. Reliability of fuel cells is unproven |

Examples of new technologies that eventually found attractive economic applications include semiconductors, genomics, stents, lasers, and wireless telephony. All these technologies eventually traversed the four steps necessary for a favorable technology innovation, as shown in Figure 5.6.

Any new attractive technology has to be feasible and manufacturable, and provide valued performance. With a sound business model and strategy, the new technology venture strives to achieve profitability in a reasonably short period.

Perhaps the greatest challenge is to develop an innovation that replaces fossil fuels [Carr, 2008]. The challenge is to find a low-cost, high-energy renewable fuel. Examples include wind, geothermal, waves, biomass, and solar technologies. These innovations must be effective, consistent, sustainable, and low cost. Solving global warming and creating green technologies will challenge innovators to develop economic innovations that can be brought to market and enable significant growth of use and scale [Krupp and Horn, 2008].

**FIGURE 5.6** Four steps to achieve a favorable technology innovation.

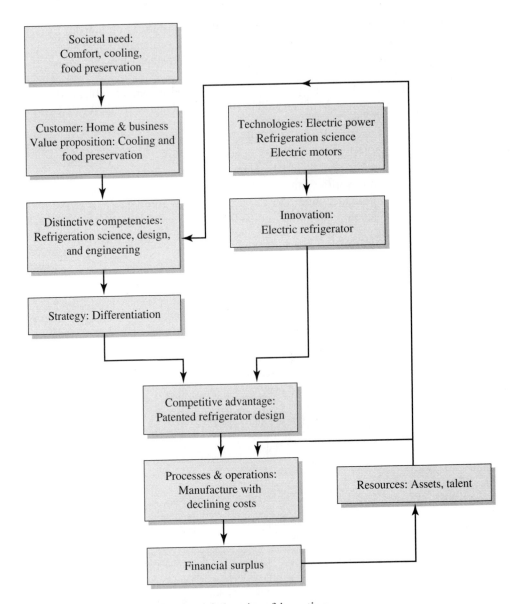

**FIGURE 5.7** The introduction of electric refrigeration.

A model of a technology innovation process is shown in Figure 5.7 using the introduction of electric refrigeration as an illustration. By the late nineteenth century, electric power, electric motors, and refrigeration science were available. With the creation of the electric refrigerator [widely available by 1915], a totally new industry was created by a discontinuous innovation. The innovation model shown in Figure 5.7 can be used to illustrate new technology applications introduced today.

Technology entrepreneurs bring together the technical world and the business world in a profitable way. Entrepreneurship is a fundamental driver of the technological innovation process [Burgelman et al., 2004]. In summary, technology entrepreneurship is about the creation of a new business enterprise that generates benefits (wealth, jobs, value, progress) for participating parties by creating unique, new arrangements of resources, including technology, to meet the needs of customers and society.

## 5.5 Spotlight on AgraQuest

Pam Maronne and several other associates formed AgraQuest in Davis, California, in 1995. AgraQuest's goal was to develop natural products for pest control that would enable farmers to avoid chemical pesticides. Maronne had already worked on natural, environmentally-friendly products and processes at Entotech, which sold to Abbott Laboratories.

Using a proprietary technology, AgraQuest's scientists analyzed tens of thousands of naturally-occurring microorganisms that could protect plants against insects, nematodes, and plant pathogens. They then selected the most promising candidates for further development. AgraQuest's innovation strategy was based upon proprietary processes and patents. The patents cover the microbe and its use as well as novel natural product compounds and mixtures.

Bayer CropScience of Germany acquired AgraQuest for $425 million in 2012. Bayer will use AgraQuest's innovations to build a technology platform for green products that control a broad spectrum of pests and diseases. Bayer intends to offer farmers integrated pest management programs to minimize the development of pesticide-resistant crops and to maximize crop yields.

## 5.6 Summary

Successful innovative firms strive to time their entry into markets. They balance a sense of urgency with a deliberate buildup to action. Working with partners—firms and individuals—most firms can enhance their capabilities and strengths for creativity, invention, and innovation. Almost all firms build an innovation strategy that strives to provide them with a sustainable action plan.

- A first-mover strategy can lead to significant benefits in an emerging market, but is not a guarantee of success.
- An innovation strategy creates a road map for continual commercialized invention.
- An ambitious venture can strive to design a product or service that is a disruptive application ("killer app.") that reshapes an industry.

> **Principle 5**
> An innovation strategy builds on creativity, invention, and technologies, acting within a value network, to effectively commercialize new products and services for customers.

## Video Resources

Visit http://techventures.stanford.edu to view experts discussing content from this chapter.

| | | |
|---|---|---|
| Three Types of Innovation | Judy Estrin | JLabs |
| Facing Competition | | |
| Through Innovation | Reid Hoffman | Linkedin |
| Out Innovate Bigger Competitors | Aaron Levie | Box.net |

## 5.7 Exercises

**5.1**  Name and describe the strategies of a company that was successful as a start-up being a first mover. Contrast that with a company that was successful being a fast follower.

**5.2**  Select an industry of interest to you and then try to find a good candidate for imitation. Describe the opportunity and tell how you will reap the benefits of imitation.

**5.3**  Go to a university's website and determine if a technology licensing office exists (e.g., http://otl.stanford.edu at Stanford University). Explore its website and featured technologies. Would you consider any of the feature technologies new-venture opportunities? Does the technology licensing office encourage innovation at this university? If so, how?

**5.4**  An inventor brings you a new design for an electric toothbrush with an oscillating head and a tilted handle that appears to meet the American Dental Association criteria. The inventor has filed a preliminary application for a patent. Also, you have tried the brush and found it easy to use. Using the factors of Table 5.3, provide a brief review of this invention. Would you recommend proceeding with commercialization?

**5.5**  Determine and describe the enabling technology used by Take-Two Interactive to develop its interactive software games (www.take2games. com). Describe Take-Two Interactive's value network as described in Section 4.7.

**5.6** Zebra Technologies Corporation provides bar-code labeling solutions for use in automatic identification and data collection systems (www. zebra.com). Describe the technology of Zebra in terms of the three dimensions of technological inventions: importance, radicalness, and patent scope.

## VENTURE CHALLENGE

1. Describe your venture in terms of timing of entry as illustrated by Figure 5.1.

2. Summarize your technology and innovation strategy.

3. Is your product or service a disruptive innovation? Why?

# PART II

# Concept Development and Venture Formation

The founders of a new venture should create a business plan. This plan should guide the team members as they think through opportunity identification, market sizing, competitive threats, product strategy, sales and distribution strategy, required resources, and potential financial outcomes. A good business plan will be a valuable tool in attracting investors and in convincing others to join the new venture. The venture team then experiments using the plan to mitigate the associated risks while using innovation to increase the chances of success. If possible, a business design will achieve economies of scale and scope.

The management team also needs to engage in product design and development. The use of prototyping and iterative product development can result in the creation of outstanding products that most closely meet customer needs. Concurrent with product development and design, the team should create and test a marketing and sales strategy. The strategy should identify target customers, the messages that best position and differentiate the product, and the distribution strategy. Marketing programs include traditional marketing, new social media, and the use of marketing analytics. The sales plan needs to determine the right mix of direct versus indirect sales as well as the mix of inside sales versus face-to-face selling.

The venture techniques described in this book can be used by entrepreneurs to build an independent business as well as a corporate venture emerging within an established firm. Large corporations can learn to confront the innovators' dilemma and build a viable corporate venture. ■

CHAPTER 6

# The Business Story and Plan

*The method of enterprising is to plan with audacity and execute with vigor.*
**Christian Bovee**

**How are ventures actually formed and what is the role of the business story and plan?**

Entrepreneurs respond to attractive opportunities by forming new ventures. In this chapter, we consider the five-step process for establishing a new enterprise. One particularly noteworthy step in the process is the development of a story and a business plan. The story is a compelling synopsis of why this venture is needed at this moment of time and how it can achieve success. We then detail the task of writing a business plan, which is a worthy effort to better prepare the venture for testing its assumptions and hypothesis with others. ■

## 6.1 Creating a New Business

Table 6.1 shows the five-step process for starting a new venture. The new venture will follow the steps to prepare a business plan that is suitable for the team as well as for the investors and business partners. This process is broad enough to apply to all types of businesses: independent or corporate, small or large, niche or broad, family or franchise, nonprofit, and one attempting a radical versus incremental innovation.

The corporate venture will also prepare a business plan suitable for review by the parent corporation to secure the necessary resources and assistance from its parent. The five-step process of Table 6.1 can be used for corporate ventures where the investors of step 5 will normally be the parent firm. We defer until Chapter 10 the discussion of corporate ventures and the appropriate legal form a venture should adopt.

A talented leader of a new venture has a vision and a plan to implement it. He or she can motivate people and manage information and resources so that the business can create a profit. The performance of a new venture is a consequence of factors that encompass the dimensions of a business as portrayed in Figure 6.1. The quality of the opportunity and its fit with the vision lead to the accumulation or creation of the distinctive competencies based on resources and capabilities. Following from a set of competencies, a business strategy is created based on novelty or innovation within an industry context. The attractiveness of the industry with respect to its business opportunities affects the profit potential and the expected return of the venture. Access to resources and the ability to attract the entrepreneurial team will depend on the attractiveness of the new venture. The industry environment will determine the amount of resources available to the venture since capital typically flows to industries in which opportunities are abundant and attractive. Furthermore, the industry-related competencies of the entrepreneurial team, which refers to the level of experience and knowledge with the industry, will lead to the expectation of greater success.

### TABLE 6.1  Five-step process for establishing a new venture.

1.  Identify and screen opportunities. Create a vision and concept statement, and build an initial core entrepreneurial team. Describe the initial ideas about the value proposition and the business model.

2.  Refine the concept, determine feasibility, and prepare a mission statement. Research the business idea and prepare a set of scenarios. Create a story and the outline of a business plan with an executive summary.

3.  Prepare a complete business plan including a financial plan and the legal organization suitable for the venture.

4.  Determine the amount of financial, physical, and human resources required. Prepare a financial model for the business and determine the necessary resources. Prepare a plan for acquiring these resources.

5.  Secure the necessary resources and capabilities from investors, as well as new talent and alliances.

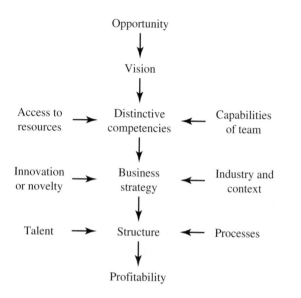

**FIGURE 6.1**  Building a new venture.

The identification and acquisition of required resources and capabilities are crucial for a firm's success. For a fast-growing firm based on continuing innovation, the intellectual resources are critical to success. The securing of the necessary resources and capabilities may occur in stages, as required.

The creation of a business plan focuses on the stages of building a business, as portrayed in Figure 6.1. The initial formation of a suitable structure for the business follows the business strategy. The ability to remain competitive and innovative while operating through an appropriate structure can lead to enduring profitability.

The greatest risk in creating a business is the failure to complete all the steps of Table 6.1 and Figure 6.1. Some entrepreneurs who possess strong technical skills and capabilities unfortunately skip the formation of a business strategy that will lead to profitability. Another risk is that the business plan may have an inadequate plan for the organizational structure and the management of processes and talent. Finally, remember the iterative and dynamic nature of this process. As an opportunity and the appropriate response to it evolve, the business plan should change, too.

## 6.2 The Concept Summary and Story

The road to success starts with the concept summary and new venture story. Once an entrepreneur has decided to act upon a business opportunity, it is important to prepare a concept summary of the new venture. This summary can be a simple statement of the problem being addressed and how the venture will solve it. Table 6.2 gives the elements of the concept summary. For example, the

**TABLE 6.2  Elements of a concept summary.**

1. Explain the problem or need and identify the customer.
2. Explain the proposed solution and the uniqueness of the solution.
3. Tell why the customer will pay for the solution.

business concept for Twitter might be summarized as "an online social network and microblogging service that enables its members to read and send short text messages."

A story is a narrative of factual or imagined events. It depicts a course of challenge, plan, actions, and outcomes in a manner similar to a plan. A good story and business plan define an opportunity, a concept, cause and effect, and an outcome all held together in a holistic way. Stories use plot lines and twists to capture the imagination and interest of the listener. They tell the ultimate goal of the venture, the ideological challenge, and the means of achieving the goal. The creation of an appealing and coherent story may be a useful form of communication for entrepreneurs in the attempt to attract interest and support for their idea and plan. To be effective, the content of the story must align with the interests and background of the listener. A well-crafted story about a new venture emphasizes the goals and merits of the venture. Telling a compelling story inspires belief in a venture's motives, character, and capacity to reach its goals [Ibarra and Lineback, 2005].

Stories are an integral part of the process by which founders start and build new ventures, acquire needed resources, and generate new wealth [Lounsbury and Glynn, 2001]. Storytelling is applicable to corporate new ventures as well as independent new ventures. Stories play a critical role in the process of the emergence of a new business. Stories can lead to favorable interpretations of the social benefits and wealth-creating possibilities of the venture, thus enabling resources to flow to the new enterprise. Furthermore, they can legitimize new ventures, thus helping to build acceptance of them. A good entrepreneurial story attracts financial and human resources, and builds industry acceptance. It can help amass support for a new venture. Successful entrepreneurs must shape interpretations of the nature and potential of their new venture for those who may supply needed resources.

As the founder of both Netscape and Healtheon, Jim Clark provided the story for both companies. In the case of Healtheon, his story was personal. Jim Clark contracted an illness in late 1995 that gave him personal experience with the health-care industry, its bureaucracy, and resulting paperwork demands. He responded with a concept: The patient would have a password and a digital record, and the doctor would use the Internet for billing forms with no hassle for both parties. He drew a diagram of the four players in the health-care industry, as shown in Figure 6.2. He then placed his new venture in the center of the diagram as the solution, via the Internet, for the entire industry. Clark told a story of critical importance to his nation, described the dramatic challenges

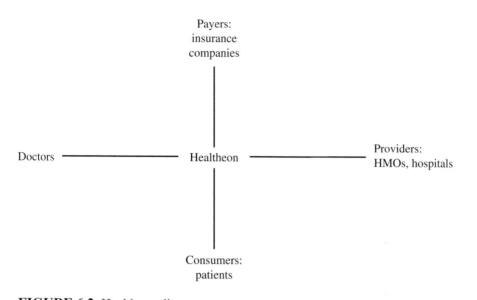

**FIGURE 6.2** Healtheon diagram.

and opportunities, and then portrayed the solution: his new venture. With his compelling story, Clark went on to raise millions of dollars to build Healtheon, which later was sold to WebMD [Lewis, 2000].

In constructing a legitimate identity for a new venture, the storyteller must forge a careful balance between emphasis on existing challenges and the potential for distinctiveness. The investor may be drawn by the credibility of the story and the storyteller to seriously consider investment.

New firms should not rely solely on a pure bullet-point presentation. A list of bullet points reduces a set of issues to a few points but provides little fabric or motivation. Bulleted lists are often generic in meaning and leave challenges and relationships unspecified. Furthermore, they often leave critical assumptions unstated. A good story includes all the challenges, relationships, and assumptions in the very fabric of the narrative. A presentation can use some bullet-point slides but should emphasize the story it wants to tell.

The new venture story consists of three elements, as shown in Table 6.3 [Shaw et al., 1998]. The first step is the setting of the stage by describing the current conditions of the industry, society, and the existing relationships and opportunities. The story should be about a person or group whose challenges the listener can relate to.

Next, the storyteller introduces the dramatic conflict by describing the challenges confronting the venture, the need for a plan to overcome the challenges at this turning point, and the critical obstacles and issues. Finally, in the third step, the storyteller describes the plan to overcome the obstacles, secure the necessary resources, and move on to resolution and success [Ibarra and Lineback, 2005].

## TABLE 6.3 Three stages of the story.

1. **Set the stage:** Define the current situation, the current players, and the opportunities coherently and clearly. Life is in a delicate balance and the listener cares about the situation.

2. **Introduce the dramatic conflict:** An inciting incident or need throws life out of balance. Describe the challenges and opportunities, and the need for a coherent plan to proceed toward success and a new balance.

3. **Reach a resolution:** Portray a coherent plan describing how the new venture can overcome the obstacles and succeed by following the plan.

As an illustration of an important story, consider the world's energy challenge. Energy is the lifeblood of industrial civilization and necessary for lifting the world's poor out of poverty. However, current methods of mobilizing energy are highly disruptive of local and global environmental conditions and processes. Thus, the challenge is to develop a new, more favorable energy system and its associated sources. The resolution of this challenge will, it is hoped, be the discovery of an energy technology that can economically convert solar energy to a locally useful form. One possibility is a solar conversion system yielding hydrogen to be used in fuel cells. Many technology ventures could exploit this opportunity favorably. It can become a great story with an important outcome.

A well-told narrative plan that shows a difficult situation and a novel solution leading to improved market conditions can be galvanizing. When listeners can locate themselves in the story, their sense of commitment and involvement is enhanced. By conveying a powerful impression of the process of succeeding, narrative plans can motivate and mobilize resources and investors.

Stories help entrepreneurs deliver direction and inspiration more powerfully than a logical argument. Facts convey information, and stories convey meaning. Entrepreneurs need facts to back up their plans and a story to convey the goals and meaning of the venture [Gargiulo, 2002].

### BET: Selling a Story

In 1980, Robert Johnson created an innovative plan to produce television programs targeted to black viewers. Johnson met with several investors and told them his story of the opportunity to serve black viewers with TV programs that resonated with their values and experiences. The power of this story generated significant financial and cable channel resources. As a result, Black Entertainment Television (BET) was launched successfully in the United States (see www.bet.com).

A great story well told, therefore, will pack enough power to be memorable. We can forget bullet points, but a great story is unforgettable. It displays the struggle of expectation and hope versus reality and challenge [McKee, 2003]. An example of an entrepreneur's story follows:

> My father and I were very close. In 1999, he exhibited congestive heart disease that appeared to be untreatable. While in the hospital for further tests, he died in the middle of the night. These tests were inadequate and failed my father and me. I have found and licensed a patent for a new blood test, but the Food and Drug Administration has been slow to respond, and the inventor has gone on to other issues. Our early tests have shown good results for a low-cost test that can illuminate paths to a cure.
>
> Our firm, Heartease, needs $1 million to complete our FDA certification. We have the data and proof of the test's efficacy. We need you on our team with your insights and money. Together, we can save your father or mother.

Ultimately, like a good book or movie, a story moves from an "inciting incident" to a good ending. With the help of colleagues and allies, the entrepreneur turns challenges and obstacles into new opportunities and an outcome desired by all.

## 6.3 The Business Plan

Once an entrepreneurial team has selected an opportunity that is attractive and feasible, it often develops a detailed plan for describing the elements of the proposed business. A **business plan** is a document that describes the opportunity, product, context, strategy, team, required resources, financial return, and harvest of a business venture. The elements of the business plan are listed in Table 6.4. This business plan can be used for many purposes, such as attracting talented individuals and resources to the venture. Of course, there is no one right way to organize and write a business plan, and many successful companies never had a formal business plan [Brinckmann et al., 2010]. Although plans are never perfect, they help entrepreneurs nail down key details by forcing them to formalize their ideas.

**TABLE 6.4  Elements of a business plan.**

- Executive summary
- Opportunity: quality, growth potential
- Vision: mission, objective, core concept
- Product or service: value proposition, business model
- Context: industry, timeliness, regulation
- Strategy: entry, marketing, operations, market analysis
- Organization: structure, culture, talent
- Entrepreneurial team: capabilities, commitment
- Financial plan: assumptions, cash flow, profit
- Required resources: financial, physical, human
- Uncertainties and risks
- Financial return: return on investment
- Harvest: return of cash to investors and entrepreneurs

As stated by Daniel Hudson Burnham (1909):

Make no little plans. They have no magic to stir men's blood and probably themselves will not be realized. Make big plans; aim high in hope and work, remembering that a noble, logical diagram once recorded will never die, but long after we are gone will be a living thing, asserting itself with ever-growing insistency. Remember that our sons and grandsons are going to do things that would stagger us. Let your watchword be order and your beacon beauty.

The business plan is a blueprint for your business. The clarity of this plan will be enhanced by the entrepreneurial team over a period of weeks or months. It enables the team to see clearly the plan of action and understand all the elements. By crafting the business plan together, the team members also can ensure that their goals and assumptions align. Potential executives and other hires will often want to see the business plan, too, in order to assess the venture and their potential fit with it.

If the venture requires outside financing, a business plan will be required by most investors. Listing the assumptions underlying the financial plan will be helpful to all participants. Most useful plans number less than 20 pages with backup and supporting material available on request.

A business plan is an important part of the business building process. Many businesses, however, start with a modestly crafted plan and build it up over the first months after launch. In other words, a business plan is a necessity, but it may not be required in a formal and complete form in order to start. In fact, in a dynamic industry such as semiconductors or nanotechnology, flexibility is key to success, and rigid adherence to a plan may be too limiting [Gruber, 2007]. Thus, the team must recognize that the plan will need to be updated periodically.

Creating a business plan teaches the team about the market, customers, and each other. A plan will probably turn up at least one or two big gaps, which can be corrected. Table 6.5 lists 10 common gaps that show up in a

**TABLE 6.5  Ten common mistakes or gaps in business plans.**

| | |
|---|---|
| ■ Solutions or technologies looking for a problem | ■ Inadequate description of revenue and profit drivers |
| ■ Unclear or incomplete business model and value proposition | ■ Limited or no description of the metrics of the business |
| ■ Incomplete competitor analysis and marketing plan | ■ Lack of focus and a sound mission |
| ■ Inadequate description of the uncertainties and risks | ■ Too many top-down assumptions such as "we will get 1 percent market share" |
| ■ Gaps in capabilities required of the team | ■ Limited confirmation of customer demand or pain |

business plan. The entrepreneurs can respond to these gaps or mistakes and seek to fix them.

A good business plan starts with a good understanding of the market, pursuit of a market that is either big or will be big, a defensible product plan that includes hard-to-copy intellectual property, and a go-to-market plan that has a cost-effective marketing and sales approach. It helps immensely if the entrepreneur has a background in previously building products for the market he or she plans to pursue. Ken Xie of Fortinet had such a plan. Ken was the founder of Netscreen, a company that was the recognized leader in remote secure-access products. He sold that company to Juniper for $3.5 billion.

As a result of the experience that Xie had accumulated with Netscreen, he had a good understanding of the security market and was aware of problems that his customers were having with securing their networks. He knew that customers were using a number of application software products to stop security threats. Managing multiple products was hard and even with those multiple products, outsiders were still penetrating corporate security. Xie felt that if he could build a single hardware network-security-appliance product that had superior capabilities in catching threats from outsiders, then corporations would be likely to buy such a solution. It would perform better, cost less, and be easier to administer than multiple software solutions. Xie knew the people who could build such a product as they had either worked for him in the past or worked for companies in the security industry that he had competed against in the past.

Xie assembled a top notch team and built Fortinet, a publicly held company on the NASDAQ that is now the recognized leader in Unified Threat Management. Not only did he bring on excellent developers, he also hired sales and marketing personnel to implement the same channel partner sales methods at Fortinet that were successfully used at his prior company.

Entrepreneurs like Xie are able to leverage their previous knowledge, relationships with customers, and relationships with domain experts to build both a great product and a successful go-to-market plan.

The business plan can be used to align the interests of all the participants of a new venture, as shown in Figure 6.3. The business plan explains how the people, the resources, and the opportunity can be linked to a deal that will hopefully benefit all stakeholders—employees, investors, suppliers, and allies [Sahlman, 1999]. See Section 2.5 for a further discussion of proper fit and alignment necessary for success.

After the business plan is outlined or drafted, the team should prepare an executive summary. The executive summary captures the essence of the business plan and, for many new ventures, it stands alone as a short business plan. When readers finish the executive summary, they should have a good sense of what the entrepreneurs are trying to do in their business. A successful executive summary captures the readers' attention and imagination, causing them to want to learn more.

An executive summary contains six elements (Table 6.6). The summary should be no longer than three pages. Most professional investors will ask to have it emailed to them. Remember that the goal of the summary is to entice someone into reading the complete plan or into a further conversation.

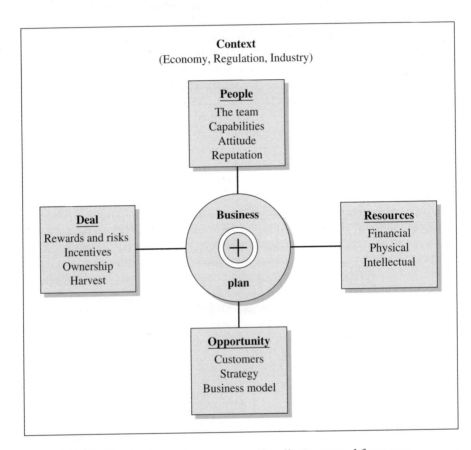

**FIGURE 6.3** The business plan serves as the alignment tool for a new business venture.

**TABLE 6.6 Elements of an executive summary.**

1.  Business concept including the problem and its solution
2.  Market, customer, and industry analysis
3.  Marketing and sales strategy
4.  Organization with key management and allies
5.  Financials including 3+ years of summary projections
6.  Financing history and investment offering

# 6.4 The Elevator Pitch

Often entrepreneurs have a chance to make their case to a potential investor or ally. The importance of a compelling story was discussed in Section 6.2. A short version of the new venture story is often called the **elevator pitch,** which gets its name from the two-minute opportunity to tell a story during an elevator ride. Chance meetings (in places such as elevators) will offer opportunities to entrepreneurs to make a brief case for their venture. A prepared short version of the venture story can be a powerful door opener.

The goal of the short story is to get approval to proceed to the next step, where the entrepreneur can tell the longer version of the story and secure new colleagues, allies, and investors. Thus, the entrepreneur must recognize that there isn't time to elaborate on many details. Instead he or she must quickly convey the essence of the opportunity in a way that invites a longer conversation. A short version of the venture story demonstrates that the entrepreneurs know their business and can communicate it effectively.

The short version of the venture story starts with an introduction, moves into a description of the opportunity, and then describes the potential benefits of the new venture. The secret of strong short stories is in grabbing the attention of listeners, convincing them with the promise of mutual benefit, and setting the stage for follow-up. The story can start with a captivating question such as: "If IKEA can provide babysitting, why can't movie theaters?"

The short version of the story should convey the vision of the venture. In the case of Genentech, it is: "We discover and make biotechnology pharmaceutical products to reduce or overcome the effects of cardiovascular, pulmonary, and cancer diseases." An important vision can provide the inspiration for the venture and its short story.

There are techniques that a storyteller can employ to make the pitch memorable. First, the storyteller should speak in terms the listeners can understand and should focus on one simple message. Second, the storyteller should use concrete images and examples rather than ambiguous or abstract phrases. Third, the storyteller should generate interest and curiosity by exposing gaps in current solutions and then filling these gaps through the proposed new venture. Ideally, this will cause an emotional response from the listeners. Finally, the storyteller must be passionate about the idea, demonstrating a genuine attachment to the problem and solution [Heath and Heath, 2007].

## Intuit's Elevator Pitch

During the founding of Intuit, Scott Cook described his venture in this way: "Homemakers need to pay the family's bills. They hate the hassle of bill collection and payment. They need a software computer program to quickly and easily pay their bills. Other programs are too slow and too hard to learn. Our solution is a fast, easy-to-use program with no instruction books needed. So the bill payer needs Quicken!"

## 6.5 An Annotated Table of Contents

Writing a complete business plan forces the entrepreneurs to crystallize key business issues in their minds. There are many ways to structure a business plan, with various references, structures, and templates available. The business plan process focuses on the venture's key success factors by posing questions that must be addressed for an idea to become a true opportunity. The remaining content of this section proposes questions for each part of a typical business plan. An example of a well-prepared executive summary is provided in appendix A. See the textbook's websites for additional plans and resources.

## Executive Summary

The executive summary is the most important part of the business plan. Many investors make their decision to proceed with further discussions (e.g., due diligence) based on this single section. The executive summary should encapsulate the key positioning and reasoning found in the rest of the business plan. Both the vision and mission statements should assist in succinctly communicating a compelling opportunity.

- Why is this a big problem and why are customers willing to pay for solutions?
- How does the venture plan on solving the customer problem or need?
- Why is this venture uniquely positioned to do this?
- How attractive are the economics? Why is this an exciting growth opportunity?
- Who is the team, and what key partnerships are in place?

## I. Opportunity and Market Analysis

Investors like funding big problems representing large opportunities. Start strong by demonstrating a solid understanding of the customer and why this problem or pain is important to him or her. Performing customer segmentation will convince readers that the venture can grow to the size of the addressable market.

- What is the problem or need being solved by the venture?
- Who is the customer or customer segment(s)?
- How large is the total addressable market, and how much is it growing?
- How is the current market context favorable or not?

Reference: Chapters 2, 3, and 4

## II. The Solution and Concept

Many product descriptions lose credibility by being too "big picture," relying on market hype, or being too product-focused in the hopes that simply explaining the technology will make it evident why the idea is valuable. It is important to balance the use of technical or industry-specific lingo with common, everyday language. In addition, detailing a value proposition and business model will ensure that the economics of the business are sound from the start.

- What is the product or service?
- Describe a "day-in-the-life" of the customer before and after adoption of the solution. What is the value proposition to a customer, and why it is compelling for the customer?
- Which customers have validated the product and are willing to pay for it?
- What is unique and defendable about the business?
- What is the business and economic model? How attractive are the financial margins?

Reference: Chapters 3, 4, 5, and 16

## III. Marketing and Sales

This section of the business plan should clearly communicate an understanding of how to successfully market and sell your product to the identified customer segments. Understanding and communicating your customer development strategy are as important as your product development strategy. It should be done in sync with product development to improve the odds of success. Your business model and pricing strategy should also extend clearly into the selected sales strategies.

- What are the most appropriate marketing mediums to reach the customer segment(s)?
- What is the most appropriate type of sales channel for the product (e.g., direct vs. indirect sales)?
- Who is the customer decision maker with purchasing power, and who influences that person to buy?
- What is the expected sales cycle length?
- Are there partnerships that can be leveraged to advertise and sell?

Reference: Chapters 4 and 9

## IV. Product Development and Operations

At this point, the reader should be convinced that the entrepreneurs have identified a compelling market and know how to generate revenue. This section focuses on product development and how the product will be made marketable. Any key technologies being leveraged for development should be explained clearly (e.g., diagrams are helpful). Demonstrate that continued revenue growth is planned by specifying long-term product goals. This section largely drives the amount and timing of cash required for the business, making it a critical component of the financial model.

- What is the current state of development of the product(s)?
- What resources will be required to finish and ship the product? Be specific about what types of resources will be required (e.g., engineering, tools, suppliers, materials, partners, and customer involvement).

- What are the planned development timelines and key milestones targeted?
- What are the key risks that will be mitigated at each milestone?
- What does the value chain look like for production and product delivery?
- Do any patents, trade secrets, or other defendable advantages exist?
- Are there any regulatory hurdles that must be cleared?

Reference: Chapters 5, 8, 13, and 14

# V. Team and Organization

Building a team is a critical part of beginning a new venture and communicating credibly to outside parties. Understanding how the current team fits into the broader venture vision will help investors and partners understand what roles remain to be filled and how they can potentially assist.

- What are the backgrounds and roles of the founders and early key employees?
- Describe the passions and skills of the team and why the team is committed to the opportunity.
- What key hires must be made to fill out the team?
- What head count levels are forecasted in each functional department?
- Does the company have advisors or board members that strengthen its story?

Reference: Chapters 10, 12, and 20

# VI. Risks

A new venture is confronted with four major types of risk: technology/product, market/competition, management/team, and financial. Many of the opportunity-specific risks are interwoven into earlier parts of the business plan. For instance, potential competitive threats should be considered when framing both the product and go-to-market sections. In addressing company-specific risks, it is important to think clearly about how each risk factor can be managed in the coming year or two. Quantitative analysis will also aid the reader. It is critical to identify which risks need to be reduced so that the reader will be confident that the entrepreneurs understand how to build a business.

- What are the key product development risks and external dependencies?
- What is being done to mitigate product execution risks?
- Who are your main competitors, and how are you differentiated from them in the marketplace?
- Can large players easily enter the market? Are there product substitutes?
- What customer, partner, or product strategies can be used to mitigate competitive threats?

Reference: Chapters 3, 4, and 7

## VII. Financial Plan and Investment Offering

Although the financial plan is considered last, the implications of financial decisions appear throughout. If the company executes successfully across product development, marketing, sales, and other company functions, the financial results must be attractive enough to make an investment. Ensure that any financial assumptions and results are feasible by citing an enterprise that is analogous to the planned venture. Investors want to know how much funding is required and what measurable milestones will be reached. Staged financing allows both investors and entrepreneurs to better manage the risk associated with a new venture. Include a timeline that integrates company sales and product milestones, planned funding events, and cash flow position.

- What is the required funding to meet the market and product milestone goals? What amount is being asked for?
- When is the venture forecasted to be cash-flow positive?
- What is the growth opportunity of the business if successful?
- What are the forecasted initial and steady-state financial margins?
- What other companies exhibit margins and growth similar to this venture?
- What are the key financial assumptions?

Reference: Chapters 15, 16, 17, 18, and 19

## Appendix. Detailed Financial Plan

A more detailed set of financial projections and assumptions are generally included in an appendix. The forecasted financials and assumptions will serve as a starting point for valuing the venture. Be sure the methodology used to arrive at the financials is transparent to the reader.

- Five-year detailed cash-flow statement, income statement, and balance sheet (monthly for first year and quarterly or yearly thereafter).
- Financial assumptions made in the construction of the financial estimates (e.g., customer penetration rates, pricing, and working capital assumptions).
- Are purchasing decisions cyclical in this industry?
- What are the largest costs of the business (e.g., engineering development, regulatory trials, manufacturing, and marketing)?
- How will product and sales costs change as volume grows?
- Has customer support and maintenance been factored in?

Reference: Chapters 16 and 17

## 6.6 Spotlight on Amazon

Jeff Bezos founded Amazon in 1994. Bezos had researched the Internet in the early 1990s and had identified a number of product categories that he thought could be sold online. He focused his attention, initially, on books. Traditionally, bookstores stocked a small percentage of all available titles and they charged a considerable markup to cover the cost of their stores and personnel. Moreover, publishers had little information on who purchased specific titles, which limited their marketing efforts. Bezos proposed Amazon as a means of cutting costs while delivering greater value to both the customers and the publishers.

In 1995, Bezos wrote his business plan, detailing the people, resources, opportunity, and deal (Figure 6.3). He hired a team, developed the website, and launched Amazon.com. Over time, Amazon has expanded its product offerings. In 2006, they launched a cloud computing service; in 2011, they entered the tablet computing market with the Kindle Fire; and in 2013, they launched the Auto-Rip digital music service. Amazon continues to execute upon its vision even as it moves into new markets.

## 6.7 Summary

Building a business is described as a process that can be learned and mastered by talented and educated entrepreneurs by developing the following descriptions:

- Opportunity, vision, value proposition, and business model.
- Concept, feasibility, and story.
- Financial plan, legal form, and business plan.
- Resource acquisition plan.
- Execution and launch process.

We note that entrepreneurs have a story and business plan that helps them to codify and communicate their roadmap to achievement. Most entrepreneurs benefit from putting this plan into written form to sharpen their thinking, communicate it well to others, and enable testing of critical assumptions and hypothesis.

> **Principle 6**
> Entrepreneurs can learn and master a process for building a new venture and they communicate their intentions by developing a story and writing a business plan.

## Video Resources

Visit http://techventures.stanford.edu to view experts discussing content from this chapter.

| | | |
|---|---|---|
| Purpose of a Business Plan | Tom Byers | Stanford |
| Make Meaning in Your Company | Guy Kawasaki | Garage |
| Work Backwards From the Customer | Diego Piacentini | Amazon |

# 6.8 Exercises

**6.1** TerraPass sells an investment in energy projects or credits used to offset environmental impact. For example, drivers of gas-guzzler cars can exhibit a decal showing their environmental investment. Create a summary for this business (www.terrapass.com).

**6.2** A new firm develops and distributes electronic games for mobile devices. These games teach children to read, recognize symbols, and perform mathematics. This new venture needs $1 million to launch a nationwide campaign for its products. Prepare a short story that will persuade a venture capitalist to support the firm.

**6.3** Evaluate the business plan from appendix A using tables and figures from Section 6.3.

**6.4** For whom is a business plan written and why? Using Figure 6.3, explain how a business plan serves as an alignment tool for key stakeholders in the business.

**6.5** What information on the competitive landscape should be included in a business plan? What frameworks would you use?

**6.6** What information on the key venture risks should be included in a business plan? Is it important for the business plan to be pessimistic or optimistic in regard to these risks?

## VENTURE CHALLENGE

**1.** Create a draft table of contents for your venture's business plan.

**2.** Describe the process you will use to create a business plan.

**3.** Write a concept summary and story for your opportunity.

# CHAPTER 7

# Risk and Return

*Our greatest glory is not in never falling but in rising every time we fall.*
**Confucius**

**How can the uncertainty and risks of ventures seeking to scale and grow be managed effectively to achieve success?**

A new venture that creates a novel solution to a problem will be subject to uncertainty of outcome. An action in an uncertain market is sure to experience a risk of delay or loss. It is the entrepreneur's task to reduce and manage all risks as much as possible.

Attractive new ventures can be designed to grow as demand for their products increases. Furthermore, it is hoped that economies of scale will be experienced so that as demand and sales grow, the cost to produce a unit of product declines. Additionally, it is desirable to have economies of scope so that costs per unit decline due to the spreading of fixed costs over a wide range of products. Many industries established on a network format exhibit network economies resulting in a reinforcing characteristic leading to the emergence of an industry standard. ∎

139

## 7.1 Risk and Uncertainty

The pursuit of important opportunities and big goals by entrepreneurs requires them to assume more risks than they might take on working for a mature company or the government. Introducing a novel product into a new market has an uncertain outcome. An outcome resulting from an action is said to be **certain** when it will definitely happen. Something certain is reliable or guaranteed. For example, it is certain that if we drop a rock, it will fall to the ground (and not float upward).

An outcome resulting from an action is said to be **uncertain** when the outcome is not known or is likely to be variable. **Risk** is the chance or possibility of loss. This loss could be financial, physical, or reputational. When Christopher Columbus embarked on his first voyage to the New World, he risked financial, reputational, and bodily harm. Farmers are a group whose fortunes are vulnerable to unpredictable outcomes due to drought, flood, or other weather conditions.

Most, perhaps almost all, people are risk-averse or risk-avoiders. Logically, an entrepreneur seeks to avoid or reduce the risk of an action. For example, farmers purchase insurance to mitigate the effects of uncertain weather. A simple measure of a person's risk aversion is provided in the following example of a coin toss game [Bernstein, 1996].

> **Example: A Coin Toss Game**
>
> You have a choice of receiving $50 for certain or an opportunity to play a coin toss game in which you have a 50 percent chance of winning $100. The $50 gift is certain, if you elect it. The game's outcome is uncertain, but the expected outcome over the long run is (if you play it many times) also $50 since, for a true coin, the probability of winning, P, is 50 percent. Surely, you would play the game if the probability, P, of winning were 100 percent. Would you play if the probability were 70 percent? What is the probability, P, that you would require to play? If your P=70 percent, you are risk-averse, and if your P=40 percent, you are a risk-seeker. Of course, your willingness to play this game will also depend on your overall wealth in relation to the $50 outcome of loss and the fun of playing this simple game. This game illustrates the fact that most people, perhaps like you, are risk-averse.

Entrepreneurs are often thought of as risk takers. Most good entrepreneurs, however, are relentless about managing risk [Gilbert and Eyring, 2010]. The ability to successfully choose the risks worth bearing is a form of human capital based on experience and good judgment [Davis and Meyer, 2000]. We usually assume that the elevated risk of an entrepreneurial venture may provide a higher return on this human capital. Entrepreneurs often have a capability to limit the downside risks of a venture by applying their skills that are useful for mitigating the risks. One mental model may be the circus high-wire walker with a strong

**TABLE 7.1  Four levels of uncertainty.**

| Uncertainty level | Level of risk | Example |
|---|---|---|
| 1. A clear, single outcome | Very low | Purchase a Treasury Bill |
| 2. A limited set of possible outcomes | Low | Set up a pushcart for selling sports items at the Olympics |
| 3. A wide range of possible future outcomes | Medium | Launch an improved product in a new market |
| 4. A limitless range of possible outcomes | High | Establish a firm to attempt to design and sell a revolutionary power source based on a fuel cell breakthrough |

net below. In fact, most successful entrepreneurial firms create value by taking calculated risks and possess the core competency to mitigate or manage these risks. All of life is the management of risk, not its elimination [Brown, 2005].

The entrepreneur is in many ways like an investment manager who chooses to pursue selected opportunities and not others. When we analyze risk, we look forward in time and try to estimate the potential outcome and the variability of that outcome. Risk is a measure of the potential variability of outcomes that will be experienced in the future. Furthermore, risk is the chance or possibility of loss. Table 7.1 lists the four possible levels of uncertainty [Courtney, 2001]. The level of return on investment we might expect should be commensurate with the level of uncertainty.

Entrepreneurs practice in the realm of opportunity in a manner similar to what financial investors do in the stock market [Sternberg et al., 1997]. They pursue opportunities with expected levels of risk and attempt to "buy low and sell high." Buying low means pursuing ideas that are not widely recognized or are out of favor. Selling high means finding a cash buyer for the successful venture, convincing the buyer that its worth will yield a significant return and it is time to harvest some of the wealth already created. As with any investment, "invest in a business you understand" is a good principle. Thus, pursuing the hope of fusion power should be left to those few who know and understand this big but risky opportunity. The entrepreneur-investor assumes the risk of the venture and should be willing to take on sizable risks with the knowledge that he or she can manage or mitigate them.

The best method for most entrepreneurs is to use a form of experimentation—trial and error. Identify possible new ventures and take a few steps toward a business and then evaluate the early feedback. If it looks good, keep going forward. Entrepreneurs will most likely take the risk of failure if the losses are constrained on the downside and the potential rewards are high on the upside [Sull, 2004]. Henry Ford said: "Failure is the opportunity to begin again, more intelligently."

Successful entrepreneurs also rank risks in order of importance and affordability, with deal killers listed first, strategic risks second, and operational risks third. They also recognize dependencies in their business plan, where a failure or delay in one area of the plan will impact other areas of the plan. With these risks identified and ranked, an entrepreneur can then work toward managing or reducing the most critical risks first—before they have invested too much in the venture [Gilbert and Eyring, 2010].

Entrepreneur-investors should consider the concept of **regret,** which we define as the amount of loss a person can tolerate. People treat regret of loss differently than the potential of possible gain. How much regret people feel varies depending on their circumstances of wealth, age, and psychological well-being. Reconsider the simple coin toss game. If you play the game and lose $50 on the first turn, it may be disturbing. Your level of regret may be $200. If you play a sequence of four turns and lose each of them, you will reach your level of regret. Thus, entrepreneurs need to evaluate their regret level of acceptable loss and limit their investment in any entrepreneurial venture. If an entrepreneur can forgo one year of income ($50,000) and willingly invest his or her savings of $60,000 in the venture, then the entrepreneur's level of regret is $110,000.

The amount of risk entrepreneurs will endure varies, but most retain some personal financial reserve so that failure will not equate to homelessness or starvation. It is easier to be a risk taker if one has some reserves to fall back on.

The risk-adjusted value of a venture, V, is

$$V = U - \lambda R \qquad (7.1)$$

where U = upside, $\lambda$ = risk-adjusted constant, usually greater than 1, and R = downside or regret. The larger the value of $\lambda$, the more risk-averse is the entrepreneur. We are neutral at $\lambda = 1$ but risk-averse at $\lambda = 2$ (Dembo and Freeman, 1998). If your regret is R = $110,000 and $\lambda = 2$, to proceed requires V > 0, or

$$U > \lambda R \qquad (7.2)$$

Therefore, the required upside, U, is U > $220,000. Entrepreneurs can use scenarios and economic analysis to estimate the potential upside, U, of a venture.

The strategic response to uncertainty is to build a venture in stages, reserving the right to adjust your core competencies and strategies, and play again at the next stage. Thus, reconsidering the case of the entrepreneur who is risk-averse with $\lambda = 2$, he or she can choose to proceed for six months with the venture so that R = $55,000, and the minimum upside required is then only $110,000. After the six months, the entrepreneur adjusts the business strategy to improve the business performance and based on a new calculation of the upside, U, proceeds to a second stage of activity or decides to terminate the venture.

To perform the kinds of analyses appropriate to high levels of uncertainty, most firms will need to enhance their strategic capabilities. Scenario-planning techniques can be useful for determining strategy under conditions of uncertainty.

Risk reflects the degree of uncertainty and the potential loss associated with the outcomes, which may follow from an action or set of actions. Risk consists of two elements: the significance of the potential losses and the uncertainty of those losses. In most new ventures, it is the significance or size of the potential losses, *hazard,* and the *uncertainty* that are estimated by new venture entrepreneurs and their investors. We propose a measure of risk as:

$$Risk = hazard \times uncertainty \qquad (7.3)$$

The hazard, H, is the size of the potential losses as perceived by the entrepreneurial team. Hazard is an entrepreneur's income forgone (opportunity cost, OC) plus the financial investment, I, which he or she will need to make. Therefore,

$$Risk = (I + OC) \times UC \qquad (7.4)$$

The uncertainty, UC, is measured by the variability in anticipated outcomes, which may be described by their estimate of the probability of loss (failure). Based on these factors, the entrepreneurial team may make a selection of a new venture (or to proceed or not), as shown in Figure 7.1. High levels of hazard may not deter entrepreneurs from choosing ventures with potentially high levels of returns. The decision to proceed or not depends on the level of risk adversity

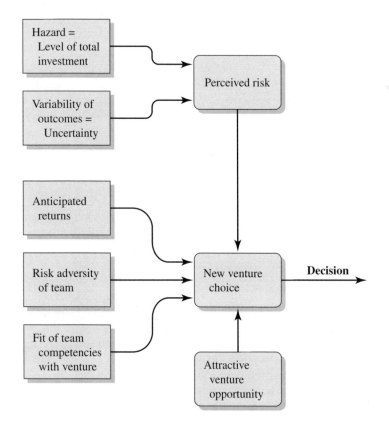

**FIGURE 7.1** Risks and new venture choice.

## TABLE 7.2  Sources of uncertainty.

1. Market uncertainties
   - Customer
   - Market size and growth
   - Channels
   - Competitors

2. Organization and management uncertainties
   - Capabilities
   - Financial strength
   - Talent
   - Learning skills
   - Strategies

3. Product and processes uncertainties
   - Cost
   - Technology
   - Materials
   - Suppliers
   - Design

4. Regulation and legal uncertainties
   - Government regulation
   - Federal and state laws and local ordinances
   - Standards and industry rules

5. Financial uncertainties
   - Cost and availability of capital
   - Expected return on investment

of the team members and their perception of the extent of the uncertainty. Of course, the estimate of potential returns is subject to the assumptions underlying their calculation.

An entrepreneurial venture is launched with a degree of uncertainty due to the novelty of the product to the market, the novelty of the production processes, and the novelty to management [Shepherd et al., 2000]. Novelty to the market concerns the lack of uncertainty of the market and customer uncertainty. Novelty of production processes is dependent on the extent of knowledge of the processes by the venture team. Novelty to management concerns the venture team's lack of the necessary competencies. Regulation and legal changes are also a source of uncertainty. Sources of uncertainty are listed in Table 7.2.

The risk of failure or poor performance is significant and should not be understated. According to the U.S. Small Business Administration, about one-half of all small businesses are acquired by another firm or leave the market within four years. Of course, being acquired at a good price may be the success one is seeking. It is fair to say that one-fourth of all business start-ups are discontinued within four years, but it is unclear at what cost or return.

Technology ventures usually have four types of risk: technology, market, financial, and team. Every effort should be made to test the assumptions about these four sources of risk. Strategic risk management involves setting up strategies that anticipate the downside of risk. The key to surviving risks is assessment and response [Slywotzky and Drzik, 2005].

Technology entrepreneurs are often less concerned with certainty than with getting into the game quickly and learning how to participate. The uncertainty and associated risks decline as the novelty in the three dimensions of market, production, and management declines. Novelty is synonymous with uncertainty, and, thus, we expect uncertainty to decline as knowledge of the market, the processes, and management competencies improves. Thus, the liability of newness declines over time. A process for managing risk and uncertainty is shown in Figure 7.2.

Whenever there is uncertainty, there is usually the possibility of reducing it by the acquisition of information. Indeed, information is essentially the negative of uncertainty. The acquisition of information and knowledge improves

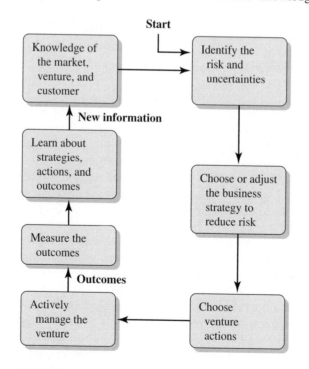

**FIGURE 7.2** Managing risk and uncertainty.

an organization's chances of adaptation and performance. An entrepreneur is continuously making decisions about highly uncertain environments and the new venture's internal structures that, in turn, modify the venture's performance outcomes. New venture managers may learn from past choices about how to perform better in the future. This learning can facilitate adaptation to changed environmental conditions since the strategic choices of managers help define the outcomes. An appropriate strategy for risk reduction uses new information learned from experience to adjust the business strategy and the actions taken to execute the strategy. Appropriate strategies might include adding new people to the team, creating new alliances, reducing costs, or improving customer relationships, among others.

Arthur Pitney received a patent in 1902 for a hand-operated postage meter that he hoped would be used to replace stamps for mass mailings. Pitney recognized his primary source of risk was regulatory since the U.S. Post Office controlled postage services. By 1918, after continual rejections, Pitney was joined by Walter Bowes to form Pitney Bowes and try again to get Post Office approval. With Bowes's persuasive skills, they finally received a license for the postage meter in 1920. They ultimately overcame the regulatory hazards to build up their firm. Many entrepreneurs underestimate their various risks.

## Symantec: Risk and Reward

In 1981, Gordon Eubanks, a software pioneer in the personal computer industry, cofounded C&E Software to develop an integrated database management and word processor product. In 1983, John Doerr, who was a venture capitalist with Silicon Valley's Kleiner Perkins Caufield and Byers (KPCB), approached Eubanks regarding one of his investments. Symantec was an artificial intelligence software firm that was struggling to stay afloat despite having an interesting technology known as natural language recognition. Doerr suggested that Eubanks merge with Symantec and incorporate Symantec's technology into C&E's product (see Figure 7.2). Although Eubanks viewed this technology integration as a compromise, he was persuaded by the upside of the deal—a substantial percentage of ownership for C&E in the merged venture, an additional cash investment from KPCB, and a chance to lead the new company. In late 1984, Symantec Corporation was reborn as a firm that developed a natural language database manager with word processing capabilities.

Even with the additional venture capital funds, it became apparent that Symantec would not survive as a one-product company. A few months before the first version of Symantec's database manager shipped in 1985, Eubanks hired Tom Byers to search for new revenue streams and to diversify the product line. Byers realized that a market existed for software utilities that added features to the then-popular Lotus 1-2-3 spreadsheet.

Symantec used a strategy similar to book publishing with these products: in return for the rights to package and sell the software, Symantec paid developers a royalty as the product's author. Eubanks had adjusted Symantec's strategy in response to risk and uncertainty.

In 1987, Eubanks further diversified the firm. Symantec purchased Breakthrough Software for minimal cash and a $10 million note payable in cash if and when Symantec went public. Timeline, Breakthrough's project management software, quickly doubled Symantec's incoming revenue, which was important because sales of Lotus utilities were rapidly declining.

Also in the late 1980s, Eubanks quickly acted on two emerging trends in the industry. First was the proliferation of computer viruses as more and more computers were connected in networks. Second was the rapid advancements of graphical user interfaces (GUI) in personal computers such as the Macintosh. To rapidly respond to these changing conditions, Eubanks made several key decisions and acquisitions.

Ted Schlein, who worked for Byers in Symantec's publishing division, identified a market for antivirus software and pushed for Symantec to publish Symantec Antivirus for Macintosh (SAM), which became the first successful commercial antivirus product. In following years, Symantec made numerous acquisitions in this category, including Peter Norton Computing; network security eventually became Symantec's core business. In addition, Eubanks acquired two Macintosh software companies whose developers were very experienced in GUI-based software. These acquisitions helped prepare the company for the subsequent decade in which all PC software was based either on Windows or Macintosh user interfaces.

Finally, marketing and distributing Symantec's growing line of software became vital to its success. Recognizing that the software industry was rapidly maturing with large corporations becoming most important, Eubanks hired John Laing as head of global sales and Bob Dykes as CFO. Laing and Dykes, experienced professionals who had worked in large business operations in the past, created the necessary systems and processes to handle the upcoming rapid growth.

In 1989, Symantec went public with 264 employees, $40 million in sales, net income of $3 million, and 15 products. During the 1990s, Eubanks focused Symantec on network security technology. In 1999, Eubanks stepped down as CEO of Symantec, and John Thompson from IBM took the reigns. Thompson continued the tradition of focusing the entire product line around enterprise security. Symantec expanded into the market of overall data reliability by merging with Veritas in 2005, becoming one of the largest software companies in the world. In 2012, Symantec had revenues of nearly $7 billion, more than 20,500 employees, operations in over 40 countries, and a market capitalization of over $12 billion. Industry leaders like Symantec continuously manage risk and return.

**TABLE 7.3  Estimating risk and reward.**

1. Describe the most likely scenario, the expected reward, and its estimated probability.

2. Describe the worst-case scenario, the expected loss, and its estimated probability.

3. Describe the best-case scenario, the expected reward, and the estimated probability of it occurring.

4. Determine how much the entrepreneurial team and their investors can afford to lose. Include their investment and opportunity costs.

Businesses can manage the problem of unpredictable customer behavior by following the ideas of portfolio management. The portfolio of customers should be diversified so as to produce the desired returns at the particular level of uncertainty the firm can tolerate. Customers are risky assets. As with stocks, the cost of acquiring them is supposed to reflect the cash-flow values they are likely to generate. The concept of risk-adjusted lifetime value of a customer has a transforming power [Dhar, 2003].

To the uninitiated, successful new ventures appear to be the right idea at the right time. However, entrepreneurs put the pieces together so that while it looks like a happenstance, it is actually the entrepreneur who makes it happen with good calculations based on sound information. As Charles Kemmon Wilson, the founder of Holiday Inn, said [Jakle et al., 1996]: "Opportunity comes often. It knocks as often as you have an ear trained to hear it, an eye trained to see it, a hand trained to grasp it, and a head trained to use it."

One way to calculate an estimate of a new venture's potential risk and reward is to answer the four questions posed in Table 7.3. In general, the entrepreneur seeks a venture where the return is expected to significantly exceed the potential losses.

## 7.2 Scale and Scope

In this section, we consider the strategic impacts of the scale and scope of a firm. The **scale of a firm** is the extent of the activity of a firm as described by its size. The scale of a firm's activity can be described by its revenues, units sold, or some other measure of size. **Economies of scale** are expected based on the concept that larger quantities of units sold will result in reduced per-unit costs. Economies of scale are generally achieved by distributing fixed costs such as rent, general and administrative expenses, and other overhead over a larger quantity, q, of units sold. This effect is portrayed in Figure 7.3. The cost per unit decreases, reaching a minimum at $q_m$. Often, the cost per unit will increase for $q > q_m$, since the complexity of coordination may increase costs per unit for a high number of units.

When significant economies of scale exist in manufacturing, distribution, service, or other functions of a business, larger firms (up to some point) have a cost advantage over smaller firms. Thus, smaller, new-entrant firms need to

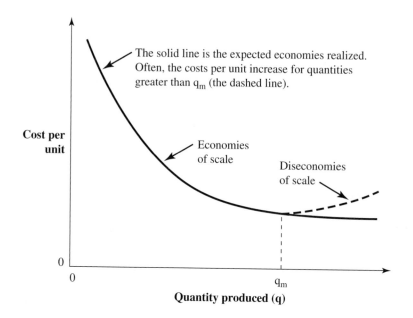

The solid line is the expected economies realized. Often, the costs per unit increase for quantities greater than $q_m$ (the dashed line).

Cost per unit

Economies of scale

Diseconomies of scale

0

0

$q_m$

**Quantity produced (q)**

**FIGURE 7.3** Economies of scale.

differentiate their product on qualities other than price. As the smaller, new-entrant firm grows in size, it can also learn to reduce its costs per unit and price competitively with larger firms.

**Google's Competitive Advantage**

Google's sustainable competitive advantage is tied to its innovation underlying its search engine. Search engines for the Web are based on computer science algorithms and need to search without failure [Hardy, 2003]. A search on Google offers great results and keeps people coming back. Revenues for Google were estimated at $100 million in 2001, only three years after starting up. In 2004, Google raised $1.2 billion in its initial public offering. By 2013, its revenues were more than $50 billion. Its primary source of revenues is from advertising on its pages. It gets its fees by selling rights to given keywords so an ad shows up first when those words are entered. With an enormous volume of searches, Google now benefits from tremendous economies of scale.

Another issue related to scale is the concept of scalability. **Scalability** refers to how big a firm can grow in various dimensions to provide more service. There are several measures of scalability. They include volume or quantity sold per year, revenues, and number of customers. These dimensions are not independent, as scaling up the size of a firm in one dimension can affect the other dimensions. Easily scalable ventures are attractive, while ventures that are difficult to grow are less so.

The consequence of growth is the necessity to respond by increasing capacity. **Capacity** is the ability to act or do something. Any firm has processes, assets, inventory, cash, and other factors that must be expanded as the company grows its sales volume. A firm that can easily grow its capacity is said to be readily scalable. For example, as a firm grows, its working capital requirements will grow. **Working capital** is a firm's current assets minus its current liabilities. Sources of working capital for an emerging firm can include long- and short-term borrowing, the sales of fixed assets, new capital infusions, and net income. The ability for a new firm to grow will be influenced by its access to new capital and assets. Managing a firm for scalability is important to its success. To preempt or match competitors, a firm must attempt to foresee increases in demand and then move rapidly to be able to satisfy the predicted demand. This strategy can be risky since it involves investing in resources before the extent of the demand is verified. The total cost, TC, of the production of units is described as:

$$TC = FC + VC \qquad (7.5)$$

where FC is the fixed costs that do not vary with the quantity of production. The variable costs, VC, do vary with the quantity produced where $VC = c \times q$, c being the cost/unit and the quantity being q. This relationship is shown in Figure 7.4. Table 7.4 describes the scalability and economies of scale for four types of businesses.

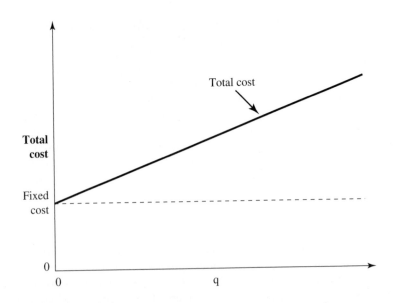

**FIGURE 7.4** Total cost as fixed costs plus variable costs.

**TABLE 7.4 Scalability and the effects of fixed and variable costs for four types of businesses.**

| Type of business | Economies of scale | Scalability | Fixed costs | Variable costs | Primary strategy | Funds required by start-up |
|---|---|---|---|---|---|---|
| 1. Based on talent; e.g., consulting | Low | Medium | Low | High | Recruit talent | Low |
| 2. Based on talent and knowledge assets; e.g., plastics, clean tech | Medium | Medium | Medium | Medium | Secure physical assets and talent | Medium |
| 3. Based on physical assets, knowledge, and materials; e.g., biotechnology, semiconductors | High | Low | High | Low to medium | Secure physical assets | High |
| 4. Based on information with few physical assets; e.g., software, movies | High | High | High | Low | Secure talent to create the software or the movie | High |

The advantage of a talent-based business such as a consulting firm is that the start-up funds are low. The firm is scalable as long as new talent can be recruited for expansion, but it has few economies of scale. The advantage of a firm based on a mixture of talent and physical assets is that it can expand as long as it can secure the necessary funds. A physical asset-based business such as a steel company must secure new plants and equipment as it grows, requiring capital infusion. An information-based business must invest funds upfront to create the software or a movie. It has low variable costs and high economies of scale.

The **scope of a firm** is the range of products offered or distribution channels utilized (or both). The sharing of resources such as manufacturing facilities, distribution channels, and other factors by multiple products or business units gives rise to **economies of scope.** For example, the cost per unit of Procter & Gamble's advertising and sales activities is low because it is spread over a wide range of products. Procter & Gamble's disposable diaper and paper towel businesses demonstrate a successful realization of economies of scope. These businesses share the costs of procuring certain raw materials and developing the technology for new products and processes. In addition, a sales force sells both products to supermarket buyers, and both products are shipped by means of the same distribution system. This resource sharing has given both business units a cost advantage compared to their competitors [Hill and Jones, 2001].

> **Facebook: Seeking Economies of Scope**
>
> Facebook is a very popular social networking site with a significant market share in the United States and Europe. Seeking to take advantage of its large user base, it launched the Facebook Platform in 2007. This allowed any developer to program small Web applications for users to add to their profiles. These applications take advantage of Facebook's comprehensive social connectivity capabilities and include games, event publicizing, gift-giving, and video sharing among others. By 2013, over 9 million applications and websites integrated with Facebook. Facebook is attracting developers seeking to take advantage of the massive economies of scope resulting from Facebook's success in the market.

The economies of scale and scope both reduce the cost per unit. For a factory, its **throughput**—the amount processed within a given time—needs to be consistently high. The introduction of the railroad to the United States and Europe reduced the travel time between markets and supply, increasing the flow of raw materials to factories. The revolution in railroad transportation and telegraph communication resulted in great increases in throughput. In 1870, the Union Pacific Railroad joined the Central Pacific at Promontory, Utah, spanning the United States [Beatty, 2001]. The United States became the world's leading industrial producer by 1929. The economies of scale and scope helped the United States to become a low-cost producer and distributor of goods.

The strategy of a new firm must incorporate plans for economies of scale and scope. Global Solar is an emerging California company whose fundamental business model depends on economies of scale. Global Solar uses a new technology, called the compact linear fresnel reflector, to gather sunlight and generate electricity by boiling water and spinning a turbine. Solar thermal power plants require large numbers of mirrors to concentrate adequate heat on the boiler. This need for a large mirror farm requires Global Solar's power plants to be much bigger than most other solar start-ups. In 2008, Global Solar constructed the first U.S. manufacturing plant for solar thermal systems to test out its new technology in the field.

## 7.3 Network Effects and Increasing Returns

In recent years, realization has been growing that network economies are an important element of competitive economics for new entrepreneurial firms. **Network economies** arise in industries where a network of complementary products is a determinant of demand (also called network effects). For example, the demand for telephones is dependent on the number of other telephones that can be called with a telephone. As more people get telephones, the value of a telephone increases, thus leading to increased demand for telephones. This process is called a positive feedback loop since as more people use the process, the value to users of the process increases and thus demand goes up, leading to more people using the process.

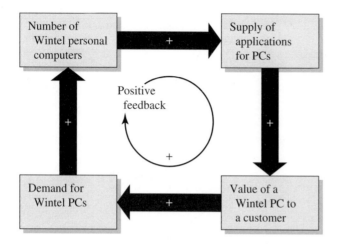

**FIGURE 7.5** Increasing demand for Wintel personal computers due to positive feedback.

The positive feedback process for Windows-Intel (Wintel) personal computers is illustrated in Figure 7.5. As the number of Wintel PCs increases, the incentive for the development of software applications increases (a complementary product). With more application software available, the PC is more valuable to a user. As the value of a PC increases, the demand for Wintel PCs increases, leading to an increased number of Wintel PCs.

Networks include telephone networks, railroad networks, airline networks, fax machine networks, computer networks, ATM networks, and the Internet, among others. The overall tendency is toward "bigger is better." As Figure 7.6 indicates, over time, a winner emerges (company A) and the competitors decline. In the PC industry, Wintel captured the largest market share, with Apple holding the rest of the market. In general, network effects exhibit reinforcing characteristics, as shown in Figure 7.7.

The value of a network, according to Bob Metcalfe [Shapiro and Varian, 1998], is approximately:

$$\text{value of network} = kn^2 \qquad (7.6)$$

where k is a constant to be determined for each industry sector and n is the number of participants in the network. Based on this model, the value of a network grows rapidly as n grows. This simple model assumes all participants are equally valuable, which is not always true. The incentive to a firm in a network economy is to secure market share, eventually taking off toward dominance. This is the underlying theory behind Amazon.com and other Internet start-ups. Of course, market share can grow rapidly while profitability may be elusive. A balance of both market share and profitability can lead an entrepreneurial firm to eventual

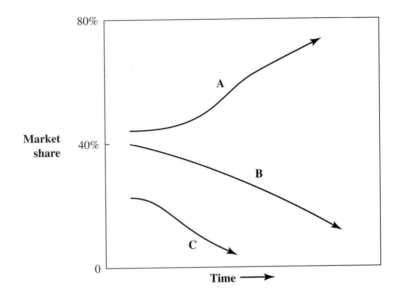

**FIGURE 7.6**  Emergence of a dominant firm, A.

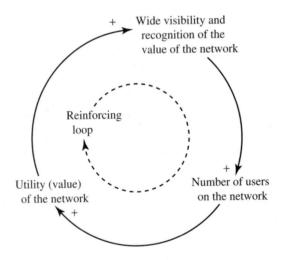

**FIGURE 7.7**  Reinforcing characteristic of a
positive loop exhibiting network effects.

success if it has a product that has high value for its customers and strong alliances with its complementors.

   Network economies work when revenues grow faster than costs. In the
late 1990s Webvan tried to become the online grocer of choice in the United
States. It had to invest in warehousing, trucks, and logistic systems, which led

to increased costs and inventories, which caused its costs to spiral upward faster than its revenues, eventually sinking the firm into bankruptcy.

**Increasing returns** mean that the marginal benefits of a good or of an activity are growing with the total quantity of the good or the activity consumed or produced [Van den Ende and Wijaberg, 2003]. Increasing returns is the tendency for a company that is ahead, firm A in Figure 7.6, to get farther ahead. The theory is that the firm that has a successful product that is increasingly becoming the standard for the industry will experience increasing returns as increasing quantities are sold. However, there is no guarantee that increasing market share (firm A in Figure 7.6) will experience profitability. Furthermore, it is not possible to predict ahead of time which firm will attain market share dominance. If a product or a company or a technology—one of many competing in a market—gets ahead by quality of offering or clever strategy, increasing returns can magnify this advantage, and the product or company or technology can go on to lock in the market. Microsoft DOS became dominant after a protracted battle with CPM and Apple for PC operating system leadership.

Many new firms may enter a new industry with the potential to eventually dominate, but only one or two survive. Products do not stand alone but depend on complementary products to make their use valuable. Examples of network industries that exhibit increasing returns are airlines and banking. As an airline expands the number of cities it serves, the value to a customer of using the airline increases. Another prominent example is eBay, which has dominant market share in online auctions. Because it was the first to connect individual buyers and sellers over the Internet in an auctionlike format, it grew very quickly. As more rare items came up for auction on eBay, more buyers were attracted to the website to bid on those items. Those extra bidders attracted still more sellers, thus leading to dominance. Note that eBay was profitable from the first year onward.

While Metcalfe's law illustrates the general idea of the value of a network, it is only an approximation to reality. The value of each node (participant) will vary. Furthermore, some links will be strong while others are weak. Customers value the number of nodes in the network but also key links in the network. A network of five nodes and eight links is shown in Figure 7.8. Note that not all nodes are connected by a link to all other nodes in this example. Consider a bank network with 100 branches. Most people do not visit many other branches except their local branch and perhaps one near their work. For many customers, the link to their account at their local branch and online or via telephone is what they value. Thus, designers of a network business must analyze their customers' needs and build their business on the best information.

Consider Southwest Airlines. It has no physical branches, nor does it use travel agents; instead it uses telephone and online links to its network. It offers strong incentives to its customers to use the online rather than the telephone links. Physical branches may be necessary for banks but not for airlines. Wells Fargo is putting minibranches in grocery stores on the theory that its customers value physical nodes (branches) as well as Internet links.

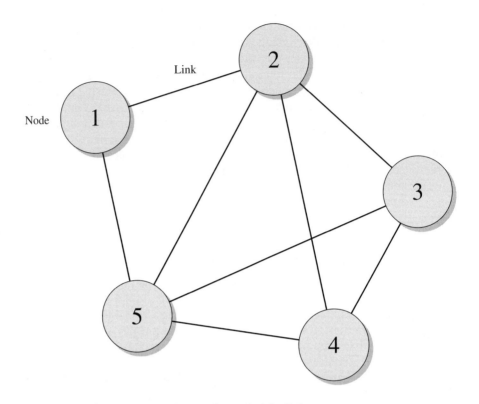

Link

Node

**FIGURE 7.8** Network of five nodes and eight links.

In general, knowledge-based products exhibit increasing returns. The upfront development costs are high, but the per-unit production costs thereafter are low. Knowledge-based products also exhibit network effects; that is, the more people who use these products, the more valuable they become.

### Network Effects at Facebook

During the early 2000s, numerous social networking sites competed for users. Mark Zuckerberg, then a sophomore at Harvard University, launched "The Facebook" in 2004. It was intended to allow students in the same class to find one another and study together. Within one month, more than half of Harvard undergraduates had registered for the service. At first, only Harvard students were allowed access. Gradually, the site expanded, extending membership to other colleges in the Ivy League. The site was then opened to all university and high school students. In September 2006, Facebook was opened to all users.

Since then, it has become the most popular social networking site on the Internet. As of 2013, Facebook had more than 1 billion users. Of people aged 18-29, 92 percent have a social network account and 89 percent of those people have Facebook accounts, according to Pew Research Center. It claims that 85 percent of U.S. undergraduate students are registered. Facebook, Orkut, and Friendster are competing for dominance in the social networking market around the globe. Which of these three companies do you expect to reap the benefits of network effects?

## 7.4 Risk versus Return

Reaching for higher returns carries higher risk. Assuming the entrepreneurs and their investors are rational beings, they will demand higher potential annual returns for higher-risk ventures. We illustrate a risk-reward model in Figure 7.9. The expected return varies as

$$ER = R_f + R \qquad (7.7)$$

where ER = expected return, $R_f$ = risk-free rate of return (T-bill), and R = risk. High-risk and high-return investments will be expected to return in excess of 30 percent annualized over a period of years, T. For a high-risk venture, T may range from three to seven years [Ross et al., 2002]. A disruptive application or radical innovation is expected to return in excess of 40 percent annualized over T years (point a in Figure 7.9).

Compact fluorescent lights (CFLs) are an energy-efficient replacement for traditional incandescent bulbs. They use up to 75 percent less power than incandescent lights and could be the replacement for most home uses. Ellis Yan, a native of China, came to the United States and recognized the opportunity presented by CFLs. He took a risk by making the bulbs when CFLs were largely ignored by large manufacturers and others in the market. CFLs are now seen as the climate-friendly solution for traditional lighting. By 2011, his company, TCP Inc. in Ohio, captured over half the U.S. market and generated revenue of almost $300 million dollars for a large return on his investment.

Often, a more moderate return and risk venture (point b in Figure 7.9) can be achieved by using an incremental technology change rather than a disruptive technology in combination with a change in the business model. A moderate change in technology plus a moderate change in business model can lead to an attractive risk-return profile [Treacy, 2004].

## 7.5 Managing Risk

New technologies and innovations create serious challenges for an existing firm. Change creates risks for competitors and a firm needs to manage risk. Business risk is growing in intensity and leaders of new ventures need to manage strategic risk: the possibility of loss due to hazards, operating risks, and competitive

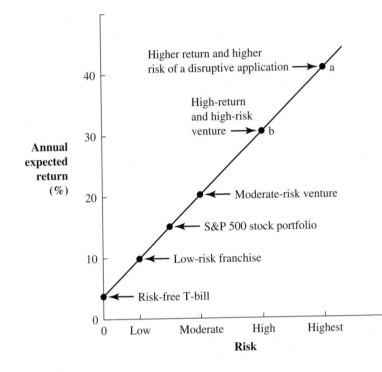

**FIGURE 7.9** Return-versus-risk model for a new venture.

challenges. Managing risk requires anticipating threats and reversing them [Slywotzky, 2007]. Risk and reward can be decoupled if a firm learns how to manage risk. Defeating risk is based on a keen knowledge of the customer, a unique value proposition, and a winning profit model.

As industrial and technological change occurs, a well-managed enterprise is ready to quickly adapt to any challenge. Firms can manage risk by constantly asking about potential challenges to product, brand, and business model. Video-rental store Blockbuster failed to manage risk and lost to a new business model introduced by Netflix. Motorola, Nokia, and Blackberry lost significant cell phone/smart phone market share beginning in 2007 when Apple introduced its iPhones. The superior design, intuitive interfaces, and embedded technologies such as cameras caused users who were less price sensitive to migrate rapidly to the Apple iPhone. Ford and General Motors experienced great risk to their business design when Toyota introduced a hybrid auto called Prius. The Prius offers energy efficiency and a high-tech brand that eclipses the inefficiencies of traditional Fords and GMs.

During an economic recession, a heightened sense of risk of capital exists. However, a new venture's success is sensitive to risks other than an economic downturn. The key factors for a start-up are the management quality of the founders and their earlier experience in the industry, as well as the timeliness of the new technology innovation. It is difficult to find investors during a recession, but they are available. Great entrepreneurs persevere by creating a frugal plan for attracting new customers with reasonably priced products.

Ultimately, innovations are the core element for the future success of a new venture in any economic conditions. Innovations can occur in times of capital market anxieties when resources are limited, as well as during good times and financial exuberance. In fact, for many entrepreneurs with a good innovation, the time to act is during difficult economic times [Graham, 2008].

## 7.6 Spotlight on Dropbox

MIT graduates Drew Houston and Arash Ferdowski founded Dropbox, Inc. in 2009. Dropbox is a cloud storage service that enables users to store photos, videos and documents. Dropbox is based on the "freemium" concept: Although the basic service is free, many subscribers pay a fee to upgrade to a premium service that provides more storage and other features. Thus, we can describe the business model of Dropbox as high-growth company that attracts "free" users and later converts them to paid services.

Dropbox has grown rapidly as existing users encourage new users to subscribe. In turn, as more and more users join Dropbox, the value of the service grows since there are more opportunities for sharing. Dropbox thus exhibits economies of scale and increasing returns. At the same time, by scaling the Dropbox service and infrastructure with demand, the Dropbox team has been able to effectively manage risk. By 2013, Dropbox had 50 million users and revenue of about $300 million.

## 7.7 Summary

A new venture is subject to risks due to uncertainty of outcomes in the marketplace. It is the entrepreneur's job to manage and reduce all risks. As ventures grow, they may experience economies of scale and scope, resulting in lower costs per unit produced. Furthermore, attractive ventures are those that can readily expand their capacity in response to demand and are said to be readily scalable. Many industries operate in a network of participants and exhibit network economies. As some network industries move toward an industry standard, a few firms may experience increasing returns. Significant value can be added to a product when a creative design leads to the embodiment of critical details.

---

**Principle 7**
The entrepreneur seeks to manage risks and attain economies of scale, scope, and networks while achieving scalability of the business.

---

## Video Resources

Visit http://techventures.stanford.edu to view experts discussing content from this chapter.

| | | |
|---|---|---|
| Realities of Consumer Internet at Scale | John Lilly | Greylock |
| Motivation Through Equity and Risk Taking | Bill Gross | Idealab |
| Risk Is a Necessity for Exploration and Growth | Peter Diamandis | X PRIZE |

## 7.8 Exercises

**7.1**   Select a well-known start-up or one that personally interests you. Use the sources of uncertainty outlined in Table 7.2 to discuss three key risk areas for that company in the immediate future. Find a recent article about the company and list which risks are discussed in that article.

**7.2**   An investor is asked to invest $10,000 in a new venture today. The expected return in three years is $28,000 with a probability of occurrence of 70 percent. Would you recommend this investment? Describe your reasoning.

**7.3**   A new entrepreneur is relatively risk-averse with a risk-adjusted constant $\lambda = 2$. Her opportunity cost is $100,000 before earning a regular salary from the venture in its second year. She also invests her savings of $50,000. Calculate the minimum annual return that will be acceptable to her in the second or third year.

**7.4**   Verify the empirical validity of Figure 7.9. Research the following questions using your favorite finance website or newspaper. What was the previous year's risk-free T-bill rate? What rate of return has the S&P 500 generated over the last year? Select a start-up that recently has gone public and determine its rate of return in the last year (or since the IPO).

**7.5**   The timing of the first revenue or customer shipment is an important milestone for any start-up. Research the average time until first revenue for Internet, biotech, and clean tech companies. If you were an investor in these three types of businesses, how would this impact your view of these businesses? How could this time risk be managed?

**7.6**   Describe how social and professional networking sites such as Facebook, Twitter, or Linkedin have leveraged network effects to expand.

**7.7**   Section 5.1 discussed the first mover's versus follower's approaches. How does the importance of this decision change in a market with network effects and increasing returns?

## VENTURE CHALLENGE

1. Describe the major risks for your venture. How can you reduce these risks?

2. What is the potential for economies of scale and scope? Is this business scalable?

3. Describe the venture's potential for creating network effects.

# CHAPTER 8

# Creativity and Product Development

*Knowledge and human power are synonymous, since the ignorance of the
cause frustrates the effect.*
**Francis Bacon**

**How can an organization establish an environment that fosters bringing new products and services to market on a consistent and timely basis?**

Over the past 20 years, product life cycles have gotten shorter and shorter—in large part due to faster technological breakthroughs. To keep pace with this environment of rapid change, companies must establish a creative environment that strongly encourages spending time on new ideas, concepts, and solutions. The creative enterprise is based on six resources, which are outlined in Table 8.1. Creative ideas, concepts, and solutions are turned into products and services through the product design and development process. This process is concerned with the concrete details that embody a new product or service. Prototypes are models of a product or service and can help a new technology venture to learn about the right form of the product for the customer. Scenarios are used to create a mental model of a possible sequence of future events or outcomes. Good product development relies on the five practices listed in Table 8.5. ∎

## 8.1 Creativity and Invention

**Creativity** is the ability to use the imagination to develop new ideas, new things, or new solutions. Creative thinking is a core competency of most new ventures, and entrepreneurs strive to have creative people on their team. Creative ideas often arise when creative people look at established solutions, practices, or products and think of something new or different. These creative ideas can underlie invention, and invention can flow to innovation. Thus, firms committed to innovation must fundamentally strive to encourage and support creativity.

The creative enterprise is based on six resources as shown in Table 8.1 [Sternberg et al., 1997]. To create something new, one needs knowledge of the field and of the domain of knowledge required. Domains are areas such as science, engineering, or marketing. Fields within a domain might be circuit design or market research.

The intellectual ability required is the ability to see linkages between things, redefine problems, and envision and analyze possible practical solutions. Creative people use inventive thinking that reflects in novel ways on a problem. A creative thinker is motivated to make something happen and is open to change. Finally, the creative person understands the context of the problem and is willing to take a reasonable risk. The person who has most of these skills is often called *intuitive*; that is, he or she has an instinctive ability to perceive or learn relationships, ideas, and solutions.

One process of creative thinking is shown in Figure 8.1. It starts with a description of a problem and rests for a period of incubation. Then, intuitive brainstorming leads to good insights and ideas that can be evaluated and tested. Finally, a prototype is built and shown to the potential customer. This process may lead the entrepreneur to revise the question or problem and to begin a second cycle through the process. The entrepreneur continues around the loop until the prototype product solves the problem.

It is also useful to think of the innovation process as involving multiple personas, each with their own skills and points of view. The first three personas occupy learning roles: the *anthropologist* observes behaviors and develops a deep understanding of how people interact with products, services, and each other; the *experimenter* prototypes new ideas continuously; and the *cross-pollinator*

### TABLE 8.1  Six resources for a creative enterprise.

- Knowledge in the required domain and fields: and knowing what is new.
- Intellectual abilities to recognize connections, redefine problems, and envision and analyze possible practical ideas and solutions.
- Inventive thinking about the problem in novel ways.
- Motivation toward action.
- Opportunity-oriented personality and openness to change.
- Contextual understanding that supports creativity and mitigates risks.

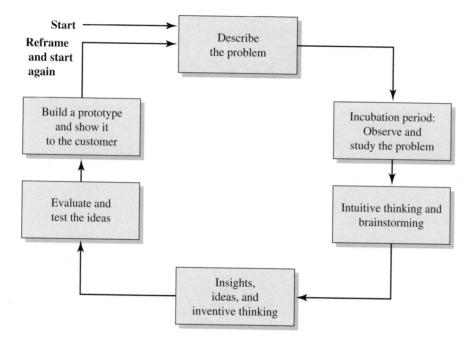

**FIGURE 8.1** Creativity process.

explores other industries and settings and borrows relevant ideas from them. The next three personas occupy organizing roles: the *hurdler* develops a knack for overcoming and outsmarting potential obstacles; the *collaborator* helps to bring diverse groups together; and the *director* gathers and inspires the team. The last four personas occupy building roles: the *experience architect* designs compelling experiences that go beyond mere functionality; the *set designer* transforms physical environments to facilitate the work of innovative team members; the *caregiver* anticipates and attends to customer needs; and the *storyteller* conveys a compelling narrative about the project [Kelley and Littman, 2005].

Figure 8.2 illustrates the "Innovation Engine," which shows how different internal and external factors can work together to enhance creativity. The three factors on the inside of the Innovation Engine are knowledge, imagination and attitude. Knowledge provides the fuel for one's imagination. Imagination, in turn, is the catalyst for transforming knowledge into new ideas. Finally, attitude sets the Innovation Engine in motion. The three factors on the outside of the Innovation Engine are resources, habitats, and culture. Resources include the assets in an entrepreneur's community. Habitat is the physical space, constraints, incentives and team dynamics that surround the entrepreneur. Finally, culture refers to collective beliefs, values and behaviors in the entrepreneur's community [Seelig, 2012].

**FIGURE 8.2** Innovation Engine.

An entrepreneur can improve all six factors to increase innovation. Imagination can be enhanced by reframing problems, connecting ideas, and challenging assumptions [Seelig, 2012]. For example, in 1954, Kay Zufall was looking for new things for children to do. She did not like the modeling clay sold for children because it was too stiff. However, her brother-in-law made a doughy mixture for cleaning wallpaper. Zufall tried it as a modeling medium and discovered it was soft and easy to mold and cut up. She and her brother-in-law reformulated it as a safe and colorful product for children, and they came up with the name: PlayDoh [Sutton, 2002].

A powerful method for enhancing the imagination is brainstorming. In a brainstorming session, diverse people come together to generate new ideas through conversation and interaction. Brainstorming sessions typically require significant planning, setup and follow-through. Table 8.2 lists eight guidelines for effective brainstorming sessions.

The mindset of the brainstorming group is particularly important. To be effective, teams should follow certain brainstorming "rules," as listed in Table 8.3. Entrepreneurs can be naturally inclined to judge ideas immediately, leading them to filter the ideas that they discuss and record in a brainstorming session. The point of a brainstorming session, however, is to stir the imagination in order to generate new ideas. Some firms, like the product-design firm IDEO, post printouts of the brainstorming rules in Table 8.3 in their work environments to remind team members of this goal.

Entrepreneurs also can act to enhance parts of the Innovation Engine beyond imagination. For example, an entrepreneur can enhance her knowledge of a market by engaging deeply with potential customers through conversations and observations. Entrepreneurs also can work to apply their knowledge across different domains. For example, Tina Seelig completed her PhD in neuroscience at the Stanford University School of Medicine. Soon after she completed her degree, she decided that she wanted to enter the world of business. She realized that her ability to do scientific research was directly applicable to the kinds of problems that she would face in the business world. Even though Seelig did not have a degree in business, she did have highly relevant knowledge. Being able to tap into your knowledge and apply it in different settings is a powerful tool for creative problem solving.

An important part of attitude lies in maintaining a willingness to experiment and the drive to push through challenges to solve problems. Research shows that the human mind is malleable and that people who maintain a positive and learning-oriented attitude have different brain activity than people who do not.

### TABLE 8.2  Guidelines for effective brainstorming sessions.

| | |
|---|---|
| Right people | A brainstorming group should be diverse, relatively small, and free of internal politics |
| Right challenge | A brainstorming group should focus on a clearly-stated challenge |
| Right mindset | A brainstorming group should adopt a creative and generative mindset, not a critical and evaluative one |
| Right empathy | A brainstorming group should focus on understanding the people who are affected by the challenge |
| Right stimulus | A brainstorming group should use questions that question assumptions, consider extreme cases, employ analogies, and explore technology scenarios and trends |
| Right facilitation | A brainstorming group should ensure that everyone participates, while keeping the conversation fresh and energetic |
| Right follow-up | A brainstorming group should have some way to ensure that ideas can later be considered and, where appropriate, implemented |

*Source:* Liedtka and Ogilvie, 2011.

### TABLE 8.3  Brainstorming "rules".

| | |
|---|---|
| Defer judgment | Don't confuse idea generation, where brainstorming excels, with idea evaluation |
| Capture all ideas | Even ideas that seem irrelevant or "bad" can be the basis of other ideas |
| Encourage wild ideas | Wild ideas can lead to insights |
| Hold one conversation at a time | Listen to other people in the group |
| Build on other people's ideas | Consider how you might modify or extend an idea into a new idea |
| Be visual | Use pictures and diagrams, not just words, to capture ideas |
| Go for volume | Emphasize quantity over quality |

*Source:* Seelig, 2009.

Finally, habitat plays an important role in facilitating creativity. Part of habitat is the physical environment itself. Many startups feature brightly-colored and open spaces with easily-movable furniture, whiteboards, and even games. These physical environments support creativity by encouraging informal interactions and by providing an easy way to capture insights.

Habitat also includes incentives and team dynamics. All firms need a culture that sustains a creative process that enables team members to engage and interact with new solutions. The leaders of a new venture can play a particularly important role in establishing this culture; firms with leaders who empower employees experience greater creativity [Zhang and Bartol, 2010]. Leaders also should work to establish a collaborative culture around creativity. Collaborative teams are more likely to achieve creative breakthroughs, since they can draw upon multiple perspectives, and are less likely to have very poor outcomes, since groups typically have a more rigorous selection process [Singh and Fleming, 2010].

Entrepreneurial leaders influence the culture by putting in place rules, rewards, and incentives that can foster innovation. Even those employees not in a leadership role with direct control over the corporate culture can indirectly influence the culture by reinforcing other parts of the Innovation Engine. For example, by increasing their imagination, they create opportunities for the organization that can ultimately impact that entire organizational culture.

Table 8.4 conveys several ideas for managing a creative work environment. These ideas can clash, however, with conventional management practices [Sutton, 2002]. A natural conflict exists between managing for creativity or exploration on one hand and implementation or exploitation on the other hand: new ideas and inventions depend upon creativity; bringing these inventions to market, however, may require routine processes [Freeman and Engel, 2007]. A small, emerging firm can accommodate both tendencies within it. As a firm grows, it needs to build a culture that reinforces the best qualities of creativity as well as efficient execution of its business processes [Brown, 2001].

**TABLE 8.4  Conventional versus creative management practices.**

| Conventional practice | Creative practice |
|---|---|
| Hire "fast learners" (of the organizational code) | Hire "slow learners" (of the organizational code) |
| Hire people who make you comfortable, whom you like | Hire people who make you uncomfortable, even those you dislike |
| Hire people you (probably) do need | Hire people you (probably) *don't* need |
| Use job interviews to screen and, especially, to recruit new employees | Use job interviews to get new ideas, not to screen candidates |
| Encourage people to pay attention to and obey their bosses and peers | Encourage people to ignore and defy their superiors and peers |
| Find some happy people, and make sure they don't fight | Find some happy people, and get them to fight |
| Reward success, punish failure and inaction | Reward success and failure, punish inaction |

## TABLE 8.4  Continued

| Conventional practice | Creative practice |
| --- | --- |
| Decide to do something that will probably succeed, then convince yourself and everyone else that success is certain | Decide to do something that will probably fail, then convince yourself and everyone else that success is certain |
| Think of some sound or practical things to do, and plan to do them | Think of some ridiculous or impractical things to do, and plan to do them |
| Seek out and be attentive to people who will evaluate and endorse the work | Avoid, distract, and bore customers, critics, and anyone who just wants to talk about money |
| Learn everything you can from people who seem to have solved the problems you face | Don't try to learn *anything* from people who say they have solved the problems you face |
| Remember and replicate your company's past successes | Forget the past, especially your company's successes |

*Source:* Sutton, 2002.

# 8.2 Product Design and Development

One of the early tasks of a new venture is the design and development of the new product. The entrepreneurial team wants to develop a new product or service that can establish a leadership position. One of the strengths of a new venture is that the leadership of the venture plays a central role in all stages of the development effort. Furthermore, the small new firm is able to integrate the specialized capabilities necessary for the development of a successful product [Burgelman, 2002].

In recent years, product complexity has dramatically increased. As products acquire more functions, the difficulty of forecasting product requirements rises exponentially. Furthermore, the rate of change in most markets is also increasing, thereby reducing the effectiveness of traditional approaches to forecasting future product requirements. As a result, entrepreneurs need to redefine the problem from one of improving forecasting to one of eliminating the need for accurate long-term forecasts. Thus, many product designers attempt to retain flexibility of the product characteristics as the development proceeds. A design and development project can be said to be flexible to the extent that the cost of any change is low. Then, project leaders can make product design choices that allow the product to easily accommodate change [Thompke and Reinertsen, 1998]. Uncertainty is an inevitable aspect of all design and development projects, and most entrepreneurs have difficulty controlling it. The challenge is to find the right balance between planning and learning. Planning provides discipline, and learning provides flexibility and adaptation. Openness to learning is necessary for most new ventures that are finding their way into the market [DeMeyer et al., 2002].

**Design** of a product leads to the arrangement of concrete details that embodies a new product idea or concept. The design process is the organization and management of people, concepts, and information utilized in the development of the form and function of a product. The role of design is, in part, to mediate between the novel concept and the established institutional needs. For example, Thomas Edison designed and described the electric light in terms of the established institutions and culture. As a result, he succeeded in developing an electric lighting system that gained rapid acceptance as an alternate to the gas lamp. A new product needs to be advanced, yet it should not deprive the user of the familiar features necessary for understanding and using the product. As new products are designed, the challenge ultimately lies in finding familiar cues that locate and describe new ideas without binding users too closely to the old ways of doing things. Entrepreneurs must find the balance between novelty and familiarity, between impact and acceptance [Hargadon and Douglas, 2001].

### Tony Fadell: Thermostat Breakthrough

Tony Fadell is a prototypical breakthrough thinker. While at Apple, he thought of the concept of a small portable digital music player, which later became the iPod. He then went on to be a key player in the development of the iPhone, which catapulted Apple to its current position as one of the world's highest valued firms. Tony looks at products and then reenvisions how these products might be made easier to use or more accessible. Fadell followed up his work at Apple by creating the Nest Learning Thermostat. The idea for a smart thermostat came to him while shopping for thermostats for an energy-efficient home he was building. At the time, Fadell realized that the state of at-home and office thermostats seemed to be caught in a time warp. These devices had few features and limited intelligence. With this knowledge, he reimagined an easy-to-use thermostat that could be controlled both manually and over the internet. His invention is upending what had been a stagnant market for decades. Nest is now vying to replace the 250 million thermostats in homes, office buildings, stores, and restaurants.

Good, effective products or services are the outcome of a methodology based on solid, proven design principles [Brown, 2008]. Innovation is powered by a thorough understanding of how people want products made, packaged, marketed, sold, and supported. The overall development process is shown in Figure 8.3 [Thompke and Von Hippel, 2002]. The overall development process can include design of the product and its architecture, its physical design, and testing. The iPad and the Tesla Roadster sports car are examples of the outcome of a creative, artistic process of design. Part of the user experience is the look and feel of a product. A good product is attractive to see and easy to use and understand. Furthermore, customers want a product that does a few things really well. Fortunately, customers can participate fruitfully in the product design process when the innovations are incremental [Nambisan, 2002]. Good designers think about the qualities of a product as well as its soft benefits such as warmth, status, and community.

Design includes aesthetics as well as basic needs. A beautiful glass must be functional as well as attractive. However, design also includes compromises and

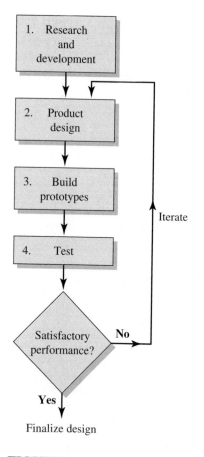

**FIGURE 8.3** Overall
development process.

limits. Even the Maglite flashlight, which is widely acknowledged for its good
design, is flawed by the spot in the middle of the light beam.

Successful product design and development requires commitment, vision,
improvisation, information exchange, and collaboration, as listed in Table 8.5
[Lynn and Reilly, 2002]. These five practices may be easy to achieve in a start-up
where collaboration is the order of the day. The product team, which may be
all of the employees of a start-up, needs to clearly understand the vision for the
product and work together effectively.

The product design process [step 2 in Figure 8.3] is shown in Figure 8.4.
The first step is to establish the goals and attributes of the product expressed
as the required performance and robustness of the product (step A). When
possible, the potential customer should be included in the design process.

**TABLE 8.5 Five practices of good product development.**

| | |
|---|---|
| ■ Commitment of senior management to the design process | ■ Improvisation and iteration to develop a prototype |
| ■ Clear and stable vision and goals for the product | ■ Open sharing of information |
| | ■ Collaboration of everyone on the team |

*Source:* Lynn and Reilly, 2002.

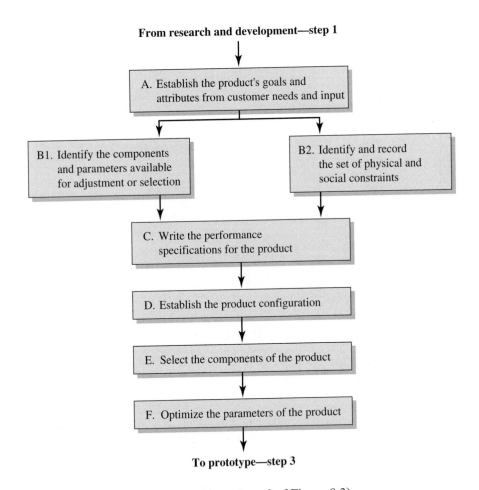

**FIGURE 8.4** Product design process (step 2 of Figure 8.3).

The voice of the customer can communicate the insights needed for the best products [Lojacono and Zacoai, 2004]. Potential customers can suggest ideas for new products and can be involved throughout the product development process to provide continuous feedback [Ogawa and Piller, 2006]. Firms may also find that *observing* potential customers, rather than simply surveying or interviewing them, can yield important information about their product needs.

In step B of Figure 8.4, the components and parameters available for adjustment are identified, and specifications for the product are agreed upon. Specifications are the precise description of what the product has to do. In addition, the set of physical and social constraints should be determined. Next, the product configuration is established, and the components of the product are recorded. Finally, the parameters of the product are optimized to achieve the best performance and robustness at a reasonable cost [Ullman, 2003]. A **robust product** is one that is relatively insensitive to aging, deterioration, component variations, and environmental conditions. Preparing a robust design implies minimizing variation in performance and quality. All designs involve trade-offs between performance, cost, physical factors, and other constraints [Petroski, 2003]. The success or failure of any design is ultimately determined in the marketplace.

**Usability** is a measure of the quality of a user's experience when interacting with a product. Usability is a combination of the five factors listed in Table 8.6. Examples of a common product with poor usability are most DVR and DVD players. New products should pass the five-minute test, which requires that the product is simple enough to use after quickly reading the instructions and then trying it for a few minutes. Information technology products with excellent usability are the iPhone, Skype, Twitter, Gmail, and Wikipedia.

Many system designs use a combination of modules within a specified architecture. A **module** is an independent interchangeable unit that can be combined with others to form a larger system. In modular designs, changing one component has little influence on the performance of others or on the system as a whole. An example is the iPod, which Apple's engineers first developed from a wide range of standard, interchangeable parts and modules. Design methods using independent modules make product design more predictable. Of course, the predictability inherent in modular design increases the chances that competitors can develop similar products.

Realistically, most products consist of modules that possess some dependency between them. For example, an automobile is a product that consists of

## TABLE 8.6  Five factors of usability.

| | | |
|---|---|---|
| 1. | **Ease of learning:** | How long does it take to learn the product's operation? |
| 2. | **Efficiency of use:** | Once experienced, how fast can the user complete the necessary steps? |
| 3. | **Memorability:** | Can the user remember how to use the product? |
| 4. | **Error frequency and severity:** | How often do users make errors, and how serious are these errors? |
| 5. | **Satisfaction:** | Does the user like operating the product? |

wheels, engine, body, and controls that are relatively interdependent. Products made up of modules with intermediate levels of interdependence are harder for competitors to duplicate and may also provide better performance than a design based on purely independent modules [Fleming and Sorenson, 2001].

Designers strive to create new products different enough to attract interest but close enough to current products to be feasible to make a market. Many new designs flow from changing the components, attributes, or integration scheme to create a new product [Goldenberg et al., 2003]. The designer asks what can be rearranged, removed, or replicated in new ways.

Over time a dominant design in a product class wins the allegiance of the marketplace. A **dominant design** is a single architecture that establishes dominance in a product class. An example is the Microsoft Windows operating system, used in the vast majority of personal computers. Eventually, a dominant design becomes embedded in linkages to other systems. For example, the use of Windows shapes hardware interfaces and the other software programs that computers can run.

A **product platform** is a set of modules and interfaces that forms a common architecture from which a stream of derivative products can be efficiently developed and produced. For example, Google's Android and Apple's iPhone seek to be the leading platform for smartphone applications. Firms target new platforms to meet the needs of a core group of customers but design them for ready modification into derivative products through the addition, substitution, or removal of features. Well-designed platforms also provide a smooth migration path between generations so neither the customer nor the distribution channel is disrupted. A good example of a platform is Hewlett-Packard's electronics and software used for its printers; although Hewlett-Packard offers a wide range of printers, these products draw upon relatively similar electronics and software.

## 8.3 Product Prototypes

Whenever possible, new business ventures should create a prototype of their product. A **prototype** is a physical model of a proposed product or service that conveys the essential features but remains open to modification. Prototypes can be used to identify and test requirements for the product by eliciting comments from designers, users, and others. New ventures can use prototypes to redefine their business models and strategies.

Prototypes can be pictures, sketches, mock-ups, or diagrams that can be collaboratively studied. They also can be physical, digital, pictorial, or some combination of media. The computer software industry uses prototypes called beta versions of software to elicit response from lead customers. For example, Microsoft made a beta version of Windows 8 available as a free download. Microsoft then used customer feedback to refine the product for its official release.

**Ford: The Power of Prototypes**

Henry Ford planned to build a horseless carriage. However, no one could be persuaded to invest in it. A key turning point came when Ford built a prototype car for the Grosse Pointe automobile races. Ford entered the races, drove the car himself, and won decisively. He repeated the feat the following year in 1902. The victory attracted investors, and the Ford Motor Company was up and running.

In the creation of a movie or play, many innovators use sketches, storyboards, and videos to describe the product. The designers of a movie or play want to see how it works and engage in a collaborative redesign. The iterative procedure for prototype development is shown in Figure 8.5. Two or three iterations of the process may be sufficient to arrive at a satisfactory prototype.

New technologies such as computer simulations can make the creation of a prototype fast and cheap. **Rapid prototyping** is the fast development of a useful prototype that can be used for collaborative review and modification. An initial prototype can be rough since it enables the team to view the product and improve it. The ability to see and manipulate high-quality computer images helps create innovative designs. BMW uses computers to help engineers visualize automobile design and the results of crash tests [Thompke, 2001]. Personal fabrication systems, clusters of tools and software that function as complete job shops, are available [Gershenfeld, 2005].

Product-development firm IDEO believes that prototypes should be "rough, ready, right." While working with Gyrus ENT to develop a better sinus-surgery tool, IDEO employees demonstrated the value of having a prototype to show customers. During one discussion, 10 surgeons struggled to explain the discomfort involved in using the existing tool. An IDEO manager picked up a film canister, a white-board marker, and a clothespin and taped them together as a prototype. The physical prototype helped to move the conversation along, allowing the surgeons to hold and adjust it. The rough prototype was just that—rough and unfinished—encouraging the surgeons to modify and tinker with it. Creating a rough prototype allowed customers to engage in the development of the product, and to enthusiastically adopt it in surgery. Eventually, IDEO and Gyrus produced a tool that is used in over 300,000 procedures in the United States every year.

It is often best to carry multiple product concepts into the prototyping phase and to select the best of those designs later in the process [Dahan and Srinivasan, 2000]. Keeping multiple product concept options open and freezing the concept late in the development process affords the flexibility to respond to market and technology shifts. It is possible to create static and dynamic virtual prototypes that are displayed at a website for review and testing by suppliers, customers, and designers. Virtual prototypes cost considerably less to build and

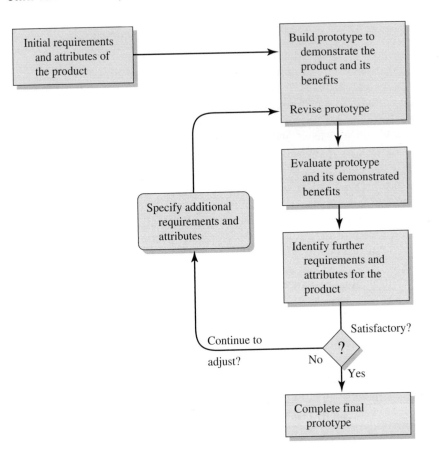

**FIGURE 8.5** Prototype development process.

test than their physical counterparts, so design teams using Internet-based product research can afford to explore a much larger number of concepts. Furthermore, Internet-based prototypes can help to reduce the uncertainty in a new product introduction by allowing more ideas to be tested in parallel.

SaltAire Sinus Relief provides a nose wash to relieve the symptoms of sinusitis and allergies (see www.saltairesinuswash.com). A bottle and pump are used to spray a supersalty solution into the sinus chamber and relieve the symptoms [Ridgway, 2003]. Two New York physicians founded the firm in 1997 and created a series of prototypes for people to try. Based on that knowledge, they showed a revised product to other physicians and launched the product in 2000. The patented dispenser bottle the firm developed won an award for innovative design.

For the innovator, a prototype is a mechanism for teaching the market about the technology and for learning from the market how valuable that technology is in that application arena. Uses for robots in situations too dangerous for people have long been imagined. Many robotics companies tried and failed to create robots that could successfully enter and explore disaster zones and other dangerous environments. For example iRobot first demonstrated a prototype of the Urbie robot in 1997. This was the first commercially available robot that was able to climb stairs. This prototype showed the market that iRobot's products had overcome many of the fundamental limitations of other contemporary robots. By 2013, iRobot's revenue was over $430 million, and its products were available in over 7,000 retail outlets. Robots like the Urbie were used to explore the rubble of the World Trade Center, and have been used by the military to explore situations that would be extremely dangerous for troops. Over three million units of the Roomba, the consumer version of the Urbie, have been sold. The Roomba uses similar technology as the Urbie to sweep and vacuum floors.

Many firms have developed their products by entering potential markets with early versions of the products, learning from the tests, and probing again. These firms ran a series of market experiments, which introduced prototypes into a variety of market segments. The initial product design was not the culmination of the development process but rather the first step, and the first step in the development process was in and of itself less important than the learning and the subsequent, better-informed steps that followed. Software products lend themselves to rapid prototyping and early tests with potential customers.

## 8.4 Scenarios

Any new venture can benefit from creating a set of scenarios to address complex, uncertain challenges as it develops its strategy. A **scenario** is an imagined sequence of possible events or outcomes, sometimes called a mental model. A few realistic scenarios based on the industrial context and a few associated possible sequences of events help a planner to plan for the future. Each scenario tells a story of how the various elements might interact under a variety of assumptions. It paints vivid narratives of the future. The goal of scenario planning is not to forecast what is going to happen but to encourage an openness of mind, a flexibility of response, and a habit of questioning conventional wisdom. As Stephen Covey and A. R. Merrill [1996] stated: "The best way to predict your future is to create it."

Scenarios lead to learning in a two-step process: constructing a scenario and using the content of the scenario to learn [Fahey and Randall, 1998]. The key elements of a scenario are shown in Figure 8.6. A scenario attempts to answer key questions and is based on a statement of the driving forces and the rationale for the story. The outcomes or results of the story lead to understanding and a useful decision. A scenario is an internally consistent picture of what the future might bring. It is not a forecast, but rather one possible outcome. Creating four or five scenarios will help portray the range of potential outcomes to core questions facing any organization.

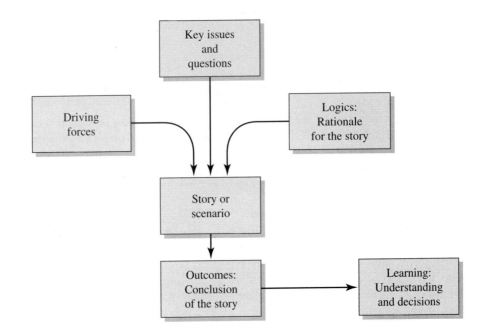

**FIGURE 8.6** Elements of a scenario.

Entrepreneurs will often weigh whether a new technology will be radical or nonlinear and have a profound impact on the marketplace. A scenario can help define the impact and time frame for a new technology.

An example of the outline of a scenario for the growth of electric auto sales is shown in Figure 8.7. The structure of the story for electric vehicles can be used to build several possible scenarios that can be used to learn about the opportunities in this market.

Scenarios can sometimes become a mirage. By 2001, the futurists—George Gilder and others—had created a scenario for the future of telecommunications that was overblown and ill-timed. This rosy, nirvana-like scenario missed the regulatory issues and the concept of excess capacity. Scenarios often can be too rosy [Malik, 2003].

## 8.5 Spotlight on Teva Pharmaceuticals

Teva Pharmaceutical Industries is an international pharmaceutical company headquartered in Israel and with facilities in Israel, North America, Europe and South America. Teva develops, manufacturers, and markets generic and proprietary branded drugs, along with active pharmaceutical ingredients. In 2012, Teva's sales totaled more than $20 billion.

Over the past 40 years, Teva has grown to become a global company through acquisitions of emerging and operating pharmaceutical firms. Thus, Teva

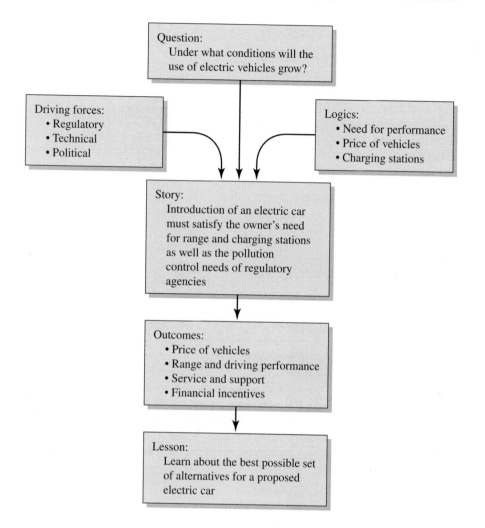

**FIGURE 8.7** Elements of scenarios for electric cars.

actively monitors research success at other companies. It then makes investments or outright acquisitions as it identifies promising products. In addition, Teva develops its own drugs. Their current offerings include drugs for treatment of multiple sclerosis and Parkinson's disease.

Teva promotes rapid prototyping through its sponsorship of Hackathons, in which engineers and developers of all sorts come together to try to design novel solutions to big problems. By focusing on innovation and by using scenarios that provide insight into possible future directions, Teva has been able to grow at a sustained 20-percent pace since the year 2000.

## 8.6 Summary

Creative thinking is a core competency of most new ventures. Entrepreneurs should design their organizations and engage in practices to encourage creativity. Product design and development builds upon creativity, turning ideas, concepts and solutions into new products or services.

Prototypes are models of a product or service and can help a new venture learn the right form and function of a product by showing it to customers and letting them observe it or try it. Furthermore, scenarios can be used to examine possible future outcomes based on specific actions.

**Principle 8**
Knowledge acquired, shared, and used is a powerful tool for the entrepreneur to build a learning organization that can design innovative products and grow effectively.

## Video Resources

Visit http://techventures.stanford.edu to view experts discussing content from this chapter.

| | | |
|---|---|---|
| Divergent Thinking | Tina Seelig | Stanford University |
| Ideas Come from Everywhere | Marissa Mayer | Google |
| Product Development Process: | David Kelley | IDEO |
| Observation | | |

## 8.7 Exercises

**8.1**   How might you influence innovation in a company using each of the six parts of the Innovation Engine from Figure 8.2?

**8.2**   What role should the "end customer" have in the product design and development process? Do customers always know what they want?

**8.3**   Capstone Turbine is a developer, assembler, and supplier of microturbine technology. Its primary customers are in the on-site power production and hybrid-electric car markets (www.capstoneturbine.com). Using the format of Figure 8.6, describe a scenario for the growth of Capstone over the next five years.

**8.4**   The advantages of the Web as a distribution platform delivering services and content are well-known. Describe a few examples of its impact on product prototyping and product design and development.

**8.5** A number of software development methodologies exist to encourage rapid design and implementation (e.g., agile software development, extreme programming, etc.). Select two of these methodologies and compare and contrast the specific product design and development processes each is attempting to address and improve.

**8.6** An *IEEE Spectrum* magazine article in September 2005 (www.spectrum.ieee.org/sep05/1685) featured some of the most expensive software failures ever. Examine the "Software Hall of Shame" and select your favorite. Describe why failure of knowledge management and the lack of behaving as a learning organization led to this result.

## VENTURE CHALLENGE

1. Examine your creative strengths and areas of improvement and share them with your team.

2. Discuss the robustness and usability of your product.

3. Discuss your plans for developing a prototype of your product.

# CHAPTER 9

# Marketing and Sales

*Successful salesmanship is 90 percent preparation and 10 percent presentation.*
**Bertrand R. Canfield**

**What is the best way to attract, serve, and retain customers?**

Marketing and sales are critical to the success of a new firm since the firm normally starts without any customers. A new business must create a marketing and sales plan, which describes its target customers for its product offering. The plan should include a product position and a mix of price, product, promotion, and distribution channels that will attract and satisfy the customer. Gaining recognition and acceptance in a target market requires the following steps in sequence:

- Describe the product offering
- Describe the target customer
- State the marketing objectives
- Gather information through market research
- Create a marketing plan
- Create a sales plan
- Build a marketing and sales staff

## 9.1 Marketing

**Marketing** is a set of activities with the objective of securing, serving, and retaining customers for the firm's product offerings. Marketing is getting the right message to the right customer segment via the appropriate media and methods. It is the task of the marketing function to help create the product and the terms of its offering as well as communicate its value to the customers. Ideally, marketing "merges the minds" of customers and product developers, facilitating the identification of unspoken but important needs [Lassiter, 2002].

The purpose of the **marketing plan** is to describe the steps required to achieve the marketing objective. The marketing plan is a written document serving as a section of a new venture's business plan and contains action steps for the marketing program for the products. Peter Drucker [2002] has said, "Because its purpose is to create a customer, the business has two basic functions: marketing and innovation. Marketing and innovation produce results; all the rest are costs."

In Chapter 3, we described the creation of a value proposition and business model for the identified customer. In Chapters 4 and 5, we described the elements of an overall business strategy and market analysis. Given these business elements, we need to develop a marketing strategy and build a marketing plan. The six elements of a marketing plan are shown in Table 9.1.

The first element of the marketing plan is a clear statement of objectives. The second element is the identification of one or more customer target segments. The goal of target segments is to carefully select the appropriate customers and to focus the marketing activities on those segments. The third element is the description of the product and the terms and conditions of its formal offering. Given our knowledge of the product and its offering, we need to determine what the response of the customer might be and how we can develop a strategy to attract and retain the customer. Next, we describe the marketing mix consisting of price, product, promotion, and place (channels). Finally, we describe plans for relating to our customer in the sales and service activities.

The marketing plan will be implemented through a marketing program. The plan will describe how we will take the product to market and attract, serve, and retain satisfied customers. The marketing and sales activity is portrayed in Figure 9.1. The new venture communicates information about its

**TABLE 9.1  Six elements of the marketing plan.**

- Marketing objectives
- Target customer segments
- Product offering description
- Market research and strategy
- Marketing mix
- Customer relationship management

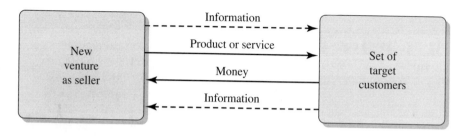

**FIGURE 9.1** Marketing and sales activity of the new venture.

product and how it sells and services the product for the customer. When the customers purchase the product, they provide some useful information about the purchase and the use of the product for the seller.

The marketing and sales plan will flow from the opportunity and the business model, as shown in Figure 9.2.

## 9.2 Marketing Objectives and Customer Target Segments

The **marketing objectives statement** is a clear description of the key objectives of the marketing program. Objectives may include sales goals, market share, profitability, regional plans, and customer acquisition goals. Objectives should be quantified and given for a time period, such as "the firm will sell 1,000 units in the initial sales phase in Texas and Oklahoma in the first year of activity."

A clear understanding of who the customers are and why they will buy is critical. Selected markets or groups of customers are often called customer target segments. A **market segment** consists of a group with similar needs or wants who reference each other and may include geographical location, purchasing power, and buying attitudes. **Market segmentation** divides markets into segments that have different buying needs, wants, and habits. Different segments will require different marketing strategies. For example, a business based on Internet sales should know that different age groups, which are one type of segment, vary dramatically in their propensity to shop and in the amount they spend [Abate, 2008]. Often a new venture identifies one target segment for its initial marketing effort, carefully describing the customer in that segment in terms of geographic, demographic, psychographic, and other variables [Winer, 2000]. Geographic variables include city, region, and type, such as "urban." Demographic variables include age, gender, income, education, religion, and social class. Psychographic variables include lifestyle and personality variables that influence a customer's wants and needs. Good segmentations identify the groups most worth pursuing and they tell companies what products to sell to these groups. Good segmentations also change over time, along with customers [Yankelovich and Meer, 2006].

**FIGURE 9.2**
Building a marketing plan and a sales plan.

**InVision: The Right Product at the Right Time**

InVision Technologies designs and manufactures electronic baggage screening systems (see www.invision-tech.com). InVision Technologies was founded in 1990 to provide airport security devices by using computed tomography to detect explosives. Its target segment, U.S. airport baggage screening, was slow growing before the terrorist attacks on September 11, 2001. It sold 250 machines in the preceding decade but 750 machines in the two years following the attacks. InVision also planned to enter the international market. InVision's success and expertise led to its acquisition by GE in 2004 for $900 million.

The target segment for Samsung's Galaxy phone can be described as young adults who crave a full-featured wireless device with cutting-edge technology. With a clear description of the target customer, a new venture can devise a plan to attract and retain them. Marketing to a segment enables the new firm to narrow the marketing strategy and put all its effort into acquiring new customers in the target market. Often, new firms reach for too many market segments in their early efforts, thus dissipating their resources before they can build up a customer base. Table 9.2 identifies four critical questions to ask about target markets.

**Redefining Flexcar's Customer Segments**

Car sharing is aimed at people who want to use a car for a short time but do not need to own it. This scheme is particularly attractive in urban centers. In general, people are moving toward access rather than ownership of autos in dense urban areas with high auto costs. Neil Peterson started Flexcar in 1999 in Seattle and then expanded to Los Angeles, San Francisco, Washington, D.C., and San Diego.

As the venture grew, the average cost of owning or leasing a new car became $625 a month. The average member in a car-sharing program spent less than $100 a month on car expenses. Flexcar initially identified its target segment as individuals. It soon discovered, however, that the biggest growth came not from individuals but from small and midsize companies that did not want to maintain their own fleets of vehicles [Stringer, 2003]. In 2007, Flexcar merged with Zipcar (see www.zipcar.com). In 2013, Avis acquired Zipcar for $500 million.

It is wise to figure out who will be your best customer and then pursue that segment. A **best customer** is one who values your brand, buys it regularly whether your product is on sale or not, tells his or her friends about your product, and will not readily switch to a competitor. Entrepreneurs identify their customer segment and position their product to serve them very well [Ettenberg, 2002].

**TABLE 9.2  Four crucial questions to ask about target markets.**

- Is there a target market segment where the company can enter the market and provide clear customer benefits at a price the customer is willing to pay?
- Do customers perceive that these benefits are superior to other solutions/options?
- How large is the target market segment and how fast is it growing?
- Will entry into the target market segment serve as a springboard for entry into other segments?

*Source:* Mullins, 2006.

## 9.3 Product and Offering Description

The product's features and primary attributes are typically described early in a business plan. If possible, a product positioning map should be developed. All products can be differentiated to some extent by communicating the most highly valuable benefit to the buyer. **Positioning** is the act of designing the product offering and image to occupy a distinctive place in the target customer's mind [Ries and Trout, 2001].

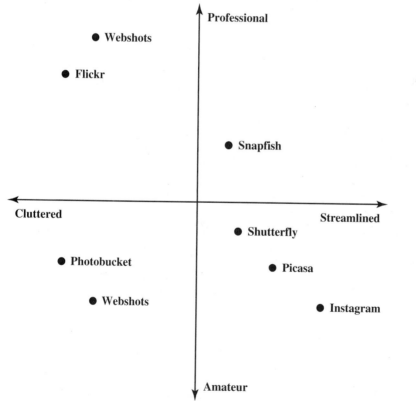

**FIGURE 9.3** Positioning map for online photo sharing showing the position of Instagram.

Positioning of a product enables the firm to differentiate it in the mind of the prospect. Volvo connotes safety, and FedEx owns "overnight." A product positioning map shows the product characteristics in relation to its competitors. Figure 9.3 shows the positioning of Instagram relative to other photo sharing applications. It would then be the task of the marketing effort to clearly communicate that position to the target customer.

Positioning a product focuses on a few key attributes of the value proposition. A positioning statement, as shown in Figure 9.4, helps to define the positioning of the product. Once we have a product position, we seek to build a powerful product offering [Moore, 2002]. A **product offering** communicates the key values of the product and describes the benefits to the customer. The **unique selling proposition** is a statement of the key customer benefit of a product that differentiates it from its competition. The unique selling proposition for Instagram could be:

> Instagram delivers an easy way to edit and share photos with your friends and family.

The unique selling proposition of FedEx is:

> We deliver your packages overnight—guaranteed.

---

- **Positioning statement**
  - **For** (target customer)
  - **Who** (statement of need or opportunity),
  - (Product name) **is a** (product category)
  - **That** (statement of benefit).                                         **(a)**

- **Differentiation**
  - **Unlike** (primary competitive alternative),
  - **Our product** (statement of primary differentiation).

---

- **Possible positioning statement for Tesla Motors**
  - For wealthy individuals and car aficionados
  - Who want an environmentally friendly and high-end car,
  - The Tesla Sedan is an electric automobile
  - That delivers unprecedented performance without damaging the
    environment.                                                              **(b)**

- **Differentiation**
  - Unlike Audis, BMWs, and Lexus sedans,
  - Our product has fantastic mileage, unparalleled performance, and no
    direct carbon emissions.

**FIGURE 9.4  (a)** Positioning statement format, and **(b)** for Tesla Motors

# 9.4 Brand Equity

As discussed in Chapter 4, the new venture should have a competitive advantage such as low cost, high quality, superior customer relationships, or increased performance. Many new technology ventures differentiate themselves from competitors by tying these product characteristics to their brand. A **brand** is a combination of name, sign, or symbol that identifies the goods sold by a firm. A brand accurately identifies the seller to the buyer and resides in the minds of consumers. Well-known brands include Intel, Philips, Sony and Samsung.

**Brand equity** is the perceived worthiness of the brand in the mind of the customer and may be portrayed as the sum of four dimensions, as shown in Table 9.3 [Aaker and Joachimsthaler, 2000]. Brand awareness or familiarity is the first step in building a brand. The perceived quality of the product and respect for the product will help build brand equity. The quality of the product and its perceived vitality will build an image of the brand. A brand association is how the customer relates to the brand through personal and emotional associations. This dimension is present in the emotional relationship that Harley-Davidson owners have with the motorcycle brand. In other words, brand loyalty responds to promises kept by the seller. Technology firms with significant brand equity include Apple, Canon, Facebook, GE, Google, and Intel.

A brand's promise of value is the core element of differentiation. This promise of value is tied to the customer, and loyalty will follow from good customer experiences. Many customers are willing to pay more for some badge of identification—Apple's logo, for example—that makes them feel they are part of a community. A strong corporate brand lets customers know what they can expect of the whole range of products that a company produces. The most successful corporate brands are universal and facilitate differences of interpretation that appeal to different groups. This is particularly true of corporate brands whose symbolism is strong enough to allow people across cultures to share symbols even when they don't share the same meaning.

The brand of a company is customer centered and focuses on the product or service offered. By contrast, the reputation of a company focuses on the credibility and respect among a broad set of constituencies such as suppliers, regulators, employees, the media, and the local communities. Brand equity depends on the delivery of a good product to customers. Reputation depends on the goodwill of communities and stakeholders [Ettenson and Knowles, 2008]. Both brand equity and reputation are important to a venture's success.

**TABLE 9.3  Four dimensions of brand equity.**

- Brand awareness or familiarity
- Perceived quality and vitality of the product
- Brand associations: connects the customer to the brand
- Brand loyalty: a bond or tie to the product

Some brands, such as Nike, Harley-Davidson, and BMW, become icons [Holt, 2003]. A brand becomes an icon when it offers a compelling story that resonates with consumers. One of the most potent stories is the depiction of a group of rebels. For example, Harley-Davidson appeals to rebel motorcycle enthusiasts who want to stand out as different from the crowd.

One approach to building a brand, therefore, is to identify the differentiating benefit that is important to the target customers and then to describe the attributes that imply this benefit. For example, Intel identifies superior quality as its benefit. Successive marketing campaigns have informed consumers that Intel integrated circuits have reliable high performance and are leading-edge products promising superior quality.

## 9.5 Marketing Mix

The four elements of the marketing mix are shown in Table 9.4. The **product** is the item or service that serves the needs of the customer. The marketing plan describes the key methods of differentiating this product. Coca-Cola, for example, differentiates regular Coke by using a distinctive trademarked bottle with ribbing. Shutterfly uses premium printing processes and shipping methods to differentiate from other online-photo competitors via high quality. Nordstrom distinguishes itself through a liberal return policy. Apple distinguishes its products through intuitive user interfaces and appealing design.

**Pricing policies** can be used to distinguish a firm's offering. Warren Buffet said it clearly, "Price is what you pay, value is what you get." For example, Amazon.com offers 30 percent off most books' list prices and free shipping for orders over $25. Price is a flexible element, and various discounts, coupons, and payment periods can be tested in test markets. The price can be initially set by estimating demand, costs, competitors' prices, and a pricing method to select the price. Effective pricing requires gathering and integrating information

**TABLE 9.4  Four elements of the marketing mix.**

| Product | Price | Promotion | Place |
|---|---|---|---|
| Product variety | List price | Public relations | Channels |
| Quality | Discounts | Advertising | Locations |
| Design | Credit terms | Sales force | Inventory |
| Features | Payment period | Direct messages | Fulfillment |
| Brand name | | | |
| Packaging | | | |
| Warranties | | | |
| Returns policy | | | |

about the firm's strategic goals and cost structure, the customer preferences and needs, and the competition's pricing and strategic intent [Nagle and Hogan, 2006]. The pricing method or strategy can seek market share, premium pricing, or maximum profit. The cost to make the product is a floor under the price, and an estimate of the total value to the customer sets a ceiling on the price (see Figure 3.3). After studying competitors' prices, the new venture can test a price on a set of test customers.

Consider the setting of a price for a textbook where the total market demand is 10,000 books per year. Competitors have established a retail price in the range $60 to $80, and the demand per year for a new textbook may be described by

$$D = 10,000 - kP \tag{9.1}$$

where D = demand in units, k = estimated sensitivity constant, and P = price in dollars. The fixed cost to produce the new book is $30,000, and the variable cost is $10 per unit. The book is differentiated from its competition by quality and clarity. What price, P, would you select within the established range of $60 to $80? To maximize market penetration, one would select the lowest price, P = $60, since this will result in the largest demand. If market research shows that the market is price-sensitive and k = 90, then when the price is set at $60, the demand is for 4,600 units. Then the gross profit = (revenues − cost of goods) is

$$GP = R - (VC \times D) = (D \times P) - (VC \times D) = (P - VC)D \tag{9.2}$$

where R = revenues and VC = variable cost. When P = $60 and k = 90, the gross profit is $276,000. As shown in Table 9.5, if you raise the price to $70, the gross profit declines. The calculation of the best price to obtain the maximum gross profit depends on the estimated sensitivity constant. If we change our assumptions so that k = 80, then we obtain the gross profit for the book as shown in Table 9.5. Note that $70 would be the price to maximize profit when k = 80. Note that k is an estimate obtained through experience and research and will change over time.

In many industries, customers demand low prices, and the competitors have little pricing power. Pricing power accrues to companies without wide competition, such as universities that raise tuition, hospitals that increase fees, or virtual monopolies such as cable-television operators. Most mature companies operate

**TABLE 9.5  Gross profit for selected values of the sensitivity constant, k, and the price of the new book (gross profit in thousands).**

|  |  | Price | | |
|---|---|---|---|---|
|  |  | **$60** | **$70** | **$80** |
| Sensitivity constant, k | **90** | $276 | $259 | $166 |
|  | **80** | $230 | $234 | $222 |

in a world of flat or falling prices due to an excess number of providers. A new venture can pick its pricing strategy from the three shown in Figure 9.5. Many new ventures use value pricing since demand will be sensitive to price and the new firms possess little brand equity. Demand-oriented approaches look at the demand for the product at various price levels and try to estimate a price that will provide a good market share and profitability for the long term. Many technology ventures with a new breakthrough product will use a premium pricing strategy.

New technology companies usually offer new, value-oriented products. A new product or service is, by its unknown nature, difficult to price. Many new products are characterized by quality and performance uncertainty. To attract customers to a new product, it may be useful to offer a warranty—a contract or guarantee of a specified performance. Another possibility is to offer quality-contingent pricing that specifies a price rebate for poor performance [Bhargava, 2003].

Since customers often use price as a signal of quality when they are unfamiliar with a product, companies should be careful not to underprice offerings [Marn et al., 2004].

Using a traditional model for growth, firms can take advantage of the demand for new goods and services by creating and marketing products that satisfy a demonstrated need in the marketplace. As their customer bases grow and the products become more and more popular, profits begin to emerge. The profits are then reinvested in projects that will provide new sources of revenue and income. A portion of the profits is retained to build brand value, which can

**FIGURE 9.5**  Three pricing methods.

be created through a variety of techniques, including aggressive pricing, savvy promotion, and advertising.

**Promotion** includes public relations, advertising, and sales methods. Selecting the message for advertising and the media for transmitting the message is a complex activity. **Advertising** is the art of delivering a sales proposition and positioning the product uniquely in the customer's mind [Roman, 2003]. The initial product message is used to attract customers to the new venture. Charles Revson, co-founder of Revlon, once said: "In our factory, we make lipstick. In our advertising, we sell hope." Many other products and services also sell hope, including weight-control products and dating services. By contrast, Twitter and Facebook sell community. Instagram sells artistic expression. Smule sells musical connectivity. Advertising can enhance brand name recognition, create value, and enhance return for a new technology venture [Ho et al., 2005].

Entrepreneurs can employ a variety of media in their advertising efforts, including websites, social networks, print media, radio and television. Table 9.6 gives a list of marketing media. Social networks have become particularly important. Facebook, for example, has one billion users–about one-seventh of the world's population. In Section 9.6 we consider the role of social media in marketing.

Social networks also can support word-of-mouth (buzz) promotion. Products that merit a buzz campaign have some unique, attractive attribute, such as the BMW's Mini Cooper or a new anticancer drug. Furthermore, they should be highly visible. The latest fashion in clothes or accessories often runs on buzz with teenage girls.

Word-of-mouth marketing is often called **viral marketing**. The concept is based on an age-old phenomenon: people will tell others about things that interest them. PayPal ran a successful viral marketing campaign by paying existing users $10 for each person they referred to the service who signed up, and by paying $10 to each new person who signed up. The marketing campaign cost PayPal tens of millions of dollars, but succeeded in generating significant buzz around the site. Hotmail, one of the first free e-mail services, included a small Hotmail advertisement at the bottom of each e-mail sent via their service. This tactic worked to quickly spread the word about Hotmail. As consumers increasingly use digital video recorders to skip commercials, listen to podcasts and downloaded music instead of radio, and use e-mail filters that prevent spam, word-of-mouth promotion will become more important.

The Tesla Roadster is a high-performance sports automobile with a retail price of over $100,000. It is a fully electric vehicle that can travel over 200 miles per charge. Tesla generated buzz around its product by getting celebrities and technology pioneers excited about it. At one point, the waiting list was over 1,000 people including Google's two co-founders, Arnold Schwarzenegger, and Matt Damon. Journalists spread photos of celebrities driving the vehicle throughout traditional and popular Internet media outlets. Despite the $5,000 deposit required to be placed on a waiting list, this "buzz" has helped Tesla Motors generate tremendous interest in this car.

**TABLE 9.6**
**Marketing media.**

Radio and podcasts

Newspapers

Magazines

Television and video

e-mail

Telemarketing

Catalogs

Infomercials

Websites

Social networks

Blogs and wikis

Presentations
  and speeches

**Place** means selecting the channels for distribution of your product and, when appropriate, the physical location of your stores. Channels of distribution are necessary to bring your product to the end user. A publisher sells a book through multiple channels such as bookstores, direct sales to the end user, and Internet bookstores, such as Amazon.com. Each industry has a some sort of distribution system. Differential advantages can accrue to sellers who creatively use different channels.

In the personal computer industry, Dell Computer sells direct to the end user via phone or the Internet, while Hewlett-Packard sells primarily via retail stores and value-added resellers. When several parallel channels are used, channel conflict can occur due to the divergence of goals between channels and domain (territory or customer) disagreements [Brynjolfsson et al., 2009].

Many technology ventures will sell their product to other manufacturers who will incorporate the product as a module or component within the final product. For example, Intel provides microprocessors that Dell incorporates into its PC.

The use of the Internet as a distribution channel has caused a shift in the relationships between consumers, retailers, distributors, manufacturers, and service providers. The Internet presents many companies with the option of reducing or eliminating the role of intermediaries and lets those providers transact directly with their customers. Before launching an e-commerce effort and bypassing its traditional distribution channels, however, a business should analyze which products are appropriate for electronic distribution. For example, information products are very compatible with electronic distribution, while bulky low-margin items like pet food may be a better fit for a traditional channel.

### "Intel Inside" Campaign

In the late 1980s, Intel decided to redirect some of its advertising efforts away from computer manufacturers to actual computer buyers. The consumer's choice of a personal computer was based almost exclusively on the brand image of the manufacturer, such as HP, Dell, and IBM. Consumers did not think about the components inside the computer. By shifting its advertising focus to the consumers, Intel created brand awareness for itself and its products, and built brand preference for the microprocessor inside the PC.

The first step was to create a new advertisement using the slogan "Intel: The Computer Inside." Second, Intel chose a logo to place on a computer—a swirl with "Intel Inside." Then it chose a name for its new microprocessor—Pentium. As a result, Intel became a leader in the PC boom of the 1990s. Intel was successful at branding a component.

Many firms are using the Internet for selling and experimentation. Procter & Gamble (see www.pg.com) is using its corporate website to invite online customers to sample and give feedback on new prototype products [Gaffney, 2001]. This approach permits P&G to test new products and their marketing mix. P&G conducts at least 40 percent of its tests online. In August 2000, when P&G was ready to launch Crest Whitestrips, a home teeth-whitening kit, it tested its proposed price of $44. P&G ran TV and magazine ads to attract people to the test. It also sent e-mails to people who had signed up to sample new products. In the first eight months, it sold 144,000 whitening kits online.

Microsoft, Yahoo, and Google have emerged as key online advertising platforms [Vogelstein, 2005]. Internet advertising is displacing newspaper and magazines. Google and others place an advertisement in front of users when they are looking to buy or research an item online.

An emerging firm has to decide how and where to spend its marketing dollars. It may have several product categories and numerous regions on which to spend its limited marketing budget. To decide, an emerging firm should collect information on each regional or international market and allocate its resources based on the regions and products that offer the best opportunity for profit [Corstjens and Merrihue, 2003].

## 9.6 Social Media and Marketing Analytics

The world has changed dramatically in recent years with the proliferation of mobile devices such as smart phones and tablets, the emergence of social networks like Facebook and Twitter, and the ability to capture and analyze massive amounts of data, including data gathered from website interactions, social networks, call centers, and devices with smart connectivity.

These changes have created both massive upheaval and massive opportunity for the companies who figure out how to best leverage these new technologies. Commerce systems and financial transaction systems are subject to disruption and new ways are rapidly emerging for how commerce will be conducted in the 21st century. Recent years have witnessed the emergence of mobile commerce, new payment platforms like eBay and Square, and new consumer platforms such as Google, Apple, and Facebook, which represent new ways for consumers to evaluate and buy products.

Customers have increased power in this new paradigm as a result of greater access to product and pricing information and an ability to get input from friends and acquaintances that have expertise or previous experience using the products or services. Customers even have new ways to buy, including group buying and rental instead of purchase. At the same time, companies also have the ability to interface with those prospects and customers in new ways that make the interaction between prospects/customers and those companies much more like an ongoing relationship than a buyer/seller paradigm. A new era is emerging where companies connect with prospects and customers on fixed and mobile devices, using the customers' communication method of choice (e.g., e-mail, Twitter, text, phone, online, or face-to-face meetings), and communicate directly to them with information tailored to their specific desires and interests.

Social media and marketing analytics are at the core of these changes. Social media networks encourage online interactions among communities. Those communities run on the Web, and the community members and Web content are accessible on both personal computers and mobile devices. Within these networks, people express and share their interests, discuss their likes and dislikes, and share input on millions of topics including music, careers, clothes, travel, education, and sports. Marketing analytics gathers and analyzes data from these various social media sites and merges the social media data with data that companies have in their corporate databases to gain deep prospect and customer insights. These analyses enable companies to create highly-successful targeted marketing and sales initiatives. Often, messages can be custom tailored down to the individual level.

Wells Fargo equity research sees a $50 billion software opportunity as companies try to bring together unified solutions for all prospect and customer facing functions of a company, including marketing, sales, customer service/support, and ecommerce. Marketing analytics enables marketing and sales people to more effectively do their jobs. In addition, it enables advertisers and publishers to target and optimize their display ads to generate the highest click-through rates and maximize online purchases. Marketing analytics tools also recommend the highest value-set of keywords for bidding, search engine optimization, and search marketing campaigns.

Table 9.7 lists representative new marketing tools that are leveraging social media and marketing analytics. With this new set of tools, the conduct of marketing and sales has changed forever. Much of the front end of the sales process, including prospect identification, brand and product awareness, providing product information, and initial customer interface is moving to the Web or to mobile devices. The role of marketing is shifting to emphasize the effective analysis of large amounts of data, gathered in large part from social media, to determine the right time to deliver product messages to interested prospects and customers. Firms are gauging customer interest through the use of marketing analytics. The marketing department's primary focus is shifting from traditional marketing activities (demand/lead generation, value proposition development, branding) to activities focused on analyzing data to figure out the right messages to get in front of prospects at the right time. Effective firms, therefore, have an integrated approach to marketing, sales, and customer support/service that enables both ecommerce and traditional commerce to be conducted in a way that provides the highest levels of prospect and customer satisfaction while enabling the highest return on investment for sales and marketing spending.

**TABLE 9.7  New marketing tools for social media and analytics with example providers.**

| | |
|---|---|
| ■ Marketing program effectiveness tools (also called Marketing Automation Platforms) | ■ Eloqua, Marketo, Hubspot |
| ■ Website analytics tools that analyze search word effectiveness | ■ Google Analytics, Omniture (part of Adobe), Webtrends, BloomReach |
| ■ E-mail marketing tools | ■ Responsys and Exact Target |
| ■ Customer experience measurement and management on websites and mobile devices | ■ Tealeaf (part of IBM), Kissmetrics |
| ■ Sales effectiveness tools | ■ Salesforce, Oracle, SAP |
| ■ Retargeting | ■ AdRoll, Criteo |
| ■ Incenting customers to tell your message | ■ Influitive and Needle |
| ■ Customer satisfaction measurement tools | ■ Comscore |
| ■ Social monitoring analytics | ■ Attensity, Salesforce (formerly Radian6) |
| ■ Text analytics | ■ Autonomy Teradata, SAS |
| ■ Social media | ■ Yammer (part of Microsoft), Jive |
| ■ Customer engagement | ■ Moxie Software, Oracle (formerly Rightnow), Lithium |

# 9.7 Customer Relationship Management

The quality of the relationship that a firm has with its customers directly influences the intrinsic value of the firm. **Customer relationship management** (CRM) is a set of conversations with the customer. These conversations consist of (1) economic exchanges, (2) the product offering that is the subject of the exchange, (3) the space in which the exchange takes place, and (4) the context of the exchange [McKenzie, 2001]. For the customer relationship to be fruitful, a firm must effectively manage the attraction of the customer, the conversion or sale of the customer, and the customer retention process. These relationships are managed through brief real-time conversations that help build a relationship. These conversations take place between the customer and the firm in a relationship space, as depicted in Figure 9.6. The first part of a conversation is the economic exchange based on a product offering that is communicated to the customer. The space in which the exchange takes place could be physical, such as a showroom, or virtual, such as a website displaying goods. The context of the exchange includes all that is known about the customer and the situation with the customer.

A necessary step to a CRM system is the construction of a customer database. Developing a database is relatively easy for banks and retail firms since they have a high frequency of direct customer interaction. It is more difficult

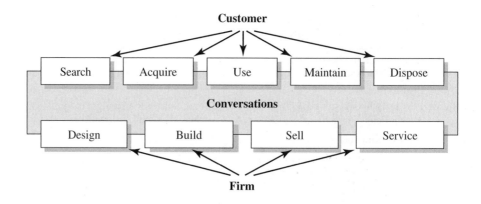

**FIGURE 9.6**  Customer-firm relationship as a conversation.

for firms that do not interact directly with the end customer, such as semiconductor and auto manufacturers [Winer, 2001]. The customer database can be used for CRM activities such as customer service, loyalty programs, rewards programs, community building, and customization.

CRM, when properly constructed, allows firms to gather customer data quickly, identify valuable customers, and increase loyalty by providing excellent service. Through the CRM process, customers become a new source of competencies engaged in building the firm's products and services as they provide ideas via the conversational process [Prahalad and Ramaswamy, 2000]. Unfortunately, too many companies distance themselves from customers by using phone loops that trap and frustrate customers seeking aid.

CRM is best operated when the customer and the CRM employee are fully engaged in conversation. A process that fully engages the customer and the employee can deliver more effective outcomes for the firm and the customer [Fleming et al., 2005].

Progressive Corporation uses CRM to relate to its customers 24 hours a day. It sells auto insurance both directly and through agents. Progressive's information systems allow customers to manage their accounts online, including paying their bills electronically and adding a vehicle or driver. It has a highly functional website, telephone call centers, and claims service available 24 hours a day, seven days a week. Progressive's claims agents travel quickly to the scene of an accident. The agents are equipped with notebook computers that communicate wirelessly to the corporate network, which lets them key in information on site.

Figure 9.7 depicts CRM and the total marketing effort. CRM helps improve marketing research, customer retention, and the marketing mix.

Firms generally strive to have repeat customers. Customers who say they are satisfied are not necessarily repeat customers because satisfaction is a measure of what people say, whereas loyalty is a measure of what they actually do. Thus, customer surveys measure opinions but are unreliable predictors of future behavior. Loyalty is not a matter of opinion [Klein, 2003]. **Loyalty** is a

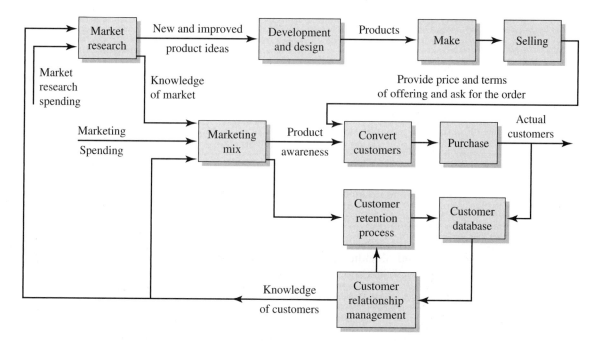

**FIGURE 9.7** CRM and the total marketing effort.

measure of a customer's commitment to a product or a company's product line. Loyalty measurements are more difficult to obtain than satisfaction measures. Good satisfaction measurement can help identify what's broken in your business today. Good loyalty measurement is a forward-looking tool that firms can use to devise strategies to hold on to customers they want to keep.

**Customization**, sometimes called one-to-one marketing, is a process that enables a product to be customized (changed) to a single customer's specifications. A firm uses a CRM system to elicit the information from each customer specific to his or her needs and preferences. Customization allows the company and the customer to learn together about the customer's needs. Customization is easy to do with digital goods such as music files, but other manufacturers can also tailor products to provide customization [Winer, 2001]. For example, BMW popularized the concept of designing and customizing a car online, which can then be purchased and delivered to a local dealership.

## 9.8 Diffusion of Technology and Innovations

Most entrepreneurial ventures have some novelty or innovation embedded in their product. Customers adopt an innovation at a particular point in time based on their perception of its advantages and its cost and risks. The **diffusion of innovations** describes the process of how innovations spread through a population of potential adopters. An innovation can be a product, a process, or an idea that is perceived as new by those who might adopt it. Innovations present the potential adopters with a new alternative for solving their problem, but they also present more uncertainty about whether that alternative is better or worse than the old way of doing things. The primary objectives of diffusion theory are to understand and predict the rate of diffusion of an innovation and the pattern of that diffusion.

Innovations do not always spread quickly. The British Navy first learned in 1601 that scurvy—a disease that killed more sailors than warfare, accidents, and all other causes of death—could be avoided through the simple solution of incorporating citrus fruits in a sailor's diet. The British Navy, however, did not adopt this innovation until 1795—almost 200 years later [Rogers, 2003]. Even the best ideas are not always quickly adopted.

The diffusion of innovations depends on a potential adopter's perception of five characteristics of an innovation, as listed in Table 9.8. The adopter's perceptions of these characteristics strongly influence his or her decision to adopt or not. Consider the introduction of black-and-white television in 1947. By 1950, 10 percent of all households had adopted this innovation, and by 1960, 90 percent of households had a TV. This rapid adoption of TV was due to its relative advantage compared to radio, its high compatibility within the home, its relatively low complexity, its easy trialability, and its ready observability in TV store windows

**TABLE 9.8  Five characteristics of an innovation.**

- **Relative advantage:** the perceived superiority of an innovation over the current product or solution it would replace. This advantage can take the form of economic benefits to the adopter or better performance.

- **Compatibility:** the perceived fit of an innovation with a potential adopter's existing values, know-how, experiences, and practices.

- **Complexity:** the extent to which an innovation is perceived to be difficult to understand or use. The higher the degree of perceived complexity, the slower the rate of adoption.

- **Trialability:** the extent to which a potential adopter can experience or experiment with the innovation before adopting it. The greater the trialability, the higher the rate of adoption.

- **Observability:** the extent to which the adoption and benefits of an innovation are visible to others within the population of potential adopters. The greater the visibility, the higher the rate of adoption by those who follow.

and friends' homes. On the other hand, consider the slow adoption of the personal computer in the home. PCs were introduced into the home market by 1982, but as late as 2007, only two-thirds of households had a PC. The complexity and cost of a PC discourages many consumers from adopting it in the home. Also, the perceived advantage is not clear to many would-be users.

As the PC example highlights, to understand a customer's likelihood of adoption, it is necessary to compare his or her current "pain" to the perceived pain of the solution presented by your company's product or service. For example, without a PC and word processor, a potential customer feels pain when typing documents that cannot be easily changed and when tracking expenses by hand rather than by using a spreadsheet. But, adoption of a PC also imposes its own perceived pain. This perceived pain is not only price, but also time and effort required to read instruction manuals, research products, wait in line, install software, and so on. Customers will adopt only when their perceived benefit exceeds this perceived pain [Coburn, 2006].

In a rapidly changing technology area, customers may wait to adopt if they think that a better technology is just around the corner. For example, many people did not purchase plasma televisions because they suspected that LCD televisions would be available soon. If customers wait long enough, they can "leapfrog" entire technologies, as with communities in Africa and Asia that skipped landlines and adopted mobile phones [*Economist*, 2006].

The adoption of any innovation usually follows an S curve, as shown in Figure 9.8. When the adoption follows the S curve, then the distribution curve of adopters follows a normal distribution, as shown in Figure 9.9, where Sd = standard deviation. The five categories of adopters are shown in Figure 9.9 and described in Table 9.9 [Rogers, 2003].

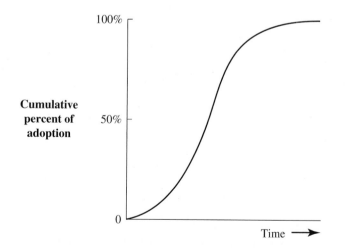

**FIGURE 9.8** S curve of adoption of an innovation.

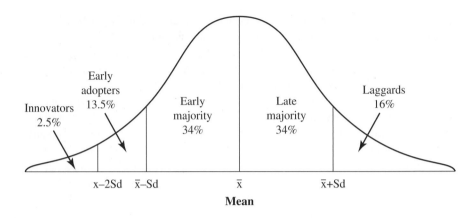

**FIGURE 9.9**  Innovation adoption categories when Sd = standard deviation.

**TABLE 9.9  Five categories of adopters of an innovation.**

- **Innovators** want to be on the leading edge of business and are eager to try new innovations. They have an ability to work with complex and often underdeveloped ideas as well as substantial financial resources to help them absorb the uncertainties and potential losses from innovations.

- **Early adopters** are more integrated with potential adopters than innovators and often have the greatest degree of opinion leadership, providing other potential adopters with information and advice about an innovation. They are visionaries.

- The **early majority** adopts just ahead of the average of the population. It typically undertakes deliberate and, at times, lengthy decision-making. Because of its size and connectedness with the rest of the potential adopters, it links the early adopters with the bulk of the population, and their adoption signals the phase of rapid diffusion through the population. They are pragmatic.

- The **late majority** is described as adopting innovations because of economic necessity and pressure from peers. While it makes up as large a portion of the overall population as the early majority, it tends to have fewer resources and be more conservative, requiring more evidence of the value of an innovation before adopting it.

- **Laggards** are the last to adopt a new innovation. They tend to be relatively isolated from the rest of the adopters and focus on past experiences and traditions. They are the most skeptical when it comes to risking their limited resources on an innovation.

## 9.9 Crossing the Chasm

The transition from the early adopters to the early majority is difficult since it requires attracting pragmatists, as shown in Figure 9.10. This large gap between visionaries and pragmatists is called a **chasm** [Moore, 2000]. The early adopters or visionaries are independent, motivated by opportunities, and quickly appreciate the nature of benefits of an innovation. However, the early majority or pragmatists are analytical, conformist, and demand proven results

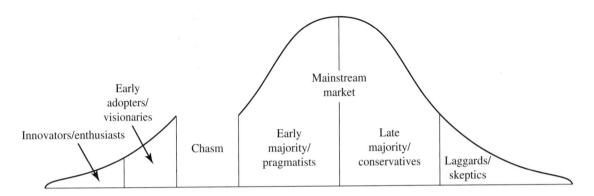

**FIGURE 9.10** Chasm model.

from an innovation. Once a product crosses the chasm, others will purchase it since they will readily observe and try it [Rohlfs, 2001]. Crossing the chasm is a challenging task for a new business. For example, the first videophone was introduced in the 1960s. It was not until Skype emerged in 2003, however, that the technology moved toward the mainstream and it took several more years to cross the chasm. Another example is three-dimensional movies that first launched in 1953. The inconveniences involved in 3-D movies have outweighed the benefits, and the adoption of 3-D stalled in the 1990s [Huntington, 2003]. With the introduction of 3-D televisions in the 2010s, manufacturers are again making a push, though 3-D technology still has not crossed the chasm.

Honda introduced hybrid cars into the U.S. market in 1999. Toyota followed in 2000 with the Prius, which enthusiasts and visionaries rapidly adopted. Hybrid cars have been increasingly popular and now have crossed the chasm into the mainstream of auto sales.

Digital photography stalled at the chasm for over a decade (1985–1995). Factors that held it were the lack of personal computers and software that could easily manipulate digital images, as well as the absence of inexpensive printers for printing photos [Ryans et al., 2000].

Many high-technology products are weak on both compatibility and complexity. They are difficult to operate and understand. Thus, they require the would-be user to learn how to use them and how they work. The firm offering these products needs to educate the user on both these matters. In some cases, this education effort can be expensive and time-consuming.

All innovators are not educators; however, anyone who adopts an innovation has to be a learner. Even the most transparent and intuitive designs present learning challenges. A bicycle looks easy to ride. But for most adults, learning to ride a bicycle is difficult because they may have to fall down before they learn how. Teaching an adult to ride a bicycle is often an exercise in frustration. Too many innovators rely not on sound training sessions and documentation to support their users' learning, but rather on Web pages that list frequently asked questions.

Encouraging users to read the directions simply does not work. Innovators often try to transfer the costs of teaching and learning to their customers. As a result, adoption of an innovation is delayed or halted.

Innovators with good ideas have no choice but to be educators, through either the medium of their innovations or teaching. They should recognize that if they really want to overcome customer resistance, they need to balance ease of use with ease of learning [Schrage, 2002].

The chasm model and the diffusion characteristics of a product help to explain the **diffusion period** required to move from 10 percent to 90 percent of the potential adopters. Table 9.10 shows the diffusion period for selected innovations.

Early on, some suggested that it could take a decade or more for the DVD format to be adopted. By 2003, purchases of DVD players and discs had doubled over the previous year. DVD sales rapidly overcame two potential hurdles: (1) the higher purchase price of DVD discs compared to VCR tapes and (2) the inability of the DVD player to record. Currently, video-on-demand delivered over broadband cable or satellite TV poses a challenge to continued growth in the DVD and Blu-ray market.

Many innovators face a chicken-and-egg problem since to use a new device, the user needs a widely available infrastructure. However, it doesn't pay others to build an infrastructure for just a few users. This is true for the world of wireless Web devices. Another example is the push for fuel-cell vehicles. Their use requires hydrogen fuel stations, but who will build them until many fuel-cell vehicles will use them? It can be difficult to get some customers to adapt innovations because markets are taking on the characteristics of networks. The interconnections among today's companies are so plentiful that often a new

**TABLE 9. 10  Diffusion period for selected innovations (U.S.).**

| Innovation | Diffusion period (Years) |
|---|---|
| Telephone | 70 |
| Automobile | 60 |
| Electric power | 40 |
| Personal computer | 30 |
| Mobile phone | 30 |
| Internet access | 18 |
| Color television | 16 |
| DVD player | 15 |
| Artery stents (for relief of heart congestion) | 8 |

*Note:* Diffusion period is the time required to diffuse an innovation from 10 percent to 90 percent of the potential adopters.

product's adoption by one player depends on its systematic adoption by other players [Chakravorti, 2004].

Other products face a dilemma in that their value depends on how many other people have a compatible product. E-mail and text messaging, for example, are only useful when a large number of other people also have e-mail and text messaging capabilities. Once these products do hit a critical mass, however, they can spread like an epidemic. Diffusion epidemics have three characteristics: contagiousness; little causes have big effects; and a big change happens, not gradually, but at one moment [Gladwell, 2000]. A **tipping point** is the moment of critical mass or threshold that results in a jump in adoption. This type of jump happens in networks where, at some point, enough people have the product that its value jumps significantly and the product takes off. The first text message was sent in 1992. In the late 1990s, when rival mobile networks allowed users to send text messages outside of the network and when mobile phones spread among young people, texting grew rapidly. As a result, 2000 was the tipping point.

An epidemic is spread by a message that is meaningful and emotional to the receiver and motivates the buyer who finds the message sticks in his or her mind and passes it on to others. The persuasive message is communicated by a trusted agent, and it moves the buyer to act. Thus, a contagious message that is memorable, motivating, and delivered by a trusted agent can take a product across the chasm by helping a product reach its tipping point.

---

**Twitter: Reaching for the Tipping Point**

Tipping points are especially evident in social networking applications. Twitter, which is a microblogging website and mobile application, experienced explosive growth in 2009. Its user base grew to over 18 million US adults who accessed Twitter on any platform, which was a 200% increase over 2008. By 2013 Twitter had over 50 million users. Awareness of the site was largely spread by trusted agents with a contagious message. For example, many *New York Times* columnists began to publish on Twitter, and some of their readers were enticed to join as well.

---

The entrepreneur is the agent who creates the vision for the new product and builds a marketing plan that will help potential adopters to understand and value the innovation and respond to the message communicated by the firm. This message must be persuasive, believable, and understandable. When new ventures have a strategy to overcome the potential adopters' lack of knowledge and understanding of the product, they are more likely to receive funds from investors and to succeed in bringing their product across the chasm.

Marketing can be described as taking the actions necessary to create, grow, and maintain your firm's place in your chosen market. To cross the chasm, the marketing strategy has to attract and retain pragmatists. They care about quality, service, ease of use, reliability, and the infrastructure of complementary products (often called the whole product).

For a small start-up, crossing the chasm may mean focusing on a particular market segment or acting locally and then expanding as sales grow. Crossing a chasm requires assembling the whole product, through alliances with partners, to satisfy the pragmatic buyer in a specific target market [Moore, 2002].

## 9.10   Personal Selling and the Sales Force

All businesses involve **selling**, which is the transfer of products from one person or entity to another through an exchange mechanism. Selling includes identifying customer needs and matching the product or solution to those needs. For many technology ventures, this process is called sales and business development. Most firms employ a sales force to make the contacts with the purchaser. In a small start-up, the salespeople may have other roles, such as product development or market planning.

New ventures should develop a selling strategy and a plan of action. Then they locate target customers and recruit, train, motivate, compensate, and organize a field sales force. They also have to manage the interactions between customers and salespeople. This dialogue is influenced by the buyer's needs and the salesperson's skills. The results of successful salesperson-customer interactions are orders, profits, and repeat customers.

Selling a technology product is difficult since the product may be less tangible than other items, like a house or a suit. Buying a technology product can take longer and may require the salesperson to inspire a buyer to take action. Thus, the technology sales person must fully understand the product offering and be able to communicate its benefits clearly.

Especially in markets where the customers are other businesses (often called B2B for business-to-business), there may be multiple decision-makers behind a purchase. The ultimate user of the technology product or service is certainly one of them. Other decision-makers, however, could include the people who make the recommendation of which solution to buy, such as the information technology staff, and the people who actually negotiate the contract, such as the purchasing agent. The involvement of multiple decision-makers can complicate and delay the sale. This length of time from the first contact until a sales transaction is completed is called the **sales cycle**. It can be as short as one day (e.g., purchasing ad listings on Google or auction listings on eBay) or many months (e.g., evaluating and choosing sophisticated MRI equipment for a hospital's radiology lab). When purchases need to be written into an organization's annual budget, it can further prolong the decision-making process. Technology ventures must estimate the length of their sales cycle as they develop their business model and financial plan.

IBM's success from 1955 to 1990 was due to a very knowledgeable, well-trained, and highly motivated sales force. IBM's salespeople had real experience with computers as well as an understanding of their clients' needs. Thomas Watson, Sr., former CEO of IBM, noted that great technological innovation combined with a powerful sales force was unbeatable.

For a new technology venture with an innovative product, the salesperson must fully understand the product and the idea of creating a solution for the customer. It is the responsibility of everyone in a new venture to (1) identify and create a purchaser, (2) offer a creative solution, and (3) make a profitable sale. In many start-ups, the staff is comfortable with steps 1 and 2 but shies away from step 3. Without actually making the sale, the start-up is destined to fail [Bosworth, 1995]. The goal is to determine the purchaser's needs (or latent pain), create a solution to meet that need, and then sell it to the purchaser.

The solution sales process is shown in Figure 9.11. The salesperson identifies the target market and makes contact with a potential purchaser. Then the salesperson determines the customer's problem and needs. Based on these needs, the solution to the customer's problem is created and presented. The benefits of the solution must be clearly communicated. Then the salesperson asks for the order and, with a positive response from the customer, confirms the order. You may have experienced this process when shopping for new clothes. The salesperson makes a contact and determines your needs. Then the salesperson shows you one or more solutions (options), and you try them on for size and appearance. The salesperson aligns his or her comments on each solution in a discussion with you. When you both see a solution, the salesperson asks for the order. If you agree, the salesperson writes the order at the cash register. This process is the same, although more complex, for a purchaser seeking a new computer system for a government agency or an electronics firm. Salespeople sell themselves, show they care, and provide proof of product, consistency of message, authority, and scarcity. The sales process rests, in part, on the skills of persuasion, as later described in Section 13.2.

New ventures often use their own people to manage the sales process but engage others, called sales representatives, under contract, to actually sell the product. The advantages and disadvantages of using company salespeople versus independent representatives are shown in Table 9.11. The choice of the right balance of company salespeople and independent representatives is a critical issue for a new business.

It is also important to grow the sales force at the right pace. Often, companies beef up their sales force capacity too early, when new products are not quite ready. A better approach is to start with a small group of salespeople who learn as much as they can about customers' responses to your product. Then, use these responses to refine the product and the sales/marketing strategy, and expand the sales force only as sales themselves accelerate [Leslie and Holloway, 2006]. A new venture needs to focus on its customer rather than product features. The sales force needs to find out what the customer needs, which will be a combination of products, services, and the product elements. They should provide a solution to the customer needs [Charan, 2007].

**FIGURE 9.11**
Solution selling process.

**TABLE 9.11**  Selling via company salespeople versus independent
representatives.

|  | Advantages | Disadvantages |
|---|---|---|
| Company salespeople | ■ Know your product well | ■ High fixed cost |
| | ■ Relatively easy to manage | ■ Low geographical dispersion |
| | ■ Provide feedback from customers | ■ Time and costs to hire and train |
| | ■ Paid salary plus commission | ■ Travel costs |
| Independent representatives | ■ Paid on commissions | ■ Sell for several firms, making it difficult to get their attention |
| | ■ Lower hiring and training costs | ■ Difficult to manage |
| | ■ Geographical dispersion | ■ Low feedback from customers |
| | ■ Have established relationships with customers | ■ May have limited understanding of your complex product |
| | ■ Low fixed costs | |

The emerging technology business may initially use a focused, direct sales force to create demand and penetrate to the primary target segment. Then, as growth accelerates, a transition to other segments and sales channels may be appropriate. It is important to clearly identify the primary target segment and key customers [Waaser et al., 2004].

New businesses encounter sales resistance due to competition and lack of knowledge of their product and its quality. One method to overcome this is to utilize trial periods, warranties, and service contracts. Many new ventures do an excellent job of building a good product and developing a solid marketing plan but then fail to make the forecasted sales.

We will cover the issues of international marketing and sales in Chapter 15.

## 9.11  Spotlight on DirecTV

DirecTV launched its satellite television service in 1994. It has become one of the largest providers of satellite television in the world and is a leading alternative to cable television.

DirecTV's target customers include households and small businesses in the U.S. and the rapidly growing middle class in Latin America. The company has worked hard to build a recognizable brand that conveys coolness, simplicity, and good customer relationships. They use Facebook, Twitter, and YouTube to engage customers and potential customers, and to reinforce their brand and product offering. They also targets their advertising and their service based upon sophisticated marketing analytics.

Since its launch in 1994, DirecTV has grown to more than 35 million subscribers. A key to DirecTV crossing the chasm was their recognition that early adopters were excited by the broad channel selection and digital picture quality, but the mainstream market cared more about good communication and specific offerings, like sports. Thus, DirecTV adjusted their marketing strategy as they grew.

## 9.12 Summary

Any new firm needs to build a marketing plan that describes how it will attract, serve, and retain the customers targeted for its products. Since a new firm normally starts without established customers, it must carefully identify the target market that will value its product. Market research can provide the information about the customers, appropriate distribution channels, and communication methods for attracting the customer.

The new firm creates a product positioning statement and selects a mix of price, product, promotion, and channels to attract and satisfy the customer. Most new firms are challenged to cross a chasm in the diffusion process that enables their product to attract the pragmatic and skeptical potential customer. The marketing process consists of describing or implementing the following elements:

- Product offering
- Target customer
- Marketing objectives
- Market research
- Marketing plan
- Sales plan
- Marketing and sales staff

---

**Principle 9**

A sound marketing and sales plan enables a new firm to identify the target customer, set its marketing objectives, and implement the steps necessary to sell the product and build solid customer relationships.

---

## Video Resources

Visit http://techventures.stanford.edu to view experts discussing content from this chapter.

| | | |
|---|---|---|
| Marketing by Surfacing the Product | Jack Dorsey | Square |
| Marketing a Start-up | Donna Novitsky | Big Tent |
| Stay on Simple Messages | Adam Lashinsky | Fortune |

## 9.13 Exercises

**9.1**   There continues to be a disparity between the advertising dollars spent on reaching TV viewers and Internet users versus the amount of time that is spent interacting with each media. A large degree of Google's success is attributable to taking advantage of this large gap. Research to determine (a) TV advertising dollars and Internet advertising dollars spent, and (b) the amount of time spent watching TV versus using the Internet. How has Google taken advantage of this disparity? What other major societal trends are forecasted that will continue to shift advertising dollars to new mediums, and why?

**9.2**   With the explosion of mobile handsets worldwide, many marketing and advertising firms are looking at how to take advantage of the ubiquity of a communications device carried everywhere. Describe why marketers view the mobile handset as such a valuable marketing platform. What types of mobile advertising challenges do you foresee arising?

**9.3**   What is viral marketing? Provide an example of a start-up using viral marketing to promote and sell its product or service. Why does it work (or not work)?

**9.4**   Facebook and Twitter are rapidly growing social networking sites based in the United States but with global reach. Prepare a positioning map for these two firms.

**9.5**   BusinessWeek with Interbrand conducts an annual worldwide brand survey ranking the top 100 global brands. Examine the most recent survey and choose a new entrant to the list. Describe that company's marketing objectives and customer target segments.

**9.6**   Powerful brands are built on innovativeness and advertising. Examine the brand value for Genentech, Merck, and Apple, and describe the reasons for their brand power.

**9.7**   HDTV is an emerging consumer electronic technology. Discuss the marketing challenges associated with HDTV (e.g., building an ecosystem of content, players, and content distribution). How do you see this influenced by (a) competing with DVD (the last technology generation) and (b) emerging substitutes like online digital content, PCs, and mobile devices? Use the categories of Table 9.9 to describe where the market response to HDTV has been to date and how long adoption has taken for each group.

**9.8**   Identify a high-tech firm that uses an indirect sales channel model. What is the model used? Why was the indirect sales model chosen for its particular products or services?

## VENTURE CHALLENGE

1.  For your venture, describe the customer and the target segment you have identified.

2.  Develop a positioning statement using the template in Figure 9.4.

3.  Briefly describe the marketing mix for your product.

4.  How will your venture sell its product and develop customer relationships?

5.  Research the industry for your venture and determine the length of its sales cycle.

# CHAPTER 10

# Types of Ventures

*Even if you are on the right track, you'll get run over if you just sit there.*
**Will Rogers**

**What forms do new businesses take and what are corporate ventures?**

The appropriate legal and organizational format used to establish a new venture will vary according to several factors such as context, people, legal and tax consequences, and cultural and social norms. In this chapter, we consider the various organizational and legal forms that entrepreneurs employ to achieve their objectives. New ventures can range from small business or consulting services to high-growth, high-impact enterprises. When entrepreneurs establish a new enterprise, they must make some critical decisions about these matters. See, for example, the first steps to establishing a new corporation in Figure 10.1. Important social enterprises are often established as nonprofit ventures.

An important contrast to these independent ventures is the corporate new venture. It emerges within larger existing enterprises and is granted autonomy so it can fulfill its promise. Corporate new ventures are an important part of the entrepreneurial world and account for many new innovations. Often constrained by existing commitments and capabilities, corporations can fail to respond to significant new opportunities. Well-planned corporate ventures, however, help refresh and strengthen large corporations. ■

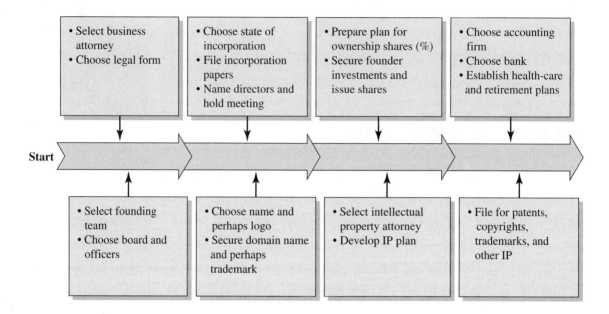

**FIGURE 10.1**  First steps of establishing a new corporation.

## 10.1  Legal Form of the Firm

When establishing a new technology venture, the entrepreneur needs to choose the legal form of the organization. The entrepreneur should choose a legal form that will facilitate the business, tax, and capital-raising objectives of the new company. The choice of a legal form depends, in part, on how the firm and its owners want to handle federal taxes. For tax purposes, we will address two types of forms: **regular taxable corporations** and **flow-through entities**. The elements of taxable corporations and flow-through entities are summarized in Table 10.1. A **corporation** is a legal entity separate from its owners. A flow-through entity, sometimes called a pass-through entity, is one that passes all losses or gains

**TABLE 10.1  Legal form of the firm.**

| Type | Taxation |
|---|---|
| 1. Regular taxable corporation: C-corporation | Taxation of the corporate profits as well as taxation of any corporate distributions to owners |
| 2. Flow-through entities<br>■ Sole proprietorship<br>■ Partnership<br>■ S-corporation<br>■ Limited liability company (LLC) | All profits or losses flow through to the owners and are not separately taxed to the firm |

through to the owners of the firm. The profits of a regular corporation are taxable to the corporation, and any distributions are taxable to the owners. This results in double taxation of any distributions such as dividends.

There are four main types of flow-through entities to choose from. They are the sole proprietorship with one owner, which is the simplest form; the partnership; the S-corporation, which is taxed much like a partnership and is named for the U.S. Internal Revenue Code subchapter that covers it; and the limited liability company (LLC).

Table 10.2 summarizes the key elements of the five types of legal form of a new business. A **sole proprietorship** is a business that is owned, and usually operated, by one person. This is a simple form of doing business but exposes the owner to unlimited liability for all debts of the business. A **partnership** is a voluntary association of two or more persons who act as co-owners of a business. Each partner who participates in management is liable for the acts of the business. Liability for all acts of a business is all-encompassing, and this factor encourages most entrepreneurs to set up a LLC or corporation.

The Hewlett-Packard Company was first established as a partnership in 1937, with William Hewlett and David Packard as equal partners. Their first sale was eight oscillators to Disney Studios in 1939 [Packard, 1995]. In 1947, Hewlett-Packard was incorporated to provide for continuity of life for the firm as well as limited liability for the owners.

In general, many businesses start out as sole proprietorships or partnerships but soon migrate to LLCs or a corporate form. With unlimited liability as a risk of the proprietor or partnership form, it may be unwise to continue in that form beyond the initial period necessary for completing a business plan. Most investors will only be willing to invest in a corporation or LLC since they wish to avoid any liability beyond the amount of their investment. The personal liability of a regular corporation or an LLC is limited to the amount of capital contributed to the entity by that person.

**TABLE 10.2  Key elements of the five types of legal forms for a new business in the United States.**

| Factors | Sole proprietorship | General partnership | Regular C-corporation | S-corporation | LLC |
|---|---|---|---|---|---|
| Owners' personal liability | Unlimited | Unlimited | Limited | Limited | Limited |
| Taxation | Proprietor's personal tax forms | Partners' personal tax forms | Profits taxed at corporation and owners pay tax on distributions | Profits or losses flow through to owners | Profits or losses flow through to owners |
| Continuity of business | Terminated by proprietor | Dissolved by partners | Perpetual | Perpetual | Varies |
| Cost of formation | Very low | Low | Moderate | Moderate | Moderate |
| Ability to raise capital | Low | Moderate | High | Moderate | Moderate |

The limited liability feature of a corporation arises from the fact the corporation is itself a legal "person," separate from its owners. If a corporation fails and proper formalities have been followed, creditors have a claim only on the corporation's assets, not on the owners' personal assets. Because they limit liability, it is often wise to use an LLC or a corporation even for sole proprietorships.

Many firms that start as a small business should consider an LLC or a subchapter S-corporation form. These forms allow the initial business losses to flow through to the owners, and these losses typically can be used to offset income from other sources.

The limited liability company offers an ideal form of ownership for small companies. It offers limited liability to the owners along with the tax advantages of a sole proprietorship or partnership. An LLC is particularly attractive to a family business that receives investments of a family's funds since it offers continuity of life, limited liability to participants, and advantages in handling tax issues.

An S-corporation is a corporation that is taxed as a flow-through entity. To qualify, a firm must meet certain requirements regarding its owners and types of stock. The S-status is established by filing with the IRS and may later be converted to a regular C-corporation. Some entrepreneurs prefer this election to the LLC. Like the LLC, the S-corporation will typically need to convert to a regular corporation to accept any venture capital or issue stock in a public market. Unlike an LLC, the S-corporation also is limited in the number of investors and it may not have foreign investors.

As the firm grows, it may be wise to consider converting the LLC or S-corporation to a regular corporation since a regular corporation has several potential advantages. It can be sold or merged into another corporation with a tax-free exchange of stock. Other factors that should be considered with an attorney are the number of owners and investors, as well as the need to raise capital, the long-term goals of the business, and issues related to accounting and health-care/retirement plans.

If the intention of the new business is to raise a significant amount of funds to start the venture and eventually to build it to a significant size, it is wise to start from the beginning as a regular C-corporation. A C-corporation provides limited liability, unlimited life, the ability to accept investments from venture capitalists and other corporations, and greater flexibility when sold.

The regular C-corporation form will be often chosen by firms that intend to seek investment from numerous professional investors and other corporations. The C-corporation allows for various classes of stock, such as common stock and preferred stock. In particular, investors in a C-corporation can purchase convertible preferred stock, which is the most common form of venture capital investment.

A corporation or LLC can be created under state laws and usually requires some legal steps. The process of forming a corporation is called incorporation. The legal steps required for incorporation in the United States are straightforward. For example, if a team selects California as the state in which to incorporate, it files articles of incorporation with the secretary of state and then pays a fee. The certificate of incorporation states the name of the corporation; the broad business purpose; the name and address of the registrant; provisions for reimbursing certain damages and expenses that directors, employees, and officers may incur on behalf of the corporation; and the authorized capital, including the total number of shares and the class of shares [Bagley and Dauchy, 2007]. The ownership shares of the corporation are called stock.

The LLC's articles of organization are similar to the articles of incorporation and establish the company's name, its duration, and the names and addresses of organizers. The operating agreement is similar to the bylaws of a corporation in that it outlines the rights of the owners and the way the LLC will operate. The owners of an LLC are called members, and their ownership interests are known as interests. These terms are equivalent to stockholders and stock in a corporation.

Usually, a company incorporates in its home state, though many companies incorporate in another state for reasons of law or ease of doing business. The majority of U.S. venture-backed companies are incorporated in a handful of states such as California, New York, or Delaware. In addition to the large number of companies resident in California and New York, relative to other states, California, New York, and Delaware have corporate laws that are well developed, stable, and transparent—all characteristics that ultimately reduce the risk to investors. Moreover, because venture capitalists and their counsel are familiar with conducting financings under such laws, the speed and efficiency with which a transaction moves will likely be improved.

Finally, some states have a new kind of corporation called a Benefit Corporation or B-corporation. A B-corporation voluntarily meets different standards of purpose, accountability, and transparency. Specifically, B-corporations include positive social and environmental benefits in their purpose; they are required to consider not only "shareholders," but also "stakeholders," including workers, the community in which they are embedded, and the environment. Additionally, they are required to issue an annual report that details their social and environmental performance. B-corporation status does not, however, affect a corporation's tax status. Thus, B-corporations can still elect to be taxed as a C-corporation or an S-corporation.

## 10.2 Independent versus Corporate Ventures

Many purposes exist for establishing new ventures in different formats. Table 10.3 describes five types of new ventures. Each of these types has a set of characteristics that distinguishes it. We can describe a **small business** as a sole proprietorship, a partnership, or a corporation owned by a few people. Examples include consulting firms, convenience stores, and local bookstores. Typically, a small business has fewer than 30 employees and annual revenues less than $3 million.

A **niche business** seeks to exploit a limited opportunity or market to provide the entrepreneurs with independence and a slow-growth buildup of the business. This business might employ fewer than one hundred employees and have annual revenues of less than $10 million. On occasion, a niche business can grow over time into a large, important enterprise.

A **high-growth business** aims to build an important new business and requires a significant initial investment to start up. A **radical-innovation business** seeks to commercialize an important new innovation and build an important new business. These enterprises are the primary focus of this textbook.

**TABLE 10.3  Five types of new ventures.**

| Type | Revenue growth | Planned-for most likely size | Description | Objective |
|---|---|---|---|---|
| 1. Small business | Slow | Small | Sole proprietorship, family business | Provide independence and wealth to partners by serving customers |
| 2. Niche | Slow to medium | Small to medium | Exploits limited opportunity or market | Provide steady growth and good income |
| 3. High growth | Fast | Medium to large | Needs large initial investment; could seek disruptive innovation | Important new business |
| 4. Nonprofit organization | Slow | Small to medium | Serves members or a social need | Serve a social need |
| 5. Corporate new venture | Medium to fast | Large | Independent unit of an existing corporation | Build an important new business unit or separate firm |

A **nonprofit organization** is a corporation or a member association initiated to serve a social or charitable purpose. Thousands of new nonprofits are organized every year to serve important social needs throughout the world. A well-known nonprofit is the International Red Cross (www.ifrc.org).

Another type of new venture is started by existing corporations for the purpose of building an important new business unit as a solely owned subsidiary or a spin-off as a separate company. This activity can be called a **corporate new venture (CNV).**

An **independent venture** is a new venture not owned or controlled by an established corporation, which includes the first four types in Table 10.3. An independent venture is typically unconstrained in its choice of a potential opportunity, yet is usually constrained by limited resources. The corporate venture is usually constrained in its choice of opportunities to those consistent with the parent business. The corporate venture, however, usually has access to the significant resources of the parent firm [Shepherd and Shanley, 1998].

While independent and corporate ventures both face the same external context, their different competencies and resources cause them to develop different strategies. The independent venture has more flexibility and potentially requires fewer resources than the corporate venture. Furthermore, the independent venture has access to a wide range of advisors, while the corporate venture is advised and controlled by the parent company. Thus, the independent venture has the advantage of flexibility, adaptability, and high incentives, while the corporate venture is usually advantaged by its access to valuable capabilities and resources.

Different types of ventures also involve different levels of novelty and different employment relationships. Table 10.4 illustrates how venture types vary along these dimensions.

The bulk of this chapter focuses on intrapreneurship corporate new ventures. First, however, we examine nonprofits and social ventures.

**TABLE 10.4  Types of entrepreneurial transitions.**

| Labor market status | Innovation and novelty | |
| --- | --- | --- |
| | Low | High |
| Self-employment | Independent contractor<br>Small business proprietor | Venture-backed startup<br>Spin-out |
| Dependent employment | Conventional employment | Intrapraneurship and<br>corporate new ventures |

*Source:* Sorensen and Fassiotto, 2011.

## 10.3  Nonprofit and Social Ventures

The purpose of a new venture is to create wealth for society. Often, wealth is seen as financial wealth. But, many entrepreneurs seek to provide social wealth for their society. The product of a nonprofit hospital is a healthy patient. Only for the tax collector does it make a difference whether the hospital is a nonprofit or for-profit. A nonprofit organization is a corporation, member association, or charitable organization that provides a service but does not earn a profit. A nonprofit organization is permitted to generate a financial surplus but may not distribute this surplus to officers, investors, or employees. Furthermore, nonprofits have no owners. Any surplus must be used for the approved nonprofit mission of the organization. Today, nonprofit organizations are often called not-for-profits or NGOs. One out of every two Americans is estimated to work as a volunteer in the nonprofit sector.

Nonprofit organizations have traditionally operated in the social sector to solve or mitigate such problems as hunger, homelessness, pollution, drug abuse, and domestic violence. They have also helped provide certain basic social goods, such as education, the arts, and health care, that society believes the marketplace may not adequately supply. Nonprofits have supplemented government activities, contributed ideas for new programs, and functioned as vehicles for private citizens to pursue their own vision of the good society.

The product of the Girl Scouts is a mature young woman who has values, skills, and respect for herself. The purpose of the Red Cross is to enable a community hit by natural disaster to regain its capacity to look after itself. In this way, the nonprofit venture forms to respond to social need.

The decision to start a nonprofit venture will be determined by the nature of the social opportunity and the creation of an innovative response that cannot or should not be performed for profit. Social functions that depend primarily on volunteers or members such as churches, museums, theaters, social clubs, industry associations, credit unions, and farmers' cooperatives are usually formed as nonprofits.

The establishment of a nonprofit organization should follow the five steps of Table 6.1. The vision for the organization is defined in terms of the creation of social value rather than economic value. Many nonprofits can be described as socially conscious service organizations. Thus, the entrepreneurial team must be committed to the social values of the new venture with its attendant risks and uncertainties.

Once the business plan is written, the required financial and human resources must be determined. How will the necessary funds be acquired and the human talent attracted? The challenge is finding donors whose special interests match those of the new nonprofit venture and who also have the expertise and commitment to provide an independent check on management judgment.

Organizations that satisfy the conditions of section 501 (c)(3) of the Internal Revenue Service Code are called charitable organizations. These organizations have religious, educational, scientific, literary, and/or charitable aims. Donations to these organizations are tax deductible for the donor and exempt from estate taxes. Noncharitable nonprofit organizations are primarily set up to serve the purposes of their members and are also tax exempt, but donations are not normally tax deductible. Nonprofits often contemplate the establishment of a related business that will generate a net surplus. Often, however, they underestimate the costs and are overly optimistic regarding revenues [Foster and Bradach, 2005].

Leading a nonprofit requires the competencies needed for most businesses mixed with a commitment to the entity's social cause. Michael Miller leads Goodwill Industries of the Portland, Oregon, area and has built it to $4 billion in revenues by using his entrepreneurial skills [Kellner, 2002]. The strategies and approaches discussed throughout this book are important for nonprofits. For example, nonprofits should leverage technology to improve the efficiency of operations and the reach of their activities. They should also consider partnering with complementary organizations. The nonprofit Women in Technology partners with IBM to cohost an engineering camp for middle-school girls, drawing on IBM's resources and expertise [Austin et al., 2007].

Nonprofits also face special challenges. For example, nonprofits often find it difficult to agree on who their customer is. For the Red Cross, is it the hospital, the blood donor, or the financial donor? Who is the ultimate beneficiary of the service? Since nonprofits are not subject to the market in the same way that traditional firms are, it is critical that they constantly evaluate the effectiveness and efficiency of their activities [Bradley et al., 2003]. Managed properly, the nonprofit sector can spawn very big, high-impact ventures in the same way that the for-profit sector does. For example, the nonprofit sector delivers much of the health care for most nations.

A unique form of nonprofit corporation is a consumer cooperative, which is a business that belongs to the members who use it. The member/owners establish policy, elect directors, and often receive cash dividends. Examples of cooperatives include credit unions, housing co-ops, food co-ops, and utility co-ops. In 1938, mountain climbers Lloyd and Mary Anderson joined with 23 fellow climbers in the Pacific Northwest to found Recreational Equipment, Inc. (REI). The group formed a consumer cooperative to supply themselves with quality gear, clothing, and footwear selected for performance and durability for outdoor recreation, including hiking, climbing, camping, bicycling, and other sports. After more than six decades, REI has grown into a supplier of specialty outdoor gear currently serving more than three million active members through 125 retail stores in the United States and direct sales via the Internet (www.rei.com), telephone, and mail.

Social ventures sometimes take the form of a nonprofit. As discussed in Chapter 4, a concern for all stakeholders, including society at large, is crucial to the long-term success of any venture. Social ventures are distinguished from other ventures by their core value proposition. The value proposition for a traditional venture is designed to create financial profit by organizing to serve markets that can afford a new product or service. Social ventures aim for large-scale and transformational value that accrues to some segment of society or to society at large. They are cognizant of financial realities. But, they do not anticipate or organize to obtain significant financial profits [Martin and Osberg, 2007]. A social venture may take the form of a nonprofit or a for-profit.

Founded by Larry Brilliant, Google.org is the charitable branch of the Internet company Google. Its mission is to fight global poverty, fund new energy solutions, and protect the environment. The organization has over $75 million in investments spread throughout a number of different enterprises. It was granted three million shares during Google's IPO and Google continually contributes 1 percent of its annual profits to Google.org. Some of its main projects include a plug-in electric vehicle that gets 100 miles per gallon and a renewable energy project designed to produce electricity at a profit from wind and solar sources. A long-term goal for the organization is to develop a way to fund Google's massive power consumption from renewable energy sources. Google's data centers consume an enormous amount of electricity every year. Google.org is hoping to fund new technologies that can help reduce this cost.

---

**David Green: Social Entrepreneur**

David Green has started several small businesses in developing countries to make inexpensive medical devices such as intraocular lenses and hearing aids. These profitable businesses make low-cost devices that meet the needs of the poor without sacrificing quality [Kirkpatrick, 2003]. See www.aurolab.com.

---

A **social entrepreneur** is a person or team that acts to form a new venture in response to an opportunity to deliver social benefits while satisfying environmental and economic values. Social entrepreneurs focus on the social welfare of their customers or clients, while remaining cognizant of the economic and environmental costs and benefits. The goal is to harness innovation for social and public good [Jackson and Nelson, 2004].

Social entrepreneurs are, first and foremost, entrepreneurs. Thus, like all entrepreneurs, they recognize and relentlessly pursue opportunities; they act boldly without being limited by their current resources; and they engage in a process of continuous innovation, adaptation, and learning [Dees et al., 2002]. Social entrepreneurs focus on the scalability of their ventures, ultimately seeking to cause a large-scale transformation and not just a local effect [Martin and Osberg, 2007]. For this reason, they are especially cognizant of the importance of technology in facilitating their efforts.

> ### Key Success Factors for Social Entrepreneurs
>
> Social ventures need to have all of the attributes of non-profit ventures and must be able to convince people of the benefits of their social impact while accomplishing more with fewer resources. Kiva Microfunds (known as Kiva.org) is a worldwide leader in providing small business loans to people in developing countries throughout the world. Kiva needed to build out an online banking platform across dozens of countries, develop a group of willing lenders, and find a set of fundable opportunities in developing countries. The company started by hiring people who were highly knowledgeable about finance, online commerce, and building communities. Next, Kiva convinced key entrepreneurs and business people, including Premal Shah from PayPal and Reid Hoffman from LinkedIn, to join them as advisors. Finally the company convinced the media and the public to rally behind its cause. Kiva is a powerful example of successful non-profit startup that has created a major market with worldwide impact.

Social entrepreneurs hold certain advantages over entrepreneurs in other circumstances. With a mission as the guiding vision, social entrepreneurs strive to organize and deploy diverse resources. They can engage volunteers, customers, partners, and investors through a sound business plan that furthers the organization's mission. In the social sector, success in the enterprise equals significance through improved lives and healthy communities. Social entrepreneurs focus on creating social value [Dees et al., 2002].

An excellent example of social entrepreneurship is the nonprofit organization Trees, Water, & People (TWP), which has an environmental and social mission. (See www.treeswaterpeople.org.) Its mission is to reforest degraded areas and plant fast-growing trees in Central America. It decided to view the problem from the demand side and seek to reduce the demand for fuel wood. TWP teamed up with Aprovecho Research Center in Oregon to introduce a fuel-efficient stove that burns 50 to 60 percent less wood than traditional open fires by using an insulated, elbow-shaped burning chamber. The Justa stove also saves lives by removing toxic smoke through a chimney. TWP gives these wood-conserving stoves to farmers in El Salvador as an incentive to reforest their land. Fuel conservation, health improvement, and reforestation are accomplished together.

## 10.4  Corporate New Ventures

A venture started by an existing corporation for the purpose of initiating and building an important new business unit or organization can be called a corporate new venture. Some people refer to this process as **intrapreneurship**. The building of the new business enterprise depends on an entrepreneurial team leading the effort. Corporate venturing is focused on the identification and exploitation of previously unexplored opportunities that utilize the resources and competencies of an existing corporation. It is usually involved with the birth of new businesses and the associated revitalization of a corporation [Wolcott and Lippitz, 2007]. We differentiate a corporate venture from a project by (1) its newness to the corporation, and (2) its independence from the existing activities, organizational units, and products of the corporation. The characteristics of a corporate new venture (CNV) are summarized in Table 10.5. Corporate new ventures are distinguished from normal product development efforts by their relative autonomy, freedom to innovate, and entrepreneurial leadership.

Table 10.6 summarizes the strengths and weaknesses of a corporate new venture. Corporate new ventures benefit from ready access to the capital, people, suppliers, technologies, and brand of the parent firm. On the other hand, CNVs may be limited by the budgetary and control practices of the parent. Furthermore, the parent firm may not have the technologies or people that the new venture requires. In fact, studies show that the resource advantages of the existing corporation do not necessarily translate into higher performance for CNVs [Shrader and Simon, 1997]. To enable a CNV to be most successful, it may be necessary to give the new unit more autonomy, separating it from the controls and limitations of its parent.

Corporate new ventures differ from independent start-ups along a variety of dimensions, as detailed in Table 10.7. The factors for success, however, are essentially the same between the two types of ventures: opportunity, vision,

### TABLE 10.5  Characteristics of corporate new ventures.

| | |
|---|---|
| ■ Newness and novelty of the product relative to the firm's existing products | ■ High potential for significant innovation |
| ■ Independence or semiautonomy from existing corporate structure | ■ Unique entrepreneurial team leadership capabilities |

### TABLE 10.6  Strengths and weaknesses of a corporate new venture.

| Strengths | Weaknesses |
|---|---|
| Ready access to capital | May be subjected to a corporate budget process |
| Access to capabilities of corporate employees | Multiple control and review levels |
| Suppliers willing to help in the design process | Limited autonomy compared to an independent venture |
| Emphasis on the marketing plan | Limited access to strong entrepreneurial talent |
| Gain from brand equity of the parent firm | |
| Access to processes and technologies of the parent | Risk-reward may be less attractive than for an independent entrepreneur |
| | Limited to parent firm's technologies and processes |

commitment, capabilities, resources, technology innovation, strategy, and execution. Success of corporate new ventures has been shown to be positively associated with growth and profitability of the firm [Phan et al., 2009]. A representation of this relationship and corporate ventures appears in Figure 10.2. Mature corporations that engage in new business venturing are innovative, proactive, and constantly renew themselves.

An entrepreneur within an existing firm, like HP or Intel, starts with a description of an opportunity, the required resources to pursue it, the value that would be created, and a plan to pursue it. The entrepreneur needs a sound understanding of the technology and the customer and forms a sound strategy to move forward [Sull and Spinosa, 2005].

There are four general models of corporate entrepreneurship. In the Opportunist Model, intrepid "project champions" toil against the odds to create new businesses inside a corporation. In the Enabler Model, employees are willing to develop new concepts if they are given adequate resources. In the Producer Model, the corporation assumes organizational ownership, but limits resources,

**TABLE 10.7  Contrasts between independent ventures and corporate ventures.**

| Dimension | Independent venture | Corporate venture |
|---|---|---|
| Team | Best in industry | Best available in firm |
| Scope | Entire company focused | Development team hands to corporate operations |
| Culture | Driven, team-oriented | Company's culture |
| "Contract" | Explicit—business plan | Increasingly explicit |
| Incentives | Equity (entire team) | Varies: bonus, career growth |
| Oversight | | |
| Who? | Board | Upper management |
| When? | Monthly | Design reviews |
| External feedback | Customers, new investors | Customers |
| Financial goal | IPO or M&A value versus investment | Break-even date, return on investment |
| Changes to plan | Quick action by the board | Multiple levels for approval |

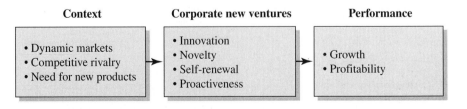

FIGURE 10.2  Corporate new venture model.

at least initially, to those provided by a specific business unit. Finally, in the Advocate Model, the corporation provides both resources and organizational ownership to encourage corporate entrepreneurship. Figure 10.3 illustrates these different approaches to corporate entrepreneurship. The appropriate model for any particular company depends on its specific objectives, such as whether the corporation is seeking change across the organization or only in a specific division [Wolcott and Lippitz, 2007].

Conventional wisdom says that large firms are weak at transforming opportunities into viable new businesses. The perceived reason for this weakness is the effects of bureaucracy and inflexibility that exist in large firms. Existing corporations are managed on a tightly defined strategy and within highly controlled boundaries. Corporate inertia can flow from successes that reinforce the rigidity of assumptions, processes, relationships, and values. All these factors are difficult to challenge: thus, the need for new independent organizations, such as new venture units, subsidiaries, or start-ups.

Often, the challenge for large firms is to protect the CNV from the pressures and controls of existing units within the firm and to "forget" the parent company's business model [Govindarajan and Trimble, 2005a]. Typically, a CNV needs to be established as a relatively autonomous unit that can accommodate the entrepreneurial team and its business plan. At the same time, large firms possess resources and capabilities that would be the envy of

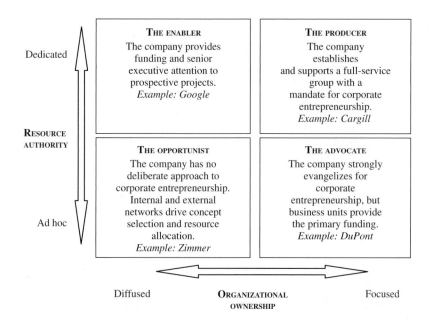

**FIGURE 10.3** Four models of corporate entrepreneurship.
*Source:* Adapted from Wolcott and Lippitz, 2007.

individual entrepreneurs trying to strike out on their own [Katila et al., 2008]. Unlike individuals, firms usually possess a share of the resources, capabilities, and knowledge necessary for innovation. Therefore, the challenge is how to achieve the right balance of integration. Table 10.8 outlines a number of methods that can be used to achieve this balance. Companies should develop strategy through disciplined trial and error, build on existing strengths, and work to integrate while preserving autonomy [Garvin and Levesque, 2006].

## Apple: Corporate Venturing

In 1985, Apple Computer faced several challenges. Its impressive Macintosh computer was in danger of being eclipsed by the new IBM PC and its clones. While Macintosh enjoyed outstanding margins, Microsoft and Intel were quickly commoditizing personal computing. Apple was hoping that its Graphical User Interface (GUI) would distinguish the company, but Microsoft introduced Windows. Another dilemma was that the Macintosh platform was dependent on Microsoft for applications. Apple did have its own modest applications group, but it could not afford to alienate other independent applications developers.

In 1986, Apple spun off its applications group and appointed Bill Campbell to lead the new venture called Claris. He decided to attack Microsoft directly and immediately recruited talented executives from within Apple. Apple transferred its applications software with some employees to Claris. Apple agreed to provide up to $20M in working capital in return for 80 percent of the new company with the option to buy back the other 20 percent anytime. The Claris culture eschewed any connection to Apple. All employees were given Claris stock options. Salaries and benefits were reduced to those befitting a Silicon Valley start-up. Claris was going to succeed or fail on its own and there would be no safety net. Claris's strategy was to create a suite of applications that worked well together. This contrasted with Microsoft's emphasis on features rather than usability. Within three short years, Claris has grown to nearly $90 million in profitable, worldwide sales through acquisitions and internal development.

Claris's ambition became its undoing. Campbell and his team were frustrated with Apple's inability to grow sales volume. Apple was more concerned with high margins than market share, and Claris had saturated the Macintosh market. As the company prepared for an IPO in 1990, Claris revealed a controversial growth strategy to enter the Windows applications market. When Apple executives became aware of this strategy, they were unhappy. They worried that Claris would make Microsoft's Windows more attractive than its own Macintosh. To Claris's dismay, Apple decided to exercise its option to spin Claris back inside. Apple tried to retain the executive team, but within months its members had all dispersed to start new companies. Claris remains a whole-owned subsidiary of Apple today with no significant impact on the greater applications market [Komisar, 2000].

In this decade, it is interesting to contrast the Claris saga with that of Apple's iPod. The recent revitalization of Apple is dependent on entrepreneurial efforts to build the iPod into an entire line of breakout consumer products, including wireless versions, portable media devices, and home network products [Sloan, 2005]. In 2008, Apple launched the iPhone. It combines the features of an iPod with a smart phone. Apple has successfully become a collection of corporate technology ventures that often redefine and recreate their business. What did Apple learn from the Claris experience?

**TABLE 10.8  Strategies for corporations to grow new businesses.**

**Balance trial-and-error strategy formulation with rigor and discipline.**

- Narrow the range of choices before diving deep.
- Closely observe small groups of consumers to identify their needs.
- Use prototypes to test assumptions about products, services, and business models.
- Use nonfinancial milestones to measure progress.
- Know when—and on what basis—to pull the plug on infant businesses.

**Balance operational experience with invention.**

- As leaders of emerging businesses, appoint successful managers who also are known for their willingness to challenge convention
- Win veterans over by asking them to serve on new businesses' oversight bodies.
- Consider acquiring select capabilities instead of developing everything from scratch.
- Force old and new businesses to share operational responsibilities.

**Balance new businesses' identity with integration.**

- Assign both corporate executives and managers from divisions as sponsors of new ventures.
- Stipulate criteria for handing new businesses over to existing businesses.
- Mix formal oversight with informal support by creatively combining dotted- and solid-line reporting relationships.

*Source:* Adapted from Garvin and Levesque, 2006.

## 10.5  The Innovator's Dilemma

As discussed in Chapters 2 and 5, disruptive innovations can revolutionize industry structure and cause existing firms to decline or fail [Christensen and Tedlow, 2000]. Incumbent firms listen to their existing customers, who usually do not express a desire for radical innovation before it is introduced. Eventually, the new innovation improves and starts to challenge the existing methods, but it may be too late for the incumbent firm to catch up. By the time the new innovative firm has improved the innovation and captured the market, the existing firm has lost market share. The disruptive innovation gets its start outside of the mainstream of the market, and then its functionality improves over time. The existing firm ignores new, potential innovations at its own peril.

Another problem for all successful firms is **cannibalization,** which is the act of introducing products that compete with the company's existing product line. When companies decline to try to cannibalize their own products, they operate under the delusion that if they do not develop the new product, no other firm will do so. When new opportunities open up, new entrants into an industry can be more flexible because they face no trade-offs with their existing activities [Burgelman et al., 2004].

Intel potentially cannibalized its business when it released the Celeron in 1998 as an alternative to its powerful Pentium line of processors. The Celeron featured lower performance, but it quickly became popular due to its reduced cost. Intel gained access to different segments of the market with little to no adverse effect on the Pentium and its successors.

For many large firms, the pursuit of innovation must take a backseat to the effective exploitation of existing competencies, the satisfaction of existing customers, and the continuous improvement of existing technologies. Innovations eventually become breakthroughs when the web of people, ideas, and technologies that surrounds them grows and evolves, forcing change farther and farther into the established systems they emerged from. Existing organizations will tend to exploit an innovation that is related to their primary forms of work when the innovation enhances their existing competencies. When an innovation is unrelated, new organizations will typically emerge to capitalize on the innovation.

Established companies can offer disruptive new products of their own that capture new customers and produce new revenue growth [Jiang et al., 2011]. Instead of waiting for the threat of new products to appear, the established firm can create its own response to an anticipated threat. It is important for existing companies to invest in the development of disruptive innovations. Several companies, such as Apple and Novartis, have learned to exploit the present and explore the future. This is a form of hedging one's bets [Bhardwaj et al., 2006].

A disruptive innovation, however, often requires new competencies, resources, and value net relationships. Moreover, its effective pursuit will likely challenge some parts of the existing business. Thus, the best strategy for an existing firm may be to establish an autonomous corporate new venture unit or subsidiary with its own mandate, vision, people, and incentives. For example, Apple and Novartis separate their new, exploratory units from the traditional ones. Each new unit has its own processes, structures, and cultures, but remains integrated into the existing senior management structure [O'Reilly and Tushman, 2004]. When IBM was confronted by minicomputers, it created an autonomous business unit in Florida to develop and sell a personal computer [Christensen and Raynor, 2003].

## 10.6   Incentives for Corporate Entrepreneurs

Mature firms should strive to build CNVs to experiment with novel, emerging, and pioneering innovation to create new dynamic growth. To pursue opportunities, firms need to identify and encourage corporate entrepreneurs [Kacperczyk, 2012]. Corporate entrepreneurs are employees of a firm who take leadership responsibility for driving a venture in the firm. Art Fry, an entrepreneur within 3M, pushed the commercialization of Post-It notes through the firm. 3M has a guideline that its researchers can spend 15 percent of their time working on an idea without approval of management.

For the effective creation of a corporate new venture, the opportunity needs a champion in the parent company [Greene et al., 1999]. The champion is an executive or leader in the parent company who advocates or provides support and resources as well as protection of the venture when parent company routines are breached. The champion helps, describes, and defends the venture and secures the necessary resources. The champion expresses confidence about the CNV, persists under adversity, and helps get the right people involved [Howell et al., 2004]. The champion enables the resource transfer process, as shown in Figure 10.4 [Lord et al., 2002].

Because leading new CNVs always carry the risk of failure, many potential corporate entrepreneurs avoid joining them. They fear the loss of status that failure can bring. In light of these individual risks, a number of features can support corporate entrepreneurship. These include rewards, explicit management support, resources, organizational structures, risk acceptance, work design, and intrinsic motivation [Marvel et al., 2007]. Incentives for corporate entrepreneurs can be stock ownership, bonuses, or promotion within the firm if the expected performance is attained.

A few years ago, an employee of Virgin Atlantic noticed some empty curb space at Heathrow Airport. In a matter of days, he secured the rights to the space and laid out a plan for Virgin to start a curbside check-in kiosk business unit. As a result, Virgin became the first airline at Heathrow to offer its business class passengers the advantage of getting a boarding pass without having to stand in a check-in line. As a result of his effort, the employee received a promotion [Hamel, 2001].

A key issue is the appropriate reward to a corporate entrepreneur. The corporate entrepreneur who is offered large financial gains is usually resented by

**FIGURE 10.4**   Resource transfer process and the champion.

**TABLE 10.9  Incentives for corporate entrepreneurs.**

| | |
|---|---|
| ■ Support for and recognition of employees who create and champion new ideas and opportunities | ■ Significant degree of autonomy |
| ■ A culture that favors individual or team initiative to create new ideas | ■ Effective rewards such as promotion, stock ownership, or bonuses |
| ■ Slack time for exploring not-yet-approved projects | |

his or her associates because the entrepreneur relies on corporate resources that he or she did not create as an independent entrepreneur would [Sathe, 2003]. A list of possible incentives for corporate entrepreneurs is provided in Table 10.9. Incentives include social incentives such as recognition, support, and a culture that favors individuals or teams to take the initiative to create new ideas and explore new opportunities. Employees can be provided slack time, which is time on the job for exploring as-yet-to-be-approved projects. A significant degree of autonomy and effective financial and promotion awards provide incentives for corporate entrepreneurs. Entrepreneurs are less risk-averse and seek independence of activity [Douglas and Shepherd, 2002]. These preferences can be exploited by CNVs.

Of course, managers, too, need to support CNVs. A firm might have a goal that it will launch four new ventures each year and expect that at least one new venture will create an important new business. To achieve this goal, the firm might tie executive compensation to initiation and support of corporate new ventures. Another incentive is to encourage ownership of stock in the firm by its executives. When executives own stock in the companies they manage, they become motivated to increase the long-term value of their firm using corporate ventures [Zahra et al., 2000].

Mature firms need to exploit opportunities for novel innovations by increasing their commitment to corporate new ventures. Larger, mature firms also need to recognize, however, the barriers to CNV: familiarity, maturity, and propinquity [Ahuja and Lampert, 2001]. Familiarity exhibits itself in a firm's tendency to favor the routine or common knowledge and ways of doing things. Maturity refers to favoring fully developed knowledge rather than novelty. Finally, propinquity refers to favoring a search for solutions similar to existing solutions. By putting in place the right incentives for employees and managers, a firm can create robust and successful CNVs.

## 10.7  Building and Managing Corporate Ventures

Can once-mighty giants of industry restore their health after they mature and decline in performance? Can these mature companies use corporate new ventures to transform their performance? Many studies point to significant difficulties in transforming large firms [Majumdar, 1999]. Structural factors, such as their intrinsic complexity, formality, and rigidity, are not conducive to either high performance or reorientation. Not only are larger firms and organizations structurally sluggish, but, with the passage of time, their culture becomes rigid and hard to change because of commitments to particular ways of doing things. As Dee Hock, founder of Visa, stated: "The problem is never how to get new, innovative thoughts into your mind, but how to get old ones out."

Large firms, however, are not entirely disadvantaged. Large firms possess sizable collections of knowledge and intellectual capital. Furthermore, these firms have many talented staff who have entrepreneurial tendencies and the ability to exploit this intellectual capital of the firm. Absorptive capacity is the ability of a firm to exploit external knowledge for the production of innovations. Thus, corporate new ventures can be based on both internal and external knowledge to the extent that the ability to absorb and exploit it is rewarded. A firm's successful use of innovations depends on its ability to exploit its existing base of knowledge while learning about technologies that lie outside its existing competencies [Cohen and Levinthal, 1990].

An existing company is wise to attempt to exploit a new opportunity through some form of a new business. The types of new business arrangements that an existing corporation can use include a new independent venture, a spin-off of a new corporation, a transfer of the opportunity to the existing company's product development department, or authorization of a small project. Figure 10.5 shows the four types of business arrangements and their relationships to operational relatedness and strategic importance. Operational relatedness refers to how the new business organization couples to the existing operational resources and competencies. Strategic importance refers to the critical nature of the long-term results of this new organization to the success of the parent firm. Internal corporate new ventures are most useful for high operational relatedness and high strategic importance (quadrant 1 of Figure 10.5). It is best to view CNVs as a source

**FIGURE 10.5** Four types of new business opportunities and the best business arrangement for each opportunity.

of insights that can inform the strategic direction of the parent company as well as provide the potential for attractive returns [Burgelman and Valikungas, 2005].

An opportunity with low strategic importance and low operational relatedness (quadrant 3) most likely calls for declining to proceed with this project or proceeding with a modest project until the strategic importance becomes clear.

An opportunity with high strategic importance and low operational relatedness may call for a spin-off to a new company (quadrant 2). A **spin-off unit** is an organization that is established within an existing company and then sent off on its own. The parent provides some resources and capabilities and sets the spin-off toward independence. Often the parent retains less than majority ownership of the spin-off. An opportunity with high relatedness and low strategic importance (quadrant 4) is a good candidate for a modest exploratory project.

Cisco Systems established a spin-off named Andiamo Systems, which makes switching gear. Cisco loaned Andiamo $42 million and committed to up to $142 million more. Cisco held 44 percent ownership of Andiamo. Cisco purchased the remaining 56 percent from the 300 employee-shareholders for $750 million of Cisco stock in 2004 [Thurm, 2002].

Corporate ventures are managed differently than traditional in-house corporate research and development. A corporate venture may be riskier and less subject to rigid management of internal costs than conventional corporate product development. Indeed, protecting venture investments from such controls is one reason that corporate new ventures and spin-offs are often housed outside the corporation's walls.

Furthermore, in corporate venturing, returns are part financial and part strategic, whereas with pure venture capital, investors' expected financial returns are paramount. Corporate ventures should follow the best practices of venture capital, but the dual objectives of financial and strategic returns must be balanced in ways that do not concern venture capitalists.

Large corporations can make room for radical, low-cost innovations by establishing a process for finding and funding new ideas [Wood and Hamel, 2002]. Table 10.10 describes a three-step process for finding, evaluating, and funding new entrepreneurial entities. First, expand the conversations about new opportunities widely throughout the firm. Then, establish a process for selecting and funding the best ideas. Finally, keep the budgetary control within the new venture and avoid letting traditional managers begin to control the budget of the new venture.

To extract maximum value from corporate new ventures, managers should adhere to the set of principles outlined in Table 10.11. The process of overseeing corporate new ventures is different from that used to maximize gains from existing business units.

Corporate entrepreneurs gain acceptance for new ideas by influencing organization members' perceptions about the nature of organizational interests. For an initiative to be accepted as part of the official company strategy, belief in the idea must be linked to corporate organizational goals. Most large companies prefer to separate new venture efforts from their core business. This

## TABLE 10.10 Establishing conditions for corporate new ventures.

- Increase the sources for innovation: New ideas tend to evolve and expand through conversation. The more people you can get involved, the more high-quality ideas you will generate.

- Establish a process for collecting and evaluating ideas: Establish a forum for assessing the merits of various proposals

to ensure that the most worthy ideas receive funding.

- Do not let traditional executives control the budget: Many executives are protecting their own departments and are unwilling to risk small amounts of resources on new and untested ventures.

## TABLE 10.11 Extracting value from corporate venturing.

1. Protect new ventures from short-term pressures

2. Recognize that not all employees who volunteer to work with the corporate new venture are a good fit for the new venture

3. Don't expect the same results from the corporate venture that are expected from the core business

4. Manage with a portfolio mindset, not a project mindset

5. Be prepared to learn, since new markets are seldom like existing ones

6. Set milestones and manage the new venture in stages

7. Stop failing ventures early—and maximize the lessons learned

8. Continually evaluate learning-transfer mechanisms to ensure that ideas and lessons are shared

*Source:* McGrath et al., 2006.

permits the new venture to focus on the new opportunity and readily gather and coordinate the necessary capabilities and resources [Albrinck et al., 2002].

The elements of a business as practiced by the parent firm and the corporate new venture are shown in Table 10.12. In general, the parent firm has developed assets, revenues, reward systems, and management practices that tend to support growth, fairness, and policies that lead to orderly progress. The corporate new venture needs to leverage its assets to create new revenue streams by rewarding entrepreneurial actions and flexibility. Furthermore, the CNV wants to attract the best talent of the parent company. Separating the CNV from the parent company enables the CNV to act quickly with flexibility to seize new opportunities.

Many companies use a portfolio strategy for holding ownership in several corporate new ventures as subsidiaries or spin-offs. These companies also use a new venture development process based on the development of a business plan and the analysis of what form of CNV is appropriate. This process is summarized in Table 10.13 [Albrinck et al., 2002]. At each step of the process, the parent company must evaluate the best next steps. Step 1 helps the CNV take shape as a venture champion and entrepreneurial team are identified. Step 2 includes the development of an initial concept statement and the outline of the

**TABLE 10.12  Elements of a business as practiced by the parent firm and the corporate new venture.**

| Element | Parent firm | Corporate new venture |
| --- | --- | --- |
| Assets | Protect and use | Leverage |
| Revenues and growth | Growth of existing revenue stream | Create new revenue stream |
| Management | Adhere to policies and procedures | Act decisively with flexibility |
| Rewards | Maintain fairness and equity | Reward entrepreneurship and performance |
| Talent—people and knowledge | Retain talent and knowledge | Attract the top talent and transfer the best knowledge from the parent company to the CNV |

**TABLE 10.13  Five-step process for establishing a corporate new venture.**

1. Identify and screen opportunities. Create a vision. Designate a venture champion and an entrepreneurial team.

2. Refine the concept and determine feasibility. Prepare the concept and vision statement. Draft a brief business plan summary or outline for review and to gather support.

3. Prepare a complete business plan. Identify the person to lead the new venture.

4. Determine the best form of the corporate new venture: internal new venture unit, spin-off, subsidiary, or internal project.

5. Establish the corporate new venture with talent, resources, and capabilities transferred from the parent company.

elements of a business plan. The next step is to complete the development of a comprehensive business plan. Step 4 is focused on selecting the best organizational form for the CNV, based on the long-range objectives of the parent [Miles and Colvin, 2002]. Finally, in step 5, the corporate new venture is established with the requisite resources, talent, and capabilities transferred from the parent company.

The selection of the appropriate form (step 4) should try to fit the needs and strategy of the parent company. For example, 3M usually incorporates the CNV within an existing or new division. Conversely, Barnes and Noble, when considering the establishment of an online unit, decided to spin off a new company into the stock market.

The Virgin Group, under the leadership of Richard Branson, has created 200 new businesses in many industries such as media, airlines, and music. Business ideas come from anywhere in Virgin Group, and Branson remains accessible to employees who have proposals. Branson also hosts gatherings for employees where they can give him their ideas. One employee proposed a bridal planning service including wedding apparel, catering, air travel, and hotel reservations. She became the CEO of Virgin Bride. Virgin Group works to start independent new ventures (see www.virgin.com).

Landmark Communications launched the Weather Channel as an internal new venture in 1981 [Batten, 2002]. With Landmark's strong corporate support and commitment, the Weather Channel became a top weather information source. With the help of Landmark's talent, knowledge, resources, and capabilities, the new venture took off through several deals with cable operators. By 1996, the Weather Channel was also available online. The Weather Channel succeeded because of the investment of significant resources of Landmark. The Weather Channel was started in the face of widespread skepticism but prevailed because of the assets and capabilities of Landmark.

Existing firms have the capability to organize a market, turning an idea into something that can be economically produced, marketed, and distributed to the customer. Entrepreneurs are able to explore new technologies quickly and effectively, and make the creative leap from technological possibility to something that meets consumer needs. Effective firms that meet the challenge of change possess people who are capable of both tasks.

Many new innovations are introduced by pioneer firms, and a learning phase is started in the market. Existing corporations can recognize these new innovations and quickly join in the innovation-commercialization phase, exploiting their capability to produce, market, and support new products.

### Guidant: A Successful Spin-off of Eli Lilly

In the early 1990s, the pharmaceutical giant Eli Lilly built a series of internal ventures focused on medical devices. By 1994, Lilly had created four internal corporate venture divisions in the medical devices area concerned with cardiac and vascular issues. By September 1994, Lilly incorporated these units into a new company, Guidant, and consummated an initial public offering of its common stock. By September 1995, Lilly disposed of its ownership, and Guidant was a separate company. Guidant is an excellent example of growing from a group of internal corporate ventures to became a leading company. By 2006, Guidant had annual revenues approaching $4 billion and was purchased by Boston Scientific for $27 billion.

The life cycle of a market, such as the telephone industry, can be portrayed by Figure 10.6. The first stage is the introduction of a disruptive technology such as Bell's telephone. Then the key application, step 2, is identified and exploited. A dominant design emerges for the product and the market starts to grow (step 3). Process innovation, step 4, occurs as the product usage grows. When the marketplace matures, experimental innovation occurs in step 5. Later in maturity, the customer relationship processes are improved. In decline, business model innovation is used (step 7). Finally, structural innovation capitalizes

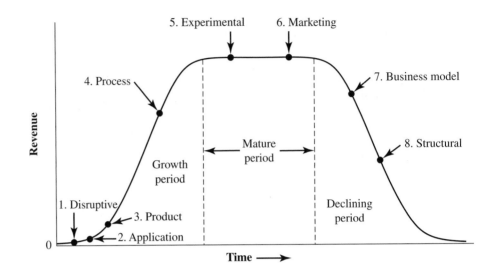

**FIGURE 10.6** Eight types of innovation and the life cycle of a market.
*Source:* Moore, 2004.

**TABLE 10.14  Eight types of innovation for periods of the market life cycle.**

| Type of innovation | Period | Description |
|---|---|---|
| 1.  Disruptive | Very early | Technological discontinuity |
| 2.  Application | Early | Technology application creates new market—Killer app |
| 3.  Product | Start of growth | Improved performance, dominant design |
| 4.  Process | Later growth | More efficient and/or effective processes |
| 5.  Experiential | Mature | Improved customer experience |
| 6.  Marketing | Mature | Improved marketing relationships |
| 7.  Business Model | Declining | Reframes the value proposition or value chain |
| 8.  Structural | Declining | Reponds to structural changes in the industry |

on disruption to industry relationships. These eight types of innovation are summarized in Table 10.14 [Moore, 2004]. In order to renew an existing firm, the leaders must choose the appropriate innovation depending upon the life cycle of the industry.

Many critics depict incumbent firms as going into decline in the face of radical technological innovation. This tendency is not universal, however, nor should it be inevitable. Corporations can respond effectively to new technological innovations when they are prepared and organized to do so.

Firms that have a history of navigating turbulence and creating loosely coupled, stand-alone divisions, and possess a critical complementary asset have a good chance of managing the challenge of a radical innovation [Hill and Rothaermel, 2003].

## 10.8 Spotlight on Twitter

Jack Dorsey created Twitter, an online social networking service, in 2006. The service quickly gained popularity, exceeding 500 million registered users in 2012. Twitter enables users to send and receive text-based messages of up to 140 characters. These messages are known as "tweets."

Twitter began as an internal project at the podcasting firm Odeo. In 2006, Jack Dorsey, Evan Williams, Noah Glass and other members of Odeo formed Obvious Corporation. They then acquired Odeo and its assets, including Twitter.com. In 2007, Twitter spun off into its own company.

As CEO, Dorsey led the firm through several rounds of capital funding by venture capital firms. Over the period 2007 to 2012, Twitter raised about $160 million. Twitter had about $100 million in 2013 revenues and is building advertising sales on its website. The emerging business model for Twitter is based upon advertising revenue from firms desiring to reach the wide range of Twitter users.

## 10.9 Summary

There are five types of new ventures: small business, niche, high-growth, nonprofit, and corporate. Important contributions have been made by small and niche businesses, especially when they later grow and extend their mission globally. High-growth ventures including radical innovation start-ups are very important to creating growth and jobs as well as providing an important service or product that makes a difference. A special form of new venture, called a nonprofit, enables an organization to meet an important social purpose.

Finally, corporate new ventures make important contributions of novelty and creativity that provide new vigor for existing large enterprises. For many firms, the pursuit of innovation and creation of a new venture independent of the existing structures may renew the vigor of the firm. A corporate new venture needs the right amount of slack, independence, and resources to create a novel business.

---

**Principle 10**

An important, vigorous new business venture can emerge from a large firm when afforded the appropriate balance of independence, resources, and people to respond to the opportunity.

## Video Resources

Visit http://techventures.stanford.edu to view experts discussing content from this chapter.

| | | |
|---|---|---|
| Sustainability for Nonprofit Organizations | Kavita Ramdas | Global Fund for Women |
| Classes of Innovations in the Product Leadership Zone | Geoffrey Moore | MDV |
| Attitude and Approach to Innovation | Sue Siegel | GE |

## 10.10   Exercises

**10.1**   Research the number of new ventures created in the last year. Try to segment the data you collect into the venture types outlined in Table 10.3. Compare the number of new ventures of each type. What growth rates have different venture types exhibited? Can this be explained by broader economic trends?

**10.2**   A partnership between a social entrepreneurship course at Stanford University and the nonprofit Light Up the World Foundation (www.lutw.org) worked to bring safe, affordable lighting to people in Mexico, China, and India. Students in engineering and business worked to design a lamp appropriate to the needs of villagers. Develop a brief plan for a social entrepreneurship project with an international nonprofit for your school.

**10.3**   Zimmer Holdings (www.zimmer.com) was incorporated in January 2001 as a wholly owned subsidiary and CNV of Bristol-Meyers Squibb Company. Zimmer designs and markets orthopedic and surgical products. The subsidiary was created from its parent in August 2001, with shareholders receiving 1 share of Zimmer for each 10 of Bristol-Meyers they owned. Study the origins of Zimmer and determine if the spin-off was the right action for Bristol-Meyers.

**10.4**   The traditional newspaper industry is in a declining phase and much has been written on how newspapers should reinvent themselves. How are various newspaper organizations addressing this challenge? What is the best next step in the newspaper industry? See Figure 10.6 for potential options.

**10.5**   Many new clean tech ventures have relied on funding and partnership from established corporations. Select a recently funded clean tech venture with corporate venture involvement. Did the funding impact the structure of the new venture? What does the new venture expect to gain from the backing by the larger corporation? What does the larger corporation expect to gain from being involved in the new venture?

**10.6** Describe the investment philosophy of a corporate venture capital firm such as Intel Capital, SAP Ventures, Eli Lilly, Google Ventures, Dow Corporate Venture Capital, or General Electric (GE) Equity. How are they synergistic with their parent companies' strategic direction?

## VENTURE CHALLENGE

1. Using Table 10.3, describe the specific type of new venture selected by your team and describe its legal form.

2. Go to the State of Delaware's website (http://corp.delaware.gov/howtoform.shtml) and download the "How to Incoporate" packet. Determine what type of incorporation best fits your venture and fill out the appropriate form(s). Alternatively, locate similar forms for another state or nation. Assuming your venture was developed as a corporate venture, describe the advantages and disadvantages of this approach using Table 10.6.

# Intellectual Property, Organizations, and Operations

The new venture organization should carefully protect its intellectual property. Trade secrets, patents, trademarks, and copyrights can provide significant long-term value for a company and create barriers to entry for possible competitors. An organizational plan that supports a collaborative, performance-based culture and a sound compensation scheme must be created to attract good talent. The acquisition of resources and capabilities and facilities will be planned for and initiated in order to build momentum for the venture. The management of operations, processes, and manufacturing will be described in an operations plan. A plan for outsourcing some activities and acquiring necessary assets and technologies will facilitate the early growth of the firm. Finally, the venture team will describe the potential for acquisitions, if any, and the plan for operating internationally in order to further stimulate growth. ■

CHAPTER $11$

# Intellectual Property

*When one door closes, another door opens; but we often look so long and so regretfully upon the closed door that we do not see the ones which open for us.*
**Alexander Graham Bell**

### How does an entrepreneur protect a new venture's intellectual property?

New ventures need to have a plan to build and protect their intellectual property. The proper array of trade secrets, patents, trademarks, and copyrights can add up to a set of very valuable proprietary assets. For many new firms built on innovation and technology, intellectual property can provide a competitive advantage in the marketplace. Licensing, including the licensing of university-developed innovations, can be an important source of new technology or a potential source of revenue. ■

## 11.1  Protecting Intellectual Property

Within intellectual assets is a subset of ideas, called intellectual property, that can be legally protected [Davis and Harrison, 2001]. Property is defined as something valuable that is owned, such as land or jewelry. Furthermore, we can distinguish real property (or physical property) from intellectual property. **Intellectual property (IP)** is valuable intangible property owned by persons or companies. A comparison of the qualities of physical and intellectual property is provided in Table 11.1.

Since knowledge and innovation are keys to competitive success, the management of intellectual property is important to most firms. For many firms, intellectual assets are the wellsprings of wealth and competitive advantage. The protection of intellectual property can lead to the possession of valuable assets; for example, the patents that cover the MP3 music/audio standard.

The protection and enforcement of legal ownership of intellectual property is more difficult than for physical property. How can a firm tell that another firm has used or taken its intellectual property? Unauthorized copying or illegal use of intellectual property can be difficult to discern and prove. The owner of a textbook or a CD purchased at a store has the right to share it with another person but is precluded by law from copying it for sale to another person.

As intellectual property is difficult to defend, it may be useful to develop a strategy to deter misappropriations [Anand and Galetovic, 2004]. Suitable strategies include overwhelming competitors by continually out innovating them and/ or licensing the IP to create cooperation with competitors.

The purpose of intellectual property law is to balance two competing interests: the public and the private. The public interest is served by the creation and distribution of inventions, music, literature, and other forms of intellectual expression. The private interest is served by rewarding people for creating and publicly disclosing these works through the establishment of a time-limited monopoly granting exclusive control to the creator.

During the course of working for a firm, an employee has an idea for a new product that is outside of the scope of the business of the firm. Who owns this intellectual property, the employer or the employee? Does it matter whether the new idea was conceived on the weekend? Entrepreneurs should avoid any

**TABLE 11.1  Comparison of physical and intellectual property.**

| Factor | Physical property | Intellectual property |
|---|---|---|
| Multiuse | Use by one firm precludes simultaneous use by another | Use by one firm does not prevent unauthorized use by another |
| Physical depreciation | Depreciates, wears out | Does not wear out |
| Protection and enforcement from encroachment | Generally can enforce and protect ownership | May be difficult or expensive to enforce and protect ownership |

potential complication or dispute of ownership of intellectual property by assiduously following the legal and moral laws of property. They should also reread all employment agreements that may state restrictions on their ownership of intellectual property. For example, most universities require that graduate students and faculty sign a form assigning intellectual property rights to the university for any invention made using university resources or made in the course of university-sponsored research.

Clearly, if a firm plans to apply for a patent or use some technical advance of a proprietary nature, the firm and its employees should not at first openly reveal the details to outsiders. Eventually, however, investors, partners, or other outsiders may require that the firm reveal more information about the proprietary asset. Moreover, firms sometimes may find that strategic revealing of intellectual property can slow down the competition by discouraging their own R&D investments [Pacheco-de-Almeida and Zemsky, 2012].

Many technology start-ups take several years to reach the market and become profitable. Typically, these new ventures are founded on a significant array of intellectual property such as patents. When the entrepreneur's personal knowledge is perceived to be a critical portion of the intellectual property, then this person will be expected to remain with the firm for several years. Arrangements such as employment agreements and stock vesting terms will help ensure this active involvement [Lowe, 2001].

Intellectual property can be extremely valuable—and violating intellectual property rights can be extremely costly. Analysts estimate the intellectual property market to be $100 billion annually. At IBM, patents and licenses represent 15 percent of revenues. Conversely, NTP received a $612.5 million settlement against Research In Motion, the makers of the Blackberry mobile phone, for patent infringement.

A useful reference for an entrepreneur is *Patent, Copyright, and Trademark,* by Richard Stim [2012]. Although it is important for entrepreneurs to educate themselves about intellectual property, they are strongly advised to seek professional legal assistance. Technology law firms sometimes advise new ventures and defer payment of cash fees until the company receives its first investment. These law firms often help the venture get under way for three to six months in return for a small percentage of equity (e.g., 1.0 percent) [Henderson et al., 2006]. If the venture succeeds like an Electronic Arts or Google, the attorneys will reap great financial benefits.

Intellectual property may be protected in a variety of different ways. The most common types of protection are trade secrets, patents, trademarks, and copyrights. We consider each type in the following sections.

## 11.2  Trade Secrets

A **trade secret** is a confidential intellectual asset that is maintained as a secret by the owner and provides the owner with a competitive business advantage. A trade secret may include knowledge, methods, ideas, formulas, or the like.

The period of life for a trade secret is potentially indefinite. The formula for Coca-Cola has been a trade secret for over a century. Trade secret protection may be lost, however, upon any unauthorized disclosure, such as theft, violation of confidentiality, independent recreation, or reverse engineering. The potential protection offered by secrecy depends on the attributes of the intellectual property and the circumstances of its use. Secrecy is valuable for formulas, algorithms, and know-how that can be implemented by a firm without its being known by other than a few people. If the knowledge must be widely shared throughout the firm, it will be difficult to protect it from those who would copy or imitate it.

Many production processes can be protected behind the walls of the firm. For example, methods for manufacturing integrated circuits are widely available, but the best production process for making them is quite complicated. Several semiconductor firms keep their competitive advantage by maintaining the secrecy of their methods as well as their processes. The risk always exists that an employee will learn the secrets of the methods and the process and decide to start a competitor firm.

A firm will have to balance the need to protect secrets with the necessity to widely share information among employees. Employees must be informed that they are dealing with secrets that are the property of the firm and they are expected to protect these secrets. For many firms, common knowledge among employees of the methods and procedures is necessary for success. For them, it is the execution of the total business process that provides the competitive advantage.

A non-disclosure agreement (NDA), sometimes called a confidentiality agreement (CA) or confidential disclosure agreement (CDA), is one method of protecting a trade secret. An NDA is a legal contract that prohibits the covered parties from sharing certain information with any third party. For example, two firms exploring a potential partnership may sign an NDA so that they can share proprietary information with one another while not releasing it to the public.

A non-compete clause in an employment agreement, sometimes called an NCC or a non-compete, can prohibit an employee from starting or working for a firm that may compete with his or her current employer. The intention of a non-compete is to prevent a former employee from exploiting technical or marketing know-how in service of a competing business [Marx, 2011]. Enforcement of non-competes, however, varies widely in the United States. In California, for example, non-competes generally are invalid except for equity stakeholders in a business.

## 11.3 Patents

Abraham Lincoln called the introduction of the U.S. system of patent laws "one of the three most important events in the world's history" [Schwartz, 2002]. A **patent** grants inventors the right to exclude others from making, using, or selling their invention for a limited period of time. In the United States, this is generally

20 years from the date of filing once the patent issues. A patent for an invention is the grant of a property right by the country in which the application is filed. Patents may be granted to new and useful machines, manufactured products, and industrial processes, and to improvements of existing ones. Patents also may be granted for new chemical compounds, foods, and medicinal products, as well as for the processes for producing them.

**Utility patents** are issued for the protection of new, useful, nonobvious, and adequately specified processes, machines, and articles of manufacture. Examples include the patent for the safety razor and the rolling bag that is widely used by air travelers.

**Design patents** are issued for new original, ornamental, and nonobvious designs for articles of manufacture. For example, Apple received a design patent for the rounded edges of the rectangular screen of the iPad. **Plant patents** are issued for certain new varieties of plants that have been asexually reproduced.

A **business method patent** is a type of a utility patent and involves the creation and ownership of a process or method, such as Amazon's "one-click" ordering process. While U.S. Courts had more broadly allowed business method patents in the past, under current law the extent to which these patents may be allowed has been curtailed.

An invention must be considered novel and useful to be considered for patentability. It must also represent a relatively significant advance in the state of the art and cannot merely be an obvious change from what is already known. Such requirements are meant to reduce the number of inventions that modify existing products in minimal ways. Patents are often granted for improvements of previously patented articles or processes if the requirements of patentability are otherwise met. However, the granting for an improvement does not by itself give the holder any rights to the underlying patent. In general, patents tend to work well in industries where the core technology is biological or chemical [Shane, 2005].

A patent is recognized as a type of property with the attributes of personal property. It may be sold or assigned to others or mortgaged, or it may pass to the heirs of a deceased inventor. Because the patent gives the owner the right to exclude others from making, using, or selling the invention, the owner may authorize others to do any of these things by a license and receive royalties or other compensation for the privilege. If anyone makes use of a patented invention without authorization, the infringement can be brought to court in a suit filed by the patent holder, who may request monetary damages as well as a court injunction to prevent further infringement [Elias and Stim, 2003]. A patent only protects what is stated in its claims. An inventor tries to make multiple claims, but must often write narrow claims to avoid conflict with a previous patent.

The patent process requires an application that includes a clear, concise description of the invention. It also defines the boundaries of the exclusive rights that the inventor claims. Furthermore, patents are territorial so inventors must apply for patents in each country where they wish to protect their inventions.

A provisional patent application is a legal document submitted to the U.S. Patent and Trademark office for the purpose of establishing an early filing date. The provisional application includes a description of the invention but does not include the formal patent claims or other detailed information. The applicant must file for a regular nonprovisional patent within 12 months, which undergoes the standard review process. Filing a provisional patent application also gives the applicant the legal right to use the term Patent Pending.

Once a patent has been issued, the owner may start a patent protection program that includes issuing notices and labeling products or services covered by the patent, monitoring uses of the patent, and pursuing known or suspected infringers of the patent. The patent provides the owner the right to exclude others from using the patent without compensation. However, the owner is responsible for enforcing that right by sending notices to infringers and possibly resorting to litigation, if necessary. Sometimes imitators are able to design new products or methods that circumvent the existing patent.

The value of patents can be high. In August 2011, Google acquired Motorola Mobility for $12.5 billion. At the time, Motorola Mobility was losing money but Google saw a tremendous amount of value in the mobile patents held by Motorola. Smart phones and tablets were becoming the preferred end-user device for interfacing with the Web. At the same time, software applications running on hosted cloud computing infrastructures were enabling a whole new set of applications to be accessed on mobile devices. The risk of not being able to participate in mobile and cloud computing was too large for Google to undertake. So it acquired a company with mobile computing know-how. Perhaps more important, Google acquired a set of patents, which would enable it to participate in the fast-growth mobile and cloud computing sectors.

Patents have proved to be very effective for inventions in the pharmaceutical and medical instruments industries: while a laptop computer may include more than 500 patented inventions held by many firms, a pharmaceutical drug will normally be covered by a single patent. Thus, drug companies typically enjoy strong intellectual property positions.

In many industries, firms eager to capture gains from their innovations are filing patent applications at an unprecedented rate. Several legal changes and court decisions in the 1980s provided more protection for patents. The 1985 case in which Polaroid won more than $900 million in damages from Kodak for instant-camera patent infringement provided strengthened precedence for patent infringement litigation. More recently, U.S. courts found that Samsung infringed several Apple patents and awarded Apple $1 billion in damages in 2012. Court decisions also have expanded patentable subject matter to include genetically modified organisms, software, and, in certain cases, business methods.

Growth in the number of patents issued in the United States in recent years is shown in Table 11.2. In 2012, the U.S. Patent and Trademark Office issued nearly 269,000 patents. Companies are increasingly building their innovation strategy around patents and intellectual property. IBM, for example, was granted more than 6,400 patents in 2012.

**TABLE 11.2  U.S. patents issued.**

| Year | 1980 | 1990 | 2000 | 2010 |
|---|---|---|---|---|
| Number issued (thousands) | 66.2 | 99.1 | 176.0 | 244.3 |

*Source:* U.S. Patent and Trademark Office, 2007.

**TABLE 11.3  Developing a patent strategy.**

1. Identify the goals of a patent portfolio.

2. Identify the intellectual assets and gather supporting documents.

3. Identify those assets most suitable for patent applications.

4. Draft invention disclosures and patent applications.

5. Develop a plan for licensing, enforcing, and enhancing patents.

*Source:* Fenwick & West LLP, R. P. Patel.

Patenting an innovation can be expensive since a single patent application can cost $20,000 in filing, examination, issuance, and attorneys' fees. In addition, the cost of infringement litigation can be very high. Proving patent infringement requires documentation and analysis of the infringing product or process. Fledgling entrepreneurs cannot expect to have the funds required to file for patent protection for all intellectual property and to litigate all possible infringements. Therefore, they must determine when it makes sense to pursue a patent. First, entrepreneurs should evaluate the core technologies that are fundamental to the business's success. If a technology is outside of this core, it probably should not be patented. Second, the entrepreneur should consider whether a competitor could easily invent an alternative to the patented technology. If so, then the technology probably should not be patented. Entrepreneurs should also consider whether other, less costly, forms of protection might be effective, such as trade secrets.

At the same time, it is important to recognize that patents can be useful when bargaining with other companies and can be a positive symbol of innovative capacity when raising money. Therefore, it is wise for entrepreneurs to consider the role of patents beyond the simple protection of a particular invention. In developing and maintaining its patent portfolio, a new venture should follow the steps outlined in Table 11.3.

## 11.4  Trademarks and Naming the Venture

A **trademark** is any distinctive word, name, symbol, slogan, shape, sound, or logo that identifies the source of a product or service. The holder of a trademark gains rights as the trademark is used. In the United States, registering a trademark gives the holder advantages in enforcement. A registered trademark is renewable indefinitely as long as commercial use is proven. A new venture should consider trademarking its company name, symbol, or logo. Commonly known trademarks include Kodak, Apple, Google, the NBC logo, and Yahoo.

The trademark owner has the right to bring legal action to halt any infringing use for damages and recovery of profits. Trademark rights are often among the most valuable assets of an emerging new venture in today's competitive marketplace. The goodwill and consumer recognition that trademarks represent have great economic value and are therefore usually worth the effort and expense to properly register and protect them [Sandner and Block, 2011].

A good trademark is an integral part of a firm's brand. To possess good value, a trademark should readily be associated with and exclusive to the firm. Excellent examples of a powerful trademark are the Apple logo and the Intel Inside logo. There were 243,000 newly registered trademarks in the United States in 2012, and there are an estimated total of 1.8 million registered trademarks in the United States. Companies may also file an "intent to use" trademark application. However, the company must actually be using the trademark before it can be registered.

A firm may lose the exclusive right to a trademark if it loses its unique character and becomes a generic name. Aspirin, thermos, and cellophane are examples of names that have become generic. Coca-Cola and Xerox have successfully protected their trademarks.

A company's name is often its most important identifier. The name of a new company should be memorable, related to the product or service, and attractive. It is also helpful if it can be used as the company's website domain address. The right name can evoke a sense of the company's character, bestow distinction, and make a powerful impression. Ideally, the name tells the prospective customer about the product. For example, Facebook is a virtual facebook or directory of individuals. LinkedIn links people into a network. Many firms such as Dell and Craigslist, are named after their founders. Other firms use creative names, such as Twitter or Uber. Some firms use a locational name, such as Silicon Valley Bank or Allegheny Technology.

The right name can deliver a subtle message about the firm's unique features. Tinker Toys evokes a spirit of play. If possible, a name will serve as a marketing tool and will be easy to remember, spell, and say. Jeff Bezos chose the name Amazon.com because it conveyed the idea of a huge entity and did not limit him to one product [Leibovich, 2002].

It is best to test the proposed name of your firm on other people since it is important to avoid negative connotations. Founder Scott Cook of Intuit, the makers of Quicken, tried and rejected the name Instinct for his new company because it sounded like "it stinks" [Taylor and Schroeder, 2003]. Once a suitable name is chosen, a name search will be required to ensure that no one else has already claimed the name. Then, the name is registered with the appropriate state office. If national and international operation is envisioned, it may be wise to register the name as a trademark with the U.S. Trademark Registry Office.

In late 1984, Leonard Bozack and Sandra Lerner initially financed their own new firm with funds from their credit cards. They named the company cisco, as in the end of the name of the city of San Francisco. Lerner designed the logo in the form of the Golden Gate Bridge. Eventually, they capitalized the name to become Cisco [Bunnell, 2000].

A new venture should make sure that the name does not translate into something embarrassing or negative in a foreign language, and that the name carries no other undesirable connotations. Another factor is its pronounceability. A software company supporting Linux chose Red Hat as a memorable name and created a red hat logo. Other examples of good, memorable, robust names include Wal-Mart, Intel, General Electric, and Microsoft. On the articles of incorporation, the name typically must include the word "Incorporated," "Corporation," or "Company." Moreover, the name usually cannot contain certain words like "insurance" or "bank" without meeting additional criteria [Bagley and Dauchy, 2007].

Once the new venture chooses its name, it should reserve a domain name for its website and e-mail address. New ventures should check into the availability of a domain name early on, because doing so often eliminates certain choices for company names and trademarks. The best situation is when you can use the same legal name and domain name. Good examples of memorable corporate names that are also used as domain names are Google and Yahoo.

## 11.5  Copyrights

A **copyright** is a right of an author to prevent others from printing, copying, or publishing any of his or her original works. For copyrights created today, the life of the copyright is for the life of the author plus 70 years after the author's death. Because copyright protection automatically attaches upon creation of a work and the process of registering a work with the U.S. Copyright Office requires only the completion of a simple form, the process of obtaining copyright protection demands very few resources.

A copyright extends protection to authors, composers, and artists, and it relates to the expression rather than its subject matter. This is important, because a copyright only prevents duplicating or using the original material. This does not prevent use of the subject matter. Therefore, software programs, books, and music are protected from copying, but the ideas in these forms may be used by others.

The protection provided by copyright is somewhat limited. In the software field, for example, courts have narrowed the scope of copyright protection. Copyright is most effective against wholesale copying of all or a significant portion of a program. It has limited protection for functional aspects of software products, such as the underlying algorithms, data structures, and protocols of multimedia technology. Copyright also may protect fanciful aspects of the graphical user interface of a program.

## 11.6  Licensing and University Technology Transfer

**Licensing** is a contractual method of exploiting intellectual property by transferring rights to other firms without a transfer of ownership. A **license** is a grant to another firm to make use of the rights of the intellectual property. This license is defined in a contract and usually requires the licensee to pay a royalty or fee to the licensor.

Many firms have a large number of unexploited or underexploited patents that a licensee may be able to exploit. IBM, for example, widely grants licenses, and its royalty income amounts to more than $1 billion each year. IBM holds more patents than any other U.S. company and licenses its software patents widely. But, while most new firms realize that intellectual property can be among their most valuable and flexible assets, they remain unaware of the earning potential of their patent holdings.

Licensing can form the core of a business model. For example, licensing is widely used to provide software to users. Microsoft derives most of its revenues from license fees for its Office suite and Windows operating system. Dolby Laboratories Inc. gets much of its revenue from licensing its products to electronics makers. It succeeds this way partly because of its well-known technology and the reasonable price it charges other firms for use of the technology.

A new venture can derive valuable income streams by licensing its intellectual property to other firms for noncompetitive, complementary uses. The benefits to the licensor include spreading the risk, achieving expanded market penetration, earning license income, and testing new products and markets. Disadvantages of licensing may include risk of infringement and nonperformance of the licensee.

A new venture also can save time and resources by licensing another firm's technology to use in its products. The terms of the license with the third-party technology owner establish what rights a start-up has to use, distribute, modify, and sublicense the licensed technology. Licensing terms often are structured in recognition of a start-up's high potential but lack of capital. For example, a firm may waive or minimize the up-front license fee charged to a start-up. But, in return, it may demand a percentage of sales revenue once the start-up releases a product.

---

**Rambus' Patent Stream**

Rambus was founded in 1990 and went public on the NASDAQ in 1997. The company derives most of its revenue from licensing patent technologies for chip interfaces. Rambus' customers include SONY, Intel, AMD, Infineon, NEC, and Toshiba. Rambus had revenues of $312 million in 2012.

---

Most start-ups founded on the basis of university-developed technologies will need a license from the university. Even if a student or professor is both the inventor and the entrepreneur who brings the technology to market, most universities own the intellectual property since they provided the lab space, salaries, and other resources to conduct the research. Thus, the inventor must obtain a license. Inventor-entrepreneurs who wish to avoid licensing their own inventions from the university must be careful to work on these inventions without using university resources.

In exchange for the license, the university typically requires some combination of an up-front fee, an annual fee, a percentage of related product revenue, and/or equity in the start-up. The university may grant an exclusive license, which provides just one company with the right to develop the invention. Alternatively, a nonexclusive license provides rights to multiple companies simultaneously. Often, universities offer a hybrid between the two approaches, such as exclusivity for one field or application area or exclusivity in one geographic region, such as North America.

Inventors of a technology that a university licenses can benefit from the marketing and intellectual property assistance that the university provides. Inventors also share in the revenue that the university collects based on the invention. A typical arrangement splits the revenue such that one-third goes to the inventor (or is split among the inventing team), one-third goes to the inventor's department, and one-third goes to the school, after patenting and administrative expenses are subtracted out.

## 11.7 Spotlight on Apple

Apple, Inc. was founded on April 1, 1976, and incorporated on January 3, 1977. The firm was established to sell the Apple I computer kit. Steve Jobs and Steve Wozniak designed and assembled the contents of the kit, which sold for $666.66. The Apple II was introduced on April 16, 1977, and incorporated VisiCalc, a spreadsheet program. On December 12, 1980, Apple went on the public market for $22 per share.

In the early 1980s, Apple encountered clones that copied the operating system of its Apple II computer. The company won a historic legal battle in 1982, when the courts determined that copyright could protect the software embedded in computer chips [Roberts, 2012]. Over time, Apple has carefully designed an intellectual property strategy that covers each product through a variety of design patents, utility patents, trademarks, copyrights, and trade secrets. In 2012 and 2013, Apple engaged in another major legal battle with Samsung over each company's mobile devices.

## 11.8 Summary

The plan to acquire, build, and protect the intellectual property of the new venture should be clear to all the participants. The proper array of trade secrets, patents, trademarks, and copyrights can come together as a strong set of valuable proprietary assets. For high-growth, technology-based companies, intellectual property can be used to build a competitive advantage. Well-protected intellectual property can be licensed to others as a source of revenue. Alternatively, ventures based upon university-developed technology will need to obtain a license. Entrepreneurs should avoid the ten mistakes with legal matters described in Table 11.4.

## TABLE 11.4 Ten mistakes with legal matters.

| | |
|---|---|
| ■ Failing to secure legal assistance | ■ Starting a business while employed by a potential competitor |
| ■ Delaying the handling of legal issues | |
| ■ Delaying the intellectual property management process | ■ Overpromising and exaggerating claims in the business plan |
| ■ Issuing founder shares without vesting provisions | ■ Failing to register the name of the firm early in the start-up process |
| ■ Failing to incorporate early | ■ Failing to develop confidentiality, nondisclosure, and noncompete agreements |
| ■ Disclosing intellectual property without a nondisclosure agreement | |

*Source:* Bagley and Dauchey, 2007.

---

**Principle 11**
The intellectual property and name of a new venture can provide a proprietary advantage leading to success in the marketplace.

---

## Video Resources

Visit http://techventures.stanford.edu to view experts discussing content from this chapter.

| | | |
|---|---|---|
| Protecting Intellectual Property | Martin Nichols | DLA Piper |
| Keeping Company Secrets | Adam Lashinsky | Fortune |
| The Role of Lawyers in the Start-up Ecosystem | Gordon Davidson | Fenwick & West |

## 11.9 Exercises

**11.1** Three friends have decided to form a firm to design and manufacture nanotechnology devices for medical applications. Michael Rogers has worked for Hewlett-Packard for 12 years and on his own has designed and submitted a patent claim for a nanotechnology manufacturing technology. Steve Allegro, a graduate student, has a software program he has developed for the design of nanotechnology medical devices. Alicia Simmons, CFO of Alletech Software Inc., is a skilled and experienced manager. Shall they incorporate immediately? What is the problem, if any, of using Rogers's patent ideas? Simmons has knowledge of several manufacturing trade secrets of Alletech. Can she use these secret methods at her new firm?

**11.2** The three founders of the new firm described in exercise 11.1 are looking for a name for their firm. One idea is Advanced Nanoscience & Technology. Another is Nanoscience Applications. What do you think of these names? Can you suggest a better name?

**11.3** Apple Inc. and Apple Corps have had trademark disputes over the use of the name Apple associated with the music business. Apple Inc. has a thriving iPod and online music store. Apple Corps is a multiarmed multimedia company formed by the Beatles in 1968, consisting of the following subsidiaries: Apple Records, Apple Electronics, Apple Films, Apple Publishing, and Apple Retail. Briefly describe the arguments for both sides. Why is it important for a company to challenge and protect its trademarks?

**11.4** Describe the advantages and disadvantages of a company name representing more than just a company name (e.g., Xerox, Kleenex, and Google).

**11.5** Mayo Clinic has filed an application for a broad method patent that gives it control over a new generation of treatments for chronic sinus inflammation (sinusitis). The patent, in effect, blocks others from selling an antifungal agent to treat the condition without Mayo's approval. Mayo will soon try to license this patent to a pharmaceutical company. Are patents helpful in the process of developing a cure for diseases?

**11.6** In 2012, Apple and Samsung engaged in a prolonged patent fight. Describe Samsung's product offerings and what Apple patents they may have been violating. How did Samsung address the patent challenge?

## VENTURE CHALLENGE

**1.** What are the key elements of the intellectual property of your firm?

**2.** What are the next steps in protecting your intellectual property?

# 12

# The New Enterprise Organization

*Two people working as a team will produce more than three working as individuals.*
**Charles P. McCormick**

**How can entrepreneurs best organize and reward the people who will lead their venture to success?**

After recognizing an opportunity and deciding it is attractive, usually one or two founders assemble a new venture team to build a plan and an organization to execute it. A key aspect of any team is that its members possess different core competencies and are able to work in a collaborative manner. A new firm also builds a set of advisors and a board of directors to provide guidance and counsel as the team moves forward. The senior leaders of the venture are identified early in the organization's development. These leaders must be capable of inspiring and motivating others to join the new venture.

As an organization grows, managers will be needed to carry out the tasks that keep the organization running well. A leader is a team's emotional guide and exhibits solid emotional intelligence. As the organization grows, the firm works to build an organizational culture and trust among team members. Leaders and teams strive to build social relationships and networks to foster collaboration.

Knowledge assets and intellectual capital are potential sources of wealth. Sharing knowledge throughout a firm can enhance the firm's processes and core competencies, thus making the firm more innovative and competitive. Most technology ventures are based on knowledge and intellectual property that must be enhanced and managed. A learning organization is skilled at creating and sharing new knowledge. ∎

## 12.1 The New Venture Team

The first step toward forming a new venture is often taken by one or two individuals who recognize a good opportunity and then develop a business concept and vision to exploit it. After a short period, it becomes clear that a team is required in order to have all the necessary capabilities in the leadership group. We define a **team** as a few people with complementary capabilities who are committed to a common objective, goals, and approach for which they hold themselves mutually accountable. In a true team, as opposed to a group, members are fully integrated and feel responsible for their collective output. Think of the contrast, for example, between a basketball team that has difficulty functioning without all players and a group of tennis players who act independently toward a common goal.

The **new venture team** is a small group of individuals who possess expertise, management, and leadership skills in the requisite areas. In knowledge-intensive, dynamic industries, entrepreneurial teams outperform single entrepreneurs, since the new venture requires more capabilities than one individual is likely to have [Beckman et al., 2007]. It is the combination of complementary capabilities that leads to success. Thus, the advantage of a team results from bringing together people with diverse characteristics, skills, knowledge, and capabilities, each of whom is an expert in his or her own particular area [Fischer and Boynton, 2005]. While some similarities between team members are important to facilitate communication and fast execution, these differences allow for greater breadth and creativity and may facilitate the identification of more market options for technology-based firms [Furr et al., 2012].

Vinod Khosla suggests that entrepreneurs should "engineer the gene pool" when they assemble a team. Often, entrepreneurs assemble a team by identifying who they know. Khosla suggests that they should instead identify what they do not know: what perspectives, skills, expertise, and experience are missing from the current team? He then recommends that entrepreneurs use tightly defined criteria for specific positions to identify the best people to fill these roles. Typically, these people will already have positions and will need to be recruited away from another firm. By filling all gaps and building a world-class team, the new venture has a much better chance of success [Khosla, 2012].

Typically, a team of two to six people is required to develop a business plan, secure the financing, and launch the firm into the marketplace. It is important to ensure that each team member contributes fully and serves a critical role. Jeff Bezos, the founder of Amazon.com, argues that team size should follow the "two pizza rule": just two pizzas should feed the entire team. Otherwise, the team is probably too big.

The capabilities of the one or two lead entrepreneurs are critical to the new venture since others are willing to join the team based on the integrity, experience, and commitment of these lead entrepreneurs. Often we call the lead entrepreneurs the **founders.** In other cases, all the members of the initial leadership team are called the founders. The founders display all the characteristics of capable entrepreneurs: passion, commitment, and vision. Entrepreneurs understand the long-term implications of the information-based, knowledge-driven, and service-intensive economy. They know what new ventures require: speed, flexibility, and continuous self-renewal. They recognize that skilled and motivated people are central to the operations of any company that wishes to flourish in the new age.

Furthermore, these new ventures must exhibit adaptability and readiness to change as the context of the business evolves. The organizational arrangement of the firm must evolve as the market and the customer change. Newly-formed firms are challenged by this necessity to change since they usually have limited resource and capability bases. Strong teams ensure that the firm is able to constantly reorganize in terms of strategies, structures, systems, and resources. The new venture team must have the skills to balance the needs for change, efficiencies, alignment of effort, and timeliness. The team must include one or more persons who can gain access to external sources of funds. One of the advantages of a new organization is that people can use active thinking rather than precedent as a basis for action [Pfeffer and Sutton, 2000].

Effective teams also depend on shared understandings and good team processes. Team members should support the team effort and engage in "buy-in" without respect to the entrepreneurial task [Shepherd and Krueger, 2002]. Each member should feel comfortable making suggestions, trying things that might not work, pointing out potential problems, and admitting mistakes. The team should create a frequent and rich flow of ideas and meet regularly to trade these ideas. The innovation process consists of the appropriate use of questioning, trust, and openness of the team [De Jong and Elfring, 2010]. The team consistently seeks the best potential product that will create significant value [Estrin, 2009]. Table 12.1 lists several characteristics of an effective team.

**TABLE 12.1  Characteristics of an effective team.**

- All members share leadership and ownership of the team's tasks.
- Communication is continuous among members in an informal atmosphere.
- Tasks and purposes are well understood.
- People listen to each other and are comfortable with disagreements within the team.
- Most decisions are reached by consensus.
- Feedback on performance is frequent.
- The division of tasks and work effort is clear.
- A collaborative effort is the norm.
- Members set their own, shared interim deadlines for project stages.
- Team members rely on each other and hold each other accountable.
- Teams learn and share their learning.

> **Migrating Geese**
>
> Geese heading south for the winter fly in a **V** formation. As each bird flaps its wings, it creates uplift for the bird immediately following it. By flying in a **V** formation, the whole flock can fly at least 71 percent farther than if each bird flew on its own. People who share a common direction can get where they are going more quickly and easily if they cooperate.
>
> Whenever a goose falls out of formation, it feels the resistance of trying to go it alone and quickly gets back into formation to take advantage of flying with the flock. Teams will work with others who are going the same way. When the lead goose gets tired, it rotates back in the formation, and another goose flies on the point. It pays to take turns doing hard jobs for the team. Perhaps the geese honking from behind are even the cheering squad to encourage those up front to keep up their speed.
>
> *Source:* Muna and Mansour, 2005.

Although entrepreneurs typically focus on full-time employees as constituting the team, it is useful to conceive of the firm's board of directors and board of advisers as part of the team, too. An incorporated firm or LLC has a board of directors. **A board of directors** is a group composed of key officers of a corporation and outside members responsible for the general oversight of the affairs of the entity. This board normally consists of the founders of the firm and one or more investor-owners. A new start-up might find a board of three owners is adequate. As other investors are added, one or more additional owners may added to the board. The board is the overseer of the corporation with responsibility to select and approve the appointment of the chief executive officer (CEO) and other officers of the corporation. Directors should possess significant knowledge and competencies in the industry of the company. The board of directors is a legally constituted group whose responsibility is to represent the stockholders. A board of five might consist of two insider executives, two representatives of investors, and one independent director. This board must approve the bylaws, officers, and annual report to the shareholders, as well as any financial offerings to investors or banking activities. The members of this board have fiduciary responsibility, meaning that they are under a legal duty to act in the best interests of the corporation and its stockholders.

The board of directors for a new venture typically differs from the board of directors for a large established corporation in several key areas, as outlined in Table 12.2. For example, members of the board of directors for a venture tend to have deep industry expertise and significant financial incentives. They can also have a greater diversity of interests, however, and may have conflicts of interests between their investment interests and the venture's performance.

**TABLE 12.2  Comparison of board of directors for ventures and public firms.**

|  | Ventures | Public firms |
|---|---|---|
| Board as internal governance mechanism | Yes | Yes |
| Principal-agent alignment | High | Low |
| Firm maturity, overall uncertainty, slack resources | Low, High, Low | High, Low, High |
| Directors' financial incentives | High | Low |
| Directors' knowledge of the sector | High | Low |
| Directors' diversity of interests, conflicts of interests | High, High | Low, Low |

*Source:* Garg, 2013.

Entrepreneurs are advised to consider how board composition should change as their venture grows.

In selecting directors, a premium should be placed on a wide range of expertise and backgrounds [Tuggle et al., 2010]. Above all, boards should include people who will seek to expose the downsides as well as the upsides of every major decision. Directors should work for better governance of the firm. They should know the strategy of the firm and keep everyone focused on the firm's innovation strategy. Savvy start-ups look for directors who are fluent in one or more of the following: audit and finance, strategy, marketing, and sales.

Compensation of directors will normally be in the form of stock. The members of the board should, after a few years, have a reasonably substantial stake in the firm. This can be achieved with the use of stock options or restricted stock that vests over a period of time.

The board of advisors, if a venture chooses to have one, is constituted to provide the firm with advice and contacts. The members have good reputations and extensive skills and knowledge in order to provide good advice. The board of advisors is nonfiduciary and does not engage in the legal or official actions of the corporation. Thus, the advisors are free from liability as long as they refrain from any legal or official role. At a minimum, advisors have their venture-related expenses, as with travel to meetings, reimbursed. They can be granted stock, too, though typically both their compensation and involvement are less than that of members of the board of directors.

Good boards are those that function and work well. They are distinguished by a climate of trust, respect, and candor. Members feel free to challenge each other's assumptions. They should feel a responsibility to contribute significantly to the board's performance. In addition, good boards assess their own performance, both collectively and individually [Sonnenfeld, 2002]. Table 12.3 lists five goals for an effective board process [Finkelstein, 2003].

> ### Lending Club's Board of Directors
>
> Lending Club is a New York based start-up looking to dis-intermediate the consumer and small business loan market. The company offers person-to-person lending with lower rates than a borrower would traditionally get from a bank. Lending Club has built a board with a number of high-profile banking and financial executives including John Mack, the former chairman of Morgan Stanley, Lawrence Summers, the secretary of the treasury during Bill Clinton's presidency, Hans Morris, the former president of VISA and CFO at Citi Markets and Banking, and Mary Meeker, one of the world's most well-known and respected technology analysts while she was at Morgan Stanley. Lending Club uses this board for advice but also to add credibility to the company and validity to the new consumer and small business lending model used by Lending Club.

**TABLE 12.3  Five goals for an effective board process.**

1. Engage in constructive conflict—especially with the CEO.
2. Avoid destructive conflict.
3. Work together as a team.
4. Work at the appropriate level of strategic involvement—avoid micromanagement.
5. Address decisions comprehensively.

*Source:* Finkelstein and Mooney, 2003.

## 12.2  Organizational Design

**Organizational design** is the design of an organization in terms of its leadership and management arrangements; selection, training, and compensation of its talent (people); shared values and culture; and structure and style. The nine elements of an organization are listed in Table 12.4. Note that the last four elements in the list can be considered as the elements of an organizational design.

The **talent** consists of the people, often called employees, of an organization. The leadership team and the firm's managers are responsible for communicating and leading the firm in the appropriate direction. The shared values and corporate culture are the guiding concepts and meanings that the members of an organization share. The structure of a firm is its formal arrangement of functions and activities. Style is the manner of working together—for example, collegially or team-oriented.

What is the best way to organize a group of people so as to maximize productivity and innovation? Most successful innovative organizations include many small units having free communication with each other, significant independence in pursuing their own opportunities, and freedom from central micromanagement.

**TABLE 12.4 Nine elements of an organization.**

1. Mission and vision

2. Goals and objectives

3. Strategy

4. Capabilities and resources

5. Processes and procedures

6. Talent

7. Leadership team and management

8. Shared values and culture

9. Structure and style

They avoid layers and bureaucratic structures, and keep communication flowing with a bias toward action [Joyce et al., 2003]. Innovation grows most rapidly under conditions of an intermediate degree of fragmentation. Excessive unity and excessive fragmentation are both ultimately harmful. The best organization design is one of teams or units that compete and generate different ideas but maintain relatively free, open communication with each other [Diamond, 2000].

Competency-based strategies depend on talented people operating in a loose-tight structure. Thus, hierarchical structures need to be replaced by networks, bureaucratic systems need to change into flexible processes, and control-based management roles need to evolve into relationships [Bartlett and Ghoshal, 2002]. Flexible organizations that effectively adapt to change are often called **organic organizations**.

Organizational performance is the result of individual actions and behavior. Successful firms have people who take the right actions in concert with others. Thus, the form of the new enterprise is often a network characterized by relationships within the firm. In general, the firm starts off as a single team, and as it grows, it evolves into a series of cross-functional teams.

Southwest Airlines has been productive due to its effective use of its major assets—its aircraft and people. Southwest uses **relational coordination** (RC), which describes how its people act as well as how they see themselves in relationship to one another [Gittell, 2003]. RC requires frequent, timely problem-solving carried out through shared goals, shared knowledge, and mutual respect. Three conditions that increase the need for RC are reciprocal interdependence, uncertainty, and time constraints—all common to new business ventures.

One model of an organizational design is shown in Figure 12.1. The three activities of an organization are operations, innovation, and customer relationship management (CRM). These activities all support the key objective of the organization to create and maintain a sustainable competitive advantage. The integration of innovation, operations, and CRM can lead to a strong competitive advantage. The Internet and an intranet can help provide low interaction costs between the three activities. Most new ventures have an advantage because their newness permits them to easily integrate the three activities shown in Figure 12.1, thus rapidly gaining a competitive edge.

A new venture normally starts out with a team or a **collaborative structure** that primarily consists of teams with few underlying functional departments. In a collaborative structure, the operating unit is the team, which may consist of 5 to 10 members. The best collaborative structures are self-organizing and adaptive. A **self-organizing organization** consists of teams of individuals that benefit from the diversity of the individuals and the robustness of their network of interactions. This collaborative effort coupled with the self-organizing behavior of the network can lead to benefits that exceed the sum of the parts—often called synergy.

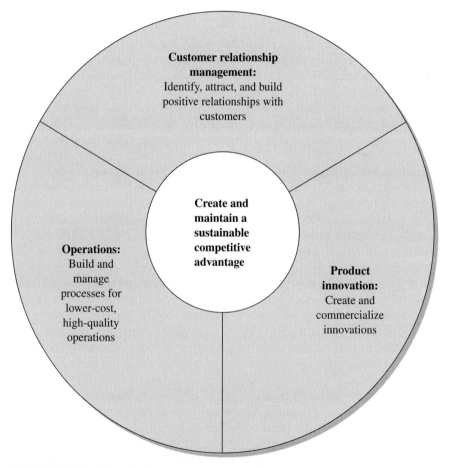

**FIGURE 12.1** Model of an innovative organization.

> **Palo Alto Network's Experienced Leadership Team**
>
> The Palo Alto Networks' founders, management team, and board of directors brought a wealth of experience to the company that enabled them to develop industry-leading products, to gain significant market share versus entrenched companies, and to scale to the point of an IPO. Nir Zuk, the founder of Palo Alto Networks, was also the CTO at Netscreen, a company acquired by Juniper Networks, and the CTO of OneSecure, a company that competed in a market similar to Palo Alto Networks. The Palo Alto Networks management team members have been key drivers of some of the most important network security companies and related technologies in the last 15+ years, including the invention of stateful inspection, hardware-based security, and intrusion prevention. The engineering team brings a proven track record in delivering high-quality networking and security solutions at companies like Check Point, NetScreen, McAfee, Juniper Networks, and Cisco. When they were getting ready to go public, Palo Alto Networks hired Mark McLaughlin, who had previously served as president, CEO and director at Verisign, another leading security company.

## 12.3 Leadership

Leadership is the process of influencing and motivating people to work together to achieve a common goal by helping them secure the knowledge, power, tools, and processes to do so. Leadership is critical to an entrepreneurial venture and is normally provided by one or two leaders of the new venture. The leader of a new venture can be thought of as a leader of a jazz band that is known for its ability to play familiar and new music while collaboratively creating and improving new variations. Mobilizing an organization to adapt its behaviors in order to thrive in changing business environments is critical. Responses to challenges reside in the collective intelligence of employees at all levels who need to use one another as resources, often across boundaries, and together learn their way to those solutions.

A good leader is hopeful about the venture's goals and can readily describe the vision of the venture. They communicate a clear vision and the value of the venture and are convinced they are the one to make it happen. Good leaders make good judgement calls about people, strategy, and challenge [Tichy and Bennis, 2007]. Judgement is the nucleus of leadership and is based on a practiced process of naming and framing the issues, exercising good judgement, and listening to the team. Most leaders are skilled at framing the issue and creating the case for change, strategy, and actions needed. Then the team proceeds to the execution phase and makes the required steps for success.

The leader of the new venture responds to routine work and challenging work issues in different ways, as shown in Table 12.5 [Heifetz and Laurie, 2001]. The

**TABLE 12.5  Leadership of routine and adaptive work.**

| A leader's role in: | Routine issues | Challenging and adaptive work |
|---|---|---|
| **Direction** | Define problems and possible solutions | Define the challenge and the issues |
| **Team and individual responsibilities** | Clarify and define roles and responsibilities | Define and discuss the necessity to adapt roles and responsibilities to changing needs |
| **Conflict** | Restore order and reduce conflict | Accept useful conflict and use it to define new approaches and strategies |
| **Norms and values** | Reinforce norms and values | Reshape norms and values |
| **Teaching and coaching** | Provide training and skill learning for existing employees | Teach and coach new people |

most important capability of the entrepreneur-leader is the ability to cultivate and make use of the competencies of the talented team members [Davidsson, 2002]. Responding to challenges and adapting the effort of the talent is the role of a leader. Challenging problems confronting an organization require the members of the organization to take responsibility for solving the problem. Thus, the leader helps the team members confront the challenges and learn new ways to solve the problems.

Leaders build companies through a blend of personal humility and professional will [Collins, 2001]. They are ambitious, but primarily for the organization, not themselves. Leaders have the drive to build great companies through the efforts of the team members. They seek sustained results and facilitate new approaches to challenging situations while maintaining clear goals and methods for routine work. Leadership is the ability to acquire new organizational methods and capabilities as the situation changes. The leader stimulates discussion so that people contribute and understand the issues, ultimately leading to a shared strategy for sustained advantage.

There are four styles of leadership, as shown in Figure 12.2 [Northouse, 2001]. A leader's behaviors are both directive (task) and supportive (relationship). *Directive behaviors* assist group members in goal accomplishment through giving directions, establishing goals and methods of evaluation, setting time lines, defining roles, and showing how the goals are to be achieved. Supportive behaviors involve two-way communication and responses that show social and emotional support to others.

The *supporting* style is used when a leader does not focus exclusively on goals but uses supportive behaviors that bring out the employees' skills around the task to be accomplished. The *directing* leader gives instructions about what and how goals are to be achieved by the subordinates and then supervises them

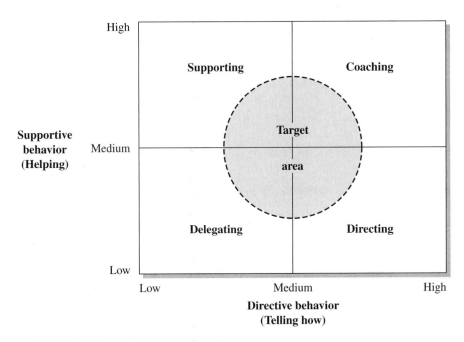

**FIGURE 12.2** Four leadership styles.

carefully. The *coaching* style calls for a leader to focus on goal achievement and give encouragement to subordinates. The *delegating* style occurs when the leader is less directive and facilitates employee confidence. Most effective leaders adopt all four styles of leadership depending on the needs of the situation and the team members while operating within the central target area. The leader of a new technology venture will most likely use a directing-supporting style in the early period of the venture. Later, in the growth period of the firm, the leader will probably use a mix of all four leadership styles.

Leaders exhibit seven traits, as recorded in Table 12.6. Leaders are seen as authentic, decisive, focused, caring, coaching, communicative, and improvement-centered [Collins and Lazier, 1992]. Leaders articulate a clear, compelling vision for the venture and stimulate the team to achieve high performance. They avoid letting talk substitute for action. They also strive to develop a sustainable competitive advantage through building new competencies and products in a timely way. Leaders have a bias for simple concepts that can be clearly understood and acted on. Informed action is their goal. The entrepreneurial leader uses a collaborative style while setting high standards and driving toward achievement.

Teaching is at the heart of leading. In fact, it is through teaching that leaders lead others. Teaching is how ideas and values are transmitted in an organization. Leading is helping others to see a situation as it really is and to understand what responses need to be taken so that they will act in ways that will move the organization toward where it needs to go. Organizational

### TABLE 12.6  Seven traits of leaders.

- Authenticity: consistent actions and words

- Decisiveness: willing to act on limited, imperfect information

- Focus: create a priority list and stick to it

- Care: build relationships and social capital

- People skills: offer helpful feedback and good coaching to all team members

- Communication: stimulate conversation and communicate vision

- Continuous improvement: keep learning and energy flowing in the firm, retain optimism

*Source:* Adapted from Collins and Lazier, 1992.

performance is the result of individual actions and behaviors. In successful companies, people do the "right things." Those companies have effective leaders who create conditions under which their people have the information, authority, and incentives to make the right decisions. When leadership is effective, behavior at all levels of the organization is both aligned and adaptable; thus, the organization performs to its potential.

Leaders have a guiding vision and passion that allow them to communicate a sense of hope to followers. The key to this communication is integrity and credibility. Leaders with both have a heightened sense of self-awareness and a strong understanding of what they believe in and value. The five functions of leadership are challenging the status quo, inspiring a shared vision, enabling people to act, modeling through personal example, and motivating people to act.

Leaders display an inner strength and a constant set of values that everyone knows and can rely on. They avoid self-aggrandizement, inspire others, and exhibit a combination of modesty and extraordinary competence.

What makes a great leader? One study reveals a key attribute: the capability to handle adversity and to learn from such experiences [Bennis and Thomas, 2002a]. Bennis and Thomas call these shaping experiences *crucibles*. These experiences make leaders stronger and more confident. Leaders are formed by a combination of their individual personalities and the events of the era in which they spend their formative years, which are then transformed in a crucible of experience and challenge. They organize the meaning of these experiences into capacities to respond to future challenges. These great leaders also evidence a capability to adapt as well as resilience, and are not necessarily charismatic [Khurana, 2002]. They are likely to be capable people with solid experiences that helped shape and build their leadership skills.

Finally, the leader is a venture's emotional guide. **Emotional intelligence** (EI) is a bundle of four psychological capabilities that leaders exhibit: self-awareness, self-management, social awareness, and relationship management, as described in Table 12.7. Self-awareness refers to the ability to understand one's moods, emotions, and motivations, as well as their effect on others. Self-management is the ability to control emotions, as well as

**TABLE 12.7  Four elements of emotional intelligence.**

- Self-awareness of one's emotions, emotional strengths and weaknesses, and self-confidence; the ability to read one's own emotions

- Self-management of honesty, flexibility, initiative, optimism, and emotional self-control; the ability to control one's emotions and act with honesty and integrity

- Social awareness of empathy and organizational currents, and recognition of the needs of followers and clients

- Relationship management achieved by communicating through inspiration, influence, catalyst actions, conflict management, and collaboration

exhibit optimism and adaptability. Social awareness is the empathetic sensing of other people's emotions and awareness of the social currents within an organization. Relationship management includes inspiration, influence, conflict management, and teamwork.

According to Goleman and colleagues, leaders and managers who possess these capabilities have a high degree of EI and tend to be more effective. Their self-awareness and self-management help to elicit the trust and confidence of their colleagues. Strong social awareness and relationship management skills can help to earn the loyalty of their colleagues. Empathetic and socially adept persons tend to be skilled at managing disputes between people, better able to find common ground and purpose among diverse constituencies, and more likely to move people in a desired direction than leaders who lack these qualities.

People with high emotional intelligence tend to (1) behave authentically, (2) think optimistically, (3) express emotions effectively, and (4) respond flexibly in their relationship styles.

Leaders and managers who are in touch with their colleagues are said to be in resonance, which is the reinforcement of emotion. People are in resonance when they are "in sync" or on "the same wavelength." Resonant leaders use their EI skills to spread their enthusiasm and resolve conflicts [Goleman et al., 2001]. Good teams work to establish group resonance by building emotional and social awareness and management.

## 12.4  Management

**Management** is a set of processes such as planning, budgeting, organizing, staffing, and controlling that keep an organization running well. Managers are concerned with the allocation of resources and may be particularly focused on routine tasks. The management of a new venture firm will work to accomplish all the tasks required to keep the company operating. Management of a new firm is important and should not be undervalued compared to entrepreneurial leadership—both are valuable. The goal of the manager is to make a business carry out activities efficiently and on time. Managers work hard on focused goals in order to implement the strategy of the firm. They make deliberate choices about resources. One reason that purposeful managers are so effective is that they are adept at conservatively leveraging resources, such as time and money.

Managers use their personal contacts when they need information or help. These informal networks create part of the social capital of an organization. People who link and connect people through a business network can be valuable managers who cross boundaries, help build subnetworks, and make the organization work [Cross and Prusak, 2002].

Managers are good at pattern recognition—making generalizations out of inadequate facts. Good managers are also prepared, frugal, and honest. Getting employees to stick to important strategic initiatives—and to give those initiatives their undivided attention over time—is crucial to competing successfully. Great organizations have outstanding strategies and capabilities as well as the mechanisms necessary for executing the firm's strategies. The follow-through with operations and processes is part of a firm's competitive advantage. Managers focus on performance, feedback, and decision-making. They are also synthesizers who bring resources together in a timely manner. Given the rate of change of the competitive environment, they can make a great contribution by managing resources on a rolling time scale of 12 to 18 months.

Managers set expectations by describing the desired outcome, not the path. Talented people will determine the best path once they know the desired result. Most good managers motivate their people by offering them the opportunity for achievement, recognition for achievement, the work itself, responsibility, and growth or advancement [Harzberg, 2003]. Managers need to tell employees the span of their job, the support they can expect, and what they will be accountable for. Then, as they achieve good results, they should receive the credit due to them [Simons, 2005].

Good managers turn the team members' talent into good performance. They engage people in decisions that directly affect them, explain why decisions are made the way they are, and clarify what will be expected of them after changes are made. They also provide feedback to team members regarding their performance. Organizations profit when employees ask for feedback and deal well with criticism. Openness to feedback is critical for all staff in a new venture facing many transitions. As people begin to ask how they are doing relative to management's priorities, their work becomes better aligned with organizational goals [Jackman and Strober, 2003].

All successful managers excel in the making, honoring, and remaking of commitments. A commitment is any action taken in the present that binds an organization to a future course of action. Managerial commitments take many forms, from capital investments to personnel decisions to public statements, but each exerts enduring influence on a company. A commitment may impede a response to changing conditions. Managers can learn to recognize when commitments have become roadblocks to needed changes. They can then replace those roadblocks with new, rejuvenating commitments. The ability to make and rewrite commitments is an important managerial skill [Sull and Spinosa, 2005].

Management of a new venture is complex. Managers balance five perspectives: reflection, analysis, contextual dynamics, relationships, and change [Gosling and Mintzberg, 2003]. Reflection can lead to understanding. Analysis can result in better organizational arrangements and performance. Understanding of context (or environment) can lead to regional and global activities. Collaboration with other firms and sound industry relationships are important. Also, a bias toward action and beneficial change is another valuable perspective for a manager. The effective manager weaves together all five perspectives into a fabric for growth and performance.

Managers can make better business decisions if they ask for evidence of the efficacy of a proposed action [Pfeffer and Sutton, 2006]. The first step is to demand evidence (facts) that proves that a proposed action or process will work. Look for gaps in logic, inference, and applicability to your situation. Managers should consider running trial programs, pilot studies, and small experiments to help provide facts and insights that can lead to better management.

## 12.5 Recruiting and Retention

New ventures serious about obtaining profits through people need to ensure that they recruit the right people in the first place. The organization needs to be clear about what are the most critical skills and attributes needed in its applicant pool. The notion of trying to find "good employees" is not very helpful. Instead, new ventures need to be as specific as possible about the precise attributes they are seeking. Technology start-ups tend to seek people who have strong technology expertise but also are flexible, and willing and able to assume a number of key roles in a new venture. The qualities of a good new member of a new venture include flexibility, experience, technical knowledge, and self-motivated creativity. The idea is to hire a "great athlete who has already run the race before." Successful new enterprises also look for people who can thrive in environments in which people trust each other and are willing to debate assumptions, share information, and express their thoughts openly.

Perhaps the most important qualities of a new hire are proficiency in the skills required and demonstrated capability to acquire the attributes needed for future situations. Thus, a raw ability to learn is critical for technology ventures [McCall, 1998]. Most new ventures benefit from attracting new people who possess stretchwork capabilities [Bechky and O'Mahony, 2006]. Stretchwork is the capability to readily bridge from proven competencies to new ones demanded by the needs of the new venture. These people possess strong competencies and quickly learn new ones.

The growth and survival of an emerging business depends on both star players and good supporting talent [DeLong and Vijayaraghavan, 2003]. Star employees can make important contributions to performance. Yet a firm's future also depends on the capable, steady performers. These steady performers bring stability and depth to an organization's resilience. Furthermore, they are more likely to be loyal to the organization.

Identifying, attracting, and retaining the best people can be difficult. To the extent possible, as a venture grows, it should recruit through existing team members and the firm's supply chain. By clearly defining what the venture seeks in terms of crucial competencies and experience, the venture can run a more focused search for candidates who are more likely to be a strong fit [Fernandez-Araoz, 2005]. Ventures also should consider approaching the hiring process as a "selling" process rather than a "buying" one. In other words, rather than focusing solely on screening applicants, they also should work to excite and entice potential employees. To be sure about the hire, new ventures often use a trial period in a consulting role or a temporary relationship on a project basis. Internships are common, too, and may turn into full-time positions if both the company and the employee see a fit.

Talented people in new ventures work hard and fast, trying to get to market with the hope of big payoffs. They remain focused on the critical factors of success and resist any tendency to drift toward peripheral goals. They are motivated by significant responsibility, participation, and the possibility of big financial gains. Their internal commitment is, in fact, aligned with the realities of their incentives, and the potential results can be motivating. Although winning is important to them, the goal is to win as an enterprise, rather than as individuals. Venture capitalist John Doerr of Kleiner Perkins says that when he looks at the business plan, "I always turn to the biographies of the team first. For me, it's team, team, team. Others might say, people, people, people—but I'm interested in the team as a whole" [Fast Company, 1997].

Attracting and retaining key employees depends on compensation, work design, training, and networks. Compensation systems include wages, incentives, and ownership options. New ventures need to provide reasonable compensation and benefits. People in a start-up may have to sacrifice financial compensation for an initial period while building the firm toward financial break-even. Thus, it is important to offer reasonable benefits and ownership opportunities. Offering health benefits may be necessary to attract people to the new venture. Most technology ventures offer health benefits to their employees before reaching break-even.

Ownership interest in the new firm will be of great interest to most new employees. Broad-based ownership, when done right, can lead to higher productivity, lower workforce turnover, better recruits, and improved outcomes [Rosen et al., 2005]. **Stock options** are offered in a plan under which employees can purchase, at a later date, shares of the company at a fixed price (strike price). Stock options take on value once the market price of a company's stock exceeds the exercise (strike) price. Stock options should vest with the recipient over a period of years. For example, a new hire may receive an option for 10,000 shares exercisable at $1 per share vesting over four years. This is equivalent to 2,500 shares each year. Stock options give employees the right to buy the company's stock in the future at a preset price, thus motivating them to work to increase productivity and innovation—and eventually, the market value of the firm.

The purpose of employee stock options is to create a noncash substitute for part of the wage compensation the firm must provide to attract and retain employees. A new, entrepreneurial firm may not be able to provide the cash compensation needed to attract outstanding workers. Instead, it can offer stock options. From the beginning, Starbucks decided to grant stock options to every employee in proportion to their level of base pay: it called these options "bean stock." Google created thousands of millionaires due to the issuance of wide stock options.

An alternative to stock options is **restricted stock,** which is stock issued in an employee's name and reserved for his or her purchase at a specified price after a period of time—say, one, two, or three years. Some call this type of stock "reserved stock." New firms need to increase the level of share ownership through such means as offering restricted shares and requiring that employees hold their shares for certain periods of time. For example, an employee could have 10,000 shares reserved to be purchased at $2 per share as they vest in two years. If the price of the stock appreciates to $10, the gain in the stock's value is $80,000. Restricted stock still has value if the stock price falls, while options expire as worthless if the stock does not appreciate. In 2003, Microsoft switched from offering stock options to restricted stock.

Although compensation is an important means of attracting and retaining employees, a simple reliance on financial rewards is insufficient. Many, if not all, technology start-ups use an organizational design and culture that involves challenging work, peer group control, and selection based on specific task abilities. Few imperatives are more vital to the success of young technology companies than retaining key technical personnel, whose knowledge often represents the firm's most valuable asset [Baron and Hannan, 2002]. Therefore, the leaders of the enterprise need to maintain their commitments to their employees and retain their trust.

Keeping faithful, loyal employees and partners in a new venture can be based on six principles, as shown in Table 12.8 [Reichheld, 2001]. A simple reliance on financial rewards is insufficient. Good communication, trust,

**TABLE 12.8  Six principles for retaining loyal people and partners.**

1. Preach what you practice. Communicate the vision, goals, and values of the organization. Practice what you preach is also required.

2. Partners must win also. Enable your vendors and partners to participate in a win-win venture.

3. Be selective in hiring. Select people with values consistent with the firm's. Membership on the team is selective.

4. Use teams of talented people. Use small teams for most tasks and give them the power to decide. Provide simple rules for decision-making so teams can act.

5. Provide high rewards for the right results. Reward long-term values and profitability. Provide solid compensation, benefits, and ownership.

6. Listen hard, talk straight. Use honest, two-way communication and build trust. Tell people how they are doing and where they stand.

*Source:* Adapted from Reichheld, 2001.

and treatment are essential to retaining talent and partners. High standards of decency, consideration, and integrity are necessary for all members of the new venture. Through loyalty to ideals, the firm becomes worthy of loyalty from its people and partners.

## 12.6  Organizational Culture and Social Capital

**Organizational culture** is the bundle of values, norms, and rituals that are shared by people in an organization and govern the way they interact with each other and with other stakeholders. An organization's culture can have a powerful influence on how people in an organization think and act. **Organizational values** are beliefs and ideas about what goals should be pursued and what behavior standards should be used to achieve these goals. Values include entrepreneurship, creativity, honesty, and openness.

**Organizational norms** are guidelines and expectations that impose appropriate kinds of behavior for members of the organization. Norms (informal rules) include how employees treat each other, flexibility of work hours, dress codes, and use of various means of communication such as e-mail. **Organizational rituals** are rites, ceremonies, and observances that serve to bind together members of the organization. Examples of rituals are weekly gatherings, picnics, awards dinners, and promotion recognition.

In innovative firms, the values and beliefs favor collaboration, creativity, and risk-taking [Jassawalla and Sashittal, 2002]. These firms employ stories and rituals to reinforce these values and beliefs.

---

### Intel: Staying Entrepreneurial

Intel, which has become the semiconductor standard for personal computers, uses three principles to keep its entrepreneurial spirit alive. First, leaders must be willing to solve complex problems. There is no substitute for deep domain knowledge in technology industries, and at Intel, new products are continually being released. Second, the most effective managers are not afraid to change the rules. The "Intel Inside" branding campaign was a highly unusual marketing idea for a company with little previous direct contact with consumers. Last, Intel rarely tries to convince anyone to take a job or assignment. Managers must display a raw enthusiasm to try something new and a clear passion of their own [Barrett, 2003].

---

Culture is expressed in community. Communities are built on shared values, interests, and patterns of social interaction. The fit of the company culture in the business environment is critical to the company's competitive advantage. Most entrepreneurial start-ups have the founders and employees bound together by strong values and norms that include long hours working closely together. The sense of solidarity of a new venture is very high. The goals are uniformly shared—survival and early success being foremost. Start-ups are often founded by friends or former colleagues and exhibit high sociability within the organization. Employees possess a high sense of organizational identity and membership. In the early days, employees of Apple thought of themselves as "Apple people." Entrepreneurial firms often sponsor social events that take on ritual significance.

Hewlett-Packard Company fostered a culture known as the "HP Way" [Packard, 1995]. This culture was captured in a statement of objectives, values, and norms regarding fairness and justice. Employees were promised opportunity for security, job satisfaction, and sharing in profit. For example, during the 1981 recession, rather than lay people off, Hewlett-Packard introduced a 10 percent cut in pay and hours across every rank.

---

### Zappos: Enjoying Work

Zappos is known worldwide for its culture. The culture not only establishes a great organizational environment for employees to work in, but also incorporates the way that Zappos interfaces with its customers. There are ten core Zappos values and they permeate the organization. Employees are trusted to do the right thing and customers are also treated with a degree of trust that is unprecedented for an online retail establishment. Every employee must spend his or her first three to four weeks working in the call center dealing with customer issues. Call center employees are empowered to do what is necessary to ensure customer satisfaction. Each manager at Zappos is expected to spend 10 to 20 percent of the time on employee team-building activities. The culture seems to be working: over 75 percent of sales at Zappos comes from repeat customers.

---

Entrepreneurial firms usually demonstrate high sociability and solidarity in their early years, built around the leadership of the founders—such as Hewlett and Packard. As the founders leave and other challenges come to bear, it may be difficult to retain the communal culture of a start-up. Yahoo built a fast-growing firm based on a set of founders who worked furiously at their jobs. Their communal culture helped them build the firm, but their insularity may have made them see the marketplace through a Yahoo lens in 2000 [Mangalindan and Hoang, 2001]. Google has succeeded due in part to a powerful culture that is maintained at all of its locations worldwide. This helps develop a consistency in user experience that customers value.

Perhaps the strongest element of an organizational culture is trust [Covey, 2006]. **Trust** is a firm belief in the reliability or truth of a person or an organization. It is critical that we can trust those with whom we work. Handy provides seven principles of trust, as listed in Table 12.9. Teams are formed with many people we already trust, but new members will need to earn our trust. By *trust,* organizations mean confidence in a person's work, competence, and commitment to the organization's goals and tasks. Every team member must be capable of self-renewal, learning, and adaptability. When trust is broken, the person needs to be reassigned, leave the organization, or have his or her boundaries constrained. Teams need to build bonds among their members to enable trust. People need to meet in person to restore the group bonds. Finally, organizations, like people, need to continually demonstrate that they are trustworthy.

Good communication is part of any sound and trusting community. Knowing when to speak up and when to keep silent is an important skill for any worker. A new enterprise needs openness and creativity to grow and become fruitful. Carefully choosing the right issues to raise and avoiding ill-chosen conflict can be important. When emotions are high and the new enterprise is challenged with obstacles, it may be wise to keep silent. Nevertheless, managers of any enterprise must welcome ideas and comments. Organizations of all sizes and arrangements must strive to keep the ideas flowing. Breaking the silence can bring an outpouring of fresh ideas from all levels of an organization—ideas that might just raise the organization's performance to a whole new level [Perlow and Williams, 2003].

In a study of successful companies, the building of a performance-based culture was central [Joyce et al., 2003]. The study showed that winning

### TABLE 12.9  Seven principles of trust.

1. Trust is not blind. It is unwise to trust people whom you do not know well, whom you have not observed in action over time, and who are not committed to the same goals.

2. Trust needs boundaries. It is wise to trust people in some areas of life but not necessarily in all.

3. Trust requires constant learning. Every individual of a team must be capable of self-renewal and learning.

4. Trust is tough. When trust proves to be misplaced because people do not live up to expectations or cannot be relied on to do what is needed, then those people must go, be reassigned, or have their boundaries severely curtailed.

5. Trust needs bonding. Teams of people need to build their own bonds.

6. Trust needs touch. Personal contact is necessary, and teams need to meet in person to renew their trust and bonds.

7. Trust has to be earned. Organizations that expect their people to trust them must continually demonstrate that they are trustworthy.

*Source:* Adapted from Handy, 1999.

### TABLE 12.10  Four principles of a performance-based culture.

1. Inspire everyone to do their best.

2. Reward achievement with praise and pay-for-performance, and keep raising the performance goals.

3. Create a work environment that is challenging, rewarding, and fun.

4. Establish, communicate, and stick to clear values.

*Source:* Joyce et al., 2003.

companies built a culture on the four principles described in Table 12.10. Examples of successful firms that have a performance-based culture include Intel, Cisco Systems, and General Electric. Just about any high-potential firm needs to focus on performance, define it, and build a culture that reinforces it.

Leaders can develop a design for a workplace that can be fulfilling for all involved. With growing trust and dignity as well as wide participation on decision-making, all employees can experience ownership and pleasure in their work [Bakke, 2005].

---

**Cisco Community Statement**

Cisco's culture requires that all employees, at every level of the organization, are committed to responsible business practices. Additionally, our business strategy incorporates our dedication to corporate citizenship, which includes our commitment to improving the global community in which we operate, empowering our workforce, and building trust in our company as a whole.

Cisco was founded on, and still thrives today in, a culture based on the principles of open communication, empowerment, trust, and integrity. These values remain at the forefront of our business decisions. We express these values through ethical workplace practices; philanthropic, community, and social initiatives; and the quality of our people.

*Source: Cisco Systems Annual Report*, 2003.

---

Effective organizational design and culture, along with good hiring and retention processes, can enable a firm to grow its social capital. **Social capital** consists of the accumulation of active connections among people in a network [Coleman, 1990]. Social capital was also considered in Chapters 1 and 4. Social capital refers to the resources available in and through personal and organizational networks [Baker, 2000]. These resources include information, concepts, trust, financial capital, collaboration, social structure, and emotional support. These resources reside in networks of relationships. Social capital depends on whom you know. Social capital, like financial capital, can lead to increased productivity, if used wisely. A firm can build up or deplete its social capital by its actions. Relationship networks are often as important as technology, land, capital, or other assets of a venture.

Social capital tends to be self-reinforcing and cumulative. Successful collaboration in one endeavor builds connections and trust—social assets that facilitate future collaboration in other, unrelated tasks. Firms that build up social capital demonstrate a commitment to retaining people and promoting from within. They also enable far-flung teams to meet in person periodically. They give people a common sense of purpose and keep their promises to people. Employees need to hear the same messages that an organization sends out to vendors and customers. Alternatively, social capital may be depleted by declining trust among a firm's people and the effects of people working off-site or on their own [Prusak and Cohen, 2001].

Social capital can be described as consisting of three dimensions: (1) structural, (2) relational, and (3) cognitive. The structural dimension concerns the overall pattern of relationships found in organizations. The relational dimension of social capital concerns the nature of the connections between individuals in an organization. The cognitive dimension concerns the extent to which employees within a social network share a common perspective or understanding [Bolino et al., 2002]. Social capital is valuable because it facilitates coordination, reduces transaction costs, and enables the flow of information between and among individuals. In other words, it improves the coordinated effort and organization.

Better knowledge sharing can lead to increased trust and better decisions. Teamwork can lead to inventiveness, creative collaboration, and a good spirit. Trust is the fuel of a social capital engine that in turn engenders more trust. When people in an organization say their firm is "political," they often mean that trust is low throughout the organization. An organization with strong capital is a community with shared values and good trust. Moreover, social capital extends to relationships outside of the company, which should also be characterized by shared values and trust.

Managing to develop and utilize social capital can lead to enhanced corporate performance [Lee and Kim, 2005]. Providing the appropriate networks that enable creativity can improve quality and productivity. Engaging the creative energies in a collaborative activity of the technologists and designers with customers can lead to improved results [Florida and Goodnight, 2005]. Social capital can provide greater access to other resources [Kalnins and Chung, 2006]. Or, it can substitute where other types of capital are lacking [Packalen, 2007]. For most technology ventures, the social network should be dense and redundant, and link external and internal sources to needs [Cross et al., 2005]. Firms should continually adapt their social capital to changing resource needs [Maurer and Ebers, 2006].

---

**IBM's Guiding Principles**

Louis Gerstner joined IBM in January 1993 to bring the firm back to its roots and success. In his first months, Gerstner created a set of principles for the firm that included [Gerstner, 2002]:

- At our core, we are a technology company with an overriding commitment to quality.
- We operate as an entrepreneurial organization with a minimum of bureaucracy.
- We think and act with a sense of urgency.
- Outstanding, dedicated people make it all happen, particularly when they work together as a team.

Gerstner also described the IBM culture this way: "In the end, an organization is nothing more than the collective capacity of its people to create value." By 2002, Gerstner left IBM as a powerful technology company focused on its entrepreneurial principles.

## 12.7   Managing Knowledge Assets

Assets are potential sources of future benefit that a firm controls or can access. Knowledge is an asset that is a potential source of wealth, as described in Chapter 1. The creation and management of knowledge leads to new, novel applications and products that can result in wealth creation. **Knowledge** is the awareness and possession of information, facts, ideas, truths, and principles in an area of expertise. Intellectual capital is the sum of the knowledge assets of a firm. These knowledge assets include the knowledge of its people, the effectiveness of its management processes, the efficacy of its customer and supplier relations, and the technical knowledge that is shared among its people. It can be thought of as best practices, new ideas, synergies, insights, and breakthrough processes. Thus, the firm's intellectual capital (IC) is the sum of its human capital (HC), organizational capital (OC), and social capital (SC), as described in Chapter 1.

From the generation of new ideas through the launch of a new product, the creation and exploitation of knowledge are core themes of the new product development process. In fact, the entire new product development process can be viewed as a process of embodying new knowledge in a product [Rothaermel and Deeds, 2004].

The knowledge creating and sharing activities of a firm can be represented by Figure 12.3 [Leonard-Barton, 1995]. The value of commercial knowledge is in its use, not its possession. The value of knowledge compounds when it is shared. Using current knowledge for cooperative problem-solving is the first of the four knowledge activities of a firm. The second is the implementation of new processes and tools within the firm. The third activity is experimenting and learning in order to build the knowledge base. The fourth activity is acquiring knowledge from outside the firm.

As a result of creating and sharing knowledge, a firm can enhance its people's skills and capabilities, as well as the knowledge embedded in its processes and managerial systems. Knowledge finds value in practice and use. The strategic approach of a new venture is linked to a set of intellectual assets and capabilities. Thus, if a firm has an opportunity that requires certain knowledge and it is not yet available, we can state that the firm has a knowledge gap. Acquiring the knowledge to fill the gap will be critical to the future success of this firm.

The knowledge of a firm encompasses (1) cognitive knowledge ("know what") or the basic mastery of a discipline, (2) advanced skills ("know how") or the ability to apply knowledge to complex real-world problems, (3) system understanding and intuition ("know why") or deep knowledge of cause-effect relationships, and (4) creativity ("care why") or the will, motivation, and adaptability for success. The first three forms of knowledge can be stored in the firm's systems, databases, or operating technologies. The fourth form often is found in the firm's culture [Quinn et al., 1996].

Knowledge can be *prepositional*, which deals with beliefs about natural phenomena, such as scientific discoveries and practical insights into the properties of materials, waves, and nature. *Prescriptive* knowledge is all about

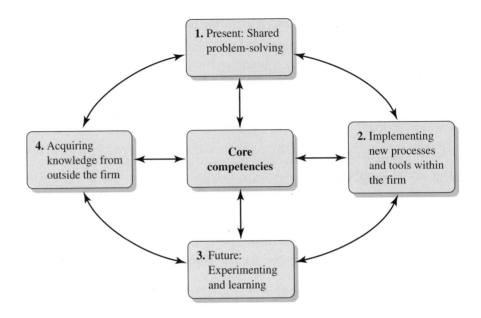

**FIGURE 12.3** Knowledge creating and sharing activities of a firm.

techniques—the manipulation of processes and formulas, such as how to write a piece of software. The growing interplay between these two forms of knowledge transformed the world economy during the twentieth century [Mokyr, 2003].

Knowledge can be seen as a source of innovation and change leading to action. Also, it provides a firm with the potential for novel action and the creation of new ventures. With increased flow of new information, firms need to develop the means to convert that information into insight [Ferguson et al., 2005]. Knowledge creates real wealth for a new venture through multiple applications, which can have breadth across an organization. The knowledge represented by patented inventions, software, marketing programs, and skillful employees makes up 70 to 90 percent of the assets held by corporations like Microsoft, Amgen, and Intel.

The growth of a new venture rests, in part, on the increasing value of a knowledge base within the emerging firm. **Knowledge management** is the practice of collecting, organizing, and disseminating the intellectual knowledge of a firm for the purpose of enhancing its competitive advantages. The four steps for managing knowledge in a new venture are given in Table 12.11. The first step is to identify and evaluate the role of knowledge in the firm. How is knowledge created, stored, and shared? The second step is to identify the expertise, capabilities, and intellectual capital that create value for a firm. Then, we examine the uniqueness and value of these intellectual assets.

**TABLE 12.11  Managing knowledge in a technology venture.**

| | | |
|---|---|---|
| 1. | **Role:** | Identify and evaluate the role of knowledge in the firm. |
| 2. | **Value:** | Identify the expertise, capabilities, and intellectual capital that create value in the form of products and services. |
| 3. | **Plan:** | Create a plan for investing in the firm's intellectual capital and exploiting its value while protecting it from leakage to competitors. |
| 4. | **Improve:** | Improve the knowledge creation and sharing process within the new venture. |

The third step is the creation of an investment and exploitation plan for maintaining and harvesting the value of the knowledge assets. Finally, the firm improves the process of creating and sharing its knowledge. Though knowledge is one of the few assets that grows when shared, the new venture needs to carefully determine what knowledge to share and what knowledge should be protected or kept secret. This is particularly true for technology ventures for which intellectual property is usually their key asset.

Part of an emerging firm's knowledge base is information about competitors. This knowledge is useful in responding to competitive changes and challenges. **Competitive intelligence** is the process of legally gathering data about competitors. Competitor intelligence may include securing data about competitors' products, services, channels of distribution, pricing policies, and other facts. Legal means of acquiring competitor intelligence include gathering company reports, news releases, and industry reports, and visiting competitor websites and trade show booths.

Knowledge is worthy of attention because it tells firms how to do things and how they might do them better [Davenport and Prusak, 1998]. The key skill for an emerging start-up is the ability to turn knowledge into products and services. Knowledge turns into action as it is embedded in the products, routines, processes, and practices of the new technology venture. Knowledge embedded in a company's activities can be a sustainable competitive advantage since imitating it is difficult for competitors.

## 12.8 Learning Organizations

New ventures grow powerful from learning and adapting to new challenges and opportunities. A **learning organization** is skilled at creating, acquiring, and sharing new knowledge and at adapting its activities and behavior to reflect new knowledge and insights. A technology venture creates and acquires knowledge and shares this knowledge among its people. As a result of this new knowledge, the organization adapts its actions and behavior. Learning organizations are skilled at five activities: systematic problem-solving, experimentation with new approaches, learning from their own experience and past history, learning from the experiences and best practices of others, and transferring knowledge quickly and efficiently throughout the organization. The learning organization is active, imaginative, and participative. It attempts to shape its future rather than react to forces. A learning firm adapts itself to its learning and increases opportunities, initiates change, and instills in employees the desire to be innovative. A learning organization confronts the unknown with new hypotheses, tests them, and creates new knowledge. Thus, the learning organization creates innovation and new knowledge that is used by the technology venture to develop new products and services.

Information that does not enable an action of some kind is not knowledge. Knowledge comes from the ability to act on information as it is presented. It truly is power, giving an organization the ability to continuously better itself. The power of knowledge depends on the company's ability to provide a supportive environment: a culture that rewards the sharing of knowledge across various barriers. The company that develops the right set of incentives for its employees to work collaboratively and share their knowledge will be successful in its knowledge management effort. Knowledge management has several benefits: it fosters innovation by encouraging a free flow of ideas, enhances employee retention rates, enables companies to have tangible competitive advantages, and helps cut costs.

The decisions of an entrepreneurial firm are the result of the firm's ability to process knowledge and learning [Minniti and Bygrave, 2001]. Knowledge acquired through learning-by-doing takes place when entrepreneurs choose among alternative actions whose payoffs are uncertain. Over time, entrepreneurs repeat only those choices that appear most promising and discard the ones that resulted in failure. Thus, entrepreneurship is based on a process of learning that allows entrepreneurs to learn from successes as well as failures. Former CEO and Chairman of General Electric, Jack Welch, described the learning process thus: "In the end I believe we created the greatest people factory in the world, a learning enterprise, with a boundary-less culture" [Welch, 2002].

Siemens, a global organization, uses the ShareNet network to enable 19,000 technical specialists in 190 countries to help each other in solving problems. In one case, a project manager in South America was trying to find out how dangerous it was to lay cables in the Amazon rain forest in order to determine the type of insurance his project needed. He posed the question on ShareNet and within hours a project manager in Senegal who had encountered a similar situation responded. Getting the right, *actionable* information before the cables went underground saved the company several million dollars in insurance costs [Tiwana and Bush, 2005].

Managers and entrepreneurs often have distorted pictures of their businesses and their environments. Busy among the trees, they can lose sight of the forest. They can review the impacts of their actions, however, and then modify their approach accordingly. The greatest asset that entrepreneurs can bring to knowledge and learning is their willingness to seek and make wise use of feedback [Mezias and Starbuck, 2003].

The entrepreneur's learning process is based on the six steps outlined in Table 12.12 [Garvin, 1993]. At each stage in the development of the new business, the entrepreneur encounters a set of challenges or problems that require

**TABLE 12.12 Entrepreneurial learning process.**

| Step | Question | Outcome or action required |
|------|----------|----------------------------|
| 1. Identify the problem or opportunity | What do we want to change? | Desired specific result |
| 2. Analyze the problem or opportunity | What is the key cause of the problem? | Key cause identified |
| 3. Generate potential solutions | How can we make a positive change? | List of possible solutions |
| 4. Select a solution and create a plan | What is the best way to do it? | Establish a criteria, select the best solution, and set a plan to accomplish it |
| 5. Implement the selected plan | How do we implement the plan effectively? | Monitor the implementation |
| 6. Evaluate the outcome and learn from the results | How well did the outcome match our desired result? | Verify that the problem is solved. Why did it work? |

resolution. A firm can use the method shown in Figure 12.4 to resolve issues and learn from its successes and failures.

Organizational learning looks at an organization as a thinking system. Organizations rely on feedback to adjust to a changing world. Thus, organizations engage in complex processes such as anticipating, perceiving, envisioning, and problem-solving in order to learn. This approach is very important for new technology ventures to adopt and improve.

Process improvement projects can produce two types of learning. *Conceptual learning* is the process of acquiring a better understanding of cause-and-effect relationships by using statistics and scientific methods to develop a theory. *Operational learning* is the process of implementing a theory and observing positive results. Conceptual learning yields know-why—the team understands why a problem happens. Operational learning yields know-how—the team has implemented a theory and knows how to apply it and make it work. It is useful to design projects that are more likely to deliver both conceptual and operational learning [Lapre and Van Wassenhove, 2002].

For learning to be widely used in a firm, knowledge must spread widely and quickly throughout the organization. Ideas carry maximum impact when they are shared broadly rather than retained by a few people. A variety of mechanisms enable this process, including written, oral, and visual reports, knowledge bases, personnel rotation programs, education, training programs, and formal and informal networks. The organization needs to foster an environment that is conducive to learning. The new venture must strive to set aside time for reflection and sharing. Furthermore, boundaries that inhibit the flow of knowledge must be reduced so that learning is shared. A new venture can profit from efforts to eliminate barriers that impede learning and to place learning high on the organizational agenda.

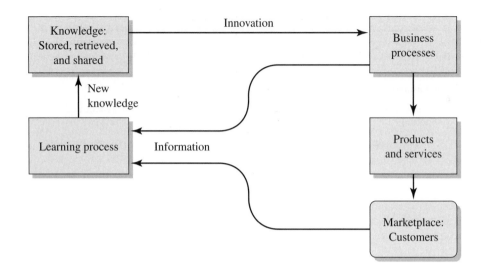

**FIGURE 12.4** Knowledge and learning within a technology firm.

Genmab was established in 1999 to develop products based on human antibodies for the treatment of a number of life-threatening and debilitating diseases. The company was founded by Lisa Drakeman, who spotted the work of a Dutch scientist while working at another company. She proposed starting a company, but American venture capitalists were unwilling to invest. Genmab was thus founded in Copenhagen, with research facilities in the Netherlands. The company had its initial public offering in 2000. In 2008, its market value reached $2.5 billion as one of the world's top 20 biotechnology firms. The venture created powerful intellectual property (knowledge) via its learning processes. It then converted that value into financial assets through an IPO.

A learning organization, properly managed, can enable a firm to meet the challenge of change by constantly reshaping competitive advantage even as the marketplace rapidly shifts. The learning organization is able to improvise or adapt to balance the structure that is vital to meet budgets and schedules with flexibility that ensures the creation of innovative products and services that meet the needs of changing markets [Brown and Eisenhardt, 1998]. The firm that is most responsive to change and capable of learning is the one that succeeds [Galor and Moav, 2002].

Knowledge is stored in documents, databases, and people's minds. Knowledge created in a learning process as depicted in Figure 12.4 is a social process that leads to increasing knowledge [McElroy, 2003]. Knowledge is shared by people and embedded within the business processes of the firm. Innovation flows through the business processes, products, and services of the firm, as shown in Figure 12.4. As the firm learns and creates new knowledge, new innovation is created and new opportunities are identified [Lumpkin and Lichtenstein, 2005].

## Genentech: Learning from Prior Experience

In the early 1970s, the biotechnology industry was just beginning to emerge when Cetus was formed by Ronald Cape (a Ph.D. biologist with an MBA), Donald Glaser (Nobel laureate in physics), Peter Farley (a physician with an MBA), Calvin Ward (a scientist), and Moshe Alafi (a venture capitalist). Being one of the first firms in the industry and having the backing of a Nobel laureate, Cetus was able to attract a star-studded advisory board. Unfortunately, neither Cetus's employees nor its advisors knew what a biotechnology firm should look like or what it should do. Therefore, Cetus took money and formed partnerships with whoever was willing, and attempted to be all things to all people. The end result was that Cetus worked on projects that spanned from health sciences to agriculture to finding better processes for making industrial alcohol. In the early 1980s, Cetus recognized that it needed to tighten its focus. It channeled 70 percent of its R&D spending into the health care field, it brought in a professional manager to run the firm, and it was more forthright with analysts and the media. Unfortunately, by then, Cetus had lost many of its supporters and investors.

Frustrated with Cetus's lack of direction and armed with their experience, several of the board members left Cetus to form their own biotechnology companies. They were convinced—and investors appeared to agree—that a more focused strategy was a better way to go. One individual who was familiar with Cetus's business strategy was Robert Swanson, a young venture capitalist with Silicon Valley's Kleiner Perkins. In 1976, just five years after Cetus was founded, Swanson approached Herbert Boyer, then a professor at University of California at San Francisco (UCSF), about starting a new biotechnology company based on work that Boyer had completed with Stanford University professor Stanley Cohen. What Boyer had scheduled as a 20-minute polite conversation turned into a 3-hour meeting as Swanson won Boyer over with his enthusiasm. By the time Swanson and Boyer left the bar, they had agreed to form Genentech (for "genetic engineering technology").

Swanson left Kleiner Perkins and delved right into learning the science and becoming a hands-on, deeply involved CEO. Boyer also became deeply involved and took a leave of absence from UCSF. Swanson and Boyer worked hard to build a creative firm that was in many ways the antithesis of traditional pharmaceutical firms. To lure postdoctoral candidates away from academia, they offered their employees stock options and structured the R&D portion of their firm to resemble an academic lab: scientists worked flexible hours, dressed casually, and were allowed to publish their research.

In 1980, Genentech became the first biotechnology firm to go public, pricing its million shares at $35 each. Within half an hour of trading, the stock hit $89 per share and closed the day at $70. Genentech's public offering broke many previous IPO records. Today Genentech continues to be a well-managed learning organization.

*Sources:* Lax, 1985; Swanson, 2001; Teitelman, 1989; Robbins-Roth, 2000.

## 12.9  Spotlight on Intuit

Intuit provides software for personal finance, small business finance and accounting. These software products enable individuals and businesses to automate financial tasks. Formed in 1984, the firm went public in 1993. Revenues reached nearly $4 billion in 2012.

Intuit has adopted a company-wide method of ongoing product reinvention. The company encourages all employees to find new opportunities. It has created an idea collaboration portal that lets employees share their concepts and receive support and suggestions. Intuit also fosters a culture of frugal experimentation. This fast experimentation method improves employee's sense of ownership and cooperation, resulting in the examination of assumptions and feedback that can lead to company-wide success. With more experimentation, the fear of failure is reduced company-wide. As a result, the corporate culture supports continual reinvention, which has led to consistent year-to-year growth.

## 12.10  Summary

A new venture depends upon a team of people with complementary capabilities who are committed to a common goal. The team includes a board of directors and a board of advisors to help monitor and enhance the growing firm. Entrepreneurs should design an organization that facilitates effective coordination and collaboration among these team members. The firm's leaders strive to motivate and inspire the team members and to foster an innovative culture. As the firm grows, managers join to build the structure and carry out the detailed tasks of the new firm. The firm also puts together a compensation scheme that emphasizes "buy-in" or ownership—normally achieved by awarding stock options or restricted stock. Effective organizations manage their knowledge assets carefully and strive to continually learn.

> **Principle 12**
> Effective leaders coupled with a good organizational plan, a collaborative, performance-based culture, and a sound compensation scheme can help align every participant with the goals and objectives of the new firm.

## Video Resources

Visit http://techventures.stanford.edu to view experts discussing content from this chapter.

| | | |
|---|---|---|
| Strength of a Team | Vinod Khosla | KPCB |
| Founding Teams | Ashwin Navin and Ping Li | BitTorrent |
| Leadership and Capability | Carly Fiorina | HP |

# 12.11 Exercises

**12.1** Examine the beginnings of a technology venture that currently has lots of "buzz". Who were the company founders? What background, capabilities, and qualities did each bring to his or her new role? Who was hired next and why?

**12.2** Genentech (www.gene.com) has a unique culture known for rigorous science, guarding of industry secrets, and rigid rules. Its key principle is: good scientists make for good science make for good products make for a good company. Describe Genentech's culture in terms of norms and rituals.

**12.3** Take-Two Interactive (www.take2games.com) develops entertainment software games. Revenues grew from $19 million in 1997 to $1.5 billion in 2008. Describe the organizational design for Take-Two, which has about 2,100 employees. Use Figure 12.1 to help determine its organizational model.

**12.4** Red Hat is the leading distributor of Linux software and services. The firm had over 4,500 employees and revenues of $1.3 billion in fiscal year 2012. Using the concepts of Section 12.6, describe the firm's organizational culture.

**12.5** 24/7 Customer provides services to U.S. firms that wish to outsource call centers and CRM activities to India. Using the information provided at the firm's website, describe the founding team of this firm. Also, describe its organizational design (www.247customer.com).

**12.6** Examine Google's corporate philosophy "Ten things Google has found to be true" (www.google.com/corporate/tenthings.html). What does Google's "do no evil" phrase mean? How are these broad truth statements translated into action within the Google organization and culture?

## VENTURE CHALLENGE

**1.** Describe the team and the organizational arrangement for your venture.

**2.** Discuss your plans to build a board of directors and a board of advisors. Name some candidates for these boards.

**3.** Describe the means of managing knowledge and learning that will be used in your venture.

CHAPTER 13

# Acquiring and Organizing Resources

*To get profit without risk, experience without danger, and reward without work,*
*is as impossible as it is to live without being born.*
**A. P. Gouthev**

**How can entrepreneurs efficiently acquire and organize the resources needed to launch their venture?**

To tap required resources, entrepreneurs need to build credibility and legitimacy in the marketplace of resources and talent. Influence and persuasion can help entrepreneurs build their case for securing scarce resources for their venture.

Both choosing a physical location and operating as a virtual organization are viable options for a firm today. We examine the benefits of joining a cluster of interconnected enterprises operating within a geographic region. All firms need to create a plan for outsourcing functions while maintaining critical functions within the firm. As firms strive to be innovative and competitive, they seek to control costs by outsourcing functions to those who can do them better and cheaper. However, these firms are challenged to retain the cohesion and coordination required to effectively manage these supplier partners. Seeking financial resources will be discussed in Chapter 18. ■

## 13.1 Acquiring Resources and Capabilities

Another definition of entrepreneurship is the pursuit of opportunity without regard to resources currently controlled [Stevenson et al., 1999]. This view stresses the idea that the entrepreneur can locate and access resources when they are needed. For example, when an entrepreneurial team needs legal counsel, it can engage a lawyer. When it needs a circuit designer, it can hire one. Resources are usually scarce, however, and the attraction of talented employees or financial investors is not easy or guaranteed. A firm's competitive advantage flows from a combination of resources and capabilities executing a unique strategy. If these resources and capabilities are scarce, then the new venture needs to compete to secure them.

The founders of a new venture attempt to acquire resources and capabilities by contacting key organizations and people and asking them to support their venture. For example, they ask their customers, suppliers, and sources of financial capital to take some risk and support the new venture, which will be further discussed in Chapter 18.

This resource-seeking activity can be represented by the cycle shown in Figure 13.1 [Birley, 2002]. The founders are asking all the participants in the credibility cycle to believe in their opportunity, vision, and story, and invest in their venture. To move forward, entrepreneurs need to persuade someone in the cycle to believe in them. If the entrepreneurs get some talented people to commit to the venture, this will help convince the suppliers. If the entrepreneurs get some customers to tentatively commit to purchasing the product, the sources of financial capital (bankers and investors) will become more interested. The entrepreneurs travel around the cycle, slowly building their credibility. In other words, the entrepreneurs demonstrate the legitimacy or truthfulness of the new venture to the members of the credibility cycle [Zimmerman and Zeitz, 2002]. Legitimacy

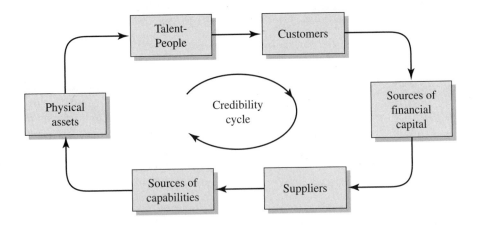

**FIGURE 13.1** Credibility cycle.

**TABLE 13.1  Sources of legitimacy.**

| | |
|---|---|
| ■ Regulatory: legal actions, accreditation, credentials | ■ Talent: known, respected people |
| ■ Social: fair treatment, endorsements, networks, image | ■ Location: within an industry cluster, favorable location, visible |
| ■ Industry: attractive, respected industry, known and understood business model | ■ Intellectual property: trade secrets, patents, copyrights |

or credibility is evidence of a social judgment of desirability and enables a new venture to access resources. The holders of scarce resources provide resources to new ventures only if they believe that the ventures are efficient, worthy, and needed, and the teams are competent. The greater the level of a new venture's legitimacy, the more resources it can access.

A new venture can build its legitimacy by tapping the sources of legitimacy listed in Table 13.1. A new venture can join industry associations, secure endorsements, and get commitments from talented, respected individuals. Patents, copyrights, and trade secrets also help to build legitimacy. New ventures should focus on actions that have the greatest payoff for legitimacy. A certain amount of legitimacy is required to make the credibility cycle build up the investment of resources. A successful new venture needs to acquire, build, and use legitimacy to secure the necessary resources to commence operations and grow successfully. Creating, building, and retaining a firm's credibility or legitimacy are critical tasks for the leadership team of any business. They determine the key influencers in your industry and reach out to them about the opportunity. For example, it has been shown that attaining the support of credible and reputed venture capitalists results in increased credibility of the venture [Hsu, 2004]. The leadership team of a new venture recognizes and creates new economic or social opportunities and makes decisions on the location, form, and use of resources [Wennekers et al., 2002].

To the extent possible, smart leaders of ventures manage the new venture like it is a public company [Carl, 2007]. They manage it with the idea that it will grow to become an important enterprise eventually. They hire a highly qualified law firm and accountant. They look for the best employees and talent available, including compelling board members and advisors. A new venture should strive to be one of the best from the beginning.

Entrepreneurs who exhibit higher social competencies and emotional intelligence greatly improve their ability to access resources. Entrepreneurs possessing a high level of social capital (e.g., a favorable reputation and an extensive social network) gain access to persons important for their success. Once such access is attained, their social competence influences the outcomes they experience [Baron and Markman, 2003].

In order to attract resources, entrepreneurs craft and use a story about their venture that they tell to potential partners and visitors [Downing, 2005]. In addition, they can use advisors and guides to help them in building their credibility and story [Chrisman et al., 2004].

---

**Aaron Levie's Quest for Resources at Box**

Aaron Levie founded Box in March 2006 to provide low-cost, cloud-based storage for documents and other content, as well as the ability to sync and use documents across a spectrum of devices, including desktop computers, laptops, and smartphones. Initially aimed at consumers, he quickly realized that Box could be sold to businesses as well. Selling to enterprises required significant capital. As of January 2013, Levie had raised $312 million in capital from private investors, venture capital firms, private equity firms, and other companies who were strategic partners of Box.

---

## 13.2 Influence and Persuasion

Influence and persuasion play a role in the entrepreneur's acquisition of resources. They are part of the process of selling, acquiring resources, and structuring deals regarding acquisitions and investments. Every entrepreneurial team ideally needs a person who is a master of persuasion [Cialdini, 1993]. Persuasion skills can often influence the behavior of others more than formal power relationships.

Persuasion is governed by basic principles that can be taught, learned, and applied [Cialdini, 2008]. The six principles of persuasion appear in Table 13.2. The principle of *liking* states that people like to please or work with those who sincerely like them also. One can uncover shared interests and bonds and offer sincere praise and compliments.

The second principle of *reciprocity* states that one can elicit the desired behavior from others by displaying it first. Offering help or information to others first can encourage them to reciprocate. The third principle of *social proof* states that people look for and respond to a display of endorsements from those they trust.

The fourth principle of *consistency* states that people stick to their verified commitments—those they make as voluntary, public statements. People who make verbal public or written commitments are likely to stay with them. The fifth principle of *authority* states that people highly regard experts. Therefore, it is useful to show and display your firm's expertise and competencies.

Finally, the principle of *scarcity* states that people want scarce or unique products. Therefore, it is necessary to explain and demonstrate the unique benefits and the chance to gain exclusive advantages.

**TABLE 13.2  Principles of persuasion.**

| 1. Liking |
|---|
| ▪ People like those who like them. |
| ▪ Uncover shared bonds and offer sincere praise and compliments. |

| 2. Reciprocity |
|---|
| ▪ People respond in kind to others. |
| ▪ Give to others what you want to receive. |

| 3. Social proof |
|---|
| ▪ People respond to a display of endorsements by people they trust. |
| ▪ Use testimonials and endorsements from trusted leaders. |

| 4. Consistency |
|---|
| ▪ People adhere to their verified commitments. |
| ▪ Ask for voluntary, public commitments. |

| 5. Authority |
|---|
| ▪ People highly regard experts. |
| ▪ Show and state your expertise. |

| 6. Scarcity |
|---|
| ▪ People want scarce products. |
| ▪ Describe unique benefits. |

*Source:* Cialdini, 2008.

## Celtel: A Better Way

As mobile telephony grew in prominence during the last several decades, the giant telecommunications companies invested in infrastructure worldwide. In this spurt of development, Africa was largely ignored due to its widespread instability, corruption, and poverty. Companies that attempted to do business in Africa faced the unpleasant reality that bribes were often a prerequisite to conducting business. In 1998, Mo Ibrahim, a consultant with experience in African telecom companies, saw this as an opportunity.

Mo founded Celtel, which is a mobile phone service provider, and took an especially effective approach to doing business. Rather than targeting individual African markets, he eliminated inefficiencies by widening the scale of operations to a pancontinental level. This large scale also increased the respectability and political clout of Celtel, allowing him to operate without paying bribes. In this way, Mo was able to utilize the positive forces of influence to achieve his goals rather than resorting to immoral pecuniary incentives.

## 13.3  Location and Cluster Dynamics

Choosing a location can have long-lasting effects on a new venture. Entrepreneurs need to choose their location with their customers, future employees, suppliers, partners, and competitors in mind. While the importance of location to start-ups in the retail and restaurant businesses is obvious, location is important to all companies. Criteria for location selection are listed in Table 13.3. Knowledge-based enterprises, especially, need to locate in an area where skilled employees and complementors are readily available. The cost of doing business will be an important factor in location selection, too. Furthermore, housing that is affordable to all employees is important. A company's home should be a place that current and future employees will like. Thus, it should have good schools, a feeling of safety, and good transportation links.

Location advantages are based on the flow of knowledge, relationships, and access to institutions. Commonly known ideas and technologies that can be accessed from anywhere will be widely available to all competitors. Thus, they cannot serve as a competitive advantage. Local advantages, such as access to knowledge and research at a university, therefore can become a unique advantage [Audretsch et al., 2005]. It is important for companies to be physically near some key constituents [Weterings and Boschma, 2009]. These constituents include customers, competitors, suppliers, and other companies in related fields, as well as venture capital firms. Moreover, ventures that are launched in regions in which their founders have lived longer tend to perform better [Dahl and Sorenson, 2012].

Companies should attempt to proactively link into strong local institutions such as trade associations, universities, and professional societies. Entrepreneurs should take advantage of opportunities in their local region. Long-term competitive advantage relies on being able to avoid imitation by competitors. Location-based advantages in innovation may prove more sustainable than implementing corporate best practices [Porter, 2001].

Firms in industrial regions benefit from localization of cost economies derived from specialist suppliers, a specialist labor pool, and the spread of local knowledge [Best, 2001]. Since new ventures benefit from new knowledge, competent suppliers, and available talent, they should consider locating in a region that offers easy access to all three factors. Entrepreneurial activity differs significantly across regions of a country. Examples of attractive locations for new ventures are provided in Table 13.4.

**TABLE 13.3  Criteria for location selection.**

| | |
|---|---|
| ■ Availability of potential employees and consultants | ■ Costs of doing business |
| ■ Availability of complementary firms | ■ Availability of suitable facilities |
| ■ Road and airplane transportation | ■ Proximity to markets |
| ■ Quality of life—education, culture, recreation | ■ Availability of support services |
| | ■ Affordable housing |

**TABLE 13.4 Selected centers of entrepreneurial activity in technology.**

**Western United States:**
Boulder, Los Angeles, Portland, Salt Lake City, San Diego, San Francisco and Silicon
 Valley, Seattle

**Eastern and Southern United States:**
Atlanta, Austin, Boston, New York, Pittsburgh, Raleigh-Durham, Washington, D.C.

**Asia:**
Bangalore, Beijing, Shanghai, Singapore, Sydney, Taipei

**Europe/Middle East:**
Berlin, Helsinki, London, Tel Aviv

**South America:**
Santiago

A **cluster** is a geographic concentration of interconnected companies in a particular field. Clusters can include companies, suppliers, trade associations, financial institutions, and universities active in a field or industry. A good example is the Hollywood cluster of firms and infrastructure coming together for the creation of movies. If a new venture wants to enter the movie industry, it is probably wise to consider locating in Los Angeles.

An excellent example of an emerging cluster is Israel's high technology cluster, often called "Silicon Wadi." In 2012, venture capitalists invested $867 million in Israel, much of which went to companies based in this cluster. Although this is only a fraction of that spent in Silicon Valley, the cluster is growing rapidly. It provides an environment conducive to technology venture formation: elite universities, venture capital firms, research centers run by big corporations, and a location attracting talented engineers.

A start-up can gain regional advantages by joining a cluster of companies that have complementary or competitive capabilities and resources. A new company within a cluster is more likely to find the employees and infrastructure that it needs [Iansiti and Levien, 2004]. Firms located in clusters have better product innovation performance, sales growth rates, and survival rates. In general, good clusters provide access to ideas, role models, informal forums, and sources of talent [Gilbert et al., 2008].

Clusters promote both competition and cooperation. Firms form alliances, recruit each other's talent, and compete, all at the same time. A cluster also provides new firms with a critical mass of talent, knowledge, and suppliers for easy entry into an industry. A cluster of independent and informally linked companies and institutions represents a robust organizational form that offers advantages in efficiency, effectiveness, and flexibility.

Furthermore, clusters are conducive to new business formation for a variety of reasons. First, barriers to entry are lower than elsewhere. Second, individuals working within a cluster can more easily perceive gaps in products or services around which they can build businesses. Finally, the formation of new ventures creates market opportunities for others. For example, a new venture may introduce a new product that creates the possibility of complementary products offered by other local enterprises.

The availability of firms providing complementary products may be critical to a new venture's success. In a tourism cluster, such as New York City, the quality of a visitor's experience depends not only on the appeal of the primary attraction but also on the quality and efficiency of complementary businesses such as hotels, restaurants, shopping malls, and transportation systems. Because members of the cluster are mutually dependent, good performance by one can improve the success of the others.

The area south of San Francisco called Silicon Valley is a cluster for electronics, medical devices, green tech, and Internet companies. Connectedness and mobility of talent and ideas are a way of life in Silicon Valley. The support structure includes entrepreneurs, venture capitalists, attorneys, consultants, board members, universities, and research centers. Technology firms in Silicon Valley prosper in a dynamic environment of novelty and innovation. This network environment is the outcome of collaborations between individual entrepreneurs, firms, and institutions focused on the pursuit of innovation and its commercialization. Silicon Valley has an openness to change and is supportive of the creative and the different [Florida, 2002]. It displays all of the cluster characteristics that support innovation, as listed in Table 13.5.

The entrepreneurial attitude prevalent in Silicon Valley is exemplified by the views of T. J. Rogers, founder of Cypress Semiconductor, who said [Malone, 2002]:

> What makes us special and different here in Silicon Valley is that we're truly capitalists. We invest. There is no safety net. You can go out of business. You can crash into the wall. There are companies, you can count them on both hands every day, that go out of business, and that's life.

Dynamic, growing industrial regions are constantly upgrading their capabilities and resources, and commercializing innovations. A cluster's boundaries may be defined by the totality of the industry participants and may reach across political boundaries. Regional clusters can be a virtuous circle leading to better opportunities, more venture capital, increasingly educated talent, and more success. The cluster of independent activities can engender dynamic flows of cost reductions and competitive advantage.

**TABLE 13.5  Cluster characteristics that support innovation.**

| | |
|---|---|
| ■ High-quality human resources | ■ Acceptance of globalization |
| ■ Research in local universities | ■ Successful role models |
| ■ Availability of investment capital | ■ Moderate regulation and taxes |
| ■ Representative customers | ■ Suppliers and complementors |
| ■ Rule of law | ■ Competitors |
| ■ Sufficient infrastructure | ■ Consultants, attorneys, and accountants |

Once the entrepreneur chooses the city or region of location, the next task is to find a suitable facility. A building must fit the needs of the organization and allow for expansion as the firm grows. For most new ventures, leasing space in an existing facility is the most economic choice. For some more established firms, owning rather than renting can pay off. However, most start-ups need to use their financial resources for innovation and marketing.

Finally, entrepreneurs should carefully consider the layout within their facility. A **layout** is the arrangement of the facility to provide a productive workplace. This can be accomplished by aligning the form of the space with its use or function. Today's facilities have replaced the private office and laboratory with public spaces and open-plan areas without walls. Since innovation is, in part, a social or collaborative activity, the work spaces are laid out to host team activity. New ventures are best started in an open facility with few walls and doors in order to promote collaboration. Studies have shown that communication between people declines at a rate inversely proportional to the distance between them. For example, we are five times more likely to communicate with someone who sits six feet away as we are with someone who sits 60 feet away [Allen, 2000]. Thus, a firm needs to avoid separate facilities for as long as possible. The center of the facility might be a public area with a coffee bar and meeting tables. The goal is to design the facility for flexibility and collaboration.

## 13.4 Vertical Integration and Outsourcing

New ventures normally have limited financial resources and are unable to internally provide all the functions required for operation of all activities. One way to identify resources and activities that have the potential for creating competitive advantages for a firm is to consider the value chain. The **value chain** of a firm is a sequence of business activities for transforming inputs into outputs that customers value, as depicted in Figure 13.2. The issue for the new venture is to decide which of the activities on the value chain will be accomplished by the firm and which activities will be provided by other firms (outsourced). A new or emerging firm will necessarily focus on a few of the activities on the value chain and outsource the others. **Vertical integration** is the extent to which a firm owns or controls all the value chain activities of a business.

In this chapter, we discuss the value chain and the decision to outsource or retain an activity within the firm. In Chapter 14, we return to the value chain and discuss how to manage and operate it to attain a competitive advantage.

The decision by a new venture to choose to carry out a value chain activity can be based on four questions, shown in Table 13.6 [Barney, 2002]. The decision to focus on a few value chain activities can be aided by an analysis of the four issues: (1) value, (2) rarity, (3) ability to be imitated, and (4) mission and organization. In many cases, it may be necessary to extend the activities of a firm to control a specific activity on the chain, especially when the firm affirmatively answers all four questions of Table 13.6.

**FIGURE 13.2**
Value chain from concept to customer.

Many companies are operating in a hypercompetitive global market where there is overcapacity in most industries. In such an environment, they are being called upon to achieve profitability by relentless cost cutting. This often entails heavy **outsourcing** to lower-cost labor and moving business abroad.

If a firm decides to outsource an activity, it plans on gaining a cost advantage or access to the supplier's superior competency or economy of scale [Weigelt and Sarkar, 2012]. Access to a supplier's superior competency and cost advantage may be favorable. For example, a new venture cannot offer its employees a superior, low-cost, internally operated cafeteria and food service. When it is cheaper and easier to conduct an activity internally, then new ventures may consider taking it on. The best reason, however, to carry out an activity on the value chain is that it is strategically critical to a firm's success. Typically, product design and marketing cannot be outsourced since they are critical to the success of most new technology ventures.

A new venture in the personal computer business may choose to control product design and marketing while outsourcing the other functions after answering the four questions of Table 13.6. On the other hand, a packaged food business would probably attempt to control all product design, manufacturing, and marketing functions while relying on other firms to provide the technology development, distribution, and service activities [Aaker, 2001].

As companies outsource more functions, the scope for competitive differentiation narrows. Almost all routine activities are of low value, not rare, easily imitated, and not central to the mission of a firm. Thus, most new ventures outsource routine services such as payroll, accounting, and other administrative services. The transaction costs of managing outsourced services have recently declined, thanks to cheaper communication and the standardization of Web-based tools. Managing a relationship with a strategically important outsourcing agent is far more complex than coping with an ordinary supplier. In many cases, a partnership arrangement with a critical outsourcing agent is required. Reasons for failure of outsourcing include poor contracts, poor control of the function, and not planning for a termination strategy [Barthelemy, 2003]. Failures can be avoided or managed if accountability for managing the outsourced functions is clearly assigned to one or two persons in the new venture.

**TABLE 13.6** Questions for selecting value chain activities that will be carried out by the new venture firm.

| | | |
|---|---|---|
| 1. | **Value** | Is the activity a primary source of product value for the firm? |
| 2. | **Rarity** | Does the activity include a resource or capability controlled by the firm and rarely available to competing firms? |
| 3. | **Ability to be imitated** | Do competing firms have a cost disadvantage when imitating the scarce resource or capability held by the new venture firm? |
| 4. | **Organizational mission** | Is the activity critical to the mission of the firm, and is the firm organized to exploit this valuable, rare, and costly-to-imitate resource or capability? |

Transaction costs with suppliers and customers can be the most important types of costs. Transaction activities are time-consuming and prone to errors. Thus, companies use technology to automate their purchasing and contracting transactions with their suppliers and providers. Discount broker Charles Schwab & Company entered the securities brokerage market by offering a transaction cost advantage for both its customers and suppliers. Another firm that offers lower transaction costs to its customers and suppliers is FedEx [Spulker, 2004].

Although a new venture should consider outsourcing many of its activities, outsourcing can lead to problems. If an activity is outsourced and the supplier fails to deliver the required result (or activity) on time and with the required performance, the new venture can experience great difficulties. The conditions that favor the internalization of the activity within the new venture are summarized in Table 13.7. If the demand forecast for an activity or component is highly uncertain, it may be better to do the task internally. If there are only a few, powerful suppliers of a service or component, the danger is that they might use their power and not meet the required needs at the agreed-upon cost and time. If a firm's technology is valuable and proprietary, it may decide to keep it internal for reasons of secrecy.

Sometimes firms are led to outsource those value-added functions in which most of their profits will be made in the future. In the struggle for a sustainable competitive advantage, it is important to retain these value-added functions that will be critical to the firm's advantage in the future [Brown, 2005].

Henry Ford built a vertically integrated facility at River Rouge near Detroit in 1917. The self-reliant plant made steel for auto bodies as well as all parts from engines to windshields. From Ford's own forests came wood for the paneling. For Ford, integration meant control of all the activities. A new firm cannot afford this type of vertical integration. It is too expensive and too risky. It is risky because with vertical integration comes control and commitment to a large investment in one way of doing things. Loss of flexibility is risky in a continuously changing economy. Today, Ford Motor Company is responsible only for the design, assembly, and marketing of its vehicles. All the modules and parts are provided by an array of suppliers and partners. In the near future, automobile companies may do only the core tasks of designing, engineering, and marketing vehicles. Everything else, including final assembly, may be done by the parts suppliers.

**TABLE 13.7  Conditions that favor an activity operating internally within the new venture.**

| Factor | Internal activity favored due to: |
|---|---|
| Costs | Total cost to produce internally is lower |
| Demand forecast | High uncertainty |
| Number of suppliers | Few powerful suppliers |
| Proprietary, nonpatented technology | Need to keep secret |
| Value-added function | Source of the firm's future sustainable competitive advantage |

**Salesforce.com's Outsourcing Model**

Salesforce.com uses the "software-as-a-service" model as an application service provider (ASP). The firm offers software for "rent" that is delivered to customers online through a Web browser [Clark, 2003b]. Salesforce.com rents software online that companies use to manage their salespeople and lets other software developers rent out their software using its computers. This is a novel idea. One obstacle to this service is that most corporations do not want sensitive information stored on any computer other than their own. Should a start-up firm use this outsourcing service?

The online grocery service offers important lessons in the challenges and rewards of vertical integration. Webvan was founded in 1999 to offer an extensive line of grocery and nongrocery items for selection online with delivery to the customer's door. Following several rounds of venture capital investments and an initial public offering, Webvan offered same-day delivery in several cities. To provide this service, Webvan found it necessary to build distribution centers. Webvan hoped to minimize costs by setting up a string of futuristic, $35 million warehouses with motorized carousels and robotic product-pulling machines. The cost of its delivery fleets, however, was enormous. With margins as low as 2 percent, Webvan struggled to make a profit. After burning its way through more than $1.2 billion, Webvan closed its doors in July 2001. Webvan failed because it had a flawed business model that required a large investment in expensive distribution and service activities, as shown on the value chain of Figure 13.2.

Tesco of Britain, by contrast, offers an online grocery service that is profitable. Tesco used a new channel, the Internet, to reach its existing customers as well as new ones. Tesco provided online ordering combined with customers picking up their selected and boxed groceries at a Tesco store or paying a delivery charge. Tesco used existing stores, while Webvan built new warehouses and extensive delivery services. This illustrates why efficient operations along the value chain are critical to profitability.

## 13.5  Innovation and Virtual Organizations

A **virtual organization** manages a set of partners and suppliers linked by the Internet, fax, and telephone to provide a source or product. In this case, the value provided by the company is primarily the networking of the participating partners and outsourcing agents. This value, however, is often not rare and can be readily imitated.

Amazon.com now operates the online stores and fulfillment activities of the online operations of Target, Toys-R-Us, and Circuit City. Amazon has become an outsourcing agent of Web services, logistics, and customer service for brick-and-mortar giants. GSI Commerce (www.gsicommerce.com) is an outsourcing company that operates e-commerce businesses for 25 sports retailers, including

Sports Authority and the Athlete's Foot. Global Sports owns the merchandise and manages inventory, fulfillment, and customer service. Each retailer, in turn, receives a single-digit percentage cut of any revenues that come from its site. To consumers, it just looks like they're buying from their local retailer. Global Sports relieves the retailers of the burden of building and maintaining costly e-commerce infrastructure. Moreover, because Global Sports bears all the costs, it allows a client's e-commerce operations to be profitable from day one. Global Sports, in turn, makes money because of the scale of its business. Moreover, Global Sports spends nothing on advertising and marketing. The retailers take care of that activity.

Companies using outsourcing and networks can pull together resources to address specific projects and objectives without having to build permanent organizations. Virtual organizations use computers and networks to build an integrated system. Software applications can be "rented" from an application service provider (ASP), for example.

---

**[24]7 Customer: An Example of Outsourcing**

As a U.S. startup, [24]7 Customer received $22 million in its first round of venture capital from Sequoia Capital in 2003. The firm initially catered to U.S. corporations that wanted to outsource backoffice operations to India. It now provides customer service and technical support via the Internet from India and other countries. [24]7 Customer is now profitable with over 8,000 employees. [Vogelstein, 2003]. (See www.247-inc.com.)

---

New innovative firms can access new ideas from an "idea" marketplace through licensing, alliances, and renting (subscriptions). Flexible, dynamic firms know that the best ideas are not always within their own boundaries. Importing new ideas is a good way to multiply the building blocks of innovation [Rigby and Zook, 2002]. Furthermore, more companies are willing to outsource their ideas and technologies to serve this market.

The creation of virtual organizations brings several challenges with it. Virtual firms encounter difficulty in building trust, coordination, and cohesion among the partner firms and outsource suppliers. New ventures need to invest time and resources in the task of keeping the coordination, trust, and synergy active in a virtual firm [Kirkman et al., 2002].

## 13.6 Acquiring Technology and Knowledge

For many firms, the effective use and management of the outsourcing function can be a competitive advantage. An open-architecture value chain can be a powerful business model [Moore, 2000]. The prudent use of other firms that provide significant contributions to a given need can be productive. For example, an electronic system might be built with an Intel microprocessor, Micron memory, and EMC storage, assembled by Solectron, and distributed by Ingram.

The new venture needs to identify which tasks are *core* and which are *context*. A task is core when its outcome directly affects the firm's competitive advantage. Everything else is context [Gottfredson et al., 2005]. The core/context ratio is a direct measure of effectiveness at generating shareholder value.

The asset base that a firm seeks to leverage through entrepreneurship has shifted over the past decade. The key assets are no longer plants and physical assets but instead are technology, science, and knowledge [Hill et al., 2002]. Entrepreneurs strive to see where new products have become feasible due to the availability of new technologies. With the advent of the new technologies, such as genomics, entrepreneurs see the opportunity for important new medical drugs, for example.

New enterprises also need to import leverage and recombine knowledge bases. Imported knowledge bases can include licensed technologies, purchased technologies, and knowledgeable employees.

A new technology venture will have developed or acquired a basic new technology. This new technology will, it hopes, be the basis for developing and contributing to countless products for many industries. For example, a new venture may possess powerful competencies and knowledge in the science and technology of superconductors. This new technology venture would look for applications in the electric power industry and the electronics industry.

The success of the embodiment of any new basic technology depends on seven characteristics, as shown in Table 13.8 [Burgelman, 2002]. These categories or characteristics are useful because they apply in all industries. For example, a powerful new superconducting technology might provide superconductivity of metal at a low initial cost to semiconductor manufacturers and be easy to use in integrated circuits with very low operating costs, highly-reliable operation and serviceability, and high compatibility with normal circuits. In this case, we would have a very powerful new technology for the electronics industry.

**TABLE 13.8  Seven key characteristics of a new basic technology.**

| | |
|---|---|
| ■ Functional performance: an evaluation of the performance of the basic function | ■ Reliability: service needs and useful lifetime |
| ■ Acquisition cost: initial total cost | ■ Serviceability: time and cost to restore a failed device to service |
| ■ Ease of use: use factors | |
| ■ Operating cost: cost per unit of service provided | ■ Compatibility: fit with other devices within the system |

> **Cisco's Acquisition Methodology**
>
> One of the most active purchasers of technology is Cisco Systems. Between 1993 and 2012, Cisco acquired 159 high-tech companies. Many of these acquisitions were made in 1999 and 2000. During this period, entrepreneurs would talk of founding companies in hopes that they would be acquired by Cisco. Although the number of acquisitions that Cisco made dropped dramatically, purchasing nascent technology remains an important part of its growth strategy.
>
> To facilitate the often arduous task of integrating the acquired firm, Cisco developed and used a documented template that focused on integrating both the people and the technology. Cisco thought carefully about who it would acquire, often taking months to make the decision. The firm did not believe in hostile takeovers and usually acquired geographically proximate companies with "market congruent" visions. Cisco preferred companies that were old enough to have a first product but still young enough that they were not entrenched in their ways or enmeshed with a broad base of customers. Second, Cisco had a no-layoff policy and made a point of keeping the acquired firm's staff. In the late 1990s, Cisco's turnover rate for acquired employees was well below 5 percent, and senior executives were often folded into Cisco's senior ranks.
>
> On the technology side, the R & D and product organizations were integrated with Cisco's other products and immediately labeled with the Cisco brand. In addition, any nonstandard technology was eliminated from the acquired firm, and its employees were given immediate access to Cisco's own infrastructure and core applications. The result was that most of Cisco acquisitions were fully integrated within 60 to 100 days.

## 13.7 Spotlight on Netflix

Netflix Inc. is a provider of DVDs by mail and of on-demand streaming media. The company was incorporated in 1997 and has its headquarters in Los Gatos, California. Today, Netflix offers its 50 million members access to a collection of more than 100,000 titles. Netflix's competitors include Amazon, Dish Network, Comcast and Hulu.

Netflix has acquired resources and capabilities continuously, as described by Figure 13.1. Its location in Silicon Valley enables it to take advantage of cluster dynamics through the availability of employees and proximity to firms with complementary resources. Netflix also outsources some functions, such as the delivery of DVDs.

In the highly-competitive market for streaming media, Netflix has attempted to acquire and employ new technologies. In 2013, it announced $100,000 in prizes to architects and developers who make improvements in cloud technology.

## 13.8  Summary

Successful entrepreneurs are good at locating and acquiring the resources they need to start and build their firm. They need capital, people, and intellectual and physical assets to launch and grow their business. They do this by building credibility and legitimacy with the sources of these scarce resources. Typically, they are good at telling persuasive stories about their vision and its potential. They use their skills of persuasion to acquire the required resources in a timely way.

Entrepreneurs also create a plan for outsourcing some functions while retaining critical functions, such as product design and marketing. They use the Internet to help communicate and manage their relationships with their partners and suppliers in a virtually integrated firm.

> **Principle 13**
> Effective new ventures use their persuasion skills, credibility, and location advantages to secure the required resources for their firm in order to build a well-coordinated mix of outsourced and internal functions.

## Video Resources

Visit http://techventures.stanford.edu to view experts discussing content from this chapter.

| | | |
|---|---|---|
| Outsourcing | Jeff Hawkins | Palm |
| Reflections on Combining Companies | Phil Libin | Evernote |
| Being Small Inside of Big | Teresa Briggs | Deloitte |

## 13.9  Exercises

**13.1**   In 2005, Google released Google Earth and the Google Map APIs. Google brought compelling mapping and visualization functions within reach of all Web developers. In 2006, Amazon released a grid storage Web service called S3 that stands for Simple Storage Service. Amazon positions the service as a highly scalable, reliable, and low-latency data storage infrastructure at very low costs. In 2007, Facebook opened up its social network to third-party widgets via Facebook developer APIs. What business models are these three companies pursuing with their products? How are these types of Web services impacting the resource acquisition strategies of new ventures? Have these strategies been successful for Google, Amazon, and Facebook?

**13.2**   Determine who is or was one of the most persuasive people you have known. Using Table 13.2, describe how this person exercised his or her sources of legitimacy.

**13.3** Four of the most popular U.S. locations for technology firms are Boston, the San Francisco Bay area, Austin, and Seattle. Using Table 13.3, determine the most attractive location for an orthopedic medical devices start-up. Do the same for a clean tech start-up.

**13.4** Identify a local start-up company (or select one from another part of the world). Does this company operate in a specific industrial cluster? What local resources or local advantages does the company leverage? Why is the company located where it is?

**13.5** There is a strong trend in many information technology fields of outsourcing to another country and continent. What are the primary motivations for this movement? How and why are start-ups participating in this trend? What are some of the risks for a new enterprise considering outsourcing?

# VENTURE CHALLENGE

**1.** How do you plan to attract talent and resources?

**2.** What location have you selected? Describe your planned use of Internet technologies and commerce.

**3.** Describe what functions your venture will outsource.

# CHAPTER 14

# Management of Operations

*Real intelligence is a creative use of knowledge, not merely an accumulation of facts.*
**D. Kenneth Winebrenner**

**How do new firms build a set of operational processes that serve to create, make, and provide the product to the customer?**

Most businesses build a chain of activities that add value at each section of the chain. Each element of the value chain has a capability that provides value added to the product. A new venture manages its value chain to provide the ultimate product to the customer. The firm also moves, stores, and tracks parts and materials to its value-adding partners and strives to ensure timely, efficient production of the service or product. Information flow along the value chain enables the coordination of the distributed tasks. An effective enterprise manages for operational excellence by trying to develop and communicate measurements of efficiency and timeliness.

Using the Internet and related technologies, new enterprises can build a powerful virtual organization. The Internet can be used to communicate with a set of partners in a quick and effective manner. This set of partners can work from a common schedule, manage how jointly assigned tasks are allocated, and see how interdependencies impact a common schedule. Interrelated tasks can be better managed across firms, which results in better forecasting and stronger chances of achieving on-time production. ■

## 14.1 The Value Chain

As discussed in Chapter 13, the purpose of a firm is to provide products that customers value. A **value chain** is a series of activities for transforming inputs into outputs that customers value. Each value chain activity adds value, as shown in Figure 14.1. Information flows back from the customer and the sales and service activities so that the value chain can maximize value for the customer. More than merely things with features, products are increasingly viewed as things with features bundled with services. Products and services are grounded in activities and relationships in a value-creating system. Furthermore, each element of the value chain has certain capabilities that can be improved over time. Capability development along the chain and the design of the chain can lead to a powerful core competency for a new venture. Furthermore, customers participate in this value-creating process by communicating their preferences and priorities. Understanding the customer enables the producer to better match the customer's needs, as described in Table 14.1.

A highly integrated company provides most, if not all, of the functions along a value chain. This approach is most suitable when proprietary interdependent activities occur at each stage of the chain. As many industries mature, the functions along the chain become independent so that modular subproducts are available at each stage. At that point in time, the value chain breaks up, and a number of independent firms participate in the activity chain. For example, today's automakers are adopting modular architectures for their mainstream models. Rather than putting together individual components from diverse suppliers, they are procuring subsystems from fewer suppliers. The architecture within each subsystem (braking, steering, chassis) is becoming progressively more interdependent as these suppliers strive to meet the auto assembler's performance and cost demands [Christensen et al., 2001].

Every industry has its own rate of evolution that erodes its competitive advantage. In a fast-changing industry, a firm must have the ability to readily redesign its value chain to find new sources of competitive advantage [Fine et al., 2002]. In designing or redesigning a value chain, each stage of the value chain can be assigned an economic value-added measure (EVA), which accounts for knowledge assets and strategic assets. Strategic assets are those in which the firm has relative competitive advantage. Strategic assets may include logistics, manufacturing, and distribution assets. Knowledge assets will exist primarily in the research design, marketing, and service functions. A new firm should retain the

**FIGURE 14.1** Value chain and information flow.

## TABLE 14.1 Understanding the customer.

| | |
|---|---|
| ■ Preferences | ■ Buyer behavior |
| ■ Purchase criteria | ■ Functional needs |
| ■ Decision-making process | |

functions with high EVA and outsource functions with low EVA. If the industry is changing rapidly, however, a firm may decide to retain a key function to strategically respond to the change internally. Often a sound approach is to retain the high EVA activities and key strategic assets while outsourcing the low EVA activities.

Vertical integration along the value chain provides firms with an opportunity to choose the value-added stages in which it will compete. Intel is a manufacturer of both integrated circuits and circuit boards, but it also assembles personal computers under an original equipment manufacturer (OEM) agreement with PC companies.

Zara, a European clothing retail firm with 1,000 stores, has retained its manufacturing capability rather than outsourcing it so it can respond quickly to changing fashion demands. Other firms may be able to make the clothes more cheaply, but the strategic asset retained by Zara is the ability to quickly deliver new fashions to its stores. Information flows (see Figure 14.1) from the store floors back to the designers, who redesign the products to fit the customers' changing ideas and tastes. In fashion, nothing is as important as time to market [Helft, 2002]. Zara has new designs arriving in its stores every two weeks. Many new designs arrive at the stores within a few days.

**Logistics** is the organization of moving, storing, and tracking parts, materials, and equipment. Logistics can be the basis of competitive advantage since a firm with fast, accurate logistics will be first to respond to customers. Logistics systems usually are based on electronic networks, such as a supply chain intranet. Companies look for unique ways to service customers quickly through improved tracking, transporting, handling, and delivery in an effort to create unique competencies. Apple is a good example of a company that uses logistics as a competitive advantage. Although Apple does some final assembly, it works through suppliers and partners to build almost all of its products. Apple benefits from not having to carry its own inventory and it can gain significant discounts from its suppliers as a result of the volume commitments that Apple can make. In addition, Apple uses its volume to gain significant reductions in transportation costs from air and land carriers. Apple is becoming so big, in fact, that the company can even get custom parts for noncustom prices because its suppliers experience economies of scale even though they are producing for only one company.

Logistics might sound like a simple business of moving things around, but it is growing more complex as customers demand timely, customized services. New technology and greater use of the Internet opened up new ways of sharing information. Companies are also trying to build only after receiving orders from customers (known as built-to-order, or BTO), rather than estimating what will be in demand and supplying it from accumulated stocks. The BTO concept tries to avoid producing any item without a firm purchase order. One of the best examples of value chain management is Modcloth. Modcloth enables customers to specify their desired product and to pay for it before Modcloth starts working with outsourced manufacturers to assemble the clothing.

The information flow along the value chain can be facilitated by the Internet [Hammer, 2001]. Using this information and working closely with partners along the chain to design and manage the activities, efficiencies can be improved. Who captures the profits from these efficiencies? Often companies such as Wal-Mart and Home Depot have assumed the distribution, sales, and service functions, and captured significant value created by their suppliers. Furthermore, by using the Internet, companies like Amazon and eBay have become electronic sales and distribution channels. At the same time, companies like Nokia are producing a seamless manufacturing, distribution, sales, and service offering. Managing the value chain is a challenging task for all new business ventures, which can ill afford to assume many of the value chain activities.

Wal-Mart has a profit margin less than 4 percent, and many supermarket chains have a profit margin less than 2 percent. Clearly, every penny saved is a penny earned. The bar code became a mainstream success after Wal-Mart adopted it in 1980. Now Wal-Mart will require that its suppliers use radio-frequency-identification (RFID) on their supply pallets. RFID relies on a computer chip to hold and convey information. Wal-Mart manages its supply chain as if it were an orchestra.

Intermediaries in a value chain make sense only for exchanges in which the parties to the transaction can save more money by hiring the intermediary than it costs. It can cost 20 percent of revenues to "hire" a retailer to sell a product. Is it cheaper to sell direct? Can the Internet reach the same customers effectively?

Value chain speed is important to any new venture. With a long lead time from design to the customer, a new firm may wind up with inventory it cannot sell. Since many products have short life cycles, the chain must be able to move fast. Zara can design and manufacture a new clothing fashion in one week, if necessary. On the other hand, a movie studio can take a year to produce a new film and introduce it into the market. One of the best examples of value chain management is the Dell Computer system, which enables customers to specify their desired product and pay for it before Dell starts assembling the computer.

---

**E-Ink Holding Systems: A Tablet and E-Reader Value Chain Participant**

E-Ink Holdings pioneered electronic paper technology that provides up to a 50-percent better contrast ratio than competing alternatives. E-Ink licenses this technology to Sony, Motorola, Amazon, and Barnes & Noble. These corporations utilize E-Ink's technology in their tablets and readers so the users of the products have an easier time reading their displays.

While goods and services flow largely down the chain, information flows in both directions. For example, information about what is demanded at successive stages passes up the chain, while information about supply conditions such as availability, pricing, time-to-manufacture, and so on passes down. Because this is an information-intensive process, the Internet holds the potential to significantly increase the amount of value created.

A new venture's lasting core competency is its ability to continuously assess industrial and technological dynamics, build value chains that exploit current opportunities, and select the high value-added activities for operation by the firm itself.

## 14.2 Processes and Operations Management

An **operation** is a series of actions, and **operations management** is the supervising, monitoring, and coordinating of the activities of a firm carried out along the value chain. Operations management deals with processes that produce goods and services. A **process** is any activity or set of activities that takes one or more inputs, transforms and adds value to them, and provides one or more outputs [Krajewski and Ritzman, 2002]. Processes are the series of operations, methods, actions, tasks, or functions leading to the creation of an end product or service. Processes are used to transfer value to customers in the form of products. A product could involve delivering a tangible good to customers, it could involve performing a service for customers, or it could be, and usually is, a combination of the two [Melnyk and Swink, 2002]. At a factory, a process transforms materials into products. At an insurance company, a process transforms client information into an insurance agreement. A network of processes helps create the value provided at each stage of the value chain. Business processes can add unique value to products and services.

The business processes of an organization should be aligned with its strategies and employee competencies, as shown in Figure 14.2. Alignment requires a continual rebalancing of strategy, business processes, and the competencies of the people to satisfy and retain the customers in order to keep the business theory as the clear driving force of the business, as shown in Figure 14.2. The business theory should be expressed as a clear statement of vision and purpose. Recall the vision statement of eBay: "We help people trade practically anything on Earth through an online system."

**Ergonomics** is making a physical task easier and less stressful to accomplish. Products should be designed ergonomically, and tools used in factories need to be ergonomic. For example, Herman Miller's Aeron chair has been acclaimed as an ergonomically pleasing product. It provides form-fitting support and maximum comfort due to a suspension system that allows for proper ventilation.

Processes bring value to customers and stakeholders. Business success comes, in large part, from a company's process performance. Therefore, a firm should strive for superior process design. An example of a simple business process is shown in Figure 14.3. Part of this process can be automated.

**Intel's Operations Strategy**

Gordon Moore and Robert Noyce founded Intel in 1968. Their first act was to recruit a director of operations. They offered the job to Andrew S. Grove, who became Intel's third employee. Though Grove had no manufacturing experience, they recognized his innate intelligence and drive. Grove was responsible for getting products designed on schedule and built within budget. The scope of this position extended into nearly every functional area at Intel, from marketing to sales to engineering. His influence, presence, and attitudes pervaded the company, and within three years of Intel's formation, it became clear that the majority of daily decision-making was passing to Grove. Operations has always played a critical role in Intel's success.

**FIGURE 14.2** Business alignment.

Functions in operations management include design of processes, quality control, capacity, and operations infrastructure. The new venture needs to plan for its operations or production function, which will be led by one member of the entrepreneurial team. Both service and product firms need to design and control operational processes to achieve efficiencies, throughput, capacity availability, inventories, capital expenditures, and productivity. Unique operational management competencies can be part of the competitive advantage of a firm [Vonderembse and White, 2004].

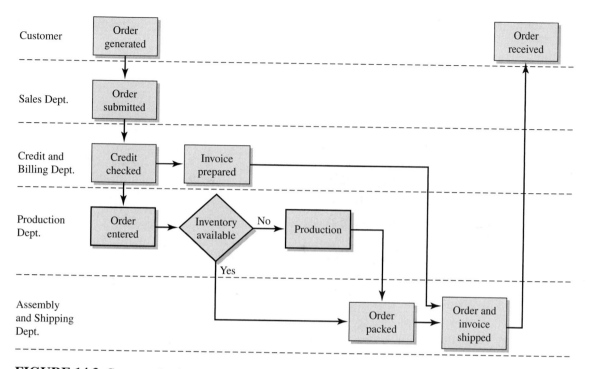

**FIGURE 14.3** Common business process.

The best return on investment is often found in companies that combine operational excellence—consistently outstanding performance for customers that is brought to the bottom line—with sustained rapid growth. Operational excellence is a necessity today. Investors mercilessly punish companies that fail to meet these expectations [Lucier and Dyer, 2003]. Operational excellence can lead to lower costs, as shown in Figure 14.4. With economies of scale, unit costs will drop, and thus, prices can decline. As prices drop, the product is more attractive and sales increase. As financial resources increase, investments in marketing and operational processes will lead to better economies of scale. This can be a very powerful self-reinforcing loop.

A firm can implement four competitive capabilities: low cost, high quality, speed, and flexibility. **Quality** is a measure of a product that usually includes performance and reliability. **Performance** is the degree to which a product meets or exceeds certain operating characteristics. **Reliability** is a measure of how long a product performs before it fails. **On-time speed** measures the pace of lead time, on-time delivery, and product development. **Flexibility** is a measure of a firm's ability to react to a customer's needs quickly. All of these goals need to be achieved while operating with the additional goal of worker safety.

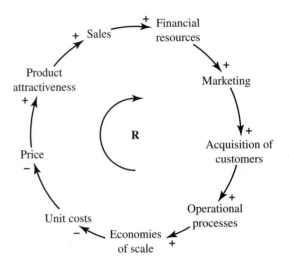

**FIGURE 14.4** Self-reinforcing growth through acquisition of customers. A self-reinforcing loop, R, is assumed to work as long as the economies of scale are actually realized through effective operational processes.

Many firms have adopted a six-sigma quality goal, with the aim of getting rid of defects in a process. Six sigma is a statistical term that measures how much a process deviates from the ideal. Six-sigma quality equals just 3.4 flaws per million. The six-sigma method attempts to build low-defect products with low costs [El-Haik and Roy, 2005]. Six-sigma quality is the result of a well-defined and structured process that is highly repeatable—a process with well-defined tasks and milestones.

Consumers often react favorably or unfavorably to the experience they have with a product or service, such as the packaging, clarity of an operating manual, or ease of use. Many customers view consumption as an experience rather than a singular purchase event [LaSalle and Britton, 2003]. Most products or services include *objective value,* such as performance, and *subjective value,* such as experience. A candle purchased for light might cost only $1, while a second candle with shape, form, and scent may provide a richer experience (and sell for $5 to $10). Thus, for many products, the value of the product involves the experience of it and the associated purchase and fulfillment process. The designer needs to identify, design, and fulfill a customer experience that will register with the customer as positive.

**Supply-chain management** is focused on the synchronization of a firm's processes and those of its suppliers to match the flow of materials, sources, and information that meets customer demand. Today, the goal is to minimize the stock of goods or inventory required to support variability of customer demand. As products become more susceptible to changing demands, the risk grows that a given product line will have disappointing sales. But if a manufacturer decides to go lean on inventories, it risks running out of stock, losing sales, and endangering relationships with its customers.

Operations systems that are designed to create efficient processes by using a total systems perspective are called **lean systems.** Flexible or lean systems aim to reduce setup times and increase the utilization of key processes. Flexible systems can quickly respond to changes in demand, supply, or processes with little cost or time penalty. They often use a **just-in-time** (JIT) approach that focuses on reducing unnecessary inventory and removing non-value-added activities. This system uses a pull method in which the customer activates production. For example, an order for an auto chassis at a carmaker activates the manufacturing processes [Liker, 2004].

The Taguchi method is used to design and improve production systems. The Taguchi method is a technique for designing experiments that converge on a near-optimal solution for a robust system. The method uses the term *noise* to describe uncontrolled variations and states that a quality product should be robust to noise factors. The design of a firm's production system is critical to its overall success [Ulrich and Eppinger, 2004].

Companies strive for quality service at low cost. Google, the Web search firm, handled 7 billion page views a day in 2012 with a hardware plant of about one million servers (computers) that cost about $2,000 apiece. When a server fails, Google pulls it and replaces it immediately. Google has no fix-it department; it just pulls and inserts. With this approach, Google saves funds and keeps its system up 99.99 percent of the time [Barroso et al., 2003].

Entrepreneurial firms seek to develop unique capabilities by fostering an interactive dynamic between their capabilities and market opportunities. The market provides them with signals of opportunities, and they respond with new products. Adjustment of business models, production capabilities, and skill formation enable the firm to respond to opportunity. In value networks, firms focus on their core competencies and use others for complementary capabilities.

The goal of operations **throughput efficiency** (TE) may be measured by the formula

$$TE = \frac{VA}{VA + NVA} \tag{14.1}$$

where VA = value-adding time and NVA = non-value-adding time. Examples of NVA are waiting in a queue or system downtime. The goal is to reduce NVA.

Gentex, which makes rearview mirrors and other auto equipment, uses technology to control costs and increase throughput (see www.gentex.com). Gentex continually adds automation and monitoring systems in its factories and increased its throughput 30 percent from 2001 to 2003 [Green, 2003].

> **Improving Operations in the Operating Room**
>
> Intuitive Surgical, Inc., makes surgical robots used in prostate and heart bypass surgeries. When they are given the opportunity to test drive the da Vinci robot, surgeons quickly become enamored with the level of control they achieve, using only a small incision. Although Intuitive Surgical's robots have been enthusiastically embraced, the company continues to find ways to make the surgical experience better for doctors and patients. The small incisions improve overall efficiency in the healthcare system, since they lead to more rapid recovery times for patients. Intuitive is looking at ways to provide ultrasound and other diagnostic images on the same screen that the surgeon uses to control the robot. This would improve efficiency by letting the surgeon focus on only one image, and improve the real-time surgical experience by making sure that all relevant information is only a click away (www.intuitivesurgical.com).

## 14.3 The Value Web

A series of business activities can be thought of as a business process, as shown in Figure 14.5. Another way to describe a process is as a set of interrelated tasks accomplished in a network of activities. Instead of a value chain—a linear series of processes—the value-creating process can be organized as a **value web.** Webs are grids with no center but allow open communication and movement of items and ideas. In a web, each participant focuses on a limited set of core competencies [Tapscott et al., 2000]. A value web is usually based on an Internet infrastructure to manage operations dispersed in many firms. The value web consists of the extended enterprise within a network of interrelated stakeholders that create, sustain, and enhance its value-creating capacity. The long-term success of a firm is determined by its ability to establish and maintain relationships within its entire network of stakeholders. It is relationships rather than transactions that are the ultimate sources of organizational wealth [Post, 2002]. The value web of a typical firm is shown in Figure 14.6.

The value web organized and operated by Amazon includes participants such as Ingram, Target, and Toys-R-Us. Amazon takes responsibility for choosing and offering the product selections, setting prices, and ensuring fulfillment. Cisco Systems leads a value web that provides its routers and computers to its customers. Cisco designs and markets the product, while others do most of the manufacturing, fulfillment, and on-site customer service. The Cisco Systems value web is shown in Figure 14.7. Recall from earlier chapters that CRM means customer relationship management. Cisco defines the goals and coordinates the integration of the value web providers. Many new ventures will use the Internet to coordinate their value web effectively.

Consider the operations management strategy of IKEA, the Swedish furniture company that has 332 stores in 38 countries. Its business strategy is to make and sell inexpensive, solid, well-designed furniture through large stores.

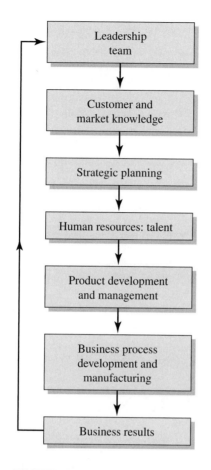

**FIGURE 14.5** A business process
is a series of activities.

Its business process starts with the identification of a needed product and the
specification of a low target price for this item. Next, IKEA determines what
materials will be used and what manufacturer will do the assembly work. IKEA
buys from about 1,800 suppliers in 55 countries. The next step is to design the
item and select the parts. After manufacture, the item is shipped disassembled
in a flat cardboard box to one of IKEA's 18 distribution centers and ultimately
to one of its stores. IKEA sells its unassembled furniture without salespeople.
The customer selects the item, gets the correct box from its rack, and brings it
to the checkout counter. IKEA implements a low-cost, quality strategy through
a far-flung value network that provides design, parts, and manufacturing in a
coordinated manner.

In the past as companies grew, they added assets. As companies grow today,
they tend to add relationships and enhance their value web. Orchestrating a value

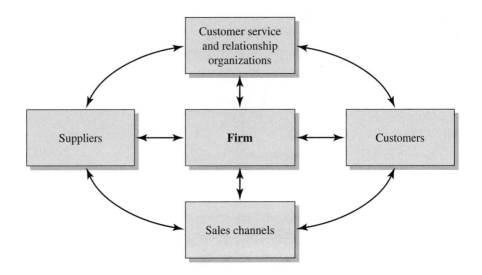

**FIGURE 14.6** Value web for a firm.

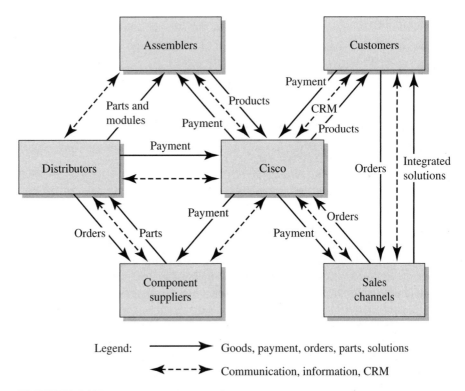

**FIGURE 14.7** Value web of Cisco Systems. (Adapted from Slywotzky and Morrison, 2000.)

web is a powerful process for growth. Of course, it is possible to develop too complex a web of partners and lose control. Effective management of the value web enterprise requires a new conception of the firm as a network, rather than a hierarchy. The key to effective implementation is recognition of value web management as a core competence.

In a fast-paced, competitive world, competitive advantage can result from the effective concurrent design of products, processes, and capabilities. Designing the product, how it is produced, and a supply chain that works harmoniously is critical to a firm's success. The firm that controls the interdependent links in the value web captures the most profit [Lawrence et al., 2005]. The coordinated product, process, and supply chain system is depicted in Figure 14.8 [Fine et al., 2002].

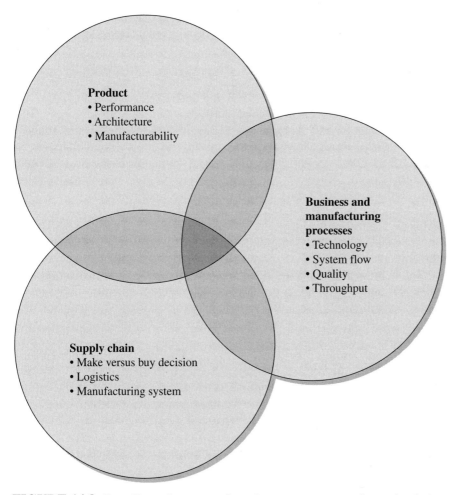

**FIGURE 14.8** Coordinated system of product, processes, and supply chain.

## 14.4 The Internet and Operations

Networks of people communicate, in part, using technological means such as the Internet. The use of the Internet is widespread. It is estimated that about 80 percent of the North American population and two-thirds of the European population used the Internet in 2013. Similar growth of the use of Internet is occurring worldwide. The Internet is ubiquitous, cheap, and standardized, and it accommodates data, voice, video, and e-mail. It readily allows individuals to search, collaborate, coordinate, and transact. Thus, a new business venture can use the Internet to synchronize its activities through different channels, through different stages of the value experience, and across different offerings [Sawhney and Zabin, 2001]. Perhaps the most revolutionary aspect of the Internet is that it gives virtually everybody access to the same information. It is this transparency that has caused the shift in power from sellers to buyers. Moreover, because transaction costs can be much lower on the Internet than in traditional channels, companies can shift some or all of their business and supplier functions onto the Web.

E-commerce involves digitally enabled commercial transactions between and among organizations and individuals. E-commerce grew using Internet technology starting in 1995. Between 1998 and 2000, venture capitalists invested $120 billion in about 12,500 Internet start-ups, often called "dot.coms." Successful dot.coms include Amazon.com, eBay, and Yahoo. Jeff Bezos founded Amazon in 1995 as an online bookstore, raising several million dollars for the venture. Amazon offers convenience, low prices, and a wide selection of books and other items. In May 1997, Amazon raised $50 million by selling its shares to the public for the first time. By 2000, it held a clear position as the world's largest online retailer. Another use of the Internet is the distribution of information in the form of a magazine or a newsletter. The *New York Times* and *The Economist* have successfully offered an online version in addition to their print editions.

The Internet also facilitates three important functions: personalization, customization, and versioning [Luenberger, 2006]. **Personalization** is the provision of content specific to a user's preferences and interests. It uses software programs to find patterns in customer choices and to extrapolate from them. For example, Amazon provides personalized book and music recommendations. **Customization** is providing a product customized to a user's preferences. Navigenics offers health-risk profiles customized to the genetics of its users. **Versioning** is the creation of multiple versions of a product and the sale of these modified versions to different market segments at different prices. *The New York Times* offers a free online version of today's newspaper but charges per article for archived material.

While personalization and customization are good ideas, there are some concerns. Many users report that they do not notice any differences in the site due to personalization. Often, attempts to customize are cumbersome and imperfect. Customization is a more powerful approach if executed well since the customer is actively involved in the selection process. Customization is a powerful tool when customers want their preferences converted directly into a specific form of product.

A central activity of the Internet is the search process that enables users to find information on a seemingly infinite set of items, ideas, terms, and issues. A buyer of airline tickets can use Expedia, Travelocity, and Orbitz to search for a bargain. Travel services appear to be an ideal service/product for the Internet since travel is an information-intensive product requiring significant consumer research.

Not all customers want to do business online; most prefer having a choice of various ways. The **hybrid model,** sometimes called "bricks and clicks" or "clicks and mortar," utilizes the best of the Internet as well as other channels. A hybrid model can extend a company's reach to new market segments as well as its global reach. Alliances are set up to combine the functions served by each company, such as the partnership between Drugstore.com and Amazon early in their corporate histories.

The advantages of e-commerce are low transaction costs, ubiquity, wide reach, and massive information. Since product and price information are readily available on the Web, the pricing power of many industries has diminished. Furthermore, many early e-commerce ventures underpriced their products in order to secure customers but never showed a profit and eventually failed. A firm must have a competitive advantage to sustain itself, and very low prices may not permit profitability. Because of the wide reach of the Internet, many competitors can imitate successful offerings, thus eroding the competitive advantage of any one firm.

Internet ventures have the potential to offer extensive selection of their wares. For niche markets, Internet providers can offer an advantage over physical stores. Amazon and the iTunes Music Store exploit this advantage [*Economist*, 2012]. Apple's iTunes has become the largest music retailer in the world.

Jeff Bezos realized that Amazon had a unique competency that could provide value to other, especially brick-and-mortar, companies. Amazon had a huge online infrastructure and significant experience with direct-to-customer shipping. With these advantages in Internet retailing, it began to offer its services to other companies. A full spectrum of retail services were rolled out for individuals, small and medium businesses, and even large corporations. Amazon Commerce is Amazon's partnership program with major traditional retailers such as Target and Sears. Through this program, Amazon creates a branded website, handles online sales, and provides customer support and order fulfillment. This arrangement yields additional revenue for Amazon and gives traditional retailers access to a critical distribution channel—the Internet.

## 14.5  Strategic Control and Operations

**Strategic control** is the process used by firms to monitor their activities, evaluate the efficiency and performance of these activities, and take corrective action to improve performance, if necessary. The goal is to keep the firm's operations on track with the performance goals of efficiency, quality, and responsiveness to customers.

To evaluate the effectiveness of their strategies, some companies develop a **balanced scorecard,** a set of measurements unique to a company that includes both financial and operational metrics. This gives managers a quick yet comprehensive picture of the company's total performance. The balanced scorecard is a strategy formulation device as well as a report of performance. A successful balanced scorecard measures the tangible objectives that are consistent with meeting an organization's goals. The business operations area indicates how the operations and processes should work to add value to customers. The customer area indicates how the company's customer-oriented strategy and operations add financial value. The financial area measures the company's success in adding value to shareholders. The learning and growth area indicates how the infrastructure for innovation and long-term growth should contribute to strategic goals. A balanced scorecard appears in Figure 14.9 [Kaplan and Norton, 2004].

To build an effective scorecard, a firm needs to determine the fundamental drivers of performance and measure them. Finding the right measures such as reliability, quality, or customer satisfaction is challenging, but can yield big payoffs. Jack Welch of General Electric promoted the company's slogan "Finding a Better Way Every Day," which addressed improvement of business processes. Then, he added the idea of measuring the performance of the processes. As a result, operating margins went from 1.2 percent in 1994 to 13.8 percent in 2000 [Welch, 2002]. Welch believed that the system of operations was the key to understanding, learning, and improving results.

---

**General Electric's Digital Cockpits**

General Electric Vice Chairman Gary Rogers created the idea for a digital dashboard—the continuously updated online display of a company's vital statistics [Tedeschi, 2003]. GE's "digital cockpits" now give hundreds of managers instant access to the company's essential data on computers, tablets, and smartphones [Tedeschi, 2003].

---

New firms also should develop diagrams and flowcharts that show how their operations work. An operational plan outlines a number of actions that will be taken in the future. To consolidate the timing of events, the firm should prepare a schedule, in chart form, of all of the important milestones that the firm

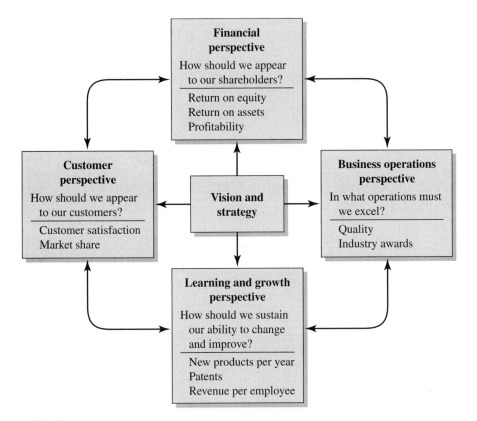

**FIGURE 14.9** Balanced scorecard. Each perspective has a question and a set of measures. [Adapted from Kaplan and Norton, 2004.]

expects to reach in the near and intermediate term. A *Gantt chart* is a way to depict the sequence of tasks and the time required for each. Gantt charts, by using shaded bars on a grid, compare what was done with what was planned over time. Timelines are a visual means of comparing the actual and planned progress of a project or activity. Timelines allow participants to envision the ending of an otherwise open-ended plan [Yakura, 2002].

Enterprises of any size can profit from using Gantt charts to depict schedules and milestones. For example, completion of task B could represent the completion of a prototype, and task C could represent the testing of the prototype (see Figure 14.10). It is important to set milestones and then strive to reach them on time.

Another, perhaps more graphic, form of representing the activities, outcomes, and schedule is to use a milestone picture, as shown in Figure 14.11. This depicts a road map to achieve a scheduled outcome.

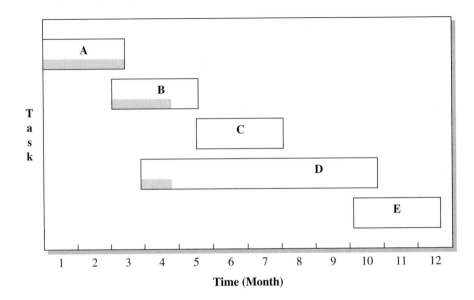

**FIGURE 14.10** Gantt chart for five tasks. The actual progress is indicated by the shaded bars.

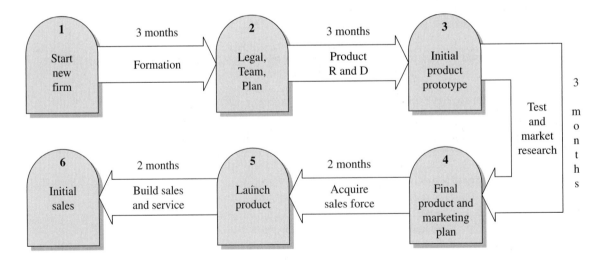

**FIGURE 14.11** Example of a road map with milestones for a new technology firm.

## 14.6  Spotlight on Clean Harbors

Clean Harbors, Inc. is a provider of environmental, energy, and industrial services, including hazardous waste disposal. Its clients include Fortune 500 companies and federal, state, provincial, and local governments. Current CEO Alan McKim founded the company in Brockton, Massachusetts, in 1980. Clean Harbors has since expanded to more than 470 locations throughout North America.

Effective management of operations has been a key to Clean Harbor's growth. For example, the company was a pioneer in the use of GPS tracking for fleet vehicles, which enabled it to reduce vehicle loss and to allocate resources more effectively. The company also has undertaken a variety of acquisitions in order to expand its resources and capabilities. For example, in the 1990s and 2000s, it acquired a number of incinerator and treatment companies; in 2009, it acquired Eveready, Inc. which provided services for the oil and gas industry; and in 2012 it acquired Safety-Kleen, which had expertise in oil recycling and re-refining.

Through these acquisitions, Clean Harbor has become increasingly vertically integrated. Since Clean Harbor is a service organization that works with physical products, it operates a large number of locations. These many locations enable Clean Harbor to have resources close to most any point of need. They also allow the company to engage with a wide range of customers and partners. In turn, these relationships enable Clean Harbor to acquire new technology and knowledge.

## 14.7  Summary

A new venture needs to design a set of operational processes that will enable it to build, store, and ship the products provided to the customer. New businesses build a supply chain of partners that add value at each stage of the assembly or manufacture of the product. Service companies use business processes to put together their service outputs. The new venture manages its value chain to effectively provide the final product or service to its customer. The firm also needs to effectively manage the logistics of parts and materials. It strives to achieve the best possible coordination of its partners as well as its internal processes.

Many firms establish a set of interrelated activities as a network facilitated by an Internet-based value web. With a common schedule, associated tasks, and synchronization, the venture can manage the value web to maintain an efficient, on-time business process.

---

**Principle 14**
The design and management of an efficient, real-time set of production, logistical, and business processes can become a sustainable competitive advantage for a new enterprise.

## Video Resources

Visit http://techventures.stanford.edu to view experts discussing content from this chapter.

| | | |
|---|---|---|
| Overcoming Some of the Early Mistakes of Dell | Michael Dell | Dell |
| Develop Operating Range | Sukhinder Singh Cassidy | Joyus |
| Navigating AgraQuest's Value Chain | Pam Marrone | Agraquest |

## 14.8  Exercises

**14.1**  Select a start-up of your choice. Using the format of Figure 14.6, prepare a value web diagram for this company.

**14.2**  Flextronics manufactures products for many electronic companies (www.flextronics.com). It is a global company, headquartered in Singapore, and has over 176,000 employees. The company's core competence is lean manufacturing. Where does Flextronix fit in the value web for Microsoft, Ericsson, and Dell?

**14.3**  China is a world-class manufacturer and has advantages of size, scale and cost. Identify a Chinese assembler of electronic products for European companies. What advantages does China have versus manufacturing in Europe?

**14.4**  The use of a new technology can bring new life to a mature industry such as the plastics industry. Logistics, supply-chain, and scheduling software enable large productivity increases in several mature manufacturing industries. Examples of such software firms are i2, Moldflow, Quad, and Keane. Select one of these firms and describe an actual application for operations productivity improvement.

**14.5**  About 5 to 10 percent of pharmaceuticals produced do not meet specifications and have to be reworked or discarded. Quality testing is done by hand, and the batch process method is widely used. A new venture has been launched to design new processes for drug makers. What new methods and approaches should it develop to sell to drug makers?

**14.6**  Describe and contrast the operational challenges faced by the following start-ups: (a) consumer Web services start-up, (b) iPhone application company, and (c) electronic device company.

**14.7**  Describe and contrast the operational challenges faced by the following start-ups: (a) new drug discovery company, (b) medical device company, and (c) biofuel company.

## VENTURE CHALLENGE

1. Describe the key business processes used by your venture.

2. Draw a road map with milestones for your venture as illustrated in Figure 14.11.

3. Draft a value web for your venture as in Figures 14.5–14.7.

# CHAPTER 15

# Acquisitions and Global Expansion

*Opportunity is rare, and a wise person will never let it go by.*
**Bayard Taylor**

**How can entrepreneurs best manage expansion via acquisitions and entry into new geographic markets?**

Entrepreneurs often create a new business by acquiring an existing firm and then improving it. The acquirers attempt to create growth and new value for the firm. Another strategy is for entrepreneurs to start and build their own firm and then expand the company by acquiring other firms. A series of successful acquisitions can help build a firm into a powerful leader in an industry. The integration of the newly acquired firm within the existing firm is a large challenge, however, especially when the cultures of the two firms differ significantly.

Most new firms develop, at the appropriate time, a plan for building an international strategy for growth. The forces for globalization are powerful, and new business ventures need to plan for them. ■

## 15.1 Acquisitions and the Quest for Synergy

An **acquisition** is when one firm purchases another. Usually the acquired company gives up its independence, and the surviving firm assumes all assets and liabilities. Acquisitions are one form of corporate entrepreneurship that can be particularly useful as an established company tries to innovate and infuse the organization with more entrepreneurial behavior or new product lines. Buying an existing business also can reduce the risk of entering a new product or market area because the business's operating history provides valuable data. Moreover, purchasing an established business has the advantages of ongoing businesses: customers, employees, products, equipment, and location(s). Obvious potential disadvantages, however, include poor location, depleted assets, obsolete inventory, depreciated brand, disenchanted customers or employees, and lack of profitability.

The three main steps for acquiring a company are (1) target identification and screening, (2) bidding strategy, and (3) integration or transition to the acquirer. Finding and evaluating an acquisition candidate can be time-consuming. Most, if not all, acquisitions are justified on the basis of an expected **synergy,** which is the increased effectiveness and achievement produced as a result of the combined action of the united firms. Suppose that you identify a firm that you determine is worth V and that a bid for its acquisition is accepted at a price of V. Then you estimate that after adding the value created by you and the entrepreneurial team, the value of the newly revitalized firm will be $V_N$. We then expect a synergy (Syn) as defined as:

$$Syn = V_N - V$$

The synergy is the expected value added by the acquiring party. The source of the synergy may be revenue enhancements and cost reductions due to capabilities and resources introduced into the firm by the new entrepreneur team. When new entrepreneurs acquire an existing business, the synergy is the added value of the entrepreneurial team that often replaces the management team of the acquired firm. The new team strives to add value to the acquired team's product that will be rarely available to competitors and is difficult to imitate. Thus, acquirers try to find firms with valuable and scarce product innovations that can be enhanced by the capabilities of the acquirer's management team.

After finding a good acquisition, the acquirer needs to arrange financing and negotiate the terms of the deal. It is a good rule to avoid bidding contests and to close the deal in a timely way.

**eBay and PayPal**

eBay and PayPal announced their combination in July 2002, promising multiple synergies. The press release added:

In a move that will help millions of Internet users buy and sell online, eBay, Inc., the world's online marketplace, today announced that it has agreed to acquire PayPal, Inc., the global payments platform. A natural extension of eBay's trading platform, the acquisition supports the company's mission to create an efficient global online marketplace. Payment is a vital function in trading on eBay, and integrating PayPal's functionality into the eBay platform will fundamentally strengthen the user experience and allow buyers and sellers to trade with greater ease, speed, and safety.

*Source:* eBay Press Release, July 8, 2002.

We will consider three common methods of valuation of a firm used by acquirers: (1) book value, (2) price-to-sales ratio, and (3) price-to-earnings ratio. The **book value** is the net worth (equity) of the firm, which is the total assets minus intangible assets (patents, goodwill) and liabilities. The price-to-sales and price-to-earnings ratios are obtained for comparative firms in a specific industry.

Consider a firm that designs and makes orthopedic devices for injured and disabled people. An accounting consultant determines that the net worth of the firm is $800,000. Annual revenues have remained at $1.2 million over the past two years. The firm has several patented products, but it has not fully exploited its marketing opportunities. Therefore, the net worth or book value sets a base value of $800,000 for the firm. With no growth in revenues, the accountant suggests a purchase value of one-half of sales, or $600,000. Earnings have held steady at $100,000 per year for the past several years. Assuming a comparable price-to-earnings ratio of nine for a zero-growth firm, the valuation could be nine times earnings, or $900,000. Assuming these three valuation methods ($800,000, $600,000, and $900,000), the buyer chooses a target price—say, $700,000—and tries to determine a suitable deal structure. One starting arrangement could be $200,000 in cash with the remaining $500,000 as a loan from the seller of the firm set at the prime rate for four years. Ultimately, the valuation and the final deal are a result of negotiation between buyer and seller.

Often acquisitions end up eroding the value of the acquired company due to difficulties with the transition to new ownership and overestimation of the value of the acquired firm. When a transition to the acquirer is attempted, difficulties may occur in working with or changing the established culture of the acquired firm. These transitions can be especially difficult when the two firms have dissimilar product offerings or operate in different geographies [Ellis et al., 2011]. Differences in trust between the buyer and seller also can challenge an acquisition's success [Graebner, 2009].

Before entering into an acquisition, a new venture should consider the technology and customer uncertainty. If the uncertainty is high, it may be appropriate to consider an alliance, which will normally cost less and limit the firm's financial exposure. If the alliance starts showing results, then a move to acquisition may be appropriate [Dyer et al., 2004].

Earl Bakken and Palmer Hermundslie founded Medtronic as a medical equipment repair company in Minnesota in 1949. The company sustained itself early on by selling and servicing medical equipment and building many of its own custom devices as well. In 1957, Medtronic developed the first wearable external cardiac pacemaker. Three years later, Medtronic purchased the exclusive rights to the first implantable pacemaker. The company grew to be a leading manufacturer of heart pacing technology worldwide. Sales outside the United States were strong, but competition was fierce. In response, Medtronic began establishing international facilities, and it gained direct control of its international operations in 1968 by purchasing the firm that had been its sales agent in Canada. Medtronic then began to acquire the firms that had been its major distributors in the United States, thus building a direct sales force to market products around the world.

In the 1980s, Medtronic purchased Johnson & Johnson's Cardiovascular Division. In addition, Medtronic acquired nearly a dozen other medical technology companies, enabling it to enter new markets. Acquisitions included a manufacturer of coronary angioplasty catheters and guiding catheters, a producer of centrifugal blood pumps, and a Dutch pacemaker manufacturer. By 1990, through a combination of internal developments and strategic acquisitions, Medtronic had successfully made the transition from a company with a limited product line to an international, diversified, medical technology corporation. By continuing to acquire market leaders with strategic mergers in the 1990s and 2000s, Medtronic has maintained its leadership with the medical technology industry.

## 15.2  Acquisitions as a Growth Strategy

Acquisitions and mergers can serve as a growth strategy in fragmented industries. A **merger** refers to the fusing together of two companies. An acquisition is when one company buys another. The difference between a merger and an acquisition is the degree of control by one of the two firms; a merger may result in 50-50 control. Mergers involve a much higher degree of cooperation and integration between the partners than do acquisitions. Most of the time, mergers occur between relatively equal-sized organizations, while, in acquisitions, one organization tends to be larger and more established. Many mergers suffer from insufficient integration of the functions and activities of the two firms.

In fragmented industries, numerous small companies are differentiated as specialists and compete for market share. Powerful forces are driving industries to consolidate into oligopolies. An **oligopoly** is an industry characterized by just a few sellers. The incentives to consolidate are significant in the technology, media, and telecommunication industries, where fixed costs are large and the cost of serving each additional customer is small.

An oligopoly, a market in which a few sellers offer similar products, is not always undesirable. It can produce efficiencies that allow firms to offer consumers better products at lower prices and lead to industrywide standards that create stability for consumers. But an oligopoly can allow some businesses to make big profits at the expense of consumers and economic progress. It can destroy the vital competition that prevents firms from pushing prices well above costs. Many industries also face large fixed costs. A typical semiconductor fabrication plant now costs between $2 billion and $3 billion, compared with $1 billion five years ago. A maker of basic memory chips must sell far more integrated circuits (chips) to justify an investment of that size. This is why makers of memory chips are eager to merge.

Industries tend to become more efficient as they undergo consolidation [Sheth and Sisodia, 2002]. In a fragmented market, the consolidator firm within the industry can realize the synergy of the economics of scale. A new venture in a mature industry ripe for consolidation can offer good opportunity. The new entrant can concentrate the resources and use them effectively in one niche and then acquire small competitor firms as resources to do so become available [Santos and Eisenhardt, 2009]. An example of a fragmented industry is the Internet service provider (ISP) sector. Every town has many independent ISPs as well as large competitors, such as Comcast and CenturyLink. As these large firms consolidate their strength, the small competitors are fading.

A merger can stimulate growth if the new conjoined firm has a sound business plan for the near-term future that includes a few key measures of profitability. The merged firms should work to redeploy unproductive assets and focus on optimizing their joint activities.

Most studies show that about two-thirds of mergers do not pay off with any synergy or gains. To realize the full value of a merger, the merged organizations must be appropriately integrated. A **horizontal merger** is a merger between firms that make and sell similar products in a similar market. The merger between Exxon and Mobil is an example of a horizontal merger. A **vertical merger** is the merger of two firms at different places on the value chain.

In 1976, Jim McCann got an urge to own his own business. He bought a flower shop in Manhattan for $10,000 and brought in a day-to-day manager. He then opened 12 new stores over the next decade. He was doing well and left his regular job to build up his floral business. He purchased the troubled 1-800-Flowers in 1987 for $2 million and assumed $7 million in debt. He then moved the firm to New York and merged it with his flower store chain. McCann then made a good set of acquisitions, enabling the expansion of 1-800-Flowers. It moved beyond flowers to include gifts in the 1990s. Flowers in 1999 acquired Great Foods, a specialty food unit. In 2001, Flowers acquired Children's Group, a maker of toys and dolls. Flowers has expanded its candy business by offering more gift-box products, higher-price candy brands, and express delivery of items. Nonfloral items such as baked goods, sweets, and jewelry are offered on its website (www.800Flowers.com).

Five different types of mergers and acquisitions appear in Table 15.1 [Bower, 2001]. The first type aims to reduce overcapacity in a relatively mature industry. The acquirer tries to close inefficient plants and reduce costs while retaining the acquired firm's technologies and customers so that economies of scale can be realized. The Hewlett-Packard and Compaq merger in 2002 was an example of this type.

**TABLE 15.1 Five different types of mergers and acquisitions.**

| Type: | Overcapacity reduction | Geographic extension (roll-up) | Product or market extension | Technology acquisition | Industry convergence |
|---|---|---|---|---|---|
| **Objective** | To reduce excess capacity and increase efficiencies | To extend a company's reach geographically and build economies of scale and scope | To extend the product line or reach a new market segment | To quickly add new technologies and capabilities | To establish a position during the convergence of an industry or sector |
| **Examples** | Daimler-Benz and Chrysler | Bank of America and Nations Bank | Tyco and Raychem | Cisco Systems acquired 159 companies between 1993 and 2012 | Sirius and XM Radio |
| | Hewlett-Packard and Compaq | Waste Management and numerous local firms | eBay and PayPal | Medtronic and numerous medical device companies | Disney and ABC-TV and radio |
| **Issues** | What to eliminate in the merged company and how to get it done in a timely way | How to merge two firms with different cultures | Merging two cultures and distribution channels | Overvaluation of the technology acquisition and loss of leaders from the acquired firm | The convergence may not materialize or may be of low value |

Many deals are based on acquiring the customers of the acquired company and reducing overcapacity. The goal is to identify the best customers and retain them while trying to attract new customers from the acquired firm's customer list [Selden and Colvin, 2003].

A geographic extension or roll-up occurs when a successful company rolls up (buys up) local or regional firms into a nationwide powerhouse. Roll-ups are designed to achieve geographic reach and economies of scale and scope.

The third category, market extension, is designed to extend a company's product line or its entry into unserved markets. To extend its product line, eBay purchased PayPal to facilitate transactions with its customers.

A technology acquisition aims to quickly acquire new technologies and capabilities by purchasing a small firm. As discussed in Section 13.6, Cisco Systems built its performance during the 1990s on a long string of acquisitions of small firms. The final type of merger is based on a perception of a future convergence of industries. Disney purchased ABC-TV and Radio, envisioning a convergence of content and media channels. Often firms use an acquisition to restore a sense of vitality to their businesses and, they hope, unleash a subsequent surge in performance [Vermeulen, 2005].

> **Boston Scientific and Scimed: A Successful Integration**
>
> The successful integration of two firms is an important goal of any merger or acquisition. For example, Boston Scientific Corporation of Massachusetts was a well-respected pioneer in the less-invasive medical device industry after its IPO in 1992. Its aggressive acquisition strategy soon led to a merger with Scimed Life Systems of Minneapolis, which specialized in angioplasty products for catheter-based treatment of cardiovascular disease. Scimed also had built a top-notch distribution system in Europe and Japan. After several challenging years working to fully integrate the two firms, Boston Scientific became an industry leader, because of the Scimed products, distribution system, and team.

The Hewlett-Packard and Compaq merger was actually based on both overcapacity reduction and technology acquisition. In this case, the surviving firm, Hewlett-Packard, acquired Compaq's technologies. The goal of the merger was to improve HP's competitiveness since the firm had largely missed out on the personal computer and Internet transitions [Anders, 2003].

Rules for integrating an acquired firm into the acquiring firm are provided in Table 15.2. The key step is to appoint an integration manager who works full-time for a period on integrating the two firms. The integration effort starts with a strategy and an integration plan. Then the goal should be to achieve a majority of the integration by a short period after the close of the deal—about six weeks. An important step is to build an integration team working with the integration manager. The team helps build the social connections for the merger and get early results. The role of the integration manager is to inject speed into the process, create a new structure, make social connections, and build success, as summarized in Table 15.3 [Aiello and Watkins, 2000]. The number of overall mergers and acquisitions varies depending on market conditions.

The leaders of entrepreneurial firms often continue to play a vital role after an acquisition by a larger firm closes. The leaders of the buying firm are often too busy with their own business to provide effective direction to the acquired employees. Moreover, they initially may not understand the acquired business well enough to make good decisions. This creates a need for the acquired managers to continue to lead their companies, even after the deal closes. Acquired leaders can add value by focusing their employees on specific goals and timelines, and by helping to resolve problems that arise as employees are assigned to their new positions and supervisors [Graebner, 2004].

Acquisitions and mergers are as likely to destroy value as to create it. They only add value if they make strategic sense, if their fair value is paid based on realistic expectations, and if management stays focused on executing the plan. For example, AOL and Time Warner failed to properly mesh the organizations and destroyed value [Klein and Einstein, 2003].

## TABLE 15.2  Rules for the acquiring firm and the acquired firm.

| Rules for the acquiring firm |
| --- |

- Use your highly valued stock as payment.
- Identify the key people of the acquired firm and get their agreement to stay.
- Decide who to keep and build relationships fast.
- Contain the tendency to act with hubris.
- Integrate the culture and the operations of the two firms.
- Appoint an integration manager or team to lead the acquisition process.

| Rules for the acquired firm |
| --- |

- Demand cash, not stock, as payment from the acquiring firm.
- Key people will agree to stay for a short period.
- Avoid signing a noncompete agreement or keep its duration short.
- Explain the benefits that accrue to the employees and managers of the acquired firm.
- Tell people who will and will not have a job.
- Restructure with respect for people.

## TABLE 15.3  Four roles of the integration manager.

1. Inject speed into the process.
   - Push for decisions and progress.
   - Set the pace.
2. Create a new structure.
   - Create joint teams.
   - Lead an integration team.
   - Provide the new structure framework.
3. Make social connections.
   - Interpret the cultures of both companies.
   - Actively be present in both companies.
   - Bring people together.
4. Build success.
   - Identify and communicate synergies.
   - Show short-term benefits.
   - Demonstrate corporate efficiency gains.

## 15.3  Global Business

A common motivation for mergers and acquisitions is to enter new markets. This strategy is often adopted by firms seeking to start doing business in a new country. The current globalization phenomenon dates from the fall of the Berlin Wall in 1989, which ended the principle division that ruled the world since 1945, when the Iron Curtain descended. If the metaphor for the Cold War period was the Berlin Wall, the metaphor for globalization is the Web. Globalization is the triumph of free-market capitalism worldwide. The previous era of globalization was built around falling transportation costs. Today's era of globalization is built around falling telecommunications costs [Friedman, 1999]. **Globalization** involves the integration of markets, nation-states, and technologies, enabling people and companies to reach around the world to offer and sell their products in any country in the world. With the Internet and globalization, a company can sell anytime, anywhere. Globalization is characterized by speed, modernization, movement, and the removal of distance.

A new venture should consider a globalization strategy, even if it is only to decide that its initial strategy is to remain local. A new design-automation firm may choose to only serve the United States in its initial years but later consider expanding internationally. We use a classification system for the strategies for globalization, as shown in Figure 15.1.

A **local** or **regional** strategy focuses all a firm's efforts locally since that is its pathway to a competitive advantage. Early-stage companies often select a regional market that they know well and attempt to be successful there first. This enables them to understand their customers and why they buy the product [Bhidé, 2008]. Thus, a new technology firm may start locally and fashion its marketing and sales methods. Then it can go global.

Another reason to remain local initially is limited availability of resources. Restaurants, retail stores, and other local opportunities are best started locally. For example, Starbucks started in Seattle and moved to other regions of the United States only after perfecting its operational capabilities locally.

| | | Low → Medium → High |
|---|---|---|
| **Pressures for cost reductions** | **High** | **Global:** Operations in many nations provide worldwide creation and coordination of products, sales, and marketing | **Transnational:** Set of coordinated subsidiaries in many nations |
| | **Medium** | | |
| | | **International:** Products are created and transferred from the home market to other nations | **Multidomestic:** Set of semi-independent subsidiaries in selected nations |
| | **Low** | | |

**Pressures for local responsiveness**

**FIGURE 15.1** Strategies for globalization.

The **multidomestic** strategy calls for a presence in more than one nation as resources permit. In this case, the firm offers a separate product and marketing strategy suitable for each nation. This strategy is not cost-efficient but does enable a firm to have independent subsidiaries in many nations. Examples of companies using a multidomestic strategy are Nokia and Sony-Ericsson.

To exploit cost economies while creating differentiated products, a **transnational** strategy can be used. This strategy rests on a flow of product offerings created in any one of the countries of operation and transferred between countries. Examples of companies using a transnational strategy are ABB and Caterpillar.

An **international** strategy tries to create value by transferring products and capabilities from the home market to other nations using export or licensing arrangements. One benefit of international activities can be the exposure to new business environments where the firm can learn about different methods, products, and innovations. Examples are Microsoft and IBM. Microsoft has regularly tried to bring the same business model to other countries.

---

**Riverbed Technologies and International Strategy**

Riverbed Technologies (a U.S.-based company) is a leading maker of wide area network (WAN) optimization devices. The WAN optimization device market grew quickly to become a multibillion dollar market in less than a decade from the time that devices were first commercialized. Most large customers were unwilling to deploy networks that used different WAN optimization devices across different countries. So, Riverbed was forced very early in its life to figure out an international strategy. Riverbed partnered early on in its existence with distributors and resellers across many countries, enabling the venture to grow rapidly. Today Riverbed has offices in over 30 countries.

---

A **global** strategy emphasizes worldwide creation of new products, sales, and marketing. The company uses facilities and organizations in several nations to create products for worldwide sales. Examples are General Motors, Intel, and Hewlett-Packard. Headquartered in Silicon Valley, Intel's international business has grown to represent 70 percent of its total revenues. Craig Barrett, Intel's former CEO, anticipates the largest growth areas in the coming years will be India, China, and Russia. These markets represent almost half the world's population and have just recently become available to U.S. technology companies [Barrett, 2003]. The advantages and disadvantages of each of the globalization strategies are listed in Table 15.4 [Hill and Jones, 2001].

> **Suzlon Energy's Global Strategy**
>
> Founded in India by Tusli Tanzi in 1995, Suzlon Energy is now a leading wind turbine manufacturer with an increasingly global strategy. The company has operations on five continents with sales, R&D, and manufacturing employees in locations like India, China, Germany, and Belgium. It offers complete wind power solutions including consulting, manufacturing, installation, and maintenance to its customers. It has rapidly grown into one of the largest multinational clean energy companies with sales exceeding $4 billion in 2012. Its leadership challenge is now formidable because the firm's many employees are separated by oceans, languages, and cultures (www.suzlon.com).

**TABLE 15.4  Advantages and disadvantages of the four global strategies.**

|  | Advantages | Disadvantages |
|---|---|---|
| ■ Multidomestic | Ability to customize products for local markets | Failure to reduce costs and to appropriate learning from other nations |
| ■ Transnational | Ability to reduce costs and learn from other nations | Difficult to implement due to many independent organizations |
|  | Ability to be locally responsive |  |
| ■ International | Transfer of unique products and competencies to other nations | Low-local responsiveness |
|  |  | Failure to reduce costs |
| ■ Global | Ability to reduce costs and gain worldwide learning | Lack of local responsiveness |
|  |  | Difficult to coordinate |

In general, a new or emerging firm should choose one of the strategies for entering the global marketplace and then determine which foreign markets to enter and when. The determinants of entry, timing, and costs will lead to the selection of a sound strategy. Some industries are local in nature, and others are international or global. A new manufacturer of integrated circuits is immediately required to quickly build an international strategy since the competitive marketplace is global. For many firms, expansion from local to regional and then to national markets will follow a natural progression. The first step is to expand to selected nations, establishing appropriate distribution channels and supply chains.

International opportunities exist in many industries. Microsoft and Intel sell their products worldwide. London's hit plays, such as *Phantom of the Opera,* move to New York and eventually on worldwide tours. Global opportunities to reduce costs, improve capabilities, and match local needs call for a transnational or global strategy. The resources necessary to mount these strategies can be large, however.

Cisco Systems uses an international strategy that enables it to derive about 50 percent of its revenues from foreign sales in 150 countries. Many new ventures or emerging firms need to consider the development of a global strategy to access unique capabilities or advantages. Reasons for considering entering the global marketplace are listed in Table 15.5. In turn, the speed at which a new venture enters the global marketplace is dependent on competitors' actions, enabling technologies, and the venture's knowledge and perceptions of foreign markets [Oviatt and McDougall, 2005].

It is conventional for a start-up to fully develop and test its product in a local or national marketplace before taking it global. Since emerging firms have limited ability to acquire information and knowledge about foreign markets and to manage foreign activities [Julien and Pamangalahy, 2003], they often limit their international efforts. Over time, as firms identify international opportunities in the same region and around the world, they must balance the pursuit of these opportunities against their available resources [Kuemmerle, 2005]. Successful entrepreneurs generalize lessons learned in one context into simple rules that apply to the next context [Bingham, 2005]. They also have the ability to improvise their execution according to the needs of different contexts [Bingham, 2009].

A successful global start-up usually has an international vision from inception, a strong worldwide network, and a unique product in demand worldwide. Consumers throughout the world increasingly demand the same selection of consumer goods, particularly automobiles, clothing, many food and beverage products, and consumer durables such as appliances and electronics. For many businesses, this global consumerism has focused sharp attention on the development of global brands, which are rapidly creating brand equity positions for their companies. Apple, Toyota, and Pfizer are global brands. New ventures need to plan for globalization in various ways. As their industry becomes global, they must plan for production, regulatory, and organizational factors [Farrell, 2004].

It can be wise to talk and act like a global company from the start, putting strategies in place early to protect your intellectual property, and being committed at a deep level to the success of the international operations. John O'Farrell, a partner at the leading venture capital firm Andreessen Horowitz, recommends that a start-up develop a separate global strategy that addresses five key areas for global operations: What is your goal in the region? What countries will you focus

**TABLE 15.5  Reasons to develop a global strategy for a new venture.**

| | |
|---|---|
| ■ Access low-cost labor or materials | ■ Access attractive markets for the firm's products |
| ■ Work around trade barriers | |
| ■ Access unique capabilities and learning from others | ■ The industry is global and all competitors are worldwide |
| ■ Obtain economies of scale | ■ Possession of a strong brand that is known worldwide |

**TABLE 15.6  Five forms of entry mode into international markets.**

| Mode | Description | Advantages | Disadvantages |
|---|---|---|---|
| 1. **Exporting** | Send goods abroad for sale in another nation | Ability to sell elsewhere | Transportation costs |
|  |  |  | Difficulties with the agent |
| 2. **Licensing** | Legal permit to use knowledge or patent to make product | Low-cost entry | Weak control of the licensee |
| 3. **Franchising** | Rights granted to a business to sell product using brand, name, and methods of operation | Low-cost entry | Lack of control over quality |
| 4. **Joint venture** | Corporation held jointly with a local entity | Access to partner's capabilities | Lack of control |
|  |  | Shared costs |  |
| 5. **Wholly owned subsidiary** | Company incorporated in another nation | Ability to act directly | High costs |

on and in what order? What products will you sell? How will you go to market? And what is the operating model for the different regions of the world? Going global can provide big returns, but only if done with good thought in advance and strong tactical operations once deployed [O'Farrell, 2011].

The mode of entry into another national or regional market offers five possibilities, as listed in Table 15.6. Exporting is an easy method to start up elsewhere. However, high transportation costs can be a disadvantage. Licensing to another party for a fee can be an inexpensive method and can provide some control of the marketing and manufacturing carried out by the licensee. Franchising is a form of licensing with an agreement to follow rules and procedures of operation. This method may result in loss of control over the quality of the product.

A joint venture with a foreign company affords access to the partner's capabilities [Yu et al., 2011]. But, it also results in diminishing control. Thus, care must be taken to align the goals of both partners and to address the imbalance in bargaining power if one partner is much larger than the other [Lu and Beamish, 2006]. A wholly owned subsidiary enables the parent company to exercise full control but it can be a costly approach.

McDonald's and Hilton Hotels are usually set up as franchises in other nations. Intel and Hewlett-Packard act through wholly owned subsidiaries. Fuji-Xerox is a joint venture between Xerox-USA and Fuji Photo Film of Japan. Harley-Davidson exports about 29 percent of its motorcycles directly to dealers and distributors in other countries.

Honeywell is a U.S.-based Fortune 100 company with large international operations. The company is involved in aerospace, environmental controls, advanced materials, transportation systems, and technologies for manufacturing and process industries. A high percentage of Honeywell's 132,000 employees lives outside the United States, and 55 percent of the company's revenues comes from outside the United States.

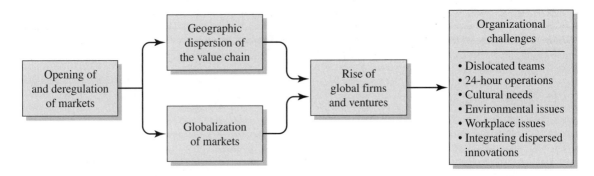

**FIGURE 15.2** Forces and consequences of globalization.

Globalization is spreading as markets open and deregulate. As shown in Figure 15.2, the forces of globalization are powerful [Barkema et al., 2002]. Many new firms will need to consider the global factors in their industry and create a plan for responding. To be successful, global entrepreneurs need to articulate a global purpose, build alliances, create and manage supply chains, and learn how to build a trusting organization across different cultures and institutional frameworks [Isenberg, 2008].

As new firms grow, they need to develop global relationships and work with people from other cultures. Cultural intelligence is the capacity of a person to adapt to new cultural settings and to interact effectively with people from different cultures [Early, 2004]. The three components of cultural intelligence are cognitive, physical, and emotional intelligences. Global business depends on good cultural relationships and smart cultural intelligence. Successful members of global teams learn to cope with different national, corporate, and vocational cultures.

Yves Doz of INSEAD defines a **metanational** company as possessing three core capabilities: (1) being the first to identify and capture new knowledge emerging all over the world; (2) mobilizing this globally scattered knowledge to out-innovate competitors; and (3) turning this innovation into value by producing, marketing, and delivering efficiently on a global scale [Fisher, 2002]. Doz identifies one example of such a firm as Nestlé. Other examples are IBM and Hewlett-Packard. Perhaps the most challenging task is to access knowledge and innovations everywhere and then integrate them into the firm's competencies and products.

Competition is similar to a three-dimensional game of global chess. The moves that an organization makes in one market may achieve goals in another market in ways that are not immediately apparent to its rivals. Where this strategic interdependence between markets exists, the complexity of the competitive situation can quickly overwhelm ordinary analysis. A firm needs an analytic process for mapping the competitive landscape and anticipating how its moves in one market can influence its interactions in others.

Leading firms in technology-intensive industries are based on core capabilities that can be leveraged globally. Innovative companies strive to draw knowledge and coordinate capabilities worldwide. They build and discover new assets and capabilities, and work to achieve economies of scale and scope. As the firms grow, they move toward a global network of capabilities and assets [Tallman and Fladmol-Lindguist, 2002].

A truly global firm, such as IBM, HP, Cisco, or Intel, operates as a single integrated organization. The firm will move people and projects anywhere in the world based upon the right set of costs, skills, and the business environment. These firms operate as a "global network enterprise"—a flexible assembly of firms that together strives to provide the best set of business processes. These networks are no longer transactional (contract-based), but rather rely on mutual cooperation and trust [Branscomb, 2008]. They rely on a cooperative network of innovation partners.

## 15.4 Spotlight on Alibaba

Founded in 1999 by Chinese entrepreneur Jack Ma, Alibaba was originally created as a portal to connect Chinese manufacturers and foreign buyers. The company has since evolved into the "Amazon of the East," with its suite of services valued at greater than $50 billion, totaling $160 billion in sales (GMV) in 2012.

Alibaba has expanded beyond China to become a global power, with portals in seven languages and customers across the world. This growth has been fueled in part by partnerships with, investments in, and acquisitions of a number of companies operating in the Chinese e-commerce space. Notable among them is Alibaba's 2005 acquisition of Yahoo! China, giving Alibaba a strong footprint in news, e-mail and search, as well as close ties to Yahoo!.

As Alibaba continues to expand its footprint in China to compete with other global technology giants, it often does so via strategic investment and acquisition, preferring to bring on existing, proven technologies rather than spend the time to develop them in-house. Examples include strategic investments in Sina Weibo, known as "China's Twitter," and AutoNavi, a well-known Chinese mapping and location-technology company.

## 15.5 Summary

A new venture may start as a purchase of an existing company by a team of entrepreneurs. Alternatively, entrepreneurs can start their new venture and acquire other small firms to grow their own company. Most, if not all, acquisitions are justified on the basis of an expected synergy, which is the increased effectiveness of the combined firms. Acquisitions can be used for efficiency improvement, geographic expansion, product or market extension, and technology acquisition.

> **Principle 15**
> All new technology business ventures should formulate a clear acquisition and global strategy.

Most new firms start with a regional or national strategy and later develop a plan to export their product internationally. As these firms grow, they may shift from exporting to establishing a wholly owned subsidiary in other nations.

## Video Resources

Visit http://techventures.stanford.edu to view experts discussing content from this chapter.

| | | |
|---|---|---|
| Reasons to Acquire a Company | Dan Rosensweig | Chegg.com |
| Getting Acquired by Google | Brett Crosby | Google |
| Building Resistance to Acquisition | Hugh Martin | Pacific Biosciences |

## 15.6 Exercises

**15.1**  An acquirer looks for a company with a good profit margin, a proven history, and a fair price. Choose an industry of interest and list five criteria for selecting candidates for acquisition.

**15.2**  A software firm is available for purchase, but it has experienced no growth for several years. The firm provides a cash return annually before taxes and owner's salary of $100,000. It has annual sales of $1 million. As a purchaser of this firm, you select a discount rate of 14 percent. Calculate the price you would offer for the firm. Assume that you can increase and maintain a growth rate of sales and cash flow of 2 percent annually.

**15.3**  Describe how the acquisition strategies of these firms have differed: Apple, Google, Microsoft, Oracle, and Qualcomm.

**15.4**  Identify a technology-based company that used acquisition to enter into an international market. Describe the strategic alignment and the executed deal terms. Was it a success? What were the key challenges that arose?

**15.5**  Interview a technology executive based in a country different from the company's headquarters. How are various company functions handled (e.g., marketing, sales, operations, development)? What are the key challenges of operating away from company headquarters?

**15.6** When Google initially expanded search services into China, there was surprise over Google's decision to censor some content per requests by the Chinese government. Why did some view this as "evil"? How did Google view the concern? What are other challenges facing companies expanding into international markets?

## VENTURE CHALLENGE

1.  Describe your venture's approach to acquisitions and mergers as a growth strategy.

2.  For your venture, discuss the plans for going global and describe your approach using Figure 15.1 and Table 15.6.

# Financing and Leading the Enterprise

The venture should have a clear revenue model and a workable path to profitability. Furthermore, there should be a plan of how the wealth created will be harvested by the owners. A comprehensive financial plan will be designed to demonstrate the potential for growth and profitability that is based on accurate and reliable assumptions. With the financial plan in hand, sources of investment capital will be explored and tested. The presentation of the total business plan will require a compelling story about the venture. Furthermore, skillful negotiations about the terms of the deal will be required. When funded and launched, the venture team must continuously and ethically implement the business plan and adapt to changing conditions. ■

# CHAPTER 16

# Profit and Harvest

*Profit is the product of labor plus capital multiplied by management. You can hire the first two. The last must be inspired.*

**Fost**

**How will a new venture generate revenue and achieve positive cash flow?**

A new firm creates a sales model describing how it will generate revenues from its customers. Then it determines a cost model and how to generate profits from its revenues. The revenue and profit engines show how the firm will create powerful value for its customers and how customers will enable the new firm to profit. Many new ventures assume that profit will flow naturally from sales but discover that profits are not guaranteed. It is difficult to operate in a market that is chronically unprofitable.

A new firm seeks positive cash flow as soon as is feasible and acts to move to profitability early in its life. Managing revenue growth is important since uncontrolled growth can lead to negative cash flow and the need to constantly raise new funds from outside investors. Furthermore, a firm needs a plan to harvest the benefits of its growing venture for all owners. Entrepreneurs must also be realistic and accept that termination of the new venture is a possibility. ■

## 16.1 The Revenue Model

A firm's **revenues** are its sales after deducting all returns, rebates, and discounts. A firm's **revenue model** describes how the firm will generate revenue; five models are listed in Table 16.1. Most firms generate revenues by selling a product in units to a customer using a **product sales model.** For example, Lenovo sells its personal computers to one customer at a time, and Intel sells its chips to electronics companies.

In the **subscription revenue model,** a business offers content or a membership to its customers and charges a fee permitting access for a certain period of time. This model is used by magazines, information and data sources, and content websites. *Consumer Reports* offers its information to magazine subscribers (members) as well as to subscribers to its online service for a fee. This model is also used by clubs, cooperatives, or other member-based organizations. Technology firms sometimes license their technology for a fee.

The **advertising revenue model** is used by media companies such as magazines, newspapers, and television broadcasters that provide space or time for advertisements and collect revenues for each use. The media entities that are able to attract viewers or listeners to their ads will be able to collect the highest fees. Facebook, Google, and Yahoo collect most of their revenues through the sale of advertising space.

Some firms receive a fee for enabling or executing a transaction. The **transaction fee revenue model** is based on providing a transaction source or activity for a fee. Examples of firms based on transaction fees are Charles Schwab, Visa, and eBay.

The **affiliate revenue model** is based on steering business to an affiliate firm and receiving a referral fee or percentage of revenues. For example, this revenue model is used in the real estate business and by companies that steer business to Amazon.

Magazines and newspapers such as the *New York Times* and *The Economist* use both the subscription and the advertising models to generate revenues. Some new ventures use a mix of the five revenue models. Amazon.com, for example, uses the product sales, transaction fee, and affiliate revenue models. Google generates revenues through advertising and licensing its technology to others for a subscription fee.

Any new business needs to determine its revenue model and test it on potential customers. Who is the customer? Will the target customer pay for the offering? Facebook started out as a social network with no viable revenue-generation model. The first model the company tried was display advertising, previously

### TABLE 16.1  Five revenue models.

| | |
|---|---|
| ■ Product sales | ■ Transaction fee revenue |
| ■ Subscription (member) fee | ■ Affiliate revenue |
| ■ Advertising revenue | |

> ### JBoss: Generates Revenue Through Service and Support
>
> JBoss of Atlanta, Georgia, has an unusual revenue model and mission: "JBoss Inc.'s mission is to revolutionize the way enterprise middleware software is built, distributed, and supported through the Professional Open Source model. We are committed to delivering innovative and high quality technology and services that make JBoss the safe choice for enterprises and software providers." Revenue is generated by supporting software that is given away for free. By providing professional service and support for open source software, JBoss found a comfortable and profitable niche within its industry. It was acquired by Red Hat in 2006. (See http://www.jboss.com.)

pioneered by Google. Facebook has generated and continues to generate significant revenue from display ads but is currently being challenged to explore other models as more of its users are viewing the site from mobile devices on which display advertising is less profitable. One new alternative is a search tool that mines Facebook's wealth of user data. The social networking giant believes this new search has the potential to challenge Google by delivering far more personal, tailored results—recommendations are driven by the likes and interests of your Facebook friends rather than an algorithm. As of 2013, the effectiveness of this model is still being tested.

Because they have yet to achieve profitability, most start-ups use a revenue metric such as revenue per employee to track their performance during their formative years. Their goal is to exceed $100,000 per employee as soon as possible. Most mature growth firms exceed $200,000 per employee, and technology companies sometimes exceed $400,000 per employee.

## 16.2  The Cost Model

A **cost driver** is any factor that affects total firm costs [Niraj et al., 2001]. Typically, costs vary over time or volume of output. The four primary types of cost drivers are: fixed, variable, semivariable, and nonrecurring. Fixed costs do not vary at all with volume. Examples include rent and management salaries. Variable costs change directly and proportionally with volume. Examples include sales commissions (which vary with sales) and materials costs (which vary with the amount of goods produced). Semivariable costs also change with volume, though not as directly and they contain elements of fixed costs. For example, a store may need to keep a minimum staff on hand (which is a fixed cost), but can add more staff as the number of customers increases (which is a variable cost). Finally, nonrecurring costs occur infrequently or irregularly. Examples include equipment and building purchases [Hamermesh et al., 2002]. In order to maximize profits, entrepreneurs must be aware of their various cost drivers and of the types of costs.

## 16.3  The Profit Model

**Profit** is the net return after subtracting the costs from the revenues. The **profit model** is the mechanism a firm uses to reap profits from its revenues. Google makes most of its profits by auctioning text-based advertisements that appear in search results and participating websites. Microsoft makes most of its profits from licensing Windows and Office software. General Electric generates one-third of its profits from its financial arm, GE Capital. Newspapers make most of their profits from classified ads. Hewlett-Packard and Xerox make most of their profits from replacement toner cartridges.

Figure 16.1 shows the value of a product and the distribution of the value to the customers and the profit captured by the firm. To remain profitable, a firm strives to reduce its costs while maintaining or increasing the value of its product to the customer. To generate profits, a firm needs to examine all its activities on the value chain and determine if its cost versus value generated is in line.

Profit accrues to a company that maintains a competitive advantage as conditions change. When the PC industry took off in the 1980s, IBM ceded the operating system's rights to Microsoft when it incorrectly determined that profit would flow to the branded integrator of hardware and software components. Key to profit capture is ownership of the unique, value-added element of the value chain or the product makeup. Examples would be ownership of an essential pipeline, control of the customer interface, or ownership of unique locations for a retail operation. It pays to hold the largest "value-added" step in a value chain or the unique innovation that no one else can match.

During the early years of a firm, the firm may be patient for growth but should be impatient for profitability. As a firm works to gain profitability, it is testing its assumptions that customers will pay for a profitable product [Christensen and Raynor, 2003].

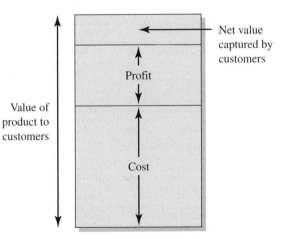

**FIGURE 16.1**  Value of a product and profit.

The revenue and profit engines are driven by the firm's business model, strategy, resources, capabilities, operations, and processes, as shown in Figure 16.2.

The best conditions for profit occur when the perceived value of a product to a customer is high and the cost to produce the product is low. Figure 16.3 shows a value grid that enables a firm to determine its potential to reap large profits. Low cost to produce a product leads to low price per unit of perceived value to the customer. The upper-right quadrant is a high profit location that many firms seek to occupy [Chatterjee, 1998].

Most start-ups initially invest time and energy in learning about their customers. Then they use that knowledge to create improved solutions for them. They lose money initially but make money after a period we will call T, as shown in Figure 16.4. Of course, it is best to keep T relatively short and the peak negative profit (NP) small [Slywotzky, 2002]. The profit curve shown in Figure 16.4 is often called a "hockey stick" expectation.

It is useful to try to estimate the attractiveness (the potential profitability) of a market segment. It is of little value to win market leadership in a market segment that is chronically unprofitable [Ryans et al., 2000].

**FIGURE 16.2**
Revenue and profit flow from the firm's operations.

**FIGURE 16.3** Value grid.

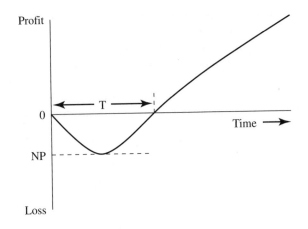

**FIGURE 16.4** Early losses of a successful start-up turn profitable after a time, T. The peak negative profit is NP.

A firm can create a metric for its profitability as

$$\text{metric} = \frac{\text{profit}}{X}$$

when x is chosen to fit the firm's goals and business model. As shown in Table 16.2, firms chose the variable x so as to illuminate their profitability performance. A commonly used profit metric is **profit margin,** which is the ratio of profit divided by revenues.

Jeff Bezos of Amazon uses a simple two-by-two matrix to describe margins, as depicted in Figure 16.5. Amazon sits in the lower-left quadrant offering products with low margins but selling a high volume of product. In order to sit in that quadrant, a company must be incredibly efficient. Amazon has structured its entire business around efficiency and has hit the margin goals outlined by Bezos.

During the telecommunications boom of 2000, companies often used poor indirect metrics of their growth such as the number of building leases. They

**TABLE 16.2** **Metric for profitability performance for selected firms.**

| x | Metric | Example firm |
|---|---|---|
| Customer | Profit / Customer | Gillette |
| Employee | Profit / Employee | Abbott Labs |
| Customer visit | Profit / Customer visit | Walgreens |
| Tons of output | Profit / Tons | Nucor |
| Revenue | Profit / Revenue | ExxonMobil |

|  | *Low Margin* | *High Margin* |
|---|---|---|
| *Low Volume* | (avoid this quadrant) | BMW |
| *High Volume* | Amazon | Apple |

**FIGURE 16.5** Combination of profit margins and volumes.
*Source:* Garrett, 2012.

reasoned, wrongly, that access to office buildings would translate to customers [Malik, 2003].

A business model designed for high customer relevance that delivers high value will have the best chance of capturing profit. The best business model helps the customer in the difficult or time-consuming areas of their purchasing process. One of the most powerful profit models is the **installed base profit model**. The supplier builds a large installed base of users who then buy the supplier's consumable products. This is the model used by HP, which sells printers at a modest price while building a large base of users who purchase consumable ink cartridges. Nine types of profit models appear in Table 16.3 [Slywotzky et al., 1999]. A new venture

**TABLE 16.3 Nine types of profit models.**

| Name | Description | Examples |
|---|---|---|
| 1. **Installed base** | Build a large installed base of customers and sell consumables or upgrades | Dropbox<br>Hewlett-Packard printers |
| 2. **Protected innovation** | Create a unique, innovative product and protect it using patents and copyrights | Merck<br>Microsoft |
| 3. **New business model** | Find unmet customer needs and build a new business model | Twitter<br>Square |
| 4. **Value chain specialization** | Specialize in one or two functions on a value chain | Nucor<br>Intel |
| 5. **Brand** | Create a valued brand for your product | Google<br>Apple |
| 6. **Blockbuster** | Focus on creating a series of big winners | Pixar<br>Schering Plough |
| 7. **Profit multiplier** | Build a system that reuses a product in many forms | Samsung<br>Virgin Group |
| 8. **Solution** | Shift from product to unique total solutions | General Electric<br>IBM |
| 9. **Low cost** | Create a low-cost product to offer low price per unit of value | Ryanair<br>Airbnb |

will wisely select its needed profit model and work hard to build its strength and resiliency in the competitive marketplace.

Managers have always known that some customers are more profitable than others. For some emerging firms, 20 percent of the firms' customers may provide most of the firms' profits. Furthermore, the firms' worst customers may be costing more than they pay for products or services. Securing profitable customers while getting rid of unprofitable customers can help to double a firm's profits. It pays to be attentive to the best customers and ignore the worst [Selden and Colvin, 2002].

Recall that profit is

$$\text{Profit} = (P - VC)Q - FC \qquad (16.1)$$

where P = price, VC = variable costs, Q = total number of units sold, and FC = fixed costs. Managing profitability can be achieved by lowering fixed or variable costs, raising units sold, or raising price. One may be able to find a customer segment that is willing to pay a higher price or purchase more units. Otherwise, costs need to be lowered.

Another important measure of performance is **cash flow,** which is the sum of retained earnings minus the depreciation provision made by the firm. Without positive cash flow, a firm may eventually use up all its cash and close its doors. A profit and cash flow model focuses attention on the nature of the driving forces of the revenue and profit models.

Amazon.com has used aggressive price discounting and free shipping to boost its revenues. However, its low operating margin (P − VC) made profitability elusive during its formative years.

Many mass-market retailers such as Wal-Mart continually lower prices by squeezing inefficiencies from their operations and sacrificing profit margins on products in favor of selling in high volumes. Hewlett-Packard tries to avoid lowering prices to keep its profit margin robust.

All entrepreneurs need to find a suitable profit model for their firm. If profitability appears to be highly elusive and at best in the distant future, it may be wise to not proceed with the venture. We discuss the matter of terminating a venture in Section 16.6.

Cemex is a Mexican cement company. In the past, it described its profit metric as yards of concrete sold each day. Cemex realized, however, that its customers value the delivery of the right amount of concrete at the right time. Using that metric, profitability increased. The profit model led to the creation of a new business model [McGrath, 2005].

## 16.4  Managing Revenue Growth

New businesses normally strive to build up revenues and profits so that they can meet their goals. Most teams are naturally inclined to grow their business rapidly, especially in technology entrepreneurship. The degree of commitment of the entrepreneurial team to the growth of the firm can be called **entrepreneurial intensity** [Gundry and Welsch, 2001].

Commitment to growth leads to sacrifices an entrepreneur is willing to make. High growth can require significant financial resources, leading the entrepreneurs to seek outside capital and often give up majority ownership of the firm. Low growth would include firms growing revenues at a rate less than 10 percent per year, and high rates of growth would exceed 25 percent per year. Many high-growth firms grow at 50 percent or higher each year for several years after founding. Technology entrepreneurs who seek a high-growth strategy will usually select a team-based organizational structure and exhibit high entrepreneurial intensity. Furthermore, they are willing to endure the burdens associated with the demands of high growth. The characteristics of a high-growth entrepreneurial team are listed in Table 16.4. High-growth entrepreneurs are willing to put aside some of their personal or family goals and make sacrifices because they are committed to the growth of their ventures.

A growing business requires cash for working capital, assets, and operating expenses. If a company grows too fast, it will need to continually raise additional cash from investors. The cash required will depend on the operations model of the firm, which depends on its accounts payable cycle, as well as assets and working capital required [Churchill and Mullins, 2001]. In general, most growing firms are unable to sustain a growth rate of sales exceeding 15 percent using their internally generated cash. Some service businesses are less asset-intense and may be able to self-finance a sales growth rate of 20 percent to 30 percent. Very few, if any, firms can self-finance if they plan to grow at 50 percent or more per year. For high growth, a financing plan for outside cash will be required.

Since service businesses are often less asset-intensive and more labor-intensive than production firms, growth typically adds to costs and may not produce economies of scale. Some companies with a heavy emphasis on service such as IBM, Starbucks, and Southwest Airlines have managed to successfully combine growth and profitability while other companies have not. Successful service-oriented companies are able to design and implement the right strategies to keep costs low, strengthen customer loyalty, and gain competitive advantage.

The profitability of a firm may be a function of the growth rate of revenues, as shown in Figure 16.6. At a low growth rate such as $G_1$, the firm is unable to meet demand and loses sales to its competitors. At a very high growth

---

**TABLE 16.4 Characteristics exhibited by entrepreneurial teams that seek high growth rates.**

- Strong entrepreneurial intensity
- Willingness to incur the costs of growth
- Willingness to use a wide range of financing sources
- Emphasis on a team-based organizational structure
- Focus on innovation

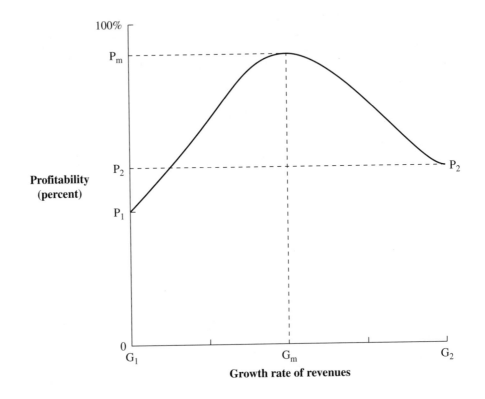

**FIGURE 16.6** Profitability of a firm as a function of the growth rate of revenues.

rate such as $G_2$, the firm is unable to efficiently manage its operations and only achieves profitability $P_2$. Growth rate $G_m$ maximizes a firm's profitability $P_m$. An emerging or new firm should try to estimate its growth rate $G_m$ that would maximize profitability. For many new firms, $G_m$ ranges between 20 percent and 40 percent.

The profitability of a firm may be represented by its return on capital or return on equity (ROE). Thus, one quick estimate of a firm's ability to grow is to state that a firm may grow organically at a rate less than its return on equity without turning to outside financial sources. We define **organic growth** as growth enabled by internally generated funds.

A more complete equation for a sustainable change in sales-to-sales ratio, $\Delta S/S$, for a firm is [Ross et al., 2012]:

$$\frac{\Delta S}{S} = \frac{PM(1+L)}{T - [PM(1+L)]} \qquad (16.2)$$

where PM = profit-to-sales ratio, L = debt-to-equity ratio, and T = the ratio of total assets to sales. If a start-up firm has no debt (L = 0), we have

$$\frac{\Delta S}{S} = \frac{PM}{T - PM}$$

For example, if PM = 0.10 and T = 0.5, then

$$\frac{\Delta S}{S} = \frac{0.10}{0.5 - 0.10} = 0.25$$

or the sustainable sales growth rate is 25 percent. Consider an asset-intensive business with T = 1.0 and determine the sustainable growth rate when PM = 0.10. Then, we have

$$\frac{\Delta S}{S} = \frac{0.1}{1 - 0.1} = 0.11$$

and the sustainable growth rate is 11 percent. If this asset-intensive firm uses debt so that L = 0.8, then

$$\frac{\Delta S}{S} = \frac{0.1(1 + .8)}{1 - [0.1(1 + 0.8)]} = \frac{0.18}{1 - 0.18} = 0.22$$

and the sustainable growth rate is 22 percent.

A start-up will need to examine its expected growth rate and its financing needs carefully. As an example, consider the growth of a fictional green-tech venture that raised $68 million via an initial public offering in 2008. Its sales were $130 million in 2010 and $518 million in 2013, growing at an annualized rate of 44 percent. During the same period, the firm's long-term debt rose from $135 million to $947 million, and its debt-equity ratio rose from 0.16 in 2010 to 0.96 in 2013. This venture was able to grow sales at a significant rate by increasing its debt (and associated risk) significantly.

Service firms require less money to start and expand than asset-intensive industries. Building and growing a service business requires adding employees. A service business has low capital and asset intensity and requires little debt. Consider Infosys, an Indian-based outsourcing company, which grew revenues from $4.2 billion in 2007 to $7.0 billion in 2012. The firm has negligible debt, thus L = 0. The profit-to-sales ratio averaged .44 during 2007 to 2012. The ratio of total assets to sales was approximately 1.01. Then, the sustainable growth in sales was

$$\frac{\Delta S}{S} = \frac{PM}{T - PM} = \frac{.44}{1.01 - .44} = 0.77$$

or the sustainable growth rate was 77.7 percent.

To grow steadily and avoid stagnation, a company should learn how to scale up and extend its business, lengthen its expansion phase, and accumulate and apply new knowledge to new products and markets faster than competitors. Entrepreneurs should choose a plan that fits with the knowledge, learning skills, and assets that the organization possesses or is developing.

## TABLE 16.5 Incentives for growth by new ventures.

- Attracting capital for market expansion
- Attracting capable team members
- Achieving economies of scale
- Development of a reputation and brand
- Growing profitability and financial rewards for owners and employees

Rapid growth and good profitability can often cover up some underlying problems of an organization. They can provide a cushion for wasteful decisions regarding the allocation of financial, human, and other resources. The excitement of growth can also veil inadequacies in leadership or management skills. Growth can mask a lack of planning or an inadequate orientation toward long-term issues. Success can disguise a variety of shortcomings while breeding a dangerous form of arrogance.

The incentives for growth by a new venture are several, as summarized in Table 16.5. One incentive to grow is to attract capital investment to expand markets and product lines. Also, growth creates a sense of pride among the employees and provides opportunity for expanding financial reward.

A firm's ability to use and coordinate new assets and activities depends on its organization and managerial capabilities. Rapid growth can challenge those capabilities severely. It is important for a publicly held company to have consistent, predictable financial growth. Consistency requires controlled growth of assets and personnel additions. Growing a staff at a rate greater than 15 percent a year will challenge any organization. Paychex, Inc., is a $2 billion payroll-processing firm that has increased revenues at an average of 18 percent for many years. With economies of scale, it has increased profits by 20 percent each year.

Most companies employ a mix of organic growth using both inside resources and external sources of resources. A balanced approach to growth attempts to break down barriers to growth and improve the company's core competencies. Some firms, such as Palo Alto Networks and Salesforce, have been successful in building revenues at 30 percent or more per year while maintaining a balanced mix of internally and externally financed growth.

Most highly innovative firms become high-growth firms [Kirchhoff, 1994]. Microsoft, for example, grew its revenues at 29 percent per year for the decade 1991 to 2001 and grew its profits at 35 percent per year for the same period.

### Microsoft's Quest for Growth

In 1975, two college students, Bill Gates and Paul Allen, began to develop and sell software that ran on the original personal computers (PCs). Soon they incorporated their firm as Microsoft. In 1981, industry giant IBM decided to enter the PC market and hired Microsoft to develop an operating system. Microsoft quickly created DOS, which became the dominant operating system in the industry. In March of 1986, Microsoft raised $61 million in its initial public offering of stock. In 1992, Microsoft announced the third version of Windows, marking the end of its collaboration with IBM on operating systems. Windows allowed Microsoft to expand its offerings of applications, greatly enhancing its revenue generation.

Applications made by Microsoft continue to generate significant revenue for the company. Products such as the Office Suite and Explorer continue to be ubiquitous on computers around the world, despite increasing challenges from open source competitors. In late 2001, Microsoft built on its strength in the computer gaming industry by entering the game console market. Eventually, Xbox 360 gained market share to become a popular console alongside Sony's PlayStation 3 and Nintendo's Wii. The Xbox allows Microsoft to gain additional revenue through the sale of games.

Microsoft has cleverly added revenue streams over the years, while continuing to play to its core competencies as a software development firm. Current forays, such as its Zune music products and smartphone technologies, seek to add revenue growth to the firm in the future. Microsoft is currently the #1 software company in the world, a position it has maintained for most of the last 30 years. In 2013, Microsoft's market capitalization was more than $290 billion with revenues of $78 billion. This valuation reflects over three decades of sustained profitability and continuous revenue growth.

Apple grew its revenues at 44 percent per year for the decade 2003 to 2012 and grew its profits at 51 percent per year for the same period.

The sources of revenue growth for a firm can include increasing brand recognition and international expansion, as shown in Table 16.6. Very few firms are able to increase prices in the face of tough competition. New offerings of valuable products are a good source of revenue growth. Successful companies active in new product offerings are Apple and Google.

The market value of a firm can be described as a result of three drivers, as shown in Figure 16.7 [Rappaport et al., 2001]. Changes in volume, price, and sales mix lead to changes in sales growth rate. Operating profit margin (profit before taxes divided by revenues) is driven by four factors, as shown in Figure 16.7. Incremental investment rate is the result of investment efficiencies. Operating leverage is the ratio of profit margin increases to upfront preproduction expenses for new product development and capacity expansion. Accounting for all these factors, a firm seeks to increase its efficient use

**TABLE 16.6  Sources of revenue growth.**

| | |
|---|---|
| ■ Increasing brand recognition | ■ Acquisition of other firms |
| ■ Intellectual property licensing | ■ Price increases |
| ■ International expansion | ■ New product offerings |

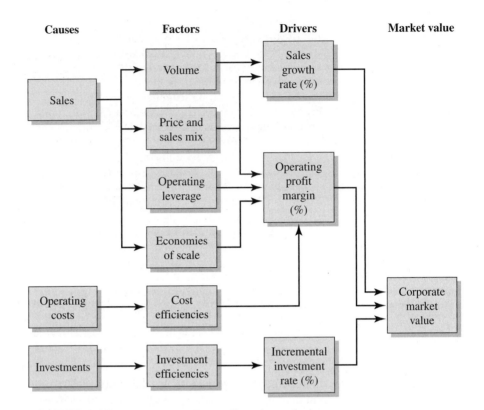

**FIGURE 16.7**  Causes and drivers of market value.

Note: Operating leverage $= \dfrac{\Delta \text{ Profit margin}}{\Delta \text{ Costs for product development manufacturing}}$

of investments and reduce its operating costs. Furthermore, increased volume, operating leverage, economies of scale, and an improved price and sales mix can lead to an improved sales growth rate and operating profit margin.

---

**Millennium Cell: When Market Value Disappears**

Wireless devices such as cell phones and laptop computers use batteries for power. The replacement for a battery could be a small hydrogen fuel cell. Stephen Tang and his team at Millennium Cell developed a miniature fuel cell and used the factors of Figure 16.6 to increase his investment and cost efficiencies and improve his product development capabilities. Unfortunately, this could not be achieved profitably. The firm ceased operations and filed for bankruptcy in 2009.

# 16.5 The Harvest Plan

Assuming a successful venture, investors will want to know the plan for providing a cash return. A **harvest plan** defines how and when the owners and investors will extract some of the economic value from the investment. Professional investors will expect a return on their investments within five to seven years. Thus, investors will expect a plan for cash liquidity for themselves. Note, however, that "harvest" does not mean the challenges and responsibility of the business are over.

For high-growth firms, the value created by an innovative venture can lead to significant returns over a four- to ten-year period, depending on the industry and market conditions. Both the founders and the investors will desire to access the financial return accrued by the new growing firm at the end of that period. This will mean some action will be necessary to yield a cash flow from the firm to the investors and owners. Table 16.7 lists five methods of harvesting a firm. The sale of the entrepreneurial firm to an acquiring firm is an attractive route for the founders and investors. Proceeds from the sale of a private company usually consist of cash, shares of the acquiring company, or a combination of shares and cash.

Fast-growing companies with annual revenues greater than $20 million may find a solution in the public stock market by using an initial public offering (IPO). If the investors are patient, the issuance of cash dividends to individuals can serve to provide cash to the investors. Of course, it is often possible to arrange a sale of the firm to the managers and employees of the firm. Finally, many owners of relatively small firms will consider passing the firm on to family successors.

The selection of a harvest strategy will depend on the interests of the founders and investors. Professional investors such as venture capitalists expect large annualized returns and normally seek the issuance of an initial public offering by year five or six. Alternately, the venture may be acquired by another, typically larger, firm that provides the liquidity sought by the professional investors.

The entrepreneurial team may describe an **exit strategy** for their venture after a specified period. This plan will be part of the negotiation with the investors.

A planned sale to employees and managers may be outlined in the business plan. This transfer can use an employee stock ownership plan (ESOP). The firm first establishes an ESOP and guarantees any debt borrowed by the ESOP for the purpose of buying the company's stock. Then the ESOP borrows money

**TABLE 16.7  Five methods of harvesting the wealth created by a new firm.**

- Sale of the firm to an acquiring firm
- Sale of the firm's stock on a public market through an initial public offering
- Issuance of cash dividends to the owners and investors
- Sale of the firm to the managers and employees
- Transfer of the firm through gifts and sales to family successors

from a bank, and the cash is used to buy the owner's stock. The shares of the firm are held by a trust, and the company makes annual tax-deductible contributions to the trust so it can pay off the loan. As the loan is paid off, shares are released and allocated to the employees. While an ESOP benefits the owner by providing a market for selling stock, it also carries with it some tax advantages that make the approach attractive to owner and employees alike.

Few events in the life of the entrepreneur or the firm are more significant than the harvest. Without the opportunity to harvest, a firm's owners and investors will be denied a significant amount of the value that has been created over the firm's life. The founders may need a harvest strategy due to a desire to retire or diversify their portfolio of assets. Investors may need to realize their returns to invest them elsewhere or benefit in other ways. The timing of the harvest may be uncertain, but a harvest strategy does help the entrepreneur team plan together for the future.

A good time to sell a company is when it is very successful. When that time arrives, it may be best to review or exercise the harvest plan. At that time, a firm needs to determine a realistic valuation for the firm and obtain advice from its board of directors. Entrepreneurs often choose to sell for personal, nonfinancial reasons, such as burnout from the long hours and high stress of running their own businesses. Entrepreneurs who have raised money from family and friends may be especially eager to sell, since they feel a heightened pressure to return their investors' money. Moreover, entrepreneurs typically have much of their personal wealth tied up in a single company, making them eager to sell so they can diversify their holdings.

When entrepreneurs decide to sell, their choice of buyers is about more than price. Company leaders often choose buyers based on "soft" criteria such as strategic and organizational fit. They care about the fate of their employees as well as whether their strategic vision will be carried on. It is rare for entrepreneurs to hold a formal auction for their company; instead, they hold informal discussions with a small number of potential buyers with whom they see a good fit [Graebner, 2004].

First Solar, Inc., was founded in 1999 by Harold McMaster. The company manufactures cadmium telluride-based thin-film solar panels. In 2006, the company became profitable and raised about $400 million in its IPO. Subsequently, the founders sold some of their shares as a way to harvest a portion of their gains. They retained ownership of the majority of their stock, but had the alternative of liquidating shares in the future.

### AIR: A Successful Harvest

In 1994, David Edwards, a postgraduate researcher in Robert Langer's lab at MIT, began working on a novel drug delivery system that used large, porous particles to deliver drugs directly to the lungs. Although the technology was entering a seemingly crowded market, Edwards's idea had the potential to perform significantly better than other forms of inhalation delivery systems. From the technology's initial stages, Langer, who had founded several successfully biotechnology firms, recognized the commercial potential of

Edwards's idea, but an attempt in 1995 to license the technology to a public drug delivery firm was unsuccessful.

Reluctant to start his own company when the technology was still preliminary, Edwards left MIT in early 1995 for a faculty position at Pennsylvania State University. While at Penn State, he continued to refine the delivery system and to visit Langer at MIT about once a month. By early 1997, Edwards and Langer were sufficiently pleased with their progress to approach Terry McGuire, a Harvard Business School graduate and recent founder of Polaris Ventures. Langer knew McGuire because he had invested in several of Langer's ideas. McGuire was initially hesitant to invest in the novel technology, which was entering a crowded market. His faith in Langer, however, as both a stellar scientist and entrepreneur, along with an influential *Science* article and external affirmation of the importance of Edwards's technology, convinced McGuire to put aside his uncertainty. In the summer of 1997, McGuire invested $250,000 in return for 11 percent of Advanced Inhalation Research (AIR) and an option to purchase an additional 9 percent. McGuire took on the role of temporary CEO, and Edwards took a leave of absence from Penn State to return to Boston to work on the idea full time. In January 1998, the company opened its headquarters in Cambridge, Massachusetts. The first three employees were all previously affiliated with the chemical engineering department at MIT.

Once established in their new offices, Langer and Edwards quickly went to work on their first human clinical trial. As their science progressed, so too did interest from outside parties. The founders decided on a two-tier strategy. For drugs that had gone off patent, they would manufacture generic versions of the drugs themselves. For newly developed drugs, AIR would partner with leading pharmaceutical firms to exclusively manufacture the drugs for the delivery system. McGuire focused on successfully closing the deals with minimal dilution of Polaris's stake in the firm.

For a nascent firm, AIR was extremely successful. It had raised capital for an early-stage technology, was making progress in its clinical trials, and was forming favorable partnerships. Nevertheless, when suitors began knocking at the door, AIR needed to consider their offers seriously. To continue to grow, AIR would need to scale up its operations significantly and move from a small R & D firm to a larger manufacturing operation. This task was extremely difficult and one that only a small handful of biotechnology firms had achieved. In late 1998, just a year and a half after AIR was founded, Alkermes offered to acquire the firm. In February 1999, AIR agreed to Alkermes's offer of an all-stock deal valued around $125 million. The deal was considered a success on all sides. Polaris received a healthy return on its investment, AIR's technology was a strategic fit for Alkermes's existing drug delivery systems, and Alkermes's savvy management team and established reputation in the industry allowed it to find the most appropriate partner for AIR.

All three founders remain active in the biotechnology industry. Edwards, now a professor at Harvard University, cofounded Pulmatrix in the spring of 2003 and has been involved in the formation of a number of nonprofit organizations. McGuire remains a general partner with Polaris Ventures and has invested in other early-stage life science firms. Finally, Langer continues to be wildly prolific. He holds more than 800 granted and pending patents, has licensed his technology to more than 100 firms, and was named one of the 100 most important people by CNN and *Time* magazine.

*Sources:* Roberts and Gardner, 2000, and www.alkermes.com.

## 16.6 Exit and Failure

A large percentage of new ventures shut down within a few years of initiation. Some terminate their efforts when they fail to achieve the original goals; others terminate when they simply run out of cash. Most entrepreneurs and investors assert that new ventures fail because of inadequate management skills, a poor strategy, and inadequate capitalization as well as poor market conditions [Zacharakis et al., 1999]. For most entrepreneurs, an inadequate team with inadequate past experience leads to failure.

Many people learn from business failure by revising their knowledge and assumptions about their skills [Shepherd, 2003]. Learning can be measured in terms of an increased understanding of why the business failed, in an effort to prevent repeating the same mistakes. Often entrepreneurs are overconfident and unrealistically overrate their knowledge and skills. They discount the risks inherent in a new venture. Furthermore, they may exaggerate their ability to control events and people. Examples of highly visible failures in telecommunications are Helio and Amp'd Mobile.

Knowing when to stop or terminate a venture may be as important as knowing when to start. The concept of **sunk** is that a cost that has already incurred cannot be affected by any present or future decisions. In other words, funds and time invested on a new venture are already gone, regardless of any action you take today or later. If a new venture has not worked out as planned, one should look at proceeding with the venture as a new decision, as shown in Figure 16.8. The decision to terminate or continue should, if possible, be looked at afresh with the information at hand. If a venture has run out of cash and the market has not responded to the new venture, it may be wise to exit the venture.

Pandesic was a joint venture of SAP and Intel designed to develop information architectures for Internet companies. Founded in 1997, Pandesic intended to create a unique e-business solution that would automate the entire business process for companies doing business on the Internet. By March 1997, Pandesic had grown to 100 employees. By the end of 1997, sales were slowly appearing, and by 1998, it was still struggling without many sales. By April 1999, the sales force was reorganized. As a result, sales started to grow, but the firm was still

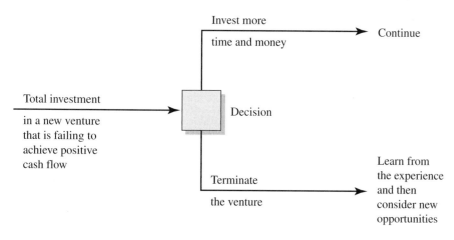

**FIGURE 16.8** Decision tree for the sunk cost dilemma.

unprofitable. Pandesic had 400 employees and 100 customers by mid-2000, but it was still experiencing negative cash flow (said to be $80 million per year) and closed its doors in July 2000. The decision to terminate was based on large cash losses and recognition that profitability was not within reach [Girard, 2000].

If the decision is to continue, it may be wise to consider the next phase of the company as a turnaround and devise deliberate interventions that increase the level of communication, collaboration, and respect among all the participants [Kanter, 2003].

If the decision is to terminate the new venture, it is best to try to learn from it. Ideally, it is the venture that failed, not the people in it. Picking oneself up, learning from the venture, and then moving on is the best process. Every exit is an entry somewhere else.

The alternative of investing more time and money should be based on rational recognition of the sunk costs and the potential for recovery and success. An entrepreneur can examine the situation afresh and decide if the opportunity still looks good enough to invest more time and money.

## 16.7 Spotlight on Baidu

Robin Li and Eric Xu incorporated Baidu in January 2000. Baidu is a Web services company headquartered in Beijing, China. It offers Internet search and information services in the Chinese language.

Baidu has a revenue model based on advertising revenues. Most of these revenues are linked to online shopping, a growing market in China. Baidu also is growing in the mobile-device and cloud-storage markets.

Baidu's 2012 revenue was 22.3 billion yuan ($3.5 billion) with profits of about 7 billion yuan ($1.1 billion). Baidu is consistently striving to build its revenue and profits by adjusting its revenue and profit models. Support from the Chinese government, increasing demand from consumers, and expansion of Baidu's services have fueled strong growth.

## 16.8 Summary

A new firm formulates its revenue model to clearly describe how it will generate and grow its revenues. Revenues are important, but positive cash flow and profitability are critical to ultimate success. Thus, cost and profit models that can be readily implemented must be created early in a firm's planning. Profit does not occur naturally but rather flows from value shared with the customer. Thus, enough customers must experience high value and find that the firm is the best, if not the only, provider of this value.

Managing revenue growth to match the growth of cash flow is important to achieve organic, internally funded growth. Otherwise, the new firm must constantly seek new financial resources from investors and lenders. Many firms find it necessary to terminate their activities when they are unable to access new sources of funds. With positive results and reasonable growth, a firm should consider a plan to harvest the rewards so that all owners receive a financial return on their investment.

---

**Principle 16**

A new firm with a powerful revenue and profit engine can achieve strong but manageable growth leading to a favorable harvest of the wealth for the owners.

---

## Video Resources

Visit http://techventures.stanford.edu to view experts discussing content from this chapter.

| | | |
|---|---|---|
| Critical Early Decisions with Long Lasting Results | Robin Li | Baidu |
| Why Google Won Desktop Search | Dan Rosensweig | Google |
| Exit Strategy for the Single Product Company | Mir Imran | InCube |

# 16.9 Exercises

**16.1** A company's value chain was discussed in Section 14.1. Figure 16.1 outlines the division between profit and cost. Extend Figure 16.1 to account for the profits and costs of partners and suppliers in a company's value chain.

**16.2** Google uses a complex revenue model and a related profit engine. It has a large base of users and advertisers, and works to link its users to retailers. Describe its revenue and profit model. Contrast this with Yahoo's revenue model. Examining recent quarterly reports will give an indication of the "split" of revenue among business units.

**16.3** Zipcar gives its members short-term, on-demand use of a fleet of cars (www.zipcar.com). Before launching this service, Zipcar proposed a pricing scheme of a $300 refundable security deposit, a $300 annual subscription fee, and $1.50 per hour plus 40 cents per mile. Create another pricing scheme that will be more profitable.

**16.4** Skype offers Voice over Internet Protocol (VoIP) telephony services and PC to PC calling. The telephone service began as a free service. Describe the revenue and profit model for Skype specifically. Has its acquisition by Microsoft changed Skype's revenue model?

**16.5** Salesforce.com sells software as a service delivered online. Corporations pay about $60 per month for each user (www.salesforce.com). Describe its revenue model and profit model.

**16.6** Research and determine the most profitable industry sector in a country of interest (profitability on either a relative or absolute basis). What is driving this high profitability? Is it sustainable, and for how long?

**16.7** Compare the most recent yearly income statements for Microsoft, Apple, Dell, Sony, and Qualcomm. Examine the gross margins for each of these companies (cost of goods sold divided by revenue). Complete a similar ratio analysis against Sales, General & Administrative (SG&A) and Research & Development (R&D). When normalized against revenues, what do these ratios tell us about the strategies and operations of these companies?

## VENTURE CHALLENGE

**1.** Describe the revenue model of your venture.

**2.** Describe the profit model of your venture.

**3.** Using Table 16.7, discuss your harvest plan.

CHAPTER $17$

# The Financial Plan

*Budgets are not merely affairs of arithmetic, but in a thousand ways go to the root of prosperity of individuals, the relation of classes, and the strength of kingdoms.*
**William E. Gladstone**

### How do entrepreneurs describe the financial elements of their new venture?

Entrepreneurs build a financial plan to determine the economic potential for their venture. This plan provides an estimate of the potential of the venture. Of course, any estimate is based on a set of assumptions regarding sales revenues and costs. Using the best available information and their intuition, entrepreneurs calculate the potential profitability of the venture. Furthermore, they need to determine the flow of cash monthly to identify the cash investments that will be required over a two- or three-year period. An income statement and a balance sheet also are required to demonstrate profitability and liquidity.

Using the estimates of sales, the venture team can determine the number of units it needs to sell to break even. Furthermore, it can calculate several measures of profitability that demonstrate the return provided by its venture for investors. The best ventures grow sales consistently and provide positive cash flow and profit early in their life. ■

## 17.1  Building a Financial Plan

A sound business plan is based on a solid vision and a business design or concept. It is an expression of the theory of the business in the form of a story. This story also needs to make sense financially as a business model. The business model tells a story about the customer and the value proposition that leads to revenue and profit. To create this value for the customer, a new firm needs to build a financial plan that describes the expected revenues, cash flows, profits, and investments necessary to achieve them. The purpose of any business is to create value for its customers and to generate a return on investment for its owners. A financial plan provides an estimate of projected cash flow and return on investment.

To create a financial plan, entrepreneurs must clearly state their assumptions about sales and costs. What resources will it take, over what time frame, to achieve expected sales and profitability? The calculation of cash flows is based on a set of assumptions, which we will call the **base case,** that portrays the most likely outcomes. It may be prudent to also determine the outcome of a situation in which the expectations are not realized as expected, called the **pessimistic case. Cash flow** is the amount of cash flowing into or out of a firm during a specific period. It is arrived at by subtracting the amount paid out in dividends from the net profit and then adding back noncash expenses such as depreciation. See Table 17.10 at the end of this chapter for a glossary of accounting and financial terms.

The entrepreneurs' goal is to develop a solid set of financial projections that will include a pro forma income statement. **Pro forma** means provided in advance of actual data. Pro forma statements are forecasts of financial outcomes. The creation of a set of financial projections starts with a sales forecast based on a set of assumptions regarding the customer and sales growth. Then the calculation of projected sales over a two- or three-year period can be developed. This is step 1, as shown in Table 17.1. The second step is to state the assumed costs of doing business in the time frame described in step 1. In step 2, the costs associated with the projected sales can be calculated. Step 3 is to calculate the expected income and cash flow forecast over the time frame based on a set of assumptions regarding the timing of sales and receipts as well as payables to vendors and others. The final step is to calculate the balance sheet on an annual basis for the two- or three-year period. The balance sheet at the starting point of the new venture will need to be described by stating the assumed starting investments and required assets. Please visit this textbook's websites for tools and links useful towards building a financial plan.

The cash flows, assets, balance sheet, and revenue projections are all interconnected through linkages. Accounting items are classified into "accounts" according to their nature, translated into monetary units, and organized in statements. The basic accounting formula is

$$\text{Assets} = \text{liabilities} + \text{equity}$$

where assets are what the company owns and liabilities are the amounts it owes to other persons or entities. Equity is the company's net worth (book value) expressed as

## TABLE 17.1 Four steps to building a financial plan.

### 1. Sales forecast

- Time frame—two or three years
- Assumptions about sales per customer, number of customers, and growth rate of sales
- Calculation of the sales forecast

### 2. Costs forecast

- Assumptions about the costs of doing business in the specified time frame
- Calculation of the costs associated with the projected sales of step 1

### 3. Income and cash flow forecast

- Assumptions about the timing of cash receivables and payables specified in the time frame
- Calculation of the income and cash flow associated with the projected sales and costs on a monthly basis over the time frame

### 4. Balance sheet

- Assumptions about the starting value of cash and assets
- Calculation based on the income and cash flows from step 3

$$\text{Equity} = \text{assets} - \text{liabilities}$$

**Equity** is the ownership of the firm, usually divided into certificates called common or preferred shares of stock. The assets and liabilities are linked to income and expenses, as shown in Figure 17.1. Assets are used to generate income, and liabilities require expenses such as rent, payments, or return of loans. Book value is the firm's net worth and is often called accounting value. Market value is share price times the number of issued shares. Note that book value is not equal to market value, which is the perceived value of the firm given its growth potential.

The financial plan is critical to the evaluation of the business model of the new venture. With sound assumptions, the projected results will help in evaluating the venture and its financial viability. The resulting financial plan is only as good as the quality of the assumptions. One reason forecasting models are so fallible is that they rely on the assumptions that the user chooses to input [Riggs, 2004].

New firms should select two or three parameters of the business that display the greatest impact on the cash flows of the business. Examples are sales growth rate and new customer acquisition rate. They should then test the changes in sales as each of these parameters is changed. Analysis of these sensitivity results will allow for more focus and better decision making.

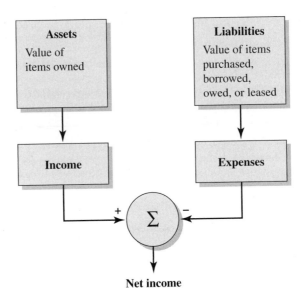

**FIGURE 17.1**  Assets generate income and liabilities lead to expenses. Net income is income minus expenses.

## 17.2  Sales Projections

Sales projections will normally be developed for a two- or three-year period on a monthly basis. The sales forecast for a new venture is often the weakest link in the financial plan. Since the new venture has not actually obtained sales, the firm can only work with assumptions based on inadequate information.

In this chapter, consider a fictitious new venture, named e-Travel. It seeks to sell travel guide books (e-books) directly to readers who will download the guides via the Internet and read them on their tablets or other devices. Short books such as these travel guides operate on a pull model, since customers order them only when they wish to read them. This is opposite of the usual push model, in which the publisher prints the book and then tries to find a purchaser. For travelers, a tablet is easier to use than several heavy books. Using keywords such as "pizza restaurant Denver," the reader can use the search function to get the information quickly.

The new venture needs to build a financial plan. The first step is to build the sales projections. e-Travel has created a network of the best authors of travel guides. These authors have signed publisher agreements and have provided e-Travel with electronic guides to over five hundred cities, regions, and leisure and recreation destinations throughout the world. All these guides are written following a common format, and key search words are identified.

e-Travel expects to sell 1,200 guides starting in the third month of operation. The price per guide is $15, paid by credit card when the guide is ordered online. Based on market research, the expected growth rate of sales is 10 percent per month. The pessimistic growth rate is 1 percent per month. The sales projections for the expected growth rate are shown in Table 17.2 for a three-year period. Assuming a 10 percent growth rate per month, sales amount to over $3 million in year three. Note the bolded ovals for key results in tables.

## 17.3  Costs Forecast

To determine the expected costs of doing business, the new venture team must examine its needs for facilities, equipment, and employees. Our example business, e-Travel, will need an office, computers, software, and office furniture. The authors will constantly update the information in their guides and will receive royalties at 12 percent of net revenues paid on the fifteenth day of the month following the sale generated by their guides. The costs of doing business will include salaries, marketing, and communication costs, and normal office utilities and supplies. The cost assumptions for e-Travel are summarized in Table 17.3. Every new venture needs to construct a set of assumptions at a similar level of detail.

## 17.4  Income Statement

The income statement reports the economic results of a firm over a time period. The income statement is calculated as shown in Figure 17.2 [Maher et al., 2004]. Note that operating expenses are often grouped into four general categories: sales and marketing, general and administrative, research and development, and depreciation. Other formats are acceptable.

The income (profit and loss) statement for e-Travel depicts the venture's expected performance over a period of time—in this case, three years. The sales, costs, and profits or losses are shown monthly. The purpose of the income statement is to show how much profit or loss is generated. Due to the online nature of e-Travel, it has no cost of goods sold. The profit and loss statement is shown in Table 17.4. The firm is profitable by the fifth month and shows a profit of $19,954 for the first year (for the base case).

## 17.5  Cash Flow Statement

The cash flow statement shows the actual flow of cash into and out of the venture. The cash flow statement tracks when the venture actually receives and spends the cash. A venture with positive cash flow can continue to operate without new debt or equity capital. If a cash flow statement reveals projected negative cash in some period, it will be necessary to plan for a new capital infusion. We define cash flow as the sum of retained earnings minus the depreciation provision made by a firm [Maher et al., 2004].

**TABLE 17.2  Sales projections for the expected growth rate (10 percent per month).**

**Year 1**

| Month | 1 | 2 | 3 | 4 | 5 | 6 | 7 | 8 | 9 | 10 | 11 | 12 | Year total |
|---|---|---|---|---|---|---|---|---|---|---|---|---|---|
| Units | 0 | 0 | 1,200 | 1,320 | 1,452 | 1,597 | 1,757 | 1,933 | 2,126 | 2,339 | 2,573 | 2,830 | 19,127 |
| Price per unit | $15 | $15 | $15 | $15 | $15 | $15 | $15 | $15 | $15 | $15 | $15 | $15 | |
| Sales dollars | $0 | $0 | $18,000 | $19,800 | $21,780 | $23,955 | $26,355 | $28,995 | $31,890 | $35,085 | $38,595 | $42,450 | $286,905 |

**Year 2**

| Month | 1 | 2 | 3 | 4 | 5 | 6 | 7 | 8 | 9 | 10 | 11 | 12 | Year total |
|---|---|---|---|---|---|---|---|---|---|---|---|---|---|
| Units | 3,113 | 3,424 | 3,766 | 4,143 | 4,557 | 5,013 | 5,514 | 6,065 | 6,672 | 7,339 | 8,073 | 8,880 | 66,559 |
| Price per unit | $15 | $15 | $15 | $15 | $15 | $15 | $15 | $15 | $15 | $15 | $15 | $15 | |
| Sales dollars | $46,695 | $51,360 | $56,490 | $62,145 | $68,355 | $75,195 | $82,710 | $90,975 | $100,080 | $110,085 | $121,095 | $133,200 | $998,385 |

**Year 3**

| Month | 1 | 2 | 3 | 4 | 5 | 6 | 7 | 8 | 9 | 10 | 11 | 12 | Year total |
|---|---|---|---|---|---|---|---|---|---|---|---|---|---|
| Units | 9,768 | 10,745 | 11,820 | 13,002 | 14,302 | 15,732 | 17,305 | 19,036 | 20,940 | 23,034 | 25,337 | 27,871 | 208,892 |
| Price per unit | $15 | $15 | $15 | $15 | $15 | $15 | $15 | $15 | $15 | $15 | $15 | $15 | |
| Sales dollars | $146,520 | $161,175 | $177,300 | $195,030 | $214,530 | $235,980 | $259,575 | $285,540 | $314,100 | $345,510 | $380,055 | $418,065 | $3,133,380 |

## TABLE 17.3  Cost assumptions for e-Travel.

- Author royalties: 12 percent of net revenues paid on the fifteenth of the month following the sale
- Credit card: 1 percent to credit card providers, paid electronically as sale is processed
- Office rent: $1,500 per month
- Physical assets (computers, furniture, etc.): $48,000 per year; four-year life
- Depreciation of equipment (monthly):

| Year 1 | Year 2 | Year 3 |
|--------|--------|--------|
| $1,000 | $2,000 | $3,000 |

- Salaries (monthly):

|                                        | Year 1  | Year 2   | Year 3   |
|----------------------------------------|---------|----------|----------|
| President                              | $4,000  | $ 5,500  | $ 6,500  |
| Vice president                         | 4,000   | 5,000    | 6,000    |
| Administrative manager (part time)     | 1,500   | 3,000    | 3,500    |
| Total:                                 | $9,500  | $13,500  | $16,000  |

- Social Security taxes and other benefits: 15 percent of total salaries
- Marketing (monthly):

| Year 1 | Year 2 | Year 3 |
|--------|--------|--------|
| $2,000 | $2,500 | $3,000 |

- Utilities, supplies, travel, communication (monthly):

| Year 1 | Year 2 | Year 3 |
|--------|--------|--------|
| $2,000 | $3,000 | $4,000 |

- Interest expense: $1,000 per month ($100,000 loan at 12 percent per year; interest paid monthly; principal to be paid at end of five years)
- Income taxes: 30 percent of income before taxes

A growing business needs cash to operate. The detailed cash flow process is shown in Figure 17.3. Firms calculate their cash on hand at the end of each month. Therefore,

$$TC(N + 1) = (CF - \text{Disbursements}) + TC(N)$$

where $TC(N + 1)$ is the cash at the end of month $(N + 1)$, $TC(N)$ is the total cash at the end of month $(N)$, and CF is the cash flow for the month.

The cash flow statement for e-Travel is provided in Table 17.5 (for the base case). It is assumed that the founders invest $140,000 in cash and obtain a bank loan of $100,000 secured by their personal assets. This $240,000 is for operations

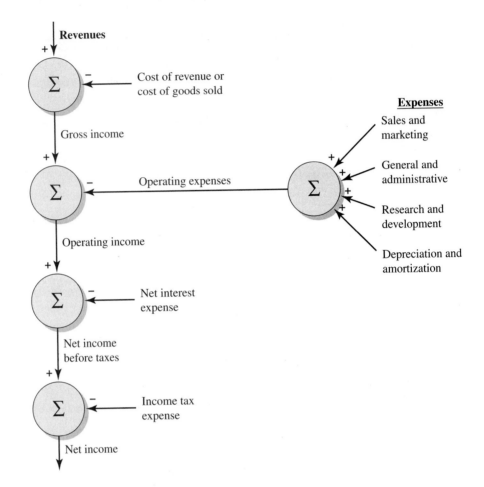

**FIGURE 17.2**  Calculation of the income statement.

and the initial purchase of long-term assets such as computers and furniture, as shown in month 1 of year 1 in Table 17.5. The initial investment of $240,000 can be considered an equity investment since the loan is personally guaranteed by the two founders. Under the base case assumption of 10 percent growth in revenues each month, the cash flow quickly becomes positive.

## 17.6  Balance Sheet

The new venture team should prepare a balance sheet at the opening of the business and for the end of each year. The balance sheet depicts the conditions of the business by displaying the assets, liabilities, and owners' equity of the business [Maher et al., 2004]. The format for a balance sheet is shown in Figure 17.4 for a business at the end of the year.

# TABLE 17.4 Profit and loss statement.

## Year 1

| Month | 1 | 2 | 3 | 4 | 5 | 6 | 7 | 8 | 9 | 10 | 11 | 12 | Year total |
|---|---|---|---|---|---|---|---|---|---|---|---|---|---|
| Revenues | $0 | $0 | $18,000 | $19,800 | $21,780 | $23,955 | $26,355 | $28,995 | $31,890 | $35,085 | $38,595 | $42,450 | $286,905 |
| Expenses: | | | | | | | | | | | | | |
| Author royalties | 0 | 0 | 2,160 | 2,376 | 2,614 | 2,875 | 3,163 | 3,479 | 3,827 | 4,210 | 4,631 | 5,094 | 34,429 |
| Credit card charges | 0 | 0 | 180 | 198 | 218 | 240 | 264 | 290 | 319 | 351 | 386 | 425 | 2,871 |
| Marketing | 2,000 | 2,000 | 2,000 | 2,000 | 2,000 | 2,000 | 2,000 | 2,000 | 2,000 | 2,000 | 2,000 | 2,000 | 24,000 |
| Depreciation | 1,000 | 1,000 | 1,000 | 1,000 | 1,000 | 1,000 | 1,000 | 1,000 | 1,000 | 1,000 | 1,000 | 1,000 | 12,000 |
| Interest | 1,000 | 1,000 | 1,000 | 1,000 | 1,000 | 1,000 | 1,000 | 1,000 | 1,000 | 1,000 | 1,000 | 1,000 | 12,000 |
| Office rent | 1,500 | 1,500 | 1,500 | 1,500 | 1,500 | 1,500 | 1,500 | 1,500 | 1,500 | 1,500 | 1,500 | 1,500 | 18,000 |
| Salaries | 9,500 | 9,500 | 9,500 | 9,500 | 9,500 | 9,500 | 9,500 | 9,500 | 9,500 | 9,500 | 9,500 | 9,500 | 114,000 |
| Social Security and benefits | 1,425 | 1,425 | 1,425 | 1,425 | 1,425 | 1,425 | 1,425 | 1,425 | 1,425 | 1,425 | 1,425 | 1,425 | 17,100 |
| Utilities, supplies, travel, communication | 2,000 | 2,000 | 2,000 | 2,000 | 2,000 | 2,000 | 2,000 | 2,000 | 2,000 | 2,000 | 2,000 | 2,000 | 24,000 |
| Profit (loss) before income tax | (18,425) | (18,425) | (2,765) | (1,199) | 523 | 2,415 | 4,503 | 6,801 | 9,319 | 12,099 | 15,153 | 18,506 | 28,505 |
| Income tax (credit) | (5,528) | (5,528) | (830) | (360) | 157 | 725 | 1,351 | 2,040 | 2,796 | 3,630 | 4,546 | 5,552 | 8,551 |
| Net profit (loss) | ($12,897) | ($12,897) | ($1,935) | ($839) | $366 | $1,690 | $3,152 | $4,761 | $6,523 | $8,469 | $10,607 | $12,954 | $19,954 |

## Year 2

| Month | 1 | 2 | 3 | 4 | 5 | 6 | 7 | 8 | 9 | 10 | 11 | 12 | Year total |
|---|---|---|---|---|---|---|---|---|---|---|---|---|---|
| Revenues | $46,695 | $51,360 | $56,490 | $62,145 | $68,355 | $75,195 | $82,710 | $90,975 | $100,080 | $110,085 | $121,095 | $133,200 | $998,385 |
| Expenses: | | | | | | | | | | | | | |
| Author royalties | 5,603 | 6,163 | 6,779 | 7,457 | 8,203 | 9,023 | 9,925 | 10,917 | 12,010 | 13,210 | 14,531 | 15,984 | 119,805 |
| Credit card charges | 467 | 514 | 565 | 621 | 684 | 752 | 827 | 910 | 1,001 | 1,101 | 1,211 | 1,332 | 9,985 |
| Marketing | 2,500 | 2,500 | 2,500 | 2,500 | 2,500 | 2,500 | 2,500 | 2,500 | 2,500 | 2,500 | 2,500 | 2,500 | 30,000 |
| Depreciation | 2,000 | 2,000 | 2,000 | 2,000 | 2,000 | 2,000 | 2,000 | 2,000 | 2,000 | 2,000 | 2,000 | 2,000 | 24,000 |
| Interest | 1,000 | 1,000 | 1,000 | 1,000 | 1,000 | 1,000 | 1,000 | 1,000 | 1,000 | 1,000 | 1,000 | 1,000 | 12,000 |
| Office rent | 1,500 | 1,500 | 1,500 | 1,500 | 1,500 | 1,500 | 1,500 | 1,500 | 1,500 | 1,500 | 1,500 | 1,500 | 18,000 |
| Salaries | 13,500 | 13,500 | 13,500 | 13,500 | 13,500 | 13,500 | 13,500 | 13,500 | 13,500 | 13,500 | 13,500 | 13,500 | 162,000 |
| Social Security and benefits | 2,025 | 2,025 | 2,025 | 2,025 | 2,025 | 2,025 | 2,025 | 2,025 | 2,025 | 2,025 | 2,025 | 2,025 | 24,300 |
| Utilities, supplies, travel, communication | 3,000 | 3,000 | 3,000 | 3,000 | 3,000 | 3,000 | 3,000 | 3,000 | 3,000 | 3,000 | 3,000 | 3,000 | 36,000 |
| Profit before income tax | 15,100 | 19,158 | 23,621 | 28,542 | 33,943 | 39,895 | 46,433 | 53,623 | 61,544 | 70,249 | 79,828 | 90,359 | 562,295 |
| Income tax | 4,530 | 5,747 | 7,086 | 8,563 | 10,183 | 11,969 | 13,930 | 16,087 | 18,463 | 21,075 | 23,948 | 27,108 | 168,689 |
| Net profit | $10,570 | $13,411 | $16,535 | $19,979 | $23,760 | $27,926 | $32,503 | $37,536 | $43,081 | $49,174 | $55,880 | $63,251 | $393,606 |

*(continued on next page)*

**TABLE 17.4  (continued)**

**Year 3**

| Month | 1 | 2 | 3 | 4 | 5 | 6 | 7 | 8 | 9 | 10 | 11 | 12 | Year total |
|---|---|---|---|---|---|---|---|---|---|---|---|---|---|
| Revenues | $146,520 | $161,175 | $177,300 | $195,030 | $214,530 | $235,980 | $259,575 | $285,540 | $314,100 | $345,510 | $380,055 | $418,065 | $3,133,380 |
| Expenses: | | | | | | | | | | | | | |
| Author royalties | 17,582 | 19,341 | 21,276 | 23,404 | 25,744 | 28,318 | 31,149 | 34,265 | 37,692 | 41,461 | 45,607 | 50,168 | 376,007 |
| Credit card charges | 1,465 | 1,612 | 1,773 | 1,950 | 2,145 | 2,360 | 2,596 | 2,855 | 3,141 | 3,455 | 3,801 | 4,181 | 31,334 |
| Marketing | 3,000 | 3,000 | 3,000 | 3,000 | 3,000 | 3,000 | 3,000 | 3,000 | 3,000 | 3,000 | 3,000 | 3,000 | 36,000 |
| Depreciation | 3,000 | 3,000 | 3,000 | 3,000 | 3,000 | 3,000 | 3,000 | 3,000 | 3,000 | 3,000 | 3,000 | 3,000 | 36,000 |
| Interest | 1,000 | 1,000 | 1,000 | 1,000 | 1,000 | 1,000 | 1,000 | 1,000 | 1,000 | 1,000 | 1,000 | 1,000 | 12,000 |
| Office rent | 1,500 | 1,500 | 1,500 | 1,500 | 1,500 | 1,500 | 1,500 | 1,500 | 1,500 | 1,500 | 1,500 | 1,500 | 18,000 |
| Salaries | 16,000 | 16,000 | 16,000 | 16,000 | 16,000 | 16,000 | 16,000 | 16,000 | 16,000 | 16,000 | 16,000 | 16,000 | 192,000 |
| Social Security and benefits | 2,400 | 2,400 | 2,400 | 2,400 | 2,400 | 2,400 | 2,400 | 2,400 | 2,400 | 2,400 | 2,400 | 2,400 | 28,800 |
| Utilities, supplies, travel, communication | 4,000 | 4,000 | 4,000 | 4,000 | 4,000 | 4,000 | 4,000 | 4,000 | 4,000 | 4,000 | 4,000 | 4,000 | 48,000 |
| Profit before income tax | 96,573 | 109,322 | 123,351 | 138,776 | 155,741 | 174,402 | 194,930 | 217,520 | 242,367 | 269,694 | 299,747 | 332,816 | 2,355,239 |
| Income tax | 28,972 | 32,797 | 37,005 | 41,633 | 46,722 | 52,321 | 58,479 | 65,256 | 72,710 | 80,908 | 89,924 | 99,845 | 706,572 |
| Net profit | $67,601 | $76,525 | $86,346 | $97,143 | $109,019 | $122,081 | $136,451 | $152,264 | $169,657 | $188,786 | $209,823 | $232,971 | $1,648,667 |

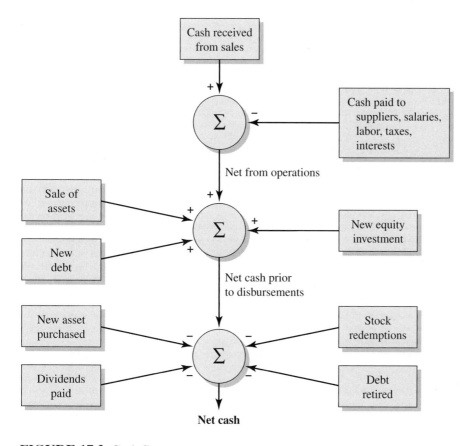

**FIGURE 17.3** Cash flow process.

The balance sheet for e-Travel is shown in Table 17.6. The balance sheet shows the assets, such as cash, equipment, furniture, and accumulated depreciation. The liabilities are the loan payable and the royalties to the authors. Total owners' equity consists of contributions of $140,000 and retained earnings. Table 17.6 shows the balance sheet at the end of month 1, year 1, year 2, and year 3. The base case balance sheet provides evidence of the financial strength of e-Travel.

## 17.7 Results for a Pessimistic Growth Rate

Any new venture needs to plan for the likely case and prepare for the worst case. For e-Travel, we will assume the pessimistic case occurs when the sales grow at a rate of only 1 percent per month. The summary of the results for the pessimistic case is shown in Table 17.7. Table 17.7A shows the sales projections for the first three years. Sales for year 3 are $284,010 for the pessimistic case, while they were estimated at $3,133,380 for the expected case. Table 17.7B shows the profit and loss statement for the pessimistic case. Note that the firm is not profitable in the pessimistic case in any year.

# TABLE 17.5  Cash flow statement.

## Year 1

| Month | 1 | 2 | 3 | 4 | 5 | 6 | 7 | 8 | 9 | 10 | 11 | 12 | Year total |
|---|---|---|---|---|---|---|---|---|---|---|---|---|---|
| *Operating activities* | | | | | | | | | | | | | |
| Net profit (loss) | ($12,897) | ($12,897) | ($1,935) | ($839) | $366 | $1,690 | $3,152 | $4,761 | $6,523 | $8,469 | $10,607 | $12,954 | $19,954 |
| Add: Depreciation | 1,000 | 1,000 | 1,000 | 1,000 | 1,000 | 1,000 | 1,000 | 1,000 | 1,000 | 1,000 | 1,000 | 1,000 | 12,000 |
| Add: Increase in royalties payable | | | 2,160 | 216 | 238 | 261 | 288 | 316 | 348 | 383 | 421 | 463 | 5,094 |
| Cash flow from operations | (11,897) | (11,897) | 1,225 | 377 | 1,604 | 2,951 | 4,440 | 6,077 | 7,871 | 9,852 | 12,028 | 14,417 | 37,048 |
| *Investing activities* | | | | | | | | | | | | | |
| Purchase of long-term assets | (48,000) | | | | | | | | | | | | (48,000) |
| *Financing activities* | | | | | | | | | | | | | |
| Bank loan | 100,000 | | | | | | | | | | | | 100,000 |
| Owners' cash contributions | 140,000 | | | | | | | | | | | | 140,000 |
| Increase (decrease) in cash | 180,103 | (11,897) | 1,225 | 377 | 1,604 | 2,951 | 4,440 | 6,077 | 7,871 | 9,852 | 12,028 | 14,417 | 229,048 |
| Beginning cash balance | 0 | 180,103 | 168,206 | 169,431 | 169,808 | 171,412 | 174,363 | 178,803 | 184,880 | 192,751 | 202,603 | 214,631 | 0 |
| Ending cash balance | $180,103 | $168,206 | $169,431 | $169,808 | $171,412 | $174,363 | $178,803 | $184,880 | $192,751 | $202,603 | $214,631 | $229,048 | $229,048 |

## Year 2

| Month | 1 | 2 | 3 | 4 | 5 | 6 | 7 | 8 | 9 | 10 | 11 | 12 | Year total |
|---|---|---|---|---|---|---|---|---|---|---|---|---|---|
| *Operating activities* | | | | | | | | | | | | | |
| Net profit | $10,570 | $13,411 | $16,535 | $19,979 | $23,760 | $27,926 | $32,503 | $37,536 | $43,081 | $49,174 | $55,880 | $63,251 | $393,606 |
| Add: Depreciation | 2,000 | 2,000 | 2,000 | 2,000 | 2,000 | 2,000 | 2,000 | 2,000 | 2,000 | 2,000 | 2,000 | 2,000 | 24,000 |
| Add: Increase in royalties payable | 509 | 560 | 616 | 678 | 746 | 820 | 902 | 992 | 1,093 | 1,200 | 1,321 | 1,453 | 10,890 |
| Cash flow from operations | 13,079 | 15,971 | 19,151 | 22,657 | 26,506 | 30,746 | 35,405 | 40,528 | 46,174 | 52,374 | 59,201 | 66,704 | 428,496 |
| *Investing activities* | | | | | | | | | | | | | |
| Purchase of long-term assets | (48,000) | | | | | | | | | | | | (48,000) |
| Increase (decrease) in cash | (34,921) | 15,971 | 19,151 | 22,657 | 26,506 | 30,746 | 35,405 | 40,528 | 46,174 | 52,374 | 59,201 | 66,704 | 380,496 |
| Beginning cash balance | 229,048 | 194,127 | 210,098 | 229,249 | 251,906 | 278,412 | 309,158 | 344,563 | 385,091 | 431,265 | 483,639 | 542,840 | 229,048 |
| Ending cash balance | $194,127 | $210,098 | $229,249 | $251,906 | $278,412 | $309,158 | $344,563 | $385,091 | $431,265 | $483,639 | $542,840 | $609,544 | $609,544 |

## Year 3

| Month | 1 | 2 | 3 | 4 | 5 | 6 | 7 | 8 | 9 | 10 | 11 | 12 | Year total |
|---|---|---|---|---|---|---|---|---|---|---|---|---|---|
| *Operating activities* | | | | | | | | | | | | | |
| Net profit | $67,601 | $76,525 | $86,346 | $97,143 | $109,019 | $122,081 | $136,451 | $152,264 | $169,657 | $188,786 | $209,823 | $232,971 | $1,648,667 |
| Add: Depreciation | 3,000 | 3,000 | 3,000 | 3,000 | 3,000 | 3,000 | 3,000 | 3,000 | 3,000 | 3,000 | 3,000 | 3,000 | 36,000 |
| Add: Increase in royalties payable | 1,598 | 1,759 | 1,935 | 2,128 | 2,340 | 2,574 | 2,831 | 3,116 | 3,427 | 3,769 | 4,146 | 4,561 | 34,184 |
| Cash flow from operations | 72,199 | 81,284 | 91,281 | 102,271 | 114,359 | 127,655 | 142,282 | 158,380 | 176,084 | 195,555 | 216,969 | 240,532 | 1,718,851 |
| *Investing activities* | | | | | | | | | | | | | |
| Purchase of long-term assets | (48,000) | | | | | | | | | | | | (48,000) |
| Increase (decrease) in cash | 24,199 | 81,284 | 91,281 | 102,271 | 114,359 | 127,655 | 142,282 | 158,380 | 176,084 | 195,555 | 216,969 | 240,532 | 1,670,851 |
| Beginning cash balance | 609,544 | 633,743 | 715,027 | 806,308 | 908,579 | 1,022,938 | 1,150,593 | 1,292,875 | 1,451,255 | 1,627,339 | 1,822,894 | 2,039,863 | 609,544 |
| Ending cash balance | $633,743 | $715,027 | $806,308 | $908,579 | $1,022,938 | $1,150,593 | $1,292,875 | $1,451,255 | $1,627,339 | $1,822,894 | $2,039,863 | $2,280,395 | $2,280,395 |

**Balance sheet of ABC Corporation, 31 December 201X**

| Current assets (liquid in less than a year) | { | Cash and equivalents<br>Accounts receivable<br>Inventories | Current liabilities (payable in less than a year) | { | Accounts payable<br>Accounts expenses<br>Short-term debt |
|---|---|---|---|---|---|
| Fixed assets | { | Property, plant, and equipment (minus depreciation) | Long-term liabilities | { | Bonds issued<br>Bank loans |
| Other assets | { | Intangibles (minus depreciation)<br>Investment securities | Shareholders' equity | { | Common stock<br>Additional paid-in capital<br>Retained earnings |

**Total assets = total liabilities + shareholders' equity**

**FIGURE 17.4** Format for a balance sheet.

We show the cash flow statement for the pessimistic case in Table 17.7C. Notice that the ending cash balance turns negative in month 1 of year 3. The company would need a cash infusion in month 1 of year 3 to continue operating.

To be a profitable venture, e-Travel needs to attain a sales growth rate that exceeds 4 percent per month over the first two years. This figure can be determined by modifying the spreadsheet calculation with various growth rates.

## 17.8  Breakeven Analysis

In the initial stages of building a financial plan, it is useful to know when a profit may be achieved. **Breakeven** is defined as when the total sales equals the total costs. Total sales (R) are

$$R = Q \times P$$

where $Q$ = number of units sold and $P$ = price per unit. Total cost (TC) is

$$TC = FC + VC$$

where $FC$ = total fixed costs and $VC$ = variable costs. Thus, breakeven is the volume of sales (Q) at which the venture will neither make a profit nor incur a loss. Sales in excess of the volume of sales needed to cover costs will result in a profit.

Total fixed costs are $221,100 for e-Travel in year 1 for the base case, and variable costs are 13 percent of sales, since royalty and credit card costs are 12 percent and 1 percent, respectively. Then, to determine Q, we have

$$R = TC$$
$$R = \$221,100 + (0.13 \times R)$$

## TABLE 17.6  Balance sheet.

| End of month 1 of year 1 | | End of year 1 | |
|---|---|---|---|
| **Assets** | | **Assets** | |
| Cash | $180,103 | Cash | $229,048 |
| Equipment and furniture | 48,000 | Equipment and furniture | 48,000 |
| Accumulated depreciation | (1,000) | Accumulated depreciation | (12,000) |
| Total assets | $227,103 | Total assets | $265,048 |
| **Liabilities** | | **Liabilities** | |
| Loan payable | $100,000 | Royalties payable | $5,094 |
| | | Loan payable | 100,000 |
| **Owners' equity** | | **Owners' equity** | |
| Owners' contributions | 140,000 | Owners' contributions | 140,000 |
| Retained earnings (deficit) | (12,897) | Retained earnings | 19,954 |
| Total owners' equity | 127,103 | Total owners' equity | 159,954 |
| Total liabilities and owners' equity | $227,103 | Total liabilities and owners' equity | $265,048 |
| **End of year 2** | | **End of year 3** | |
| **Assets** | | **Assets** | |
| Cash | $609,544 | Cash | $2,280,395 |
| Equipment and furniture | 96,000 | Equipment and furniture | 144,000 |
| Accumulated depreciation | (36,000) | Accumulated depreciation | (72,000) |
| Total assets | $669,544 | Total assets | $2,352,395 |
| **Liabilities** | | **Liabilities** | |
| Royalties payable | $15,984 | Royalties payable | $50,168 |
| Loan payable | 100,000 | Loan payable | 100,000 |
| **Owners' equity** | | **Owners' equity** | |
| Owners' contributions | 140,000 | Owners' contributions | 140,000 |
| Retained earnings | 413,560 | Retained earnings | 2,062,227 |
| Total owners' equity | 553,560 | Total owners' equity | 2,202,227 |
| Total liabilities and owners' equity | $669,544 | Total liabilities and owners' equity | $2,352,395 |

## TABLE 17.7A  Sales projections for the pessimistic growth rate (1 percent per month).

| Year 1 | | Year 2 | | Year 3 | |
|---|---|---|---|---|---|
| | Year total | | Year total | | Year total |
| Units | 12,550 | Units | 16,804 | Units | 18,934 |
| Price per unit | $15 | Price per unit | $15 | Price per unit | $15 |
| Sales dollars | $188,250 | Sales dollars | $252,060 | Sales dollars | $284,010 |

## TABLE 17.7B  Profit and loss statement for the pessimistic growth rate (1 percent per month).

| Year 1 | Year total | Year 2 | Year total |
|---|---|---|---|
| Revenues | $188,250 | Revenues | $252,060 |
| Expenses: | | Expenses: | |
| Author royalties | 22,590 | Author royalties | 30,246 |
| Credit card charges | 1,883 | Credit card charges | 2,521 |
| Marketing | 24,000 | Marketing | 30,000 |
| Depreciation | 12,000 | Depreciation | 24,000 |
| Interest | 12,000 | Interest | 12,000 |
| Office rent | 18,000 | Office rent | 18,000 |
| Salaries | 114,000 | Salaries | 162,000 |
| Social Security and benefits | 17,100 | Social Security and benefits | 24,300 |
| Utilities, supplies, travel, communication | 24,000 | Utilities, supplies, travel, communication | 36,000 |
| Profit (loss) before income tax | (57,323) | Profit before income tax | (87,007) |
| Income tax (credit) | 0 | Income tax | 0 |
| Net profit (loss) | ($57,323) | Net profit (loss) | ($87,007) |

| Year 3 | Year total |
|---|---|
| Revenues | $284,010 |
| Expenses: | |
| Author royalties | 34,079 |
| Credit card charges | 2,840 |
| Marketing | 36,000 |
| Depreciation | 36,000 |
| Interest | 12,000 |
| Office rent | 18,000 |
| Salaries | 192,000 |
| Social Security and benefits | 28,800 |
| Utilities, supplies, travel, communication | 48,000 |
| Profit before income tax | (123,709) |
| Income tax | 0 |
| Net profit (loss) | ($123,709) |

## TABLE 17.7C  Cash flow statement for the pessimistic growth rate (1 percent per month).

### Year 1

| Month | 1 | 2 | 3 | 4 | 5 | 6 | 7 | 8 | 9 | 10 | 11 | 12 | Year total |
|---|---|---|---|---|---|---|---|---|---|---|---|---|---|
| *Operating activities* | | | | | | | | | | | | | |
| Net profit (loss) | ($18,425) | ($18,425) | ($2,765) | ($2,609) | ($2,452) | ($2,295) | ($2,138) | ($1,982) | ($1,812) | ($1,643) | ($1,473) | ($1,304) | ($57,323) |
| Add: Depreciation | 1,000 | 1,000 | 1,000 | 1,000 | 1,000 | 1,000 | 1,000 | 1,000 | 1,000 | 1,000 | 1,000 | 1,000 | 12,000 |
| Add: Increase in royalties payable | 2,160 | 22 | 21 | 22 | 21 | 22 | 23 | 24 | 23 | 24 | | | 2,362 |
| Cash flow from operations | (17,425) | (17,425) | 395 | (1,587) | (1,431) | (1,273) | (1,117) | (960) | (789) | (619) | (450) | (280) | (42,961) |
| *Investing activities* | | | | | | | | | | | | | |
| Purchase of long-term assets | (48,000) | | | | | | | | | | | | (48,000) |
| *Financing activities* | | | | | | | | | | | | | |
| Bank loan | 100,000 | | | | | | | | | | | | 100,000 |
| Owners' cash contributions | 140,000 | | | | | | | | | | | | 140,000 |
| Increase (decrease) in cash | 174,575 | (17,425) | 395 | (1,587) | (1,431) | (1,273) | (1,117) | (960) | (789) | (619) | (450) | (280) | 149,039 |
| Beginning cash balance | 0 | 174,575 | 157,150 | 157,545 | 155,958 | 154,527 | 153,254 | 152,137 | 151,177 | 150,388 | 149,769 | 149,319 | 0 |
| Ending cash balance | $174,575 | $157,150 | $157,545 | $155,958 | $154,527 | $153,254 | $152,137 | $151,177 | $150,388 | $149,769 | $149,319 | $149,039 | $149,039 |

### Year 2

| Month | 1 | 2 | 3 | 4 | 5 | 6 | 7 | 8 | 9 | 10 | 11 | 12 | Year total |
|---|---|---|---|---|---|---|---|---|---|---|---|---|---|
| *Operating activities* | | | | | | | | | | | | | |
| Net profit (loss) | ($8,234) | ($8,064) | ($7,895) | ($7,712) | ($7,529) | ($7,346) | ($7,164) | ($6,981) | ($6,798) | ($6,615) | ($6,432) | ($6,237) | ($87,007) |
| Add: Depreciation | 2,000 | 2,000 | 2,000 | 2,000 | 2,000 | 2,000 | 2,000 | 2,000 | 2,000 | 2,000 | 2,000 | 2,000 | 24,000 |
| Add: Increase in royalties payable | 23 | 23 | 24 | 25 | 25 | 26 | 25 | 25 | 25 | 25 | 25 | 27 | 298 |
| Cash flow from operations | (6,211) | (6,041) | (5,871) | (5,687) | (5,504) | (5,321) | (5,138) | (4,956) | (4,773) | (4,590) | (4,407) | (4,210) | (62,709) |
| *Investing activities* | | | | | | | | | | | | | |
| Purchase of long-term assets | (48,000) | | | | | | | | | | | | (48,000) |
| Increase (decrease) in cash | (54,211) | (6,041) | (5,871) | (5,687) | (5,504) | (5,321) | (5,138) | (4,956) | (4,773) | (4,590) | (4,407) | (4,210) | (110,709) |
| Beginning cash balance | 149,039 | 94,828 | 88,787 | 82,916 | 77,229 | 71,725 | 66,404 | 61,266 | 56,310 | 51,537 | 46,947 | 42,540 | 149,039 |
| Ending cash balance | $94,828 | $88,787 | $82,916 | $77,229 | $71,725 | $66,404 | $61,266 | $56,310 | $51,537 | $46,947 | $42,540 | $38,330 | $38,330 |

(continued on next page)

**TABLE 17.7C** (continued)

## Year 3

| Month | 1 | 2 | 3 | 4 | 5 | 6 | 7 | 8 | 9 | 10 | 11 | 12 | Year total |
|---|---|---|---|---|---|---|---|---|---|---|---|---|---|
| *Operating activities* | | | | | | | | | | | | | |
| Net profit | ($11,416) | ($11,220) | ($11,024) | ($10,829) | ($10,633) | ($10,424) | ($10,216) | ($10,007) | ($9,799) | ($9,589) | ($9,380) | ($9,172) | ($123,709) |
| Add: Depreciation | 3,000 | 3,000 | 3,000 | 3,000 | 3,000 | 3,000 | 3,000 | 3,000 | 3,000 | 3,000 | 3,000 | 3,000 | 36,000 |
| Add: Increase in royalties payable | 27 | 27 | 27 | 27 | 27 | 29 | 29 | 29 | 29 | 28 | 29 | 29 | 337 |
| Cash flow from operations | (8,389) | (8,193) | (7,997) | (7,802) | (7,606) | (7,395) | (7,187) | (6,978) | (6,770) | (6,561) | (6,351) | (6,143) | (87,372) |
| *Investing activities* | | | | | | | | | | | | | |
| Purchase of long-term assets | (48,000) | | | | | | | | | | | | (48,000) |
| Increase (decrease) in cash | (56,389) | (8,193) | (7,997) | (7,802) | (7,606) | (7,395) | (7,187) | (6,978) | (6,770) | (6,561) | (6,351) | (6,143) | (135,372) |
| Beginning cash balance | 38,330 | (18,059) | (26,252) | (34,249) | (42,051) | (49,657) | (57,052) | (64,239) | (71,217) | (77,987) | (84,548) | (90,899) | 38,330 |
| Ending cash balance | ($18,059) | ($26,252) | ($34,249) | ($42,051) | ($49,657) | ($57,052) | ($64,239) | ($71,217) | ($77,987) | ($84,548) | ($90,899) | ($97,042) | ($97,042) |

or

$$0.87R = \$221,100$$

Therefore,

$$0.87 \,(Q \times \$15) = \$221,100$$

or

$$Q = 16,943$$

Therefore, after selling about 17,000 e-Travel guides, the firm is profitable.

## 17.9  Measures of Profitability

Investors in new ventures are interested in measures of annual return on their investment. The **return on invested capital** (ROIC) is the ratio of the net income earned each year expressed as a percentage of the total invested capital in the firm. The ROIC for 3M and IBM in 2013 was 20 percent and 32 percent, respectively. It is also called **return on investment** (ROI). The ROI for a venture is

$$\text{ROI} = \frac{\text{net income}}{\text{investment}}$$

where the income is distributed or allocated to the investor. As a firm grows, it may not actually distribute cash to its investors for some period. In that case, it is retaining earnings and using the retained cash earnings as investment capital. Then the retained earnings are added to the original equity investments to yield the owners' equity [Riggs, 2004].

Net income and owners' equity can provide the ratio called **return on equity** (ROE). ROE is calculated as

$$\text{ROE} = \frac{\text{net income}}{\text{owners'equity}}$$

Using Tables 17.4 and 17.6, the return on equity for e-Travel for year 2 is

$$\text{ROE} = \frac{\$393,606}{\$553,560} \times 100\% = 71.1\%$$

Investor (owner) returns can be calculated at the time of distribution of cash or when the equity is priced in a public market. Assuming the ownership held by the original investors in e-Travel can be sold at the end of year 3 for $720,000, the multiple (M) achieved by the investor group is

$$M = \frac{\$720,000}{\$240,000} = 3.0$$

**TABLE 17.8  Top 10 accounting principles for entrepreneurs.**

1.  The fundamental equation inherent in financial statements is: Assets = Liabilities + Owners' Equity.

2.  The balance sheet is a *snapshot* at a moment in time of *financial position,* while the income statement reports on *financial performance* for a *period.*

3.  An income statement details changes in retained earnings for the period.

4.  Accounting disputes turn almost solely on valuation and timing.

5.  Five key principles govern valuation: Realization (accrual), conservatism, consistency, materiality, and historic cost.

6.  Since valuations require judgments, financial statements are necessarily only estimates.

7.  "Book values" seldom equal "market values," particularly for long-term assets and owners' equity.

8.  The lifeblood of any operation is cash.

9.  Ratios are the key tool for drawing meaning from financial statements.

10. A company's ability to finance its growth internally is a function of its return on equity.

*Source:* Riggs, 2006.

The annual compound return over the three years is 44.2 percent, since $(1.442)^3 = 3.0$. Therefore, the annual return on investment (ROI) is

$$ROI = 44.2\%$$

Note that we designate the investment as \$240,000 since that original investment is the total investment by the founders. We consider a loan countersigned personally as equivalent to an equity investment by the founders.

Table 17.8 summarizes ten core accounting principles for entrepreneurs. Table 17.10 defines various financial terms and ratios at the end of this chapter. See appendix C for sources to compare financial performance metrics.

## 17.10  Spotlight on SolarCity

SolarCity is a provider of energy sources to home owners, businesses, and government organizations. SolarCity provides the design, installs the system, and provides analysis of energy needs and services to meet the needs of the customers. The largest barrier to adoption of solar power systems is the up-front cost, which can reach \$30,000 for a system for a larger home. SolarCity employs a unique financial model to overcome this barrier. First, they rely on U.S government grants to pay for a significant portion of the total costs for installation of a system. Second, SolarCity leases the system to the customer rather than relying solely upon up-front payments. As the solar systems operate, customers normally receive more power than needed and can sell this surplus power back to their utility company. These surplus sales, in turn, can cover part of the lease payment to SolarCity. Thus, many customers may find that their total cost is almost negligible.

SolarCity's financial model is attractive; however, SolarCity experienced large costs in recent years. It could be a decade before SolarCity is able to achieve a strong return on its investment.

## 17.11 Summary

The entrepreneurial team builds its financial plan to determine the economic potential for its venture and demonstrate it to potential investors. The plan uses projected figures based on the underlying assumptions of the business venture. This plan shows the profit and loss statement and the cash flow statement, which can be used to draw up the balance sheet. Monthly figures are used for the first year or two and quarterly figures for the next two or three years. Furthermore, a calculation of the sale of the required number of units for breakeven will be useful. The best new ventures are able to grow sales consistently and show positive cash flow and profit early in their lives.

> **Principle 17**
> A sound financial plan demonstrates the potential for growth and profitability for a new venture and is based on the most accurate and reliable assumptions available.

## Video Resources

Visit http://techventures.stanford.edu to view experts discussing content from this chapter.

| | | |
|---|---|---|
| Exponential Growth | Peter Thiel and Max Levchin | PayPal |
| Extreme Relevance of Cash Flow | Elizabeth Holmes | Theranos |
| No Patience for Profitability | Thomas Prescott | Align Technology |

## 17.12 Exercises

**17.1** Define the interrelationships between an income statement, balance sheet, and cash flow statement. Which one is most important for a new venture and why?

**17.2** Examples of revenue models are listed in Table 16.1. Select three of these revenue models and explain the challenges in creating sales projections for each.

**17.3**  Viscotech, Inc. is planning to enter the field of electro-optical systems for automated optical inspection and detection of defects in manufacturing components and modules. The projected financial revenues and income are shown in Table 17.9. The firm plans to start with an equity investment of $1,000,000 and a five-year loan of $500,000. Determine the cash on hand at the end of each year. Also, determine the return on equity and the return on investment for each year.

**17.4**  A new venture is launched with an initial investment (cash on hand) of $80,000. It generates sales of $40,000 each month. It has monthly operating costs of $36,000. The firm purchases equipment costing $30,000 each month for the first 4 months. Calculate the return on investment at the end of 12 months. Determine if the cash on hand remains positive at the end of each month. What, if any, investment is required and when?

**17.5**  A software firm has fixed costs of $800,000 and variable costs of $12 per unit. Calculate the breakeven quantity (Q) when each unit sells for $50. If the firm sells 50,000 units in a year, what is its profit for that year? Assume the tax rate for the firm is 20 percent.

**17.6**  A new firm, Sensor International, is preparing a plan based on its new device to be used in a security network. The cost of manufacturing, marketing, and distributing a package of six sensors is 14 cents, and the price to the distributor is 68 cents. The firm calculates its one-time fixed costs at $121,000. Determine the number of units required for breakeven.

**17.7**  Continuing exercise 17.6, Sensor International sells 300,000 packages in its first year and 400,000 in its second. If the investor's original investment was $100,000, determine the return on investment in year 1 and year 2 for the firm. Assume a tax rate of 20 percent.

**17.8**  Reconsider the firm described in exercise 17.6 when it is determined that fixed costs have declined to $30,000 and the firm has determined by studies that it can only expect to sell 60 units this year. What price should it select to ensure selling the 60 units profitably?

**TABLE 17.9  Viscotech projections.**

|                             | Year 1 | Year 2 | Year 3 | Year 4 | Year 5 |
|-----------------------------|--------|--------|--------|--------|--------|
| Revenues                    | 1,500  | 3,400  | 5,900  | 10,600 | 15,400 |
| Income after tax            | (500)  | (100)  | 200    | 400    | 600    |
| Depreciation                | 250    | 300    | 350    | 400    | 400    |
| Average shareholders' equity | 1,000 | 700    | 600    | 800    | 1,400  |
| Long-term debt              | 500    | 450    | 300    | 200    | 100    |

*Note:* All figures are in thousands of dollars.

# VENTURE CHALLENGE

1.  Describe the assumptions for your venture that will be used to create the financial projections. Sketch out an income statement.

2.  How long will it take to break even and become cash-flow positive?

3.  How much cumulative cash is necessary to reach that point?

## TABLE 17.10 Glossary of accounting and financial terms.

**Assets**—The value of what the company owns.

**Asset velocity**—The ratio of revenues to net assets, which include plant and equipment, inventories, and working capital.

**Balance sheet**—The financial statement that summarizes the assets, liabilities, and shareholders' equity at a specific point in time.

**Book value**—The net worth of the firm, calculated by total assets minus intangible assets (patents, goodwill) and liabilities.

**Cash flow**—The sum of retained earnings minus the depreciation provision made by the firm.

**Depreciation**—The allocation of the cost of a tangible, long-term asset over its useful life. The reduction of the value of asset from wear and tear.

**Discount rate**—The rate at which future earnings or cash flow are discounted because of the time value of money.

**Dividends**—A distribution of a portion of the net income of a business to its owners.

**Earnings per share**—The ratio of net income to shares of stock outstanding.

**Equity**—The firm's net worth (book value).

**Financial statement**—A report summarizing the financial condition of a business. It normally includes a balance sheet and an income statement.

**Income statement**—A financial statement that summarizes revenues and expenses.

**Liabilities**—The amounts owed to other entities.

**Net income**—Total income for the period less total expenses for the period.

**Pro forma**—Provided in advance of actual data.

**Retained earnings**—Represents the owner's claim on the earnings that have not been paid out in dividends.

**Return on invested capital**—The ratio of net income to investment.

**Return on investment**—The ratio of net income to investment.

**Return on revenues**—The ratio of net income to revenues.

**Return on stockholders' equity**—Net income divided by average stockholders' equity.

*(continued on next page)*

## TABLE 17.10  Continued

**Revenues**—Sales after deducting all returns, rebates, and discounts.

**Statement of cash flows**—The statement that summarizes the cash effects of the operating, investing, and financing activities for a period of time.

**Working capital**—Current assets minus current liabilities.

CHAPTER $18$

# Sources of Capital

*Capital is to the progress of society what gas is to a car.*
**James Truslow Adams**

**What are the sources of capital that a new venture can use to finance the start and growth of its company?**

Entrepreneurs can estimate the capital required for their new business by reviewing the financial projections they prepare using the methods detailed in Chapter 17. In examining the projections and the cash flow statement, it becomes clear how much capital will be needed and when. This chapter addresses the task of attracting investors to a new business and creating an investment offering that will meet the firm's needs and the investors' requirements for an attractive return. In this chapter, we describe the funds that may be available from various sources.

The entrepreneurs may provide some of the required capital, and friends and family may help with modest investments. Government grants, bootstrapping, and crowdsourcing can also supply necessary capital to launch a venture. Debt financing from a bank or other financial institution may be another option.

Most high-growth ventures that expect to grow to a significant scale will need outside capital from professional investors, such as angels and venture capitalists. Typically, several stages of investment will be required over the life of the business. The entrepreneurs also must determine what percentage ownership is offered to the investors. This determination is based on the valuation of the new business at each stage.

Mature ventures use an initial public offering (IPO) to raise additional growth capital and to offer early investors a means of harvesting the value created in an emerging firm. Preparation for an IPO can be an important milestone for a firm with solid growth potential. ∎

## 18.1 Financing the New Venture

The financial projections for a new venture, as described in Chapter 17, provide the entrepreneur with an estimate of the cash flow over the first two or three years. From the cash-flow statement, the entrepreneurial team can determine the requirement for financial capital. Some new firms are cash-flow positive almost immediately, and the entrepreneurs themselves can provide the necessary start-up funds. On the other hand, if it takes one year or more for the cash flow to become positive, a sizable investment may be necessary. Most high-tech firms take several years to become cash-flow positive.

A new venture may require capital to purchase fixed assets such as computers and manufacturing equipment. Furthermore, a new business needs capital to operate while building a customer base. **Working capital** is the capital used to support a firm's normal operations and is defined as current assets minus current liabilities. Financial capital is necessary to permit a new venture to purchase assets and provide working capital. As a firm grows, its needs for financial capital will normally increase. Choosing the right sources of capital for a business can be as important as choosing the team members and the location of the business. This decision will influence the future of the firm.

Finding the capital they need can be a difficult and time-consuming task for entrepreneurs. Getting and completing an investment agreement can take 3 to 12 months. For many entrepreneurs, it may be wise to first secure a few customers and then seek investors. Finding the right financial backer takes a good business plan and time. It will be necessary to continuously tell the venture's story and answer myriad questions. However, revealing proprietary information understandably makes entrepreneurs uneasy. The chance that important information will leak to competitors is real.

Many investors take a long time reviewing the plan and interviewing the team, only to turn down the proposal in the end. The entrepreneur must assume a tentative deal will not be consummated and keep looking for investors even when one investor is seriously interested. While it's tempting to end the hard work of finding money, continuing the search not only saves time if one deal falls through but also strengthens the negotiating position.

Financial capital for new ventures is available, but the key is knowing where to look. Entrepreneurs must do their homework before attempting to raise money for their ventures. Understanding which sources of funding are best suited to the various stages of a company's growth and then learning how those sources operate are essential to success.

The issue of how much money to seek is difficult to resolve. Entrepreneurs wish investors to provide all the money necessary before positive cash flow. However, most investors want to divide their investment into several milestone-based stages. Furthermore, most investors will be wary of the pro forma projections and tend to accept only the pessimistic projection or variations. Investors attempt to factor uncertainty into their calculations, while entrepreneurs are, by nature, more optimistic.

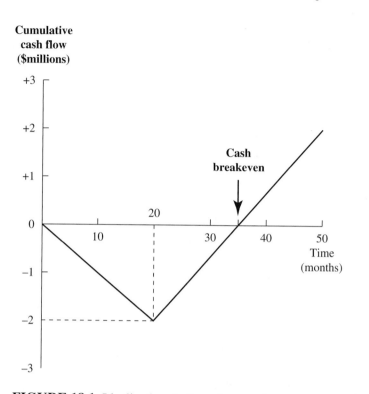

Cumulative
cash flow
($millions)

**FIGURE 18.1** Idealized cash-flow diagram for a new enterprise.

An idealized cash-flow diagram appears in Figure 18.1. This new enterprise has a burn rate of $100,000 per month in the first 20 months. It starts generating positive cash flow in the twenty-first month and reaches cumulative cash break-even in the thirty-fifth month. This firm would require an investment of at least $2 million.

Uncertainty of venture outcomes can lead to a wide range of estimates of results. Breaking a firm's development into several stages can help investors build confidence in the firm over time. However, staged investments require the entrepreneur to raise funds several times—a potentially distracting and risky effort.

Often the two parties, investors and entrepreneurs, possess asymmetrical information. The investors may know more about the industry, while the entrepreneurs may know more about the venture. Furthermore, it may be difficult for the investors to determine an appropriate value for the intellectual property of the new venture. What is a patent worth? Often one finds out in court. Furthermore, how does one value the team's capabilities? Finally, the dynamics of the marketplace are difficult to assess. For example, the marketplace can provide different valuation multiples as the market mood changes. In turn, these factors lead to different perceptions by investors and entrepreneurs [Gompers and Lerner, 2001].

**TABLE 18.1  Five factors that lead to the different perceptions of investors and entrepreneurs.**

| | |
|---|---|
| ■ Uncertainty of projected outcomes | ■ Dynamics of the industry and the financial marketplace |
| ■ Asymmetrical information | ■ Concentration of wealth risk in a venture by entrepreneurs while investors will have risk spread through a diversified portfolio |
| ■ Assigning a value to intellectual property and intangibles | |

The process of securing investment capital requires that the effects of the first four factors shown in Table 18.1 be reduced in the negotiation stage. This can be done through a full discussion of the risks, clear goals, and value of intellectual property and the leadership team. Also, the investors must build confidence that the entrepreneurs can properly manage the firm within the dynamics of the marketplace. Finally, it should be recognized by both parties that the entrepreneurs will have their financial wealth concentrated in the venture, while the investors will have a set of diversified investments. The goal of the entrepreneurs is to find investors who examine the factors in Table 18.1 and eventually align their view of these factors with those of the entrepreneurs. These investors can be called *aligned investors*. Entrepreneurs are advised to find aligned investors since they will be responsive to changing needs and provide required flexibility. Identifying aligned investors who have industry, operating, and team-building expertise is a critical first step in successful fundraising.

## 18.2  Venture Investments as Real Options

Investors make capital investments in opportunities for future cash returns. Professional investors often think of an opportunity as an *option*. **Options** may be defined as rights but not obligations to take some action in the future. Investments in new ventures can be viewed as investments in opportunities with an uncertain outcome. An investment carries many risks, but also may create unforeseen opportunities that can be exploited in the future. A sound practice in the early stages of developing a business is to keep a number of options open by committing investments only in stages while exploring multiple business paths. Once uncertainty has been reduced to a tolerable level and widespread consensus exists within the organization on an appropriate path, the full commitment to that path can be made. Investors who hold options are given the right to make a decision now or at a later date. They can exercise that right to either proceed to the next step or cut their losses and decide to cease investing.

In financial terms, an option is simply the right to purchase an asset at some future date and at a predetermined price. A **real option** is the right to invest in (or purchase) a real asset—shares in a start-up at a future date.

---

### Columbus and the Queen's Option

Christopher Columbus developed a plan to seek out exotic spices and develop a spice trade to Europe. He was an Italian sea captain who submitted, without success, a proposal to the king of Portugal in 1484 to reach Asia by sailing west. Over the next four years, he made inquiries of many European courts. Finally, in 1492, after years of fruitless effort, he received the support of Queen Isabella and the Court of Madrid. The court invested 1.4 million maravedis (currency of the time), and Columbus invested 250,000 maravedis, mostly obtained from friends and family. The contract called for his designation as "admiral" and receipt of one-eighth of the profits from of all gold, silver, gems, and spices produced or mined in "his dominions."

Columbus returned to Spain in March 1493 with some gold and a large amount of information about how to get to the "New World." Queen Isabella had purchased a real option on the discovery of gold and spices. As a result of Columbus's discovery, she decided to exercise the option and send Columbus, Pizzaro, and Cortez to the New World to find and bring back fabulous wealth to Spain. Centuries later, today's professional investors are often purchasing a real option in a start-up. For example, Peter Thiel, Jim Breyer, and Accel partners did so in Facebook in 2005.

---

Let's consider a simplified mathematical model of venture investments. Intellectual capital (knowledge) can be transformed to economic capital to increase or create cash flows as well as strategic capital to exploit new opportunities. We state that economic capital is the intrinsic value of the series of cash flows. The strategic capital is the option value (OV).

We may state broadly that the value of an investment in a new venture (V) is

$$V = IV + OV$$

where IV = intrinsic value and OV = option value.

**Net present value** (NPV) is the present value of the future cash flow of a venture discounted at an appropriate rate (r). The intrinsic value of a venture is the net present value (NPV) of the venture using a discount rate (r), equal to the expected return for the venture.

The net present value of a series of cash flow ($c_n$) is

$$NPV = \sum_{n=0}^{N} \frac{c_n}{(1 + r)^n} \qquad (18.1)$$

where n = 0, 1, 2, $\cdots$ N. For example, the NPV for a new firm over the first two years might be

$$NPV = -100,00 + \frac{65,000}{(1 + r)} + \frac{35,000}{(1 + r)^2}$$

where r = 0.15, the discount rate for this firm. The initial cash flow is negative since an investment of $100,000 was required at n = 0. Then we may calculate NPV as

$$NPV = -100,000 + 65,000(0.870) + 35,000(0.756)$$
$$= -100,000 + 83,000$$
$$= \$ - 17,000$$

The intrinsic value (IV) is equal to NPV for this case. However, the investor has an option to reinvest in the firm after the first two years. Typically, investors establish options by making an initial investment in a venture, which grants them an option to invest again later. As information flows over time, the uncertainty is reduced, and later-stage investments can be seen as less risky.

The value of an option (OV) is a function of four factors: the life of the option (T), the volatility (or uncertainty) of the price of the underlying asset ($\sigma$), the discount rate (r), and the level of the exercise price (X), relative to the current price of the venture (P). Clearly, having a longer period of time in which to decide whether or not to exercise an option increases the likelihood that the value of the firm will in the future exceed the exercise price. The higher the degree of uncertainty about the future value of the venture, the more one should be willing to pay for the option. The uncertainty can be represented by the standard deviation of the firm's price ($\sigma$) over the period (T). If the firm is new, use the standard deviation for comparable firms.

The value of an option increases with the relative value of the current stock price at the initial investment (P) to the exercise price of the option (X). The ratio P/X increases as X declines. The value of an option also increases with the discount rate (r), since higher discount rates make the option more valuable due to high discounting of future cash flows. The four factors contributing to the value of a real option are summarized in Table 18.2.

The value of the option is based on the four factors T, $\sigma$, P/X, and r. Then the option value is

$$OV = f (T, \sigma, P/X, r)$$

which increases as all four factors increase. The option value can be calculated using the Black-Scholes formula [Boer, 2002]. An option calculator is available at http://www.mystockoptions.com/black-scholes.cfm. Another valuation model for options is the binomial model, which uses a decision tree format [Copeland and Tufano, 2004].

### TABLE 18.2  Value of a real option based on four factors.

| | |
|---|---|
| ■ Increases with the level of uncertainty measured by the standard deviation ($\sigma$) | ■ Increases with the ratio of the current stock price (P) to the exercise price (X) or P/X. |
| ■ Increases with the length of time (T) the person holding the option has to decide whether or not to exercise it | ■ Increases with the discount rate (r) |

As a first approximation, we can use a linear approximation for OV as follows:

$$OV = k_1 T + k_2 \sigma + k_3 (P/X) + k_4 r$$

where $k_i$ are unspecified constants. For high relative values of the four factors, the option value will be significant and may exceed the NPV for the first few years of a high-risk venture.

Let us again consider the hypothetical case discussed in the preceding paragraphs. An initial investment of $100,000 was made with an option to invest in a second round of investments after two years. Thus, consider the case where the period (T) is relatively long, the uncertainty (standard deviation) of the firm's value is high, and the discount rate is relatively high. Also, if the current stock price of the firm at the time of the initial investment is low, say $10, and if the exercise price is preset at $5, then the option value is quite high. Perhaps an investor could estimate an option value of $50,000. Then the value of the investment is

$$V = IV + OV$$
$$= -\$17,000 + 50,000 = +\$33,000$$

Often, the value of an investment in a start-up is largely its real option value.

---

**Early Valuation of Genentech**

Genentech was founded in 1976 by entrepreneur Robert Swanson and biochemist Herbert Boyer to use gene-splicing technology to develop new pharmaceuticals. Genentech, devoid of profit, went public in 1980 and raised $35 million. With little cash flow in sight, Genentech was valued solely on its potential to exploit its intellectual property and introduce important new pharmaceuticals. Genentech became profitable in 1993, and was acquired by Roche in 2009.

---

Investments that appear overly risky from a purely financial view may be viable once the opportunities for future action are taken into account. However, as soon as an option no longer promises to provide future value, it should be abandoned. Entrepreneurs sometimes allow unpromising ventures to drag on far too long, in part because teams can develop a collective belief that their venture will succeed, regardless of overwhelming skepticism and harsh realities [Carr, 2002].

## 18.3 Sources and Types of Capital

The four financial steps for building a successful firm appear in Figure 18.2. The initial funds used to launch a new firm are usually called **seed capital**. The first round of capital needs may be limited, and funds may be readily available from the founders, their families, and friends. Most technology firms, however,

**FIGURE 18.2**
Four financial steps in building a successful firm.

require significant capital in the growth phase and turn to professional or wealthy investors. Finally, an initial public offering or an acquisition provides additional capital and a financial return to investors, founders, and other team members.

Many sources of financial capital exist for a new business. See a list of sources in Table 18.3. There are two broad types of capital: equity and debt. **Equity capital** represents the investment in ownership through purchase of the stock of the firm. The holders of equity shares are called stockholders. **Debt capital** is money that a business has borrowed and must repay in a specified time with interest, as with a bank loan. Debt capital usually does not include any ownership interest in the new firm. Usually, a new business can obtain debt capital only after some period in the marketplace and success evidenced by growing revenues and accounts receivable. In some cases, however, the government may help to secure or guarantee the loan, as with the Small Business Administration in the United States.

Equity financing of a venture at formation often begins with the funds provided by founders, friends, and family. Other equity investors and lenders, in fact, may expect the entrepreneur team to invest a significant amount of its own capital in the new business since doing so signals commitment and a firm belief in the possibility of success [Ogden et al., 2003]. A useful indicator of the quality of a new venture proposal for funding is the proportion of the entrepreneurs' wealth that is committed to the new venture [Prasad et al., 2001].

Investments from family and friends are an excellent source of seed capital and can get a start-up far enough along to attract money from private investors or venture capital companies. For example, in 1995, Mike and Jackie Bezos invested $300,000 in their son Jeff's start-up, Amazon.com. Today, those shares are worth many times what they paid for them. Family and friends are often willing to invest because of their relationships with one or more of the founders. However, family and friends should receive and review all the financing documents. Furthermore, they should be able to afford losing their investment and be comfortable with the risk.

Entrepreneurs also can approach professional investors who may make equity investments in very promising, high-growth ventures. Wealthy individuals who invest in new ventures are often called **angels**. It is estimated that angels personally

**TABLE 18.3  Sources of capital.**

| | |
|---|---|
| ■ Founders | ■ Leasing companies |
| ■ Family and friends | ■ Established companies |
| ■ Small business investment companies | ■ Public stock offerings |
| ■ Small Business Innovation Research grants (United States) | ■ Government grants and credits |
| ■ Wealthy individuals (angels) | ■ Customer prepayments |
| ■ Venture capitalists | ■ Pension funds |
| ■ Banks | ■ Insurance companies |

**TABLE 18.4  Comparison of major selected sources of growth capital.**

| Source of capital | Amounts | Advantages | Disadvantages |
|---|---|---|---|
| 1. **Individuals (Angels)** | $10,000 to $1 million with low to medium levels of patience and expertise | Create little dilution for the venture; can move fast because of minimal negotiation and due diligence requirements | Lack sufficient funds for capital-intensive opportunities; can lack long-term perspective; may not provide good advice |
| 2. **Venture capital firms** | $1–$20 million with high levels of patience and expertise | Possess large sums of money to deploy; provide recruiting assistance and other services; enhance venture's reputation and credibility immediately | Require larger percentage ownership of the venture; expect significant role in making major decisions; perhaps too active a role in building executive leadership team |
| 3. **Corporations** | $5–$50 million with medium to high levels of patience and expertise | Generate moderate dilution for the venture; provide opportunity for distribution and product development assistance | Create problems with other potential relationships (e.g., corporation's competitors); can put the venture's intellectual property at risk |

invest in more than 50,000 firms annually in the United States. **Venture capitalists** are professional managers of investment funds. They invest in more than 5,000 firms annually in the United States. The number of investments made by these two groups increases in good economic conditions. Angels and venture capitalists have invested about the same total amount in recent years.

Corporations also can invest directly in a venture. Sometimes, a corporation and a new venture establish a joint development agreement, which is aimed at developing a technology and which can provide capital to the new venture. Many technology corporations also have their own venture capital groups, which may invest directly in new ventures in exchange for partial ownership. Table 18.4 details the pros and cons of several different sources of equity capital for the growth phase.

Eventually, an enterprise can seek financial capital through an initial public offering on the stock market. Only the most qualified and experienced teams with an outstanding opportunity, however, are able to make an initial public offering in the early stages of building a business.

The overall investment process for a new firm can be portrayed as shown in Figure 18.3. The seed round of financing may be achieved with friends and

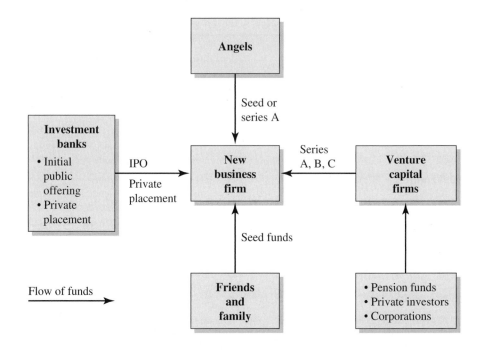

**FIGURE 18.3** Potential for the flow of funds to a new business firm.

family. Angels may supply the series A round of financing. Venture capitalists will often supply the funds in series A, B, and C. Then investment bankers will facilitate an IPO or large private placement. We elaborate on these different funding options in the following sections.

## 18.4  Bootstrapping and Crowdsourcing

Launching a start-up with modest funds from the entrepreneurial team, friends, and family is often called **bootstrap financing**. The term originated from the saying "they pulled themselves up by their bootstraps." To bootstrap a venture means to start a firm by one's own efforts. For many ventures with modest potential returns to investors, it is best to attract investors who are known to the entrepreneurs. Professional investors can only back a small fraction of the number of firms that start up each year. They seek to invest in large opportunities with high rates of return that have defensible competitive advantages, well-defined business plans, and well-known, proven founders.

Bootstrap companies start small and build their experience and know-how as they go. Thus, bootstrap entrepreneurs often start with a modest business plan and a focused opportunity and look for a quick route to breakeven and positive cash flow. They also take steps to reduce costs and to improve cash flow, such as obtaining advance payments from customers; delaying payments to suppliers; and sharing employees, assets, or business space with other firms [Winborg and Landström, 2000].

Many businesses fit the model of a bootstrap opportunity. Although actual figures are difficult to obtain, up to 75 percent of start-ups are bootstrapped, self-financed firms. They keep costs low, seek out markets that competitors are ignoring, and build the business one step at a time. Entrepreneurs often must fund a significant portion of their new business since investors and lenders are reluctant to provide the required capital [Quadrini, 2001].

Eventually, these modest beginnings can turn into large successes as the firms find new opportunities. For example, Princeton Review was launched to compete with local private tutors of uneven quality. Eventually, the firm found its place on the national scene and competed with the Kaplan chain.

Pierre Omidyar launched AuctionWeb on Labor Day, 1995, while he was still employed full-time at a software firm. As an experienced software developer, Omidyar was intrigued by the opportunities to build a business on the Internet, and his vision was to provide a "perfect market" for buyers and sellers on an Internet auction site [Cohen, 2002]. He wrote the program for the site over the Labor Day weekend using his personal website provided by his home Internet service provider. The best domain name he could find was eBay.com. Throughout the fall of 1995, AuctionWeb had hosted thousands of auctions. By February 1996, Omidyar's Internet service provider started charging him a commercial rate for his website, so he started charging a small fee for each sale. Starting in February, AuctionWeb was profitable. By April, AuctionWeb took in $5,000 in sales. In June, when revenues doubled to $10,000, Omidyar decided to quit his day job. He quickly attracted a friend, Jeff Skoll, to join him, first as a consultant and then full time by August 1996. By late 1996, they moved out of Omidyar's house to an office building in Campbell, California. By October 1996, AuctionWeb had a total of four employees when it hosted 28,000 auctions that month. AuctionWeb was dedicated to thriftiness and controlling costs. By January 1997, the site had hosted 200,000 auctions, and AuctionWeb was projecting revenues of over $4 million for 1997. Early in 1997, Omidyar and Skoll wrote their first business plan and started looking for investors. By June 1997, Benchmark Capital, a venture capital firm, paid $5 million for 21.5 percent of the company, which then changed its name to eBay.

An advantage of bootstrap financing is the ability to make some mistakes and yet keep going. Many entrepreneurs underestimate the time and the marketing costs entailed in overcoming customer inertia and conservatism, especially with respect to new, unproven products. Entrepreneurs who are unsure of their markets or who do not have the experience to deal with investor pressure may be better off without other people's capital, even if they can somehow get investors to overlook their limited credentials and experience.

Bootstrap companies usually follow five rules: (1) start small and probe the market, (2) learn from your customer and adjust the business model, (3) adjust the revenue and profit engine, (4) keep costs to a minimum, and (5) start expanding the company, once the new venture starts growing, while keeping the cost curve below the revenue curve. Often, using bootstrap financing can instill long-term frugality and financial discipline in a new venture. In fact, some research indicates that an overabundance of resources can harm a new venture [Hvide and Møen, 2010]. Thus, Guy Kawasaki tells entrepreneurs that they "should bootstrap even if you have $10 million in the bank." The advantages and disadvantages of bootstrap financing are listed in Table 18.5.

Increasingly, entrepreneurs also can use crowdfunding to obtain financing at an early stage. **Crowdfunding** refers to the joining together of a collection of individuals – the "crowd" – each of whom contributes a small amount to help fund a business. In exchange for their funding, these contributors may receive preferred treatment by the business, copies of products that a business creates, or, in some cases, a small amount of equity in the business.

Legal restrictions on ownership through crowdsourcing models vary by country. In the United States, the JOBS Act of 2012 provided an exemption for U.S. companies offering securities through an SEC-registered crowdfunding platform. Under the exemption, investors do not have to be accredited and they do not count toward the 2,000-investor threshold. Companies can sell up to $1 million of securities in a 12-month period.

---

**Founding of Siebel**

Tom Siebel earned an MBA and an MA in computer science from the University of Illinois. After graduation, he joined Oracle in 1982. By 1992, Siebel and Pat House had founded Siebel Systems. For the first 18 months, everyone worked for no salary but received equity shares. Siebel stated: "This was never about making money. It was never about going public; it was never about the creation of wealth. This was about an attempt to build an incredibly high-quality company." [Malone, 2002]. Siebel Systems was purchased by Oracle for $6 billion in 2005.

---

**TABLE 18.5 Advantages and disadvantages of bootstrap financing.**

| Advantages | Disadvantages |
|---|---|
| ■ Flexibility | ■ Unable to fund growth phase |
| ■ Maintain ownership | ■ Lack of funding commitment for future |
| ■ Operating control by founders | ■ Loss of advice from professional investors |
| ■ Little time spent on finding investors | |

Crowdsourcing websites, such as AngelList and Kickstarter, serve as forums that bring together individual investors with businesses looking for funding. These websites are growing rapidly. AngelList, for example, grew from 1,500 companies in 2010 to 38,000 companies in 2012 [Bergl, 2013]. Kickstarter grew from $28 million pledged to projects in 2010 to $320 million pledged in 2012 [Popper, 2013].

## 18.5 Debt Financing and Grants

New ventures with sales and cash flow can consider recurring short-term or long-term debt financing. Debt provides financial leverage to a firm and enables the firm to increase its return on equity. The principle of financial leverage works as long as a firm's earnings are consistent and are larger than the interest charged for the borrowed money. Of course, if the firm's net earnings should drop below the interest cost of borrowed money, the return on owners' equity will decrease. Thus, most new ventures avoid financial leverage until they achieve stable growth.

Debt financing can be easier to arrange, and often cheaper, than equity for profitable companies. Any profitable firm can borrow money if it's willing to pay high enough interest rates. Borrowers have the advantage of not giving up ownership or control of the firm—unless they cannot pay the interest. Also, the tax deduction for interest paid cuts the effective cost of debt. Using debt, however, exposes a company to another kind of risk: If it does not show enough profit to cover debt payments, its existence may be in peril.

When underwriting a traditional loan, a lender first looks to a business's cash flow. For new ventures with limited cash flow, obtaining a loan can thus be difficult. Sometimes, however, a new venture can secure a bank loan guaranteed by the Small Business Administration. Under the program, the federal agency does not actually make the loan—banks do, mostly—but it guarantees about 75 percent of the loaned amounts, and it covers losses on the guaranteed portions from a combination of fees charged to banks and taxpayer funds. SBA loans in 2012 amounted to $21.9 billion to about 54,000 borrowers.

Asset-backed borrowing also may be available to new ventures. Ventures can borrow money using the liquid assets of the company, such as accounts receivable, or fixed assets, such as equipment or property, as collateral. An asset leaseback is another way to secure cash by giving up an asset such as a building or equipment. Of course, leasing equipment or property directly is a form of borrowing, too.

Finally, many new ventures arrange for a line of credit, which is a form of short-term borrowing. The firm pays a fee for the right to access borrowed funds as needed.

While debt financing must be paid back over time, some new ventures are eligible for grants, which require neither payback nor giving up equity in the venture. A number of grants are available to support very early-stage and small-scale business efforts. The Small Business Innovation Research (SBIR) program is a U.S. government program, coordinated by the Small Business Administration, through which several government agencies make grants to small businesses. A "small business" is defined as an American-owned for-profit business with fewer than 500 employees. A similar program, the Small Business Technology Transfer (STTR) program, is focused on developing partnerships between small businesses and nonprofit U.S. research institutions such as universities. Together, SBIR and STTR grants to small businesses total $2 billion each year. Information about both programs can be obtained from the Small Business Administration (www.sba.gov).

The National Collegiate Inventors and Innovators Alliance (NCIIA), also based in the United States, offers Advanced E-Team Grants to move products or technologies from the idea stage to prototype and to help collegiate innovators secure intellectual property. Advanced E-Team Grants range in size from $1,000 to $20,000; the grant period is 12 to 18 months. More information can be obtained from NCIIA (www.nciia.org). Other examples in the United States include DARPA grants and various grants from state and local agencies.

As high-growth ventures often are based upon cutting-edge research, including research conducted at universities, entrepreneurs also are wise to investigate research funding opportunities from national and regional government agencies and foundations. While these grants typically are intended to support research activities only, such funding can be critical for start-ups in areas such as energy, nanotechnology, and medicine, where the technical challenge is the primary risk faced by a new venture.

## 18.6  Angels

**Angels** are wealthy individuals, usually experienced entrepreneurs, who invest in business start-ups in exchange for equity in the new ventures. The term *angel* was originally used for a person who backed a new theater production on Broadway. Angels are people who share the vision of the new venture and provide support, advice, and money. In a sense, angels can provide wings for lifting a new creation. Angels often have personal experience and interest in the industry that the start-up is entering.

Angel investing is a fast-growing segment of the new business financing industry, and it is often ideal for start-ups that have outgrown the capacity of investments from bootstrap financing but are still too small to attract the interest of venture capital companies. For example, after raising the money to launch Amazon.com from family and friends, Jeff Bezos turned to angels, attracting $1.2 million from a dozen angels. Angels funded about 66,000 U.S. ventures for a total of about $22.5 billion in 2011.

Angels often invest because they understand the industry and are attracted by the opportunity as well as the potential return. They may place less emphasis on an early return strategy and enjoy working with new entrepreneurs. Angels serve as investors, advisers, and mentors for the new ventures they support. They may help new entrepreneurs create and refine a business model, find top talent, build business processes, test their ideas in the marketplace, and attract additional funding. Angels tend to invest close to home and limit their investments to early-stage companies. Most of the new ventures they fund are recommended to them by a business associate or an angel group. A summary list of typical characteristics of angel investments is provided in Table 18.6.

Angels can be helpful investors, but sometimes they can be overbearing and have a negative impact. Choosing the right angel is important, so entrepreneurs should check references and capabilities of potential investors carefully.

In some geographical regions, angels join together to form groups of angels. These groups work together to screen investment opportunities. For example, the Band of Angels meets monthly in Silicon Valley to hear a few presentations by start-ups (see www.bandangels.com). Angel groups are established in many cities.

Incubators, such as Y-Combinator, can also function like angels in some ways. Y-Combinator provides seed money of $14,000, advice, and connections. In exchange, it takes an average stake of about 6 percent in a venture. Y-Combinator also arranges for each venture to receive a convertible note. A **convertible note** is a type of bond that the investor can convert to stock in the new venture or cash of equal value, at an agreed-upon price. A convertible note is a hybrid security with debt- and equity-like features. For the new venture, it has a lower rate than nonconvertible debt and thus reduces payment amounts. For the investor, the note offers the potential upside of conversion into equity while protecting the downside with cash flow from the payments.

In 1976, Steve Wozniak and Steve Jobs designed the Apple I computer in Jobs's bedroom and built the prototype in Jobs's house. To start the company, Jobs sold his Volkswagen and Wozniak sold his Hewlett-Packard calculator, which raised $1,300. With that capital and credit from local electronics suppliers, they set up their first production line. Jobs met Mike Markkula, a former marketing manager at Intel and wealthy angel who invested $91,000 in cash and personally guaranteed a bank line of credit for $250,000. Jobs, Wozniak, and Markkula each held one-third ownership of Apple Computer [Young, 1988]. Markkula, the angel investor, became chairman of the company in 1977.

**TABLE 18.6 Typical characteristics of angel investments.**

| | |
|---|---|
| ■ Within the industry in which the angel has experience | ■ Entrepreneurs with attractive personal characteristics such as integrity and coachability |
| ■ Located within convenient driving distance of the angel | ■ Good market and growth potential |
| ■ Recommended by trusted business associates | ■ Seeking an investment of $50,000 to $1 million that offers minority ownership of about 10 to 40 percent |

The founders of Google, Sergey Brin and Larry Page, approached Sun Microsystems cofounder Andy Bechtolsheim in 1998, who wrote a $100,000 check after a 15-minute pitch describing the new business. By 1999, the venture capital firms of Kleiner Perkins, Caufield and Byers, and Sequoia also invested.

## 18.7  Venture Capital

**Venture capital** is a source of funds for new ventures that is managed by investment professionals on behalf of the investors in the venture capital fund. The people who manage the venture capital fund are called venture capitalists. These funds typically invest in new ventures with high potential returns. The private venture capital firm is seeking equity participation in these high-potential firms. Each year in the United States, over 5,000 new ventures receive funding from venture capital. The venture capital firm engages in careful screening and due diligence before investing. It brings in other venture capital investors in a financing round and prepares contracts and restrictions (called term sheets) with the new venture. Venture capital firms are usually interested in technology ventures with high potential.

A typical venture capital firm will have enough investments so as to diversify its portfolio. Each venture capital firm will have several partners who are experienced, full-time investors. Their knowledge of finance and technology enables them to judge the potential investment. Normally, venture capital investing is staged financing. Thus, new information and risk reduction at each stage enables the venture capitalist to make better decisions. The new ventures are held to staged goals, milestones, and deadlines for achieving these milestones. A typical investment from a venture capital firm is $1 million to $5 million in the first financing round, with a total commitment of $10 million to $20 million. Over several stages with several venture funds involved, the combined funding potential can be $50 to $100 million.

Many entrepreneurs start up a new business to gain independence, only to find that they have a new set of partners—the venture capitalists acting as members of the board of directors. Most venture capitalists look for ventures that can become profitable and attain at least $100 million a year in revenues in the next five or six years. Entrepreneurs that envision high-growth ventures will likely turn to venture capitalists [Florin, 2005].

A venture capital portfolio of 20 to 30 new ventures may achieve an overall annual return of 30 percent. Perhaps one-half of the new ventures fail or provide a low return. Fortunately, two or three new ventures may provide an overall annual return of 50 to 100 percent. Therefore, venture capital firms are looking for new ventures that potentially can return at least 50 percent annually. Obviously, these candidates must be firms with high-potential growth.

The four investment requirements used by the venture capitalists in the selection process are (1) the industry is well-known to them; (2) the amount of the investment is greater than $1 million; (3) the company is at the appropriate stage of progress; and (4) the potential return is 40 percent or more annually. The track record of the team is also critically reviewed [Gompers and Sahlman, 2002].

Venture capitalists prefer staged or phased financing. The four stages are shown in Table 18.7. Venture capital is normally available in the development and growth stages, with greatest emphasis on the development stage. Venture capital money is limited-term money [Zider, 1998]. Typically, a venture capital firm wants to harvest (realize) its return on investment within five to 10 years. The harvest is usually facilitated through an initial public offering in the public equity markets or an acquisition by an established company.

---

**Workday and Amazon: Raising Capital Multiple Ways**

New enterprises that can become large businesses may require larger capital investments as these companies grow. Workday raised $250 million from venture capitalists and more than $600 million from its 2012 IPO. Amazon, by contrast, raised $8 million from venture capitalists and $54 million from its 1997 IPO. Amazon, unlike Workday, has chosen to use debt to fund growth, including raising a $1.25 billion debt offering in 1999.

---

Venture capitalists expect the new business to achieve certain outcomes, often called milestones, at the end of a funding stage before proceeding to the next stage. Examples of the milestones for each stage are shown in Table 18.8. Each milestone functions as a miniplan about each stage for the venture. For example, a milestone may state that a working prototype will be available in six months. Entrepreneurs should therefore develop a series of milestones and then tie their fundraising activities to the completion of them [Berkery, 2008].

A special type of venture capital is available for social benefit firms or those that promise to create products that lead to sustainable resources or environments. For example, the Silicon Valley Social Venture Fund, known as SV2, is dedicated to funding organizations in the San Francisco Bay area that facilitate social change

## TABLE 18.7 Investment stages.

1. Seed or start-up stage: Complete the team, formalize the plan, complete initial arrangements. Financial capital from angels, friends, and family.

2. Development stage (series A): Product development and prototype, ready for launch. Financial capital from venture capital funds.

3. Growth stage (series B or C, and others as required): Launch and growth phase. Financial capital from venture capital firms and corporations.

4. Competitive or maturity stage (initial public offering): Mature firm in a competitive context. Financial capital from offerings in the public equity markets.

**TABLE 18.8  Milestones for each stage of funding.**

| Stage | Milestone = expected outcome |
|-------|------------------------------|
| Seed | Formation of initial team and completion of business plan |
| Series A | Product development completed and early customer interest |
| Series B | Product test and customer acceptance proven with some revenue |
| Series C | Rapid sales growth and international expansion |

(www.sv2.org). The Omidyar Network provides funding to for-profit ventures that attempt to make a positive impact on society (www.omidyar.net).

Venture capital funds concentrate on attractive, disruptive, and high-growth industries. They concentrated on computers and biotechnology in the 1980s and communications and the Internet in the 1990s. Areas of concentration in the 2000s included biomedical devices, genomics, energy, and mobile computing. In the 2010s, the focus includes mobile, enterprise software, big data, networking,  and medical devices. By investing in emerging industries, venture capitalists hope to build great companies in important industries while reaping large rewards.

Venture capitalists carry out one or more of five functions: (1) they provide  capital for start-ups, (2) they evaluate projects for other participants in the venture, (3) they provide expertise for the development of the firm, (4) they serve as the central coordinator for all the participants involved during a firm's infancy, and (5) they provide introductions to potential customers, employees, collaborators, and future financiers. The importance of these nonfinancial contributions cannot be overstated [Bertoni et al., 2011]. Venture-capital-backed firms demonstrate more responsive product development, greater collaborative activity, and more agile management decision making [Hsu, 2006; Arthurs and Busenitz, 2006].

In a typical deal at the development stage, a group of two or three venture capitalists will invest $5 million to $15 million in exchange for 40 percent to 60 percent preferred-equity ownership. The preferred class of stock provides the venture capitalists with preference over common stock held by founders, family, friends, and other first-stage investors. They will hold a liquidation preference on rights to assets. Furthermore, venture capitalists seek voting rights over key decisions such as the sale of the firm or the timing of an IPO. They usually require seats on the board of directors.

Venture capital is high-risk and high-return capital available to high-potential firms. Venture capitalists plan for a 50 percent or more annualized return on their investment with built-in protections and controls over the new venture. The structure of most venture capital contractual agreements (deals) favors the venture capitalist and may place the entrepreneur at a disadvantage if plans do not work out. However, the venture capital firms bring sizable money, industry knowledge and contacts for recruiting and customer assistance, and a pathway to a public stock offering [Fitza et al., 2009]. The risk-and-reward profiles for

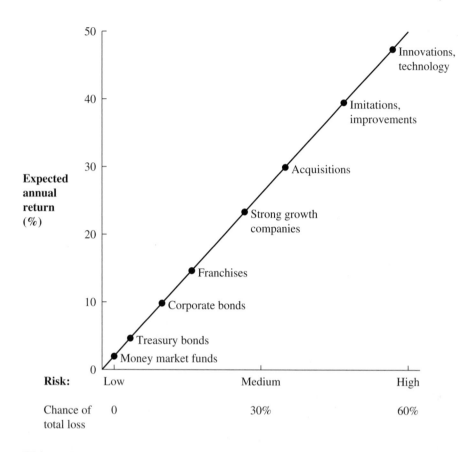

**FIGURE 18.4** Risk and reward profile for various investments.

various types of investments are shown in Figure 18.4. Venture capitalists work at the high risk–high return end of the profile.

Venture capital investments in the United States in new emerging ventures averaged $25 billion per year over the decade of the 1990s and peaked at $106 billion in 2000 during the dot.com and telecommunications boom. Worldwide venture capital investments in technology ventures are shown in Table 18.9 for 2003 to 2012 along with IPOs.

Venture capitalists get good rates of return by buying shares of private companies early and then helping management use that cash to turn the start-ups into businesses with growing revenues, profitability, and cash flow. The big winners generally have to earn 10 times the original investment money in four to five years. A multiple of 10 times is equivalent to a 58.5 percent annual return over the five years. If a venture capitalist invests $5 million in the early stage of a company, his or her ownership stake is hoped to be worth $50 million in five years. If the venture capitalist owns one-half of the company, the firm must be worth $100 million after five years for that return.

**TABLE 18.9  Worldwide technology venture capital investments and initial public offerings (IPOs) 2003-2012 ($ millions).**

| Year | VC Deals | VC Investment | Number of IPOs | Offering Amount |
|------|----------|---------------|----------------|-----------------|
| 2003 | 5,550 | $26,321 | 5 | $276 |
| 2004 | 5,752 | $32,218 | 25 | $4,300 |
| 2005 | 5,646 | $33,682 | 29 | $1,495 |
| 2006 | 6,124 | $39,119 | 22 | $1,923 |
| 2007 | 5,870 | $39,978 | 83 | $8,509 |
| 2008 | 5,723 | $41,690 | 9 | $961 |
| 2009 | 4,130 | $31,700 | 22 | $1,952 |
| 2010 | 4,813 | $39,700 | 65 | $5,570 |
| 2011 | 5,300 | $56,100 | 54 | $7,532 |
| 2012 | 6,813 | $50,300 | 65 | $2,1554 |

*Source:* Preqin Private Equity Database and Thompson Venture Economics.

**TABLE 18.10  Characteristics of a company attractive to venture capital.**

- Potential to become a leading firm in a high-growth industry with few competitors
- Highly competent and committed management team and high human capital (talent)
- Strong competitive abilities and a sustainable competitive advantage
- Viable exit or harvest strategy
- Reasonable valuation of the new venture

- Outstanding opportunity
- Founders' capital invested in the venture
- Recognizes competitors and has a solid competitive strategy
- A sound business plan showing how cash flow turns positive within a few years
- Demonstrated progress on the product design and good sales potential

The characteristics of a good venture capital deal from the venture capitalists' view and the venture founders' view appear in Table 18.10. A large opportunity in a fast-growing industry led by a very competent, experienced team is attractive to venture capitalists. With a good venture capitalist partner, the founders of the new venture can realize their dream of a well-capitalized venture that can make a big difference in their industry. At the same time, the founding team must recognize that venture capital has a high cost of potential loss of ownership, control, and even changing the vision of the venture. A list of U.S. venture capital firms is available from the National Venture Capital Association (www.nvca.org).

New and emerging businesses can use the five steps outlined in Table 18.11 to secure venture capital. Every one of these steps takes time, and the whole

**TABLE 18.11 Five steps for a venture capital deal.**

1. Determine the amount of cash needed and its use.

2. Locate appropriate venture capital investors and secure a referral to them.

3. Determine which risks are to be reduced in this financing round:
   - Team risk—recruiting great people
   - Capital risk—enough cash to achieve milestones
   - Technology risk—prove that the science works in a product
   - Market risk—deliver extraordinary customer experiences

4. Agree on valuation and ownership structure.

5. Agree on a contract (term sheet) describing the deal and its terms.

process may take 3 to 12 months. Given the extent of technology markets, there is no scarcity of good ideas. What is scarce is experienced, competent, and committed entrepreneurs. Investors strive to find the best entrepreneurial teams to invest in.

---

**FedEx: Getting the Product Right**

Fred Smith used his family money to found Federal Express (FedEx) in 1973. He leased several planes and built a 25-city network. He knew additional funding would depend on a solid start, so he tested the system for two weeks, shipping empty packages cross-country. In mid-April 1973, he opened for business. To expand his network, Smith turned to venture capitalists. He got his venture capital because he reduced the perceived risk with his own money in FedEx. By 1975, FedEx was profitable. FedEx sold stock to the public with an IPO in 1978.

---

## 18.8 Corporate Venture Capital

Large companies such as Intel, Microsoft, and Cisco Systems also invest in external start-ups [Chesbrough, 2002]. **Corporate venture capital** is the investment of corporate funds in start-up firms that are not part of the corporation. In this case, established corporations are acting as venture capitalists. Corporations can have different goals for their venture capital programs [Gaba and Meyer, 2008]. In some cases, the corporate venture capitalist is looking for new ventures that can offer synergies between itself and the new venture. Another approach is for the corporate venture capitalist to make an investment to use its industry knowledge and then generate a high return on investment. Corporate investors often use external venturing programs as a mechanism for identifying and monitoring promising acquisition targets, too.

An example of a strategic investment is one of many made by Microsoft in start-up companies that could help advance its Internet businesses. Other firms make corporate venture capital investments in firms that will drive and expand the markets for the investors' own products. For example, Intel has invested in over 1,100 companies for a total of $7.9 billion, with the goal of nurturing technologies that could stimulate demand for Intel's chip business. Another possible corporate investment strategy is to support start-ups that may be valuable to the corporation's future operations, as with investments by Applied Materials and TEL into semiconductor materials ventures. A summary of the four forms of corporate venture capital investments is in Figure 18.5.

Many start-ups can benefit from corporate venture capital, since corporations can make good partners and the cost of the capital can be less than that from regular venture capitalists. Moreover, corporate venture capital may increase the credibility of the new venture and may not exert as much control as regular venture capitalists. Perhaps the most important benefit of corporate venture capital is the potential for a strong partnership between the corporate venture capitalists and the new venture that strategically couples the strength of the large corporation with the innovation of the start-up [Mason and Rohner, 2002]. Corporate venture capital is especially beneficial to a new venture when it requires specialized complementary assets or operates in a highly uncertain environment [Park and Steensma, 2012].

At the same time, entrepreneurs need to be cautious. Corporate venture capitalists often invest to be aware of potential competitors and their interests may not be aligned with those of the start-up. Moreover, corporate venture capitalists can take longer to close a deal and may ask for distribution rights. They can also stymie a new venture's attempts to partner with a competitor to the corporation and may limit exit opportunities.

|  | Advances strategy of<br>current business:<br>Microsoft | Allows exploration of<br>potential new business:<br>Agilent |
|---|---|---|
| Tie to existing operational capabilities — Tight | | |
| Tie to existing operational capabilities — Loose | Complements strategy of<br>current business:<br>Intel | Provides potential financial<br>returns:<br>Dell |
|  | Strategic | Financial |

**Corporate investment objective**

**FIGURE 18.5** Four forms of corporate venture capital.

*Source:* Burgelman et al., 2004.

# 18.9 Valuation

The **valuation rule** is the algorithm by which an investor such as an angel or venture capitalist assigns a monetary value to a new venture. For many operating businesses, determining the net present value (NPV) is the best method for selecting alternative projects. A new enterprise, however, has uncertain pro forma cash flows, and investors find it difficult to use projected cash flows as reliable measures. Even if potential cash flows can be estimated, how does an investor decide what is a fair price for a share of ownership in a new firm?

Theoretically, the value of a company is equal to the present value of all dividends or cash disbursements paid now or later. A new firm has no historical results to use to project future cash flows. Furthermore, fads and social dynamics play a role in the determination of the value of a firm. Thus, determining the value of a start-up is difficult.

Discounted cash flow is a method of calculating the present value of a future stream of cash flow based on discounting back future flows at the end of a number of years using a discount rate (r) (see equation 18.1). Let us start with the discounted cash flow rules for the valuation of a new venture, ABC Inc., with projected cash flows as shown in Table 18.12. Since the firm is not yet operating, these projections are subject to wide uncertainty. With a potential investment in the first year, what percentage of the firm's ownership should the investor require? What is the discount rate for this calculation?

The **discount rate** is the rate (r) at which future earnings or cash flow is discounted because of the time value of money. The discount rate (r) for a firm is its cost of capital. Therefore, the discount rate for investors will be their expected return on investment. Thus, the new enterprise may propose using a discount rate of 15 percent, while the investors may demand a return of 30 percent or more. McNulty and colleagues [2002] have shown the real cost of capital of a biotech start-up is about 35 percent. Also, for the cash flow calculation, we need to have an estimate of the cash flow for year 6 and later. However, it is unreasonable to project into year 6 and later since the estimates become less reliable for later years. Thus, investors will be reluctant to use the discounted cash flow method for valuation of a firm that is yet to provide any reliable cash flow.

Venture capitalists or angels will want a harvest of the value of the firm by an IPO, acquisition, or buyout of their share of ownership by year 5 to 10. Realistically, the IPO or acquisition may not happen until year 6 or later. Examining the

**TABLE 18.12 Projected cash flow and profit for ABC Inc.**

| Year | 1 | 2 | 3 | 4 | 5 | 6 |
|---|---|---|---|---|---|---|
| Sales | 0 | 1,000 | 2,500 | 5,000 | 8,000 | 10,000 |
| Profit | −600 | −10 | 400 | 650 | 1,000 | 1,200 |
| Cash flow | −1,100 | 0 | 500 | 1,200 | 1,500 | 1,800 |

*Note:* All figures in thousands of dollars.

projections for ABC Inc. shown in Table 18.12, a valuation can be determined for year 5 using a method favored by venture capitalists and angels.

The **new venture valuation rule** uses the projected sales, profit, and cash flow in a target year (N) and the projected earning growth rate (g) for five years after year N. The investment by the investor (I) is made at the beginning of year 1. The investor requires an annual return (gain) on investment of G for N years. Thus, investors expect a capital return after N years of $(1 + G)^N$ times their original investment I. Therefore,

$$CR = (1 + G)^N \times I$$
$$= M \times I \tag{18.2}$$

where CR is the capital return and M is the multiple of the investment. Thus, if an investor invests $1.1 million in the series A stage and the expected annual return on investment, G, is 45 percent over a five-year period, then

$$M = (1 + 0.45)^5 = 6.41$$

Therefore,

$$CR = M \times I = 6.41 \times 1.1 = 7.05$$

The percentage ownership (PO) demanded by the investor will then be

$$PO = \frac{CR}{MV} \times 100\%$$

where MV is the expected market value of the new venture in year N.

To calculate the expected market value in year 5, we may use the price-to-earnings or **PE ratio** of comparative firms to estimate the market value in year N. Then,

$$MV = PE \times EN \tag{18.3}$$

where EN is the earnings in year N. In this case for year 5, EN = $1,000,000 (Table 18.12). The comparative PE ratio is obtained by looking at the industry PE ratios while accounting for the expected growth rate of earnings over the ensuing years. In this case, we might expect a growth rate of earnings of 20 percent for several years following year 5. Then, examining the industry data, we estimate the appropriate PE ratio is 16. Therefore, we have

$$MV = 16 \times \$1,000,000$$
$$= \$16,000,000$$

Then the required percentage ownership is

$$PO = \frac{CR}{MV} \times 100\%$$

$$= \frac{7.05}{16.00} \times 100\% = 44.0\%$$

The market value can also be calculated using a price-to-sales ratio (PS) for comparative firms. If comparative firms have a PS = 2.3, then the market value would be MV = PS × S. The market value for ABC Inc. is

$$MV = PS \times S$$

$$= 2.3 \times \$8,000,000 = \$18,400,000$$

Then the percentage ownership required by the investor would be

$$PO = \frac{CR}{MV} \times 100\% = \frac{7.05}{18.40} \times 100\% = 38.3\%$$

Using these calculations, the investor may reasonably expect to receive 40 to 50 percent of the firm's ownership in a series A investment. Given the uncertain nature of the sales and profit projections, the valuation of a firm is a function of the potential of the firm to achieve a big success in a short time. This simple example assumes no more stages of investment are required in the five-year period.

Netscape was formed by Marc Andreesen, coauthor of the University of Illinois Mosiac Web browser, and Jim Clark, founder and former CEO of Silicon Graphics. As an angel investor, Clark invested $3 million in Netscape in May 1994. Clark, a respected entrepreneur, then offered a $6.4 million investment opportunity (at $2.25 per share) for 15 percent of the firm to Kleiner Perkins, a premier venture capital firm. Netscape's sales rocketed to $16 million in the first six months of operations [Lewis, 2000]. Just before the IPO, Clark owned 30.0 percent of the firm and Andreesen owned 12.3 percent. The multiple (M) they achieved in 17 months on their seed shares was 37.3. Kleiner Perkins achieved a multiple of 12.4 on its first-stage investment in just 13 months. Netscape was the first high-profile venture of the Internet boom companies and reaped large rewards. A summary of the valuation of Netscape (price per share) is shown in Table 18.13.

When a firm makes an offering, it raises an investment (INV). The firm hopes to set its **pre-money** (before the investment) **value** as PREMV. Then

$$\text{post-money value} = \text{pre-money value} + \text{investment}$$

When **post-money value** is POSMV, we have

$$POSMV = PREMV + INV$$

**TABLE 18.13  Netscape valuation at four stages.**

| Stage | Date | Price per share | New investment ($ millions) |
|-------|------|-----------------|-----------------------------|
| Seed | 4/94 | $0.75 | $3.1 |
| Series A | 7/94 | $2.25 | $6.4 |
| Series B | 6/96 | $9.00 | $18.0 |
| IPO | 8/96 | $28.00 | $160.0 |

The percentage of the company sold to the investors is

$$\frac{\text{INV}}{\text{POSMV}} \times 100\%$$

Consider a firm, EZY Inc., with a series of investments as shown in Table 18.14. A set of venture capital investors invests in the firm in the series A round at 90 cents per share and owns 40.0 percent of the firm after the purchase. At the next stage, series B, the firm has missed its milestones, and the investors offer a reduced share price of 50 cents per share. This round (or stage) of financing is called a *down round,* since the price per share goes down. As a result, the venture capitalists receive a significant increase in their ownership in the firm. If the firm meets or exceeds its milestones, subsequent rounds can be "up priced." Also, entrepreneurs are urged to seek new investment rounds from other investors to check market pricing conditions.

LinkedIn's experience through its various stages of venture capital investments is shown in Table 18.15.

The timing of staged investments is a critical issue for the CEO and CFO of a new company that has not yet achieved breakeven. This company is using cash from an earlier investment to cover its negative cash flow. We define **burn rate** as cash in minus cash out, on a monthly basis. Thus, if a firm has $800,000 cash in the bank and has a burn rate of $100,000 per month, it will run out of cash in eight months unless it can reduce its burn rate.

### TABLE 18.14  Financial stages of EZY Inc.

| Stage | Investors | Price per share | Ownership by FFF* | Ownership by venture capitalists |
|-------|-----------|-----------------|-------------------|----------------------------------|
| Seed | Founders Friends Family | $0.10 | 100% | — |
| Series A | Venture capital group | $0.90 | 60.0% | 40.0% |
| Series B | Venture capital group | $0.50 | 34.0% | 66.0% |

*Founders, friends, and family.

### TABLE 18.15  Venture capital financing of LinkedIn.

| | Date | Investment (in millions) | Valuation (in millions) |
|---|------|--------------------------|-------------------------|
| Series A | November 2003 | $5 | $15 |
| Series B | October 2004 | $10 | N/A |
| Series C | January 2007 | $13 | N/A |
| Series D | June 2008 | $53 | ~$1000 |
| Series E | October 2008 | $23 | ~$1000 |
| IPO | May 2011 | $202 | $4250 |

*Source:* Various websites and filings.

Some companies spend money unwisely, so a good rule is to keep the burn rate as low as possible. The alternative is to raise new rounds of new financing, if that is possible, to avoid dilution.

---

**Amp'd Mobile's Bankruptcy**

A big opportunity often leads to success, but not always. Founded by Peter Adderton in 2005, Amp'd Mobile was a Mobile Virtual Network Operator (MVNO) located in Los Angeles, California. As a mobile phone service provider targeting youth and young professionals, it sought to use emerging technologies like 3G wireless internet to provide phones capable of delivering media content and advanced social networking capabilities. In 2007, Amp'd filed for Chapter 11 bankruptcy. It had spent all $360 million it had acquired over five rounds of funding. The company had acquired only 175,000 subscribers, which did not produce enough revenue to cover its operating costs. Some worried that Amp'd did not perform proper credit checks on subscribers, perhaps contributing to its large percentage of nonpayers.

---

## 18.10  Initial Public Offering

The first public equity issue of stock made by a company is referred to as an **initial public offering** (IPO). The newly issued stock is sold to all interested investors of the general public in a cash offer. In the United States, the IPO is a sale of a portion of the company to the public by filing with the Securities and Exchange Commission (SEC) and listing its stock on one of the stock exchanges. The offer is managed via a financial intermediary, an investment bank, which aids in the sale of the securities. The investment bank performs such services as formulating the method of issuance, pricing, and selling the new securities. Conditions that favor an IPO are shown in Table 18.16.

The new venture firm has three possible reasons to issue an IPO: (1) raise new capital, (2) liquidity, and (3) image or brand. Many fast-growing firms will need large capital infusions of more than $30 million, and the public market is suitable for these larger amounts. Second, liquidity—the ability to easily convert ownership to cash—is facilitated by the IPO. The third reason is to help build brand reputation by allowing public ownership of the new venture firm. These advantages of issuing an IPO are summarized in Table 18.17.

The disadvantages of issuing an IPO include: (1) offering costs, (2) disclosure and scrutiny, and (3) perceived short-term pressures. Table 18.17 lists

**TABLE 18.16  Conditions that favor an IPO.**

| | |
|---|---|
| ▪ Market value greater than $200 million | ▪ Profit margins greater than 14 percent |
| ▪ Sales greater than $100 million | ▪ Return on capital greater than 14 percent |

## TABLE 18.17  Advantages and disadvantages of issuing an IPO.

| Advantages | Disadvantages |
|---|---|
| ■ Raising new capital with the possibility of later, additional offerings | ■ Offering costs and effort required |
| ■ Liquidity: Ability to convert ownership to cash, potential of harvest for investors and founders | ■ Disclosure requirements and scrutiny of operations |
| ■ Visibility: Build brand and reputation | ■ Perceived pressures to achieve short-term results |
| | ■ Possible loss of control to a majority shareholder |

## TABLE 18.18  The process for an initial public offering (IPO) in the United States.

1. Examine the state of the stock market and the potential for an IPO. Consider Sarbanes-Oxley compliance costs.

2. Interview several investment banking firms and select two or three. Select a law firm with IPO experience.

3. Conduct organization meeting and set the schedule for preparation.

4. Draft the registration statement and conduct due diligence. Prepare the issuer to become a public company, including attention to board and committee composition, disclosure controls, and internal controls.

5. Complete the registration statement, provide it to a financial printer, and file the registration statement with the SEC (Securities and Exchange Commission). Gain clearance from NASDAQ or other stock exchange.

6. Receive initial comments from the SEC. Revise registration statement and file first amendment to registration statement. Submit comment response letter to the SEC.

7. Management prepares for "road show" presentations and travel.

8. Receive additional comments from the SEC. File second amendment to registration statement. Resolve SEC comments and print preliminary prospectus and begin marketing efforts.

9. Conduct "road show" of presentations to potential large investors.

10. Resolve final issues with SEC. File final amendment to registration statement and request that the SEC declare the registration statement effective. Price the offering and commence the sale of the stock.

these disadvantages. For small offerings, less than $25 million, the costs can amount to 15 percent of the offering. For a typical offering raising $50 million, the total cost to the firm for issuance of new securities can be up to 10 percent of the funds raised. The time required to prepare all the documents also can be onerous to a small, emerging firm. Moreover, the disclosure and scrutiny can be burdensome and any misstep can cause havoc with the company's share price. Furthermore, many new firms find the pressure for reports of quarterly earnings improvements difficult to satisfy. Finally, firms with a market value of less than $300 million are often thinly traded and illiquid. Table 18.19 lists the conditions that favor an IPO.

The process for an IPO is in Table 18.18. The timing of the IPO is critical since the IPO market can be very volatile. The IPO market was favorable in the periods 1998 to 2000 and 2006 to 2007, and very unfavorable in the periods 2001 to 2003 and 2008 to 2009. See Table 18.9 for IPO data for 2003 to 2012.

### The IPO That Ignited the Dot Com Boom

During the late 1990s, Silicon Valley became known for IPOs that "popped." High demand resulted in shares that were so sought after on the first day of public sale, the share price closed at much higher values than at the opening trades. In 1995, this was an uncommon occurrence. In August of that year, a start-up called Netscape went public at the young corporate age of 16 months. Demand for the shares was so high that trading was delayed for over two hours as a clearing price was sought. On that day, the stock eventually rose to nearly three times its initial price, closing at nearly double its initial offering price. Netscape had raised $140 million in its IPO, but its impact was even larger. Netscape is credited with making the Internet accessible to mainstream users around the world. By launching the hottest IPO market in history, it stimulated the appetite of investors for technology IPOs for several years. Google had a similar effect with its IPO in 2004.

Determining the appropriate offering price is the most important thing the lead investment bank must do for an initial public offering. The issuing firm faces a potential cost if the offering price is set too high or too low. If the issue is priced too high, it may be unsuccessful and be withdrawn. If the issue is priced below the true market price, the issuer's existing shareholders will experience an opportunity loss.

The IPO marketing process includes a road show. Road shows involve the lead underwriters and key firm managers marketing the firm to prospective investors (institutional investors) via presentations in major cities and one-on-one meetings with mutual fund managers.

Amazon.com was founded in July 1994 and opened for business in July 1995. It reached sales of $16 million for the quarter ending March 31, 1997, but was operating at a loss. Jeff Bezos hoped to take the company public to raise additional funds and to build public recognition for it. The total investment from Bezos, angels, and Kleiner Perkins amounted to approximately $9 million. Amazon selected Deutsche Morgan Grenfell (DMG) as its investment bankers in February 1997 and began the process of preparing the necessary documents for submission to the Securities and Exchange Commission (SEC). Amazon went public with an IPO selling three million shares at a price of $18 each. Its revenue per share, pre-IPO, was approximately $1.80. Thus, the firm went public at a price-to-sales ratio of 10. The firm increased its sales from $148 million in 1997 to $1.5 billion in 2002, but was unprofitable in those years.

If a firm intends to eventually issue an IPO when the market is favorable, it is wise to work from the beginning to be in position to qualify its IPO. Thus, a firm planning to go public needs to meet all the regulatory requirements in place when creating the prospectus. This means having audited financial statements, a complete management team, a sustainable competitive advantage, and an independent board of directors. Some firms also engage in "prestige enhancement" leading up to an IPO, hiring prestigious executives and directors with the hope that doing so will raise the share price [Chen et al., 2008].

The prospectus or selling document is a part of the information provided to the SEC for approval. The information in the prospectus must be presented in an organized, logical sequence and an easy-to-read, understandable manner to obtain SEC approval. Some of the most common sections of a prospectus are: prospectus summary, description of the company and its business, risk factors, use of proceeds, dividend policy, capitalization, dilution, management, owners, and the financial statements.

The cover page of the prospectus for LinkedIn's IPO in 2011 is shown in Figure 18.6a. The table of contents for the LinkedIn prospectus is shown in Figure 18.6b. The summary page, the offering page, and the summary financial data are provided in Figures 18.6c, d, and e, respectively. LinkedIn raised $352.8 million from the offering of 7.8 million shares.

# *7,840,000 Shares*

## Linked in.

*Class A Common Stock*

---

*LinkedIn Corporation is offering 4,827,804 shares of its Class A common stock and the selling stockholders are offering 3,012,196 shares of Class A common stock. We will not receive any proceeds from the sale of shares by the selling stockholders. This is our initial public offering and no public market currently exists for our shares of Class A common stock.*

---

*Following this offering, we will have two classes of authorized common stock, Class A common stock and Class B common stock. The rights of the holders of Class A common stock and Class B common stock will be identical, except with respect to voting and conversion. Each share of Class A common stock will be entitled to one vote per share. Each share of Class B common stock will be entitled to ten votes per share and will be convertible at any time into one share of Class A common stock. Outstanding shares of Class B common stock will represent approximately 99.1% of the voting power of our outstanding capital stock following this offering, and outstanding shares of Class B common stock held by our co-founder and board Chair, Reid Hoffman, will represent approximately 21.7% of the voting power of our outstanding capital stock following this offering.*

---

*Our Class A common stock has been approved for listing on the New York Stock Exchange under the symbol "LNKD."*

---

*Investing in our Class A common stock involves risks. See "Risk Factors" beginning on page 13.*

---

*PRICE $45.00 A SHARE*

---

| | Price to Public | Underwriting Discounts and Commissions | Proceeds to LinkedIn | Proceeds to Selling Stockholders |
|---|---|---|---|---|
| *Per Share* | $45.00 | $3.15 | $41.85 | $41.85 |
| *Total* | $352,800,000 | $24,696,000 | $202,043,597 | $126,060,403 |

*We have granted the underwriters the right to purchase up to an additional 1,176,000 shares of Class A common stock to cover over-allotments.*

*The Securities and Exchange Commission and state securities regulators have not approved or disapproved of these securities or determined if this prospectus is truthful or complete. Any representation to the contrary is a criminal offense.*

*The underwriters expect to deliver the shares of Class A common stock to purchasers on May 24, 2011.*

---

**MORGAN STANLEY**  **BofA MERRILL LYNCH**  **J.P. MORGAN**

*ALLEN & COMPANYLLC*  *UBS INVESTMENT BANK*

*May 18, 2011*

(a)

**FIGURE 18.6**

## TABLE OF CONTENTS

---

You should rely only on the information contained in this prospectus or contained in any free writing prospectus filed with the Securities and Exchange Commission. Neither we, the selling stockholders nor the underwriters have authorized anyone to provide you with additional information or information different from that contained in this prospectus or in any free writing prospectus filed with the Securities and Exchange Commission. We and the selling stockholders are offering to sell, and seeking offers to buy, our Class A common stock only in jurisdictions where offers and sales are permitted. The information contained in this prospectus is accurate only as of the date of this prospectus, regardless of the time of delivery of this prospectus, or of any sale of our Class A common stock.

**Through and including June 12, 2011 (the 25th day after the date of this prospectus), all dealers that effect transactions in these securities, whether or not participating in this offering, may be required to deliver a prospectus. This is in addition to the dealers' obligation to deliver a prospectus when acting as underwriters and with respect to their unsold allotments or subscriptions.**

For investors outside the United States: Neither we, the selling stockholders, nor any of the underwriters have done anything that would permit this offering or possession or distribution of this prospectus in any jurisdiction where action for that purpose is required, other than in the United States. You are required to inform yourselves about and to observe any restrictions relating to this offering and the distribution of this prospectus outside of the United States.

(b)

## PROSPECTUS SUMMARY

*The following summary highlights information contained elsewhere in this prospectus and does not contain all of the information that you should consider in making your investment decision. Before investing in our Class A common stock, you should carefully read this entire prospectus, including our consolidated financial statements and the related notes included in this prospectus and the information set forth under the headings "Risk Factors" and "Management's Discussion and Analysis of Financial Condition and Results of Operations."*

## LINKEDIN CORPORATION

We are the world's largest professional network on the Internet with more than 100 million members in over 200 countries and territories. Through our proprietary platform, members are able to create, manage and share their professional identity online, build and engage with their professional network, access shared knowledge and insights, and find business opportunities, enabling them to be more productive and successful.

We believe we are the most extensive, accurate and accessible network focused on professionals. We believe we are creating significant value for professionals, enterprises and professional organizations worldwide by connecting talent with opportunity at massive scale. We believe that our members and the enterprises and professional organizations that use our platform are only beginning to leverage the power and potential of our network and its underlying database of professional information.

Our comprehensive platform provides members with solutions, including applications and tools, to search, connect and communicate with business contacts, learn about attractive career opportunities, join industry groups, research organizations and share information. At the core of our platform are our members, who create profiles that serve as their professional profiles and are accessible by any other member, as well as (unless a member chooses otherwise) anyone with an Internet connection. We believe that our platform allows our members to compete more effectively, make better decisions faster and manage their careers to achieve their full potential.

The cornerstone of our business strategy is to focus on our members first. We provide the majority of our solutions to our members at no cost. We believe this approach provides the best way to continue to build a critical mass of members, resulting in beneficial network effects that promote greater utilization of our solutions, higher levels of engagement and increased value for all of our members.

We provide enterprises and professional organizations of all sizes with solutions designed to identify specific talent within our global network, enable their employees to be more productive and successful, build brand awareness, and market their products and services. Enterprises and professional organizations that use our solutions include large corporations, small- and medium-sized businesses, educational institutions, government agencies, nonprofit organizations and other similar entities. Our current products for enterprises and professional organizations include hiring solutions, marketing solutions and premium subscriptions. Our hiring solutions are transforming the talent acquisition market by providing unique access not only to active job seekers but also to passive candidates who are not actively looking to change jobs. Our marketing solutions enable enterprises to reach a large audience of influential and affluent professionals and connect them to relevant products and services.

We generate revenue from enterprises and professional organizations by selling our hiring solutions and marketing solutions offline through our field sales organization or online on our website. We also generate revenue from members, acting as individuals or on behalf of their enterprise or professional organization, who subscribe to our premium services. We strive to ensure that our hiring solutions, marketing solutions and premium subscriptions provide both a high level of value for our customers and also a high degree of relevance for our members.

(c)

**427**

We believe this monetization strategy properly aligns objectives among members, customers and our entire network and supports our financial objective of sustainable revenue and earnings growth over the long term.

We have achieved significant growth as our network has scaled and as we have expanded our product offerings. From 2009 to 2010, net revenue increased $123.0 million, or 102%, net income increased $19.4 million, or 487%, and adjusted EBITDA increased $33.3 million, or 227%. In the three months ended March 31, 2011, net revenue increased $49.2 million, or 110%, net income increased $0.3 million, or 14%, and adjusted EBITDA increased $4.2 million, or 46%, over the three months ended March 31, 2010. See "Adjusted EBITDA" below for a definition of adjusted EBITDA and a reconciliation of adjusted EBITDA to net income (loss).

## Our Mission

Our mission is to connect the world's professionals to make them more productive and successful. Our members come first. We believe that prioritizing the needs of our members is the most effective, and ultimately the most profitable, way to accomplish our mission and to create long-term value for all of our stakeholders. We will continue to concentrate on opportunities we believe are in the best interests of our members. Our long-term approach enables us to invest, innovate and pioneer in unexplored segments of our industry to increase the value proposition of our proprietary platform and extensive data.

Our solutions are designed to enable professionals to achieve higher levels of performance and professional success and enable enterprises and professional organizations to find and connect with the world's best talent.

## Our Vision

Our vision is to create economic opportunity for every professional in the world. This vision not only applies to each of our employees, but every LinkedIn member, each of whom has the ability to create economic opportunities for others. We believe this is the fundamental power of our network.

## Our Opportunity

We believe we are transforming the way people work by connecting talent with opportunity at massive scale. Our goal is to provide a global platform capable of mapping every professional's experience, skills and other relevant professional data to his or her professional graph, including connections with colleagues and business contacts.

We provide the following key benefits to our members:

- *Ability to Manage Their Professional Identity.* Through online professional profiles of record that members create, manage and control, we are reshaping how members present their professional identity. Our members share, at their own discretion, information about themselves in their profiles, and this data is available, accessible and searchable online. The accuracy of our members' public profile information combined with our search engine optimization technology often enables their professional profiles to appear at or near the top of search engine results, increasing the awareness of our members' professional identities.

- *Enhanced Ability to Build and Engage with Their Professional Network.* We enable members to build their professional networks by linking their professional profiles with those of others to whom they are directly and indirectly connected, creating an ever-expanding professional graph. In this manner, members use our platform to engage with other members whom they trust and value and reach people who are not known to them but who are known and trusted by others within their professional network. Through access to rich, up-to-date profile and professional graph data, members can make better use of their existing and new connections.

(c)

- ***Access to Knowledge, Insights and Opportunities.*** We believe we are a trusted source for comprehensive and rich, real-time news, opinions and other professional intelligence. Our proprietary platform provides solutions that enable our members to search and access insights and opportunities generated by our community of professionals, enterprises and professional organizations. The information and opportunities presented to each member are personalized based on his or her profile and professional graph, thereby providing our members with compelling and relevant information designed to make them more productive and successful.

In addition, enterprises and professional organizations also utilize our solutions to receive numerous benefits, such as attracting new talent and more fully understanding, retaining and engaging with their employees and other professionals. We provide the following key benefits to enterprises and professional organizations:

- ***Matching Talent with Opportunity.*** With the world's largest online professional network, we provide enterprises and professional organizations the ability to connect with the global professional talent pool at scale. Our extensive hiring solutions allow enterprises and professional organizations to leverage the insights from our platform to source and develop a pipeline of active and passive talent, including the ability to automate talent matching, post jobs, engage and educate candidates, streamline applications and validate information. We believe our solutions are both more cost-effective and more efficient than traditional recruiting approaches, such as hiring third-party search firms, to identify and screen candidates.

- ***Efficient Marketing Channel.*** Through our marketing solutions, enterprises and professional organizations are able to create, promote and control their corporate identity and enhance their brand awareness. Our marketing solutions, including our self-service platform, allow enterprises and professional organizations to pursue business-to-business marketing, prosumer marketing and marketing to mass consumers. Marketers use our solutions to create an online brand and corporate identity, disseminate trade publications and collateral, engage in highly targeted marketing campaigns and gain rich customer insights, all at scale and on a cost-effective basis, which is particularly attractive to small- and medium-sized enterprises and professional organizations that have limited resources. Our proprietary platform is designed to leverage viral actions, social media, trusted recommendations and our rich user-generated data to efficiently connect members, enterprises and professional organizations to relevant products and services.

- ***Targeted Advertising.*** Our member base constitutes one of the most influential, affluent and highly educated audiences on the web. According to The Nielsen Company @Plan data released in December 2010, U.S. visitors to our website represent more decision makers, have higher average household incomes and are comprised of more college or post graduates than U.S. visitors of many leading business websites. Our marketing solutions provide advertisers with the ability to target audiences based on our members' profile information, including title, function, company name, company size, industry and geography. In addition, our detailed advertising campaign reports provide advertisers with insights to further maximize the return on their advertising budget.

- ***Increase Employee Productivity.*** We serve as the central hub of a professional's online network and a platform for enterprises and professional organizations to share knowledge and professional insights with their employees and thereby increase their productivity. For example, employees who are members are able to join groups for sharing information, articles, links, conversations and opinions. In addition, sales professionals who are members are able to accelerate their sales processes by accessing information to identify leads and decisionmakers, request introductions to other members through a common connection and thereby increase their efficiency and potentially eliminate cold calls altogether.

(c)

**Our Competitive Strengths**

We believe the following strengths provide us competitive advantages in realizing the potential of our opportunity:

- *Exclusive Focus on Professionals.* As a result of our exclusive focus on professionals, we have built the world's largest professional network on the Internet with over 100 million members. We have developed a strong brand as a trusted database for profile information and provider of applications and tools for professionals to more effectively present their professional identity online, build and engage with their professional network more efficiently, and access insights to be more productive and successful.

- *Large and Growing Global Member Base.* Our member base continues to grow rapidly, adding approximately one new member every second, primarily through "word-of-mouth" and the network effects of our platform. While it took us nearly 500 days to reach our first one million members, during the second half of 2010, on average, we added more than a million new members every 10 days. Between 2008 and 2010, our member base increased at a 76% compounded annual growth rate. As of January 2011, over 50% of our members were from outside of the United States and our member base included executives from all of the companies in the 2010 Fortune 500.

- *Business Model with Powerful Network Effects.* The size and growth of our member base, the number of enterprises and professional organizations that use our platform, and the amount of rich and accurate information generated by our members increase the value we deliver to all participants in our network. A larger member base provides more opportunities to form professional connections for members, as well as increased opportunities to identify and attract talent for enterprises and professional organizations. At the same time, an increasing number of enterprises and professional organizations accessing our network enhances the relevance for members who stand to benefit from professional insights and opportunities. We believe the breadth and depth of our network would be difficult to replicate and represents a significant competitive advantage.

- *Robust and Trusted Source of Relevant Professional Data.* Our proprietary platform processes, filters and indexes a vast and growing amount of user-generated content, including updates to members' professional profiles, connections, activities and recommendations. The more data our members choose to share, the more value they receive from the network. As a result, members are more willing to share accurate and detailed information about themselves. We use the information that flows through our platform to provide more relevant searches and information to enhance productivity. Since our formation, we have provided easy-to-use controls that enable our members to select what they would like to share with whom, and we believe we have been clear and consistent on how we use this information to the benefit of members, enterprises and professional organizations.

- *Large Customer Base.* Thousands of enterprises and professional organizations use our hiring solutions and marketing solutions. In 2010, our hiring solutions were used by nearly 3,900 companies. As of March 31, 2011, our hiring solutions were used by nearly 4,800 companies, including 73 of the Fortune 100. Our customers also include many small- and medium-sized businesses using our platform to leverage their limited recruiting resources. In 2010, our marketing solutions were used by more than 33,000 customers. This broad customer base provides us with not only diversification but also market validation for additional new customers.

- *Proprietary Technology Platform.* Our proprietary software applications and technologies enable us to perform large scale real-time data and computational analyses that support our solutions. We categorize and query large sets of structured and unstructured data to personalize relevant information. For example, one of our key personalized recommendation features typically involves the processing of over 75 terabytes per day, and nearly two billion people searches were performed on our website in 2010.

(c)

## THE OFFERING

| | |
|---|---|
| Class A common stock offered | |
|     By us | 4,827,804 shares |
|     By the selling stockholders | 3,012,196 shares |
|     Total | 7,840,000 shares |
| Class A common stock to be outstanding after this offering | 7,840,000 shares |
| Class B common stock to be outstanding after this offering | 86,658,627 shares |
| Total Class A and Class B common stock to be outstanding after this offering | 94,498,627 shares |
| Over-allotment option to be offered by us | 1,176,000 shares |
| Use of proceeds | We intend to use the net proceeds from this offering for working capital and general corporate purposes, including further expansion of our product development and field sales organizations, and for capital expenditures. In addition, we may use a portion of the proceeds from this offering for acquisitions of complementary businesses, technologies or other assets. We will not receive any of the proceeds from the sale of shares to be offered by the selling stockholders. See "Use of Proceeds" on page 35. |
| Risk factors | See "Risk Factors" beginning on page 13 and the other information included in this prospectus for a discussion of factors you should carefully consider before deciding to invest in our Class A common stock. |
| NYSE symbol | "LNKD" |

The total number of shares of our Class A and Class B common stock outstanding after this offering is based on no shares of our Class A common stock and 89,670,823 shares of our Class B common stock (including preferred stock on an as converted basis) outstanding, as of March 31, 2011, and excludes:

- 16,221,375 shares of our Class B common stock issuable upon the exercise of outstanding options as of March 31, 2011 granted pursuant to our Amended and Restated 2003 Stock Incentive Plan at a weighted-average exercise price of $5.86 per share;

- 1,559,080 shares of our Class B common stock issuable upon the exercise of outstanding options granted after March 31, 2011 pursuant to our Amended and Restated 2003 Stock Incentive Plan at an exercise price of $22.59 per share;

- 13,659,553 shares reserved for future grant or issuance under our 2011 Equity Incentive Plan, which will become effective upon the completion of this offering, consisting of:

  - 2,000,000 shares of our Class A common stock reserved for future grant or issuance under our 2011 Equity Incentive Plan, and

(d)

- 11,659,553 shares of our Class B common stock reserved for future grant or issuance under our Amended and Restated 2003 Stock Incentive Plan (after giving effect to the grant of options to purchase 1,559,080 shares of our Class B common stock after March 31, 2011, and an increase of 13,000,000 shares in the number of shares reserved for issuance pursuant to our Amended and Restated 2003 Stock Incentive Plan after March 31, 2011), which shares will be added to the shares of our Class A common stock to be reserved under our 2011 Equity Incentive Plan upon its effectiveness; and

- 3,500,000 shares of Class A common stock reserved for issuance under our 2011 Employee Stock Purchase Plan, which will become effective upon the completion of this offering.

Unless otherwise stated, information in this prospectus (except for the historical financial statements) assumes:

- the reclassification of our common stock into an equivalent number of shares of our Class B common stock and the authorization of our Class A common stock;

- that our amended and restated certificate of incorporation, which we will file in connection with the completion of this offering, is in effect;

- the automatic conversion of all shares of our outstanding preferred stock into an aggregate of 45,647,201 shares of our Class B common stock immediately prior to the completion of this offering; and

- no exercise of the underwriters' over-allotment option.

(d)

# SUMMARY CONSOLIDATED FINANCIAL DATA

The following tables summarize the consolidated financial data for our business. You should read this summary consolidated financial data in conjunction with "Management's Discussion and Analysis of Financial Condition and Results of Operations" and our consolidated financial statements and related notes, all included elsewhere in this prospectus.

We derived the consolidated statements of operations data for the years ended December 31, 2008, 2009 and 2010 and the consolidated balance sheet data as of December 31, 2009 and 2010, from our audited consolidated financial statements included elsewhere in this prospectus. The unaudited consolidated statements of operations data for the three months ended March 31, 2010 and 2011, and the unaudited consolidated balance sheet data as of March 31, 2011, are derived from our unaudited consolidated financial statements included elsewhere in this prospectus. We have prepared the unaudited financial information on the same basis as the audited consolidated financial statements and have included, in our opinion, all adjustments, consisting only of normal reclining adjustments, that we consider necessary for a fair presentation of the financial information set forth in those statements. Our historical results are not necessarily indicative of the results that may be expected in the future, and our interim results are not necessarily indicative of the results to be expected for the full fiscal year.

| | Year Ended December 31, | | | Three Months Ended March 31, | |
|---|---|---|---|---|---|
| | 2008 | 2009 | 2010 | 2010 | 2011 |
| | (in thousands, except per share data) | | | | |
| **Consolidated Statements of Operations Data:** | | | | | |
| Net revenue | $ 78,773 | $ 120,127 | $ 243,099 | $ 44,716 | $ 93,932 |
| Costs and expenses: | | | | | |
| Cost of revenue (exclusive of depreciation and amortization shown separately below) | 18,589 | 25,857 | 44,826 | 8,305 | 16,783 |
| Sales and marketing | 16,986 | 26,847 | 58,978 | 10,454 | 29,361 |
| Product development | 29,366 | 39,444 | 65,104 | 12,141 | 24,735 |
| General and administrative | 12,976 | 19,480 | 35,064 | 6,672 | 13,614 |
| Depreciation and amortization | 6,365 | 11,854 | 19,551 | 3,940 | 8,159 |
| Total costs and expenses | 84,282 | 123,482 | 223,523 | 41,512 | 92,652 |
| Income (loss) from operations | (5,509) | (3,355) | 19,576 | 3,204 | 1,280 |
| Other income (expense), net | 1 277 | 230 | (610) | (346) | 449 |
| Income (loss) before income taxes | (4,232) | (3,125) | 18,966 | 2,858 | 1,729 |
| Provision (benefit) for income taxes | 290 | 848 | 3,581 | 1,043 | (349) |
| Net income (loss) | $ (4,522) | $ (3,973) | $ 15,385 | $ 1,815 | $ 2,078 |
| | | | | | |
| Net income (loss) attributable to common stockholders | $ (4,522) | $ (3,973) | $ 3,429 | $ — | $ — |
| Net income (loss) per share attributable to common stockholders: | | | | | |
| Basic | $ (0.11) | $ (0.10) | $ 0.08 | $ 0.00 | $ 0.00 |
| Diluted | $ (0.11) | $ (0.10) | $ 0.07 | $ 0.00 | $ 0.00 |
| Weighted-average shares used to compute net income (loss) per share attributable to common stockholders: | | | | | |
| Basic | 42,389 | 41,184 | 42,446 | 41,966 | 43,726 |
| Diluted | 42,389 | 41,184 | 46,459 | 44,228 | 51,459 |
| Pro forma net income per share attributable to common stockholders [1] (unaudited): | | | | | |
| Basic | | | $ 0.17 | | $ 0.02 |
| Diluted | | | $ 0.17 | | $ 0.02 |
| Pro forma weighted-average shares used to compute pro forma net income per share attributable to commonstockholders [1] (unaudited): | | | | | |
| Basic | | | 88,091 | | 89,373 |
| Diluted | | | 92,104 | | 97,106 |
| **Other Financial and Operational Data:** | | | | | |
| Adjusted EBITDA [2] | $ 5,461 | $ 14,651 | $ 47,959 | $ 9,078 | $ 13,282 |
| Number of registered members (at period end) | 32,307 | 55,111 | 90,437 | 64,177 | 101,528 |

(e)

Stock-based compensation included in the statements of operations data above was as follows:

|  | Year Ended December 31, | | | Three Months Ended March 31, | |
|---|---|---|---|---|---|
|  | 2008 | 2009 | 2010 | 2010 | 2011 |
|  | (in thousands) | | | | |
| Cost of revenue | $ 298 | $ 370 | $ 439 | $ 89 | $ 183 |
| Sales and marketing | 513 | 657 | 1,225 | 250 | 1,098 |
| Product development | 1,214 | 2,346 | 3,248 | 690 | 1,603 |
| General and administrative | 2,580 | 2,779 | 3,920 | 905 | 959 |
| Total stock-based compensation | $ 4,605 | $ 6,152 | $ 8,832 | $ 1,934 | $ 3,843 |

|  | As of December 31, | | As of March 31, 2011 | | |
|---|---|---|---|---|---|
|  | 2009 | 2010 | Actual | Pro Forma [1] | Pro Forma As Adjusted [2] |
|  |  |  | (in thousands) | | |
| **Consolidated Balance Sheet Data:** | | | | | |
| Cash and cash equivalents | $ 89,979 | $ 92,951 | $ 106,060 | $ 106,060 | $ 304,304 |
| Property and equipment, net | 25,730 | 56,743 | 65,782 | 65,782 | 65,782 |
| Working capital | 71,885 | 66,734 | 64,629 | 64,629 | 262,873 |
| Total assets | 148,559 | 238,188 | 265,332 | 265,332 | 463,576 |
| Redeemable convertible preferred stock | 87,981 | 87,981 | 87,981 | — | — |
| Convertible preferred stock | 15,413 | 15,846 | 15,846 | — | — |
| Total stockholders' equity | 9,082 | 36,249 | 46,530 | 134,511 | 332,755 |

(e)

## 18.11 Spotlight on Tesla

Tesla Motors is a California company that designs, manufactures, and sells electric cars and power train components. Tesla has two cars available: the Tesla Roadster, an sports car, and the Model S, a luxury sedan. Elon Musk, the firm's CEO, envisions Tesla as an independent maker of electric vehicles.

Tesla is a fine example of a firm using many sources to obtain early growth capital. It raised nearly $200 million through five rounds of financing. Musk, who cofounded PayPal, put much of his own money into Tesla. He was joined by other wealthy individuals, such as Google cofounders Sergey Brin and Larry Page, and by a number of venture capital firms. In 2009, Germany's Daimler AG invested an additional $50 million into Tesla. The company also received $465 million in loans from the United States Department of Energy. In 2010, Tesla entered the public stock market (symbol TSLA) and raised $226 million on the NASDAQ.

Tesla's 2012 revenue exceeded $400 million. By 2014, Tesla may break even.

## 18.12 Summary

The entrepreneur leaders of new ventures create a set of financial projections that can be used to estimate the cash investments that will be needed as well as when they will be needed. Using that information, the entrepreneurs seek out investment capital. For the first or seed round, they may rely on their own funds and investments from friends and family. Eventually, most technology firms will need a significant investment from others such as wealthy individuals called angels or professional investors called venture capitalists.

Several stages of investment may be the best means of acquiring investments based on performance milestones for each stage. A real option is the right to purchase an asset at a future date. Thus, venture capital investors often use staged investing with milestones to exercise their investment opportunity with a start-up. For high-impact start-ups, it is the future value of the potential of the new firm that is most attractive.

Using venture capital valuation methods, a start-up's value is established and an agreement on the division of ownership may be obtained. As the expected growth of revenues and profitability are achieved, the firm and the investors may wish to exercise an option to sell. This could be an initial public offering (IPO) or the acquisition of the start-up by a larger firm in order to harvest the value that has been created by the partnership of investors, founders, and employees.

> **Principle 18**
> Many kinds of sources for investment capital for a new and growing enterprise exist and should be compared and managed carefully.

## Video Resources

Visit http://techventures.stanford.edu to view experts discussing content from this chapter.

| | | |
|---|---|---|
| Corporate Backed "Smart" Money | Lisa Lambert | Intel Capital |
| Start on Your Own | Mark Suster | GRP Partners |
| Understanding Early Valuations | Dana Mead | KPCB |

## 18.13  Exercises

**18.1**   Viscotech Inc. is described in exercise 17.3. Determine the percentage ownership an angel group may demand for investing the $1 million sought by Viscotech at the start of year 1. If Viscotech is unable to obtain the bank loan of $500,000, it will need an equity investment of $1.5 million. What ownership percentage will the angel group demand for this investment? Assume the annual interest payment on the loan was planned to be 10 percent of the principal.

**18.2**   Glenn Owens' attractive technology start-up requires $10 million to launch. Projections show earnings of $10 million and sales of $80 million in the fifth year. The venture capital firm expects a return of 50 percent per year for the five-year period prior to an IPO. What valuation would you assign to the new venture? What ownership portion should the venture capitalist expect to receive? Perform a sensitivity analysis on the valuation and rate of return.

**18.3**   DGI, a new firm in formation, has developed a set of projections shown in Table 18.19. The expected and pessimistic cases are shown. The harvest of the firm is planned for the fifth year. The firm is seeking an initial investment of $1 million before launching in year 1 as well as a commitment for $1 million at the end of year 2 for expansion. Acting

**TABLE 18.19**  **Revenues and net income for the expected case and the pessimistic case for DGI.**

| Year | 1 | 2 | 3 | 4 | 5 |
|---|---|---|---|---|---|
| Expected revenues | 0.84 | 2.82 | 5.44 | 8.35 | 11.55 |
| Expected net income | 0.18 | 1.25 | 2.67 | 4.17 | 5.86 |
| Pessimistic revenues | 0.42 | 1.41 | 2.72 | 4.18 | 5.78 |
| Pessimistic net income | (0.11) | 0.26 | 0.77 | 1.25 | 1.81 |

*Note:* All figures in millions of dollars.

as an adviser to a venture capital firm, prepare an offer of investment to submit to DGI. Assume the PE (price earnings) ratio in DGI's industry is 15.

**18.4** The CEO of an early-stage software company is seeking $5 million from venture capitalists. The reasonable projected net income of $5 million in year 5 can be valued at a PE of 20. Furthermore, the sales in year 5 are projected at $25 million. Assuming no dilution from additional financings, what share of the company would the venture capitalists expect if their anticipated rate of return were 50 percent? The company has one million shares outstanding before the venture capitalists purchase shares. What price per share should the venture capitalists pay?

**18.5** Consider a new firm in the nanotechnology field that seeks a second round of financing. This year, it has revenues of $2 million and projects profitability of $200,000 next year on revenues of $3 million. It is raising $1 million from a new set of investors. What share of the company should it offer to the new investors? Assume it can increase profits at a rate of 25 percent per year over the next five years.

**18.6** Explain the purpose and value of staged financing (a) for investors and (b) for the entrepreneur(s).

**18.7** In what types of ventures and circumstances could grants be viewed as an especially attractive source of capital? When would a grant be less appropriate?

# VENTURE CHALLENGE

**1.** What sources of capital will you use for your venture?

**2.** Why did you select these sources?

**3.** How much capital is needed initially and for what purpose?

**4.** What percentage of your venture do you plan to offer to outside investors?

# CHAPTER 19

# Deal Presentations and Negotiations

*Leadership involves remembering past mistakes, an analysis of today's achievements,*
*and a well-grounded imagination in visualizing the problem of the future.*
**Stanley C. Allyn**

**How does the new venture present its vision and story and negotiate a deal with investors?**

The creators of a new enterprise need to tell their story about the future of their business. Establishing credibility and trust through presentations of the new venture's plan for a novel solution to an important problem can lead to an investment.

The integrated story and the business plan should show how the business solution would be profitable within a reasonable period. The investors are interested in a favorable return. They also want to sense that they will be partners with trustworthy and capable entrepreneurs.

The negotiation of a deal with an investor is an important part of the process. One can cement the relationship or destroy it through the negotiation process. The terms of the agreement should address the means of achieving potential return and the allocation of risk between the investors and the entrepreneurs. ■

## 19.1 The Presentation

The new venture team will be expected to verbally present its business plan to investor groups, angels, potential employees, allies, and suppliers. The purpose of these meetings will be to persuade them to cooperate, support, and participate in the new venture. Effective persuasion is a negotiating and learning process that leads to a shared vision. Formerly, people thought of persuasion as a simple process: state the position or plan, outline the supporting arguments, and then ask for the action or deal being sought. Today, most would-be investors and allies want to engage in a dialogue with the new venture team. The team outlines the venture and invites feedback and alternative solutions. Persuasion involves compromise and the development of relationships with the investors and other participants.

New enterprises must be able to sell their story to potential investors and clients. Selling ideas is intrinsically difficult. Clients and investors are naturally risk-averse when it comes to large projects that call for large investments with payoffs that are many years in the future. They are even more risk-averse when the projects do not originate from within their own organization.

Effective persuaders first establish credibility. Second, they frame their goals in a way that establishes common ground with their listener. Third, they offer solid evidence to support their plan. Finally, they build a good relationship with their potential investors or allies. This four-step method of persuasion is summarized in Table 19.1 [Conger, 1998].

Credibility and trust are established through experience over time. Thus, the shortest path to credibility is with someone the entrepreneurs already know. Otherwise, the entrepreneurs' track record and references will be particularly important. The expertise required for the venture must always exist on the venture team. It may be useful to review the concept of influence and persuasion as described in Section 13.2.

The business plan must appeal strongly to the potential investor or ally. Framing the unique benefits of the new venture so that they match the goals of the investor or ally is critical. The next step is to provide solid evidence supporting the business plan. Here, a vivid story or analogy will help to bring the plan alive. Finally, it is important to build a good relationship with the investor or ally. In this step, the entrepreneurs demonstrate their commitment to the plan and show some of their passion for the project.

**TABLE 19.1  Four-step method of persuasion.**

| | |
|---|---|
| 1. Establish credibility with the investor, ally, client, or talent. | 3. Offer solid, compelling evidence to support the plan. |
| 2. Frame the goals of a new venture to be consistent with the goals of the investors or allies; describe the unique benefits of the venture. | 4. Build a good relationship with the investor or ally. |

**TABLE 19.2  Four-part pitch.**

| | |
|---|---|
| 1.   This is a painful problem that customers need to solve. | 3.   This company has a profitable and proven solution for this problem. |
| 2.   Many customers who have this painful problem have money to spend on alleviating the pain. | 4.   The team has implemented and planned effectively in the past, and they can execute well in the future. |

**TABLE 19.3  Nine questions to answer in the business plan presentation.**

1.   What is the product and what problem is it solving?
2.   What are the unique benefits of the product?
3.   Who is the customer?
4.   How will the product be distributed and sold?
5.   How many people will buy it in the first and second years?
6.   How much will it cost to design and build the product?
7.   What is the sales price?
8.   When will you break even?
9.   Who are the key team members and how are they qualified to build this business?

A good presentation captures the listeners and lets them respond to the problem the entrepreneurs propose to solve. A four-part "pitch" appears in Table 19.2.

In presenting the business plan, it is also useful to answer the nine questions listed in Table 19.3. Of course, there will be other issues, but these nine are almost always part of a presentation. It is helpful to rehearse the presentation with an audience of a few trusted colleagues or friends who can respond with suggestions.

Trust is the basis of nearly every enduring business relationship. The presentation of a business plan is a vehicle for winning trust from potential investors and allies. Trust, confidence, and relationships are built over time. Speaking about the venture's goals needs to invoke a positive response from the listener. What difference will it make if the venture is successful? Will people live better lives or enjoy new alternatives?

A successful pitch often follows the 10/20/30 rule: 10 slides, 20 minutes, and 30-point-font text. Most presentations to investors or allies should be based on about 10 slides that the entrepreneur can present in 20 minutes. Each slide should use at least 30-point font text. This will keep the listener interested and the presenters focused on their key points [Kawasaki, 2004]. Table 19.4 outlines a sample 10-slide presentation. Renowned venture investor Guy Kawasaki elaborates upon these slides on his website: blog.guykawasaki.com.

**TABLE 19.4  A sample 10-slide presentation.**

1. Company name, presenter name, contact information

2. Description of the problem: the need and the market

3. Solution: the product and its key benefits

4. Business model and profitability

5. Competition and strategy

6. Technology and related processes

7. Marketing and sales plans

8. Leadership team and prior experience

9. Financial projections summary

10. Current status and funds required

In any presentation, the speaker should convey a sense of urgency about the problem and a strong commitment to make the solution robust. Good speakers highlight the unique benefits on slides 3 and 4.

Listeners are swayed by the quality of the ideas but even more by evidence that the presenters are creative and innovative. The investors or new team members seek to be part of a creative collaboration. The listeners look for passion and evidence that the proposed solution is a big change or discontinuity. Furthermore, the goal is to engage the listeners so that they become part of a creative collaboration. The best outcome is that the presenters successfully project themselves as creative types and get their listeners to view themselves as creative collaborators in the process of building the new venture [Elsbach, 2003].

## 19.2 Critical Issues

We discussed the development of the business plan in Chapter 6 and the presentation of the business plan in Section 19.1. After the business plan is presented to a few potential investors, their suggestions and criticisms may require adjusting the business model of the firm or other parts of the business plan. Perhaps it is necessary to revise the product to make it more compelling. Offering a product that is "nice to have" for a customer is different than offering a product that the customer "must have." It is nice to have vitamins available, but when one has a headache, an aspirin or Advil is a "must have." Is the product a complete solution to the customer's problem, or is it only a part of the solution?

Investors will ask: Are the people committed to the business, and is the opportunity large in potential? Can we identify and reduce the risks? Will there be a way we can harvest our return on investment? Is the estimated growth rate attractive?

The plan in written form and the verbal presentation should hold together as an integrated whole. Use of outdated or incorrect data will leave doubts for investors. Unsubstantiated assumptions can also hurt a plan. For example, does the plan include an honest recognition of the competitors? Although passion is important, preparedness is critical, too [Chen et al., 2009].

The business plan and its associated story can be viewed as the keystone of an arch as shown in Figure 19.1. All the elements of the business model and leadership join together to form a whole.

It will help to validate a business plan if the new venture has paying customers at the time of presentation. It is even better if the firm is about to become profitable. Investors need reasons to invest. They want to see advance orders, a letter of intent, or a customer list. That shows proof of customers and how they value the firm's product.

In 1968, Gordon Moore and Robert Noyce left Fairchild Semiconductor to found Intel. They brought along Andy Grove and several other colleagues. At Intel, they saw an opportunity to make a silicon transistor and later an integrated circuit. Moore and Noyce were leaders in their field and knew Arthur Rock, a San Francisco venture capitalist. They asked him one day if he could raise $3 million to start Intel. Rock had secured commitments for the $3 million investment by that evening. Moore and Noyce were well-known and could command a good deal for their new business. Most entrepreneurs are not as fortunate.

If the leaders of a new venture do not have a big, proven network and background, they will need to work hard to find and satisfy investors. They must convince them that their business is a one-time chance to get involved with something important that will exploit a big opportunity.

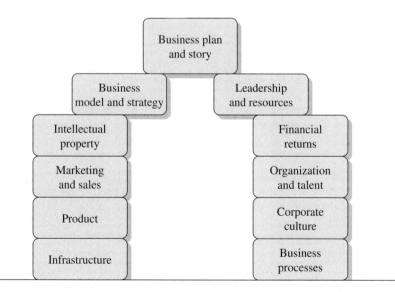

**FIGURE 19.1** Integral nature of the business plan and the firm.

## 19.3  Negotiations and Relationships

After the presentation of the business plan and follow-on discussions, an investor or ally may be confident about the potential venture but hold different expectations about it. Thus, the investor and the venture team may have different views regarding the valuation for the firm and the appropriate terms of the deal. A good deal is one that fairly meets the needs of the new venture while enabling the relationship between the firm and the investor to flourish in the future. Thus, the pricing and terms of the deal must be balanced with the future of the relationship. If possible, the new venture needs to have alternatives to closing a bad deal. With a good alternative, the new venture can walk away from a bad deal. A new venture should understand its own interests and its own no-deal alternatives or options. Price, control, and ownership percentage are usually key factors for negotiation.

Investors will normally engage in a process of checking the backgrounds of the venture team, the market data, and the key elements of the business plan. This process, called **due diligence,** consists of verifying facts and data provided in a business plan before making a commitment to the terms of an investment deal.

Negotiating a fair deal is a skill that can be learned. Most entrepreneurs have limited experience negotiating a fair deal with investors. **Negotiation** may be defined as a decision-making process among interdependent parties who do not share identical preferences. Consider an example of a manager and an employee negotiating a pay raise for the employee. They are interdependent but have different preferences for the outcome. The employee wants a raise, and the manager wants improved performance.

The best negotiation should produce an efficient, wise agreement and not damage the relationship between the parties. A wise or good agreement is one that meets the legitimate interests of both parties, resolves conflicts fairly, and is durable [Fisher and Ury, 1991]. One should try to avoid parties locking into a position, but rather reach for common ground.

The process of negotiating a good deal for all concerned can be based on four principles, as summarized in Table 19.5. Try to take personalities out of the discussion and avoid locking into positions. Try to get everyone working for a fair deal. Talk about the interests and goals of each party and avoid taking rigid positions, then generate several possible solutions that advance the interests of all parties. Finally, select the best solution with measurable outcomes.

Often negotiations will stall, and it may be best to try to reshape the scope and sequence of the negotiation. One or more of the parties can scan widely to identify elements outside of the deal currently on the table that might create a more favorable structure. For example, they may introduce new parties and terms to the deal and try to satisfy all parties [Lax and Sebenius, 2003].

Investors tend to have goals based on return on investment and time horizons to receiving this return. Entrepreneurs tend to have goals based on growth, success, and achievement, as well as return on investment. Both parties need to help generate some good options so that there is room to adjust the deal. Finally, the

**TABLE 19.5  Four principles of negotiations.**

1.  Focus on describing the problem (task or deal) and take the people out of the discussion.
    **Goal:** All participants are solving the problem.
2.  Focus on the interests of the parties, not their original positions.
    **Goal:** Each party states what they seek and the associated goals.
3.  Generate a variety of options or possibilities that advance the interests of the parties.
    **Goal:** Several real solutions.
4.  Create a final deal based on fair and objective standards.
    **Goal:** Real, measurable standards.

*Source:* Fisher and Ury, 1991.

parties select a deal that gives them a fair solution to their needs. This deal should include measurable outcomes and adjustments in ownership or other factors if the agreed-to outcomes are not realized. Also, the deal should be a good start on a long, cooperative alliance between the investor and entrepreneur [Ertel, 2004].

In any negotiation, each side must choose between two options: accepting a deal or taking its best no-deal option. Typically, the best no-deal option is to move on to find and negotiate with a new investor. Thus, the new venture team evaluates the deal offered versus seeking another potential investor. Similarly, the other side is also considering losing the deal versus losing the deal's advantages. The investor will accept the deal if it meets its own interests better than its own no-deal option, which is to lose a good opportunity. Thus, the negotiation problem for the venture is to understand and shape the investors' perceptions and help them choose in their own interest what you want [Sebenius, 2001].

When making an investment, a venture capitalist considers three forms of risk: the market risk (establishing customers for the product), the technological risk (the extent to which the technologies or concepts are well developed and not threatened by potential competitors), and the management risk (the team's technical and leadership competencies to develop the new firm). The venture capitalist wants to know which risks will be reduced by its investment in this stage.

Differences regarding valuation and ownership usually reflect differences in estimations of the future performance of the new venture. Therefore, it may be wise to develop an agreement that includes terms that are contingent on the outcome of designated measures or events. Using contingent terms in a contract enables the parties to bet on the future rather than argue about it. An agreement with contingences could include terms on such measures as revenues, profits, or number of customers achieved in an agreed-upon period. The actual outcome could lead to an agreed-upon ownership percentage. For example, the investor agrees to an ownership split of 70-30 (firm versus investor) if the goals and milestones are met but will reserve shares for a readjusted split of 60-40 if they are not met. Investors and new ventures can come to very different conclusions about many kinds of future events, such as sales and market shares as well as competitors' moves. Whenever such a difference exists, so does an opportunity to craft a contingent contract that both sides believe to be in their best interests. A contingent contract results in the two parties sharing the risk.

A common way to reach agreement about ownership is to offer a deal with warrants to tie ownership to actual performance. A warrant is a long-term option to acquire common shares, usually at a nominal price. For example, an investor may receive a warrant to receive an agreed-upon number of shares if certain performance levels are not achieved by a certain date [Smith and Smith, 2004].

A factor that can complicate negotiations is the matter of dilution of ownership by the founder group. Investors will usually seek an antidilution clause to protect themselves from dilution of their ownership percentage. This antidilution clause will usually be triggered by a lower price (down round) in a subsequent financing round. All the terms of an investment agreement, often called a **term sheet,** should be reviewed by the new venture firm's attorney. A term sheet is a funding offer from a capital provider. It lays out the amount of an investment and the conditions under which the investors expect the entrepreneur to work using their money. The key is to remember that it's just an offer, and the entrepreneur can counter that offer and negotiate all the terms before finally accepting the funds. An excellent reference on these matters is *The Entrepreneur's Guide to Business Law* [Bagley and Dauchy, 2011].

## 19.4   Term Sheets

The terms of the investment transaction are critical to the entrepreneur. The issues that concern professional investors are provided in Table 19.6. Clearly, trust and integrity are necessary between investors and the founding entrepreneurs. The venture capitalist willing to pay the highest price is not necessarily the person whom the entrepreneur will want most in the deal. Another venture capitalist who is not willing to pay quite as much may be a better partner in growing the business. Normally, founders are required to earn (vest) their stock over time.

Professional investors will normally want **preferred stock,** which has claims on dividends, if ever paid, and assets before common stock owners. With all the factors involved in completing a financing deal, the new venture needs to engage an attorney to review any terms of the deal and perhaps represent the new venture in negotiating the deal.

The terms of any deal should reflect the likelihood that the firm will require more capital at a later time. Many deals make it difficult to raise capital later at an attractive price. Often the deal is written on an assumption that everything works out as planned—an unlikely outcome. The plan for the future should include reasonable methods of obtaining new capital infusions. Often the provisions for protecting the investors in the current deal will be onerous if another capital infusion is required.

The terms of the deal should address the means of achieving potential return and the allocation of risk between investor and new venture. It may be better to lower the price to the investor and get the investor to share the future risk with

**TABLE 19.6 Issues to be resolved within the terms of the deal.**

| | |
|---|---|
| ■ Percent ownership for the investor group | ■ Type of security |
| ■ Timing of investment | ■ Reservation of ownership for future employees (stock option pool) |
| ■ Control exerted by investor | |
| ■ Vesting periods for ownership by the entrepreneur team | ■ Antidilution provisions |
| | ■ Milestones of achievement, if there are multiple stages or steps to the investment ("traunches") |
| ■ Rights to require an IPO and registration rights | |
| | ■ Stock option plans |

the new venture. The investors are looking for protection of their investment, and the new venture is seeking a capital infusion but needs to retain the right to pursue future capital infusions. Excessive protection clauses need to be traded away for lower prices for stock purchased by the investor. If possible, the investor should pay less for ownership and share the risk with the new enterprise.

## 19.5 Spotlight on LinkedIn

LinkedIn was founded in December 2002 and launched on May 2003. It is a social networking website used primarily by professionals. The site is available in many languages and has more than 200 million users in more than 200 countries.

The LinkedIn team gave a number of presentations to investors and partners as they grew the company. After examining the critical issues, Sequoia Capital funded LinkedIn in 2003. Greylock, Bain Capital and other funds joined Sequoia in subsequent rounds in 2005 and 2006. LinkedIn made a public offering of its shares on May 19, 2011. It's IPO prospectus can be found in Chapter 18.

LinkedIn has a strong business model based on enabling professionals to gain favorable network connections and to build these connections into a sales and referral network. Today, the company is expanding its mobile presence and its use of interest groups and related media.

## 19.6 Summary

The potential for a successful new business can be communicated through a formal presentation. Through this presentation of the entrepreneur's story, the potential investor, new employee, or partner learns to understand the opportunity and to recognize the competencies of the team members.

As investors become interested in the new enterprise, negotiations about valuation and performance milestones will commence. Conducting negotiations that retain and enhance the rapport with the investor is essential. The entrepreneur is negotiating until the moment of execution of an agreement and a transfer of funds. All the negotiations continuously address issues surrounding product, team, processes, business model, and intellectual property.

> **Principle 19**
> The presentation of a compelling story about a venture and the resulting skillful negotiations to close a deal with investors are critical to all new enterprises.

## Video Resources

Visit http://techventures.stanford.edu to view experts discussing content from this chapter.

| | | |
|---|---|---|
| Tips for a Good Pitch | Heidi Roizen | Mobius |
| Make a Great Pitch | Guy Kawasaki | Garage |

## 19.7 Exercises

**19.1**   Business plan contests offer an opportunity for entrepreneurs to present their business plans. The Clean Tech Open in the United States offers one such contest (www2.cleantechopen.org). Visit the site, study the description of the winners, and prepare a brief report on an enterprise that most interests you.

**19.2**   As the CEO of a new technology venture, you and your team have set a valuation for your firm of $10 million (pre-money) and found a willing venture capital firm. The venture capital firm, however, has set a valuation of $6 million. Revenues next year are projected to be about $6 million, and the firm will be profitable next year. Identify a negotiation approach for achieving a reasonable compromise valuation.

**19.3**   What is a term sheet? Specify three items in the term sheet that you would want as an entrepreneur. Specify item clauses in the term sheet that you would want as a venture capitalist. Why is it important for both parties to be happy with the resulting deal?

**19.4**   What are the key factors a venture capitalist uses to value a new venture? Describe what value parameters are most likely different between an investor and an entrepreneur.

**19.5**   What deal terms can a venture capitalist suggest to ensure entrepreneur incentives are aligned in both good and bad times for the firm? What are deal terms an entrepreneur would suggest to ensure investor incentive alignment in both good and bad times for the firm?

**19.6**   Why is it critical that the investor and the entrepreneur are both happy with the final deal? Who loses if this is not the case?

# VENTURE CHALLENGE

1.  State your venture's elevator pitch (see Section 6.4).

2.  Provide an outline of the presentation for describing your venture to investors.

3.  Sketch out a term sheet outlining your venture's capital needs, the amount of the company you are interested in selling (e.g., number of shares or what percentage of the total shares), and any other negotiation terms you consider important.

CHAPTER 20

# Leading Ventures to Success

*Well done is better than well said.*
**Benjamin Franklin**

**How do successful entrepreneurs transition from a solid business plan to an operating enterprise?**

Creating a business plan for a new enterprise is important, but implementing the plan successfully is essential. Execution of a plan is a discipline for connecting strategy with reality by aligning goals and the firm's people to achieve the desired results. Execution is about turning a concept into a great business.

New businesses move from start-up to growth to maturity in stages. Managing a new business through these stages requires different skills and organizational arrangements. Start-ups need to plan for having the right people in the right positions as they grow.

Organizations, like people, need to learn and adapt to change. Organizing for recognizing and responding to challenges can build resilience in a start-up firm. The ability to adapt to change may be a firm's only truly sustainable advantage. Furthermore, to achieve long-term success, a firm needs an ethical base for action. ■

## 20.1 Execution

**Execution** is a discipline for meshing strategy with reality, aligning the firm's people with goals, and achieving the results promised [Bossidy and Charan, 2002]. Often, the unique difference between a successful company and its competitors is the ability to execute its plan.

Execution is fundamentally about turning a concept into a business. For example, after a short period, the firm may find its prototype has defects or the sales channel is not as attractive as originally thought. Every new venture runs into trouble before reaching fruition. Then it is time to reconfirm the vision and recommit to the execution task. This requires persistence and follow through.

While admitting mistakes, the team's focus must stay on the long-term strategy. For example, Cyrus Field tried four times to lay a telegraph cable across the Atlantic Ocean to link the United States and Britain. Using a new steamship, the *Great Eastern,* he succeeded in 1866 after nine years of effort. The saga was a dramatic example of Thomas Edison's maxim, "genius is 1 percent inspiration, 99 percent perspiration." More recently, eBay faced a major challenge in the late 1990s as fraud started to increase. In 1998, the online auction firm created an antifraud campaign. Nonpaying bidders would receive only one warning before receiving a 30-day suspension from bidding. At the same time, eBay offered free insurance through Lloyd's of London. As a result, eBay became more successful.

Execution is following through on the strategy of the business plan. Execution is also the process of determining how well a firm is performing, acting to improve performance, and ensuring accountability. Any firm with a sound business model and strategy is still only as good as the implementation of the strengths of the model and strategy. In preparing the plan, all the team members have built their expectations and strategies into an agreed-upon, coherent road map. Thus, the team knows what needs to be done. The next step is to agree on how it is going to happen. Who does what and by when? The team sets short-term goals and priorities, and then assigns tasks to individuals. Rewards and recognition are linked to on-time performance. Since missing deadlines is costly, it is important to set realistic deadlines. Ultimately, the success of a new enterprise can be attributed to the execution skills of many team players rather than the decision-making skills of an omniscient entrepreneur. Six questions that can be effectively used to achieve solid implementation are provided in Table 20.1.

The new venture needs team members with a wide set of capabilities so that one person can be responsible for several tasks. If necessary, additional people

**TABLE 20.1  Six questions for implementation.**

| | |
|---|---|
| 1.  **Why** is the objective a priority? | 4.  **Who** is on the team and will be accountable? |
| 2.  **What** is the action and the expected outcome? | 5.  **When** will the activity be completed? |
| 3.  **How** will the action be achieved? | 6.  **Where** will it be accomplished? |

may be engaged to accomplish unique or difficult tasks. These tasks flow seamlessly from the firm's strategy. The team describes where it wants to be by a certain time and then divides the tasks among its members. It is critical to have a realistic assessment of the effort and time required for major tasks. The elements of an operating plan are tasks, milestones, and objectives. The leadership team will need to make trade-offs between tasks and goals so that the operating plan is realistic. A specific, written operating plan will help the firm to move forward efficiently. Also, reviews of accomplishments and the operating plan will help keep the team on task. Good execution is based on clear priorities, good assumptions, and constantly monitored performance [Mankins and Steele, 2005].

Execution is hard work. The setting of goals and deadlines should be the task of those who need to accomplish the work. The establishment of priorities is a big part of sound execution. Tasks can be divided into "must do," "should do," and "nice to do," with the priority kept on the "must do" activities as much as possible. The use of measurable goals and lists of necessary tasks and deadlines is very helpful.

The culture of the firm, if widely understood and shared, can help the firm's people to execute the strategy. With a strong culture and less formal direction, employees can take ownership over their actions and execute well. New firms can recruit, train, and reward people to take responsibility for their actions [Chatman and Cha, 2003].

During the initial period following launch of the firm, one of the primary goals will be to build and grow revenues. One key measure that can help in the early stages of a business is the ratio of revenues to expenditures plus assets employed, called a business index (BI), where

$$BI = \frac{revenues}{expenses + assets}$$

The goal of a business is to steadily increase the BI ratio by growing revenues faster than expenses and assets.

As a new firm grows, one useful measure of sound execution is the sales-per-employee ratio. A successful technology company will have at least $200,000 sales per employee. For example, as an emerging firm, storage and data management company NetApp had about $500,000 sales per employee (www.netapp.com).

An emerging new venture often needs help on operational issues as it moves toward growth. One source of help is the new firm's suppliers and customers. They have in-depth capabilities that can often be accessed by a new firm. Often these large firms want the new firm to succeed and will lend a hand on tough issues.

Truly great companies execute flawlessly. They deliver products that *consistently* meet customer expectations. Furthermore, they empower customer representatives on the frontline to respond to varying customers' needs. Few competitive advantages cannot be quickly imitated. Therefore, a new firm needs to out-execute its competitors. It must continually under-promise and over-deliver. The goal is to achieve almost perfect operational execution by constantly improving processes, training staff, and eliminating inefficiencies, as evidenced by such firms as United Technologies [Joyce et al., 2003].

## TABLE 20.2   Seven steps to building great companies.

1. **Leadership:** Be ambitious about the firm, possess strong will and resolve, desire sustained results, and retain personal humility.

2. **People:** Choose the right people, put them in the right positions, create a road map for success, and communicate it to everyone.

3. **Success:** Show unwavering faith that the firm will prevail in the long run, confront the realities and facts, and respond.

4. **Organizing principle:** Act on your passion, competencies, and economic engine to create a core principle for the business.

5. **Culture:** Build a culture of discipline where everyone is responsible for results; stay focused.

6. **Technology:** Select a technology application that will accelerate the firm's momentum.

7. **Momentum:** Build momentum slowly and consistently for the long run by constantly creating entrepreneurial projects.

*Source:* Collins, 2001.

Execution can go wrong for several reasons such as (1) allowing the strategy to shift over time and (2) failing at synchronization, which is getting the right product to the right customer at the right time [Hrebiniak, 2005]. Companies often make a chain of mistakes that undermine the firm [Mittelstaedt, 2005]. Failing to recognize mistakes and fix them can be fatal. According to Paul Graham, a successful investor at Y-Combinator, common mistakes that "kill start-ups" include hiring poorly, choosing the wrong platform, launching too early, spending too much money, and not narrowing in on a specific user problem [Graham, 2006].

Ultimately, a new venture can be said to execute well if its performance for its customers exceeds their expectations through all stages of its business. Ventures that execute well are organized around a prevailing vision and a well-understood long-term strategy. They continually invest in new ideas, products, and people [McFarland, 2008].

Jim Collins [2001] describes seven steps to building a great business, as summarized in Table 20.2. If the new enterprise can implement most of these very well, it will have a good chance for success.

### GE: Entrepreneurial Leadership

Jack Welch, retired CEO of General Electric, is perhaps the best-known U.S. executive of 1985 to 2000. He advocated many principles and methods of execution, such as "work-out" and "bullet-train speed." The key to Welch's popularity was his plain but powerful rhetoric about the critical issues of the firm [Lowe, 2001]. He insisted that each operating unit be number one or two in its market. While GE is a very large company, it retains today the atmosphere of entrepreneurial activity. General Electric meets all seven criteria for Collins's great companies.

A new venture can be said to execute well if its performance for its customers exceeds their expectations through all stages of its business. It is organized around a prevailing vision and a well understood long-term strategy. It keeps investing in new ideas, products, and people [McFarland, 2008].

## 20.2 Stages of an Enterprise

A new business venture is expected to grow over time, normally following an S-curve, as shown in Figure 20.1. The five stages of a firm are: start-up, take-off, growth, slowing growth, and maturity. During the start-up period, the firm is organizing itself, accumulating the necessary resources, and launching its product. The second stage is take-off, when revenues start to grow. The growth stage is often the most profitable and is a time when management must look toward expanding its product offerings and/or serving new types of customers and/or geographic areas, as illustrated in Figure 20.2. This stage also entails high risk since the firm has little experience with these new goals [Roberts, 2003].

Eventually, the firm's growth slows in the slow-growth period. Finally, the firm reaches the mature phase. Figure 20.1 shows a high-growth trajectory and a slow-growth trajectory for two businesses. The high-growth firm experiences

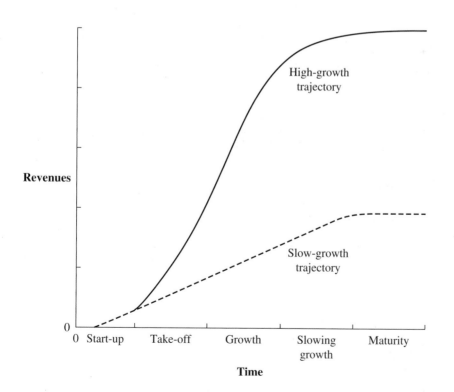

**FIGURE 20.1** Growth trajectories for two businesses.

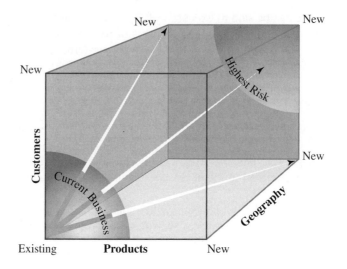

**FIGURE 20.2**  Alternative growth strategies for a venture.
*(Source:* Roberts, 2003.)

a growth rate of 40 percent per year or better. The slow-growth firm will have a growth rate of 10 percent per year.

Technology ventures needing a long start-up period for developing their product have to show success and raise their funds in stages. They continually must demonstrate credibility and engender trust in the investor community. The CEO and leadership of the venture must execute a creative and truthful strategy that unites the interests of the investors and the employees [Kleiner, 2003].

Moving from start-up to take-off may engender serious stresses as management practices that were appropriate for a smaller size and earlier time no longer work and are scrutinized by frustrated managers. High-growth ventures encounter fluid situations where rapid change and chaos occur. New products are released and marketing plans are in flux. New hires join the firm and decision-making is too slow. With high growth rates, new capital infusions may be necessary. Long working days become common and the potential for employee burnout increases. As ventures progress through stages, the original founders may depart and be replaced by management professionals.

A new venture on a slow-growth trajectory will enjoy a less-demanding workweek and fewer competitors but may be faced with limited profitability and access to capital.

As a new firm enters the take-off phase, the need for additional capital, resources, and employees will lead to more regularized processes and increasingly formalized communication. The take-off phase requires management skills and budgeting, accounting, and purchasing capabilities. By the time a firm enters the growth phase, the company moves toward decentralization and delegation of tasks. At this time, the firm may add midlevel managers for such tasks as purchasing, fulfillment, and sales. In the phase called "slowing growth," the challenge of flat revenues calls for new innovation and entrepreneurial leadership with an emphasis on renewal. The stages and their respective goals appear in Table 20.3.

When the business grows, the founding team is incredibly busy. Rapid growth puts an enormous strain on them. The business outgrows its production facilities and management capabilities. Typically, the management crunch hits in the third or the fourth year. That is when firms tend to outgrow their management base with quality falling, delivery dates missed, and customers not paying on time. As the firm enters the growth phase, the leadership team needs to ask: what does the business need at this stage? Founding CEOs tend to depart when their firm reaches a rapid growth phase and needs strong managerial expertise [Boeker and Karichalil, 2002].

**TABLE 20.3  Stages and respective goals of a business.**

| | |
|---|---|
| **Start-up** | Design, make and sell; exploit the opportunity |
| | Create the product and go to market |
| | Build the core business |
| | Emphasis on creativity |
| | Informal communication |
| **Take-off** | Efficiency of operations |
| | Refine and strengthen the business model |
| | New procedures introduced |
| | Invest in quality and customer service |
| | Communication becomes formal |
| | Emphasis on direction from leaders |
| **Growth** | Expansion of revenues and market share |
| | Expand product lines |
| | Hierarchy in place |
| | Move to decentralized structure |
| | Emphasis on delegation of tasks and responsibilities |
| **Slowing growth** | Consolidation and renewal of the firm |
| | Leadership required |
| | New initiatives needed |
| | Manage working capital |
| | Emphasis on coordination and renewal |
| **Maturity** | New innovation |
| | Strong culture and history |
| | Development of personnel seeking new opportunities |
| | Emphasis on collaboration and renewal |

eBay is an example of a successful move from take off to the growth stage. With a successful site and venture capital investment, eBay began looking in 1997 for a capable manager to take it into the growth phase. It found Meg Whitman, an experienced manager with solid marketing credentials. Whitman arrived in 1998 and immediately worked on the execution of an IPO for September 1998. In the first quarter of 1998, more than $100 million in goods changed hands on eBay, and the company's revenues exceeded $3 million a month [Cohen, 2002]. eBay sold about 9 percent of the company and raised $62.8 million. The IPO price was $18 per share, which jumped to $53 on the opening of the market. Whitman built a brand that promised trustworthy trading for buyer and seller. By the time she stepped down as CEO in 2008, eBay was a powerful auction firm with about $8.5 billion in revenues and an operating margin of 25 percent.

During the growth phase, it is also important to have financial leadership from a chief financial officer who manages the operating cash (cash flow), the capital expenditures, and any increases in working capital. The timing and form of capital investments can have a salutary effect on the profitability of the firm.

During the growth phase, competition heats up, too. Most products that are technology-based experience competition driven by technology, as shown in Figure 20.3 [Hirsh et al., 2003]. As technologies like solar panels and hybrid engines improve, customers demand better performance with lower life-cycle

**FIGURE 20.3** Technology drives most competition toward excellence and lower life-cycle cost of ownership. In this figure, automobiles are used to illustrate the principle.

cost of ownership. Firms that respond to these demands will continue to succeed.

As the firm approaches the slowing growth phase, a move toward hierarchy may be appropriate. Hierarchy helps us to handle the complexity of large organizations. Furthermore, people envision career ladders, readily understand the system, and identify with one subunit. Hierarchical structures provide rewards of power and status, and may be the most appropriate structures for managing activities in large firms. The leader strives to keep the best of the small firm—empowerment, teamwork, and shared leadership—while accepting the benefits of hierarchy in the large firm [Levitt, 2003].

Often an analogy helps us understand the management of transitions. We can envision the early-stage company in the start-up and take-off as a jazz band playing in a jam session with wonderful improvisations. The jazz band has fewer than 20 members, and all the members know each other and seamlessly take their turn leading. Several players can play several instruments. When a firm grows to more than 50 people, it starts to shift to acting like an orchestra with its separate sections—strings, wind instruments, and percussion. The orchestra needs one coordinating leader, called a conductor. He or she has a score they follow and describes a strategy the orchestra will follow for each piece of music. They act as one—as should a growing firm.

Changes in a company's growth also can challenge its management and structure. Cisco Systems's revenue growth between 1995 and 2000 was over 50 percent per year, primarily through acquisitions of small companies in exchange for Cisco stock. By 2002, Cisco's growth stalled, and the goal became managing expenses. The motto for the old Cisco was: faster, more sales. The motto for the new Cisco became: slower, better, profitable. From mid-1999 to late 2000, Cisco doubled its payroll from 22,000 to 44,000 employees. In 2001, the growth abruptly ended when businesses stopped buying. Telecommunications companies discovered they had massively overbuilt and ordered too much equipment from Cisco. Revenue fell for the first time in Cisco's history. By the summer of 2001, sales plunged one-third from their level six months earlier. By 2003, Cisco was cutting 8,500 workers [Thurm, 2003].

## Iridium: Getting It Wrong

In 1991, Motorola founded a spin-off called Iridium, LLC, to build a set of 66 low-orbit satellites that would allow subscribers to make phone calls from any global location. In 1998, the company launched its service, charging $3,000 for a handset and $3 per minute for calls. By 1999, Iridium filed for bankruptcy. As the CEO of Iridium said [Finkelstein and Mooney, 2003]: "We're a classic MBA case study in how not to introduce a product. First, we created a marvelous technological achievement. Then we asked how to make money on it." Not all ventures successfully reach the growth stage and beyond.

Downturns and recessions happen in every industry. The response of an emerging business to these tests can lead to renewal and success or disarray and failure. Few companies have soared as high, sunk as low, and struggled to survive as the software maker Novell. Founded in 1983, it has a formidable competitor, Microsoft, and has experienced large swings in revenues and profitability. In the early 1980s, Novell pioneered the market for network operating systems. In the early 1990s, Novell missed the shift to the Internet and lost market share. Drifting through the 1990s, Novell tried for a renewal with Eric Schmidt as leader. Trying to move out of the slowing growth phase, Schmidt attempted to renew the innovation of the firm. By 1999, Novell launched new software for networks and the Internet. However, Novell revenues remained flat and profitability remained elusive. Schmidt became CEO of Google in 2001.

Managing a downturn is as challenging as managing a period of fast growth. In a recession, customers are slow to pay their bills, and suppliers become weak. Furthermore, the availability of new capital dries up. If possible, an emerging new venture can reformulate a positive agenda for managing through the down period by renewing and tightening its strategy while avoiding overreacting. An economic downturn is an opportunity to clean the slate and get back to economic reality. Every downturn is a chance to rebuild the core business. Contrary to conventional wisdom, downturn winners avoid diversification. Instead, companies that successfully handle a downturn refocus on their core business and renew their strategy. They maintain a long-term view and strive to earn the loyalty of employees, suppliers, and customers [Rigby, 2001]. For example, in 2011, after several years of modest growth, Cisco elected to leave several consumer-related businesses and to return to its core switch and router business selling solutions for the data center. With a focus on the core business and a renewed strategy, the firm can see beyond the bad times. While managing costs, the firm prepares for the next upturn. If a firm has the resources, selected acquisitions may be a wise step for the future.

### What Ventures Need in CEOs

As an author and partner at Kleiner Perkins, Randy Komisar [2000] describes a start-up as requiring three types of CEOs at successive stages of its development. He cleverly uses descriptors in terms of dogs. The first CEO of a start-up is the "retriever." This CEO assembles the core team and the product to fit the original vision, and proceeds to access the necessary resources. The second CEO is the "bloodhound," who must sniff out a trail and find the right market and profitable customers. The third CEO is the "husky," who executes well and pulls the established firm steadily forward.

Another organizational issue flowing through the stages of a company's life is the matter of executive succession. As the firm moves through the stages, it often must change its CEO, particularly during periods of either very low or very high growth [Boeker and Wiltbank, 2005]. As needs change, the board of directors and the investors ask whether the incumbent has the skills to manage the firm through and into the next stage.

Entrepreneurs often are unable to make the transition from start-up mode and consequently struggle to become effective managers of a high-growth firm. The habits and skills that make entrepreneurs successful at launching and initially growing ventures can undermine their ability to lead larger organizations. Entrepreneurs tend to focus on details and tasks in the early stage of a firm, as they should. As the firm grows, a leader needs to work on leading a larger, more complex organization. Entrepreneurs can learn to grow with their firm by developing their relationship, networking, and strategic capabilities and by moving from a task-oriented approach to a coordinating approach [Hamm, 2002]. Still, fewer than 40 percent of founder CEOs make it past the second round of venture capital financing and fast-growing firms are especially likely to replace the founders [Bailey, 2003].

Founders are more likely to remain if they hold a sizeable percentage ownership of the firm [Boeker and Karichalil, 2002]. The founder of the firm that addresses a solid opportunity and grows rapidly, however, faces a dilemma. On the one hand, the company can grow faster with the necessary funds from the investors and subsequently increase the monetary value of the firm. On the other hand, the founders find it necessary to cede control and important decisions to their investor group. If the founders want to get rich, they will probably need to cede control. If the founders want to retain control, they should consider a slower growth path and require less investment. Most founding CEOs initially want both wealth and control. Choosing between money and power forces entrepreneurs to decide what success really means to them [Wasserman, 2012].

Successful executive succession can lead to superior organizational performance [Dyck et al., 2002]. Succession planning is necessary for any new venture as it moves from one stage to another. The four factors of a good succession are sequence, timing, technique, and communication, as described in Table 20.4. Using the analogy of a relay race, there is a positive relationship between successful passing of the leadership baton and organizational performance. Succession is facilitated when the incumbent and the successor have a shared understanding of the timing and technique of the hand-off.

**TABLE 20.4 Four factors of executive succession and the relay race analogy.**

| | | |
|---|---|---|
| 1. | **Sequence** | Ensuring the successor has the appropriate skills and experience to lead the organization in the next stage. The executive in the start-up stage should have strong entrepreneurial skills, while the executive in the take-off phase should have good organizational skills. |
| 2. | **Timing** | Ensuring the leadership baton is passed in a timely and expeditious way from incumbent to successor. |
| 3. | **Baton-passing technique** | Methods used for passing the baton. Ensuring that the baton is handed off as expected and the incumbent lets go of the power. |
| 4. | **Communication** | Harmonious cooperation and clear communication between incumbent and successor. |

## 20.3 The Adaptive Enterprise

No business plan survives its ultimate collision with reality. Changes in the marketplace and competition require any firm to react to change. Entrepreneurs must determine whether the assumptions on which their organization was built match the current reality.

Successful entrepreneurs have a deep knowledge of their customers and products, and know a great deal about what priorities matter to their customers at particular historical junctures. Furthermore, they learn from their experience and make rapid adjustments [Koehn, 2001].

Flexcar (www.flexcar.com) was started by Neil Peterson in Seattle to provide a time-sharing plan for automobile users. He thought that customers would use cars provided by his time-sharing company rather than buy their own car. This approach was popular in Europe, but Peterson found that people in Seattle still wanted their own car. He quickly learned to switch his marketing campaign to steer it toward universities and businesses, where the business model and sales pitch made more sense [Thomas, 2003]. Flexcar merged with Zipcar in 2007.

One of the biggest tasks for leaders in growing firms is to repeatedly mobilize their companies to change to meet new opportunities and competitive challenges. The leadership team needs to reinvent its strategy through a process of continuous renewal in a time of constant change; this process is called strategic learning [Pietersen, 2002]. A key capability of leaders in adaptive organizations is the capacity to adapt. These leaders do not get stopped by tough challenges they encounter but learn new lessons and go on. Aldous Huxley [1990] stated it as: "Experience is not what happens to a man. It is what a man does with what happens to him." **Strategic learning** is a cyclical process of adaptive learning using four steps: learn, focus, align, and execute. This adaptive process of learning and executing, if done well, may be one of a company's sustainable competitive advantages. Challenging discontinuities are new technologies, globalization, the Internet, deregulation, convergence, and channel disintermediation. A firm's strategy defines how it will respond to its challenges. Thus, the leadership team needs to focus the firm's resources on the best opportunities in the shifting context of the business world.

A **learning organization** captures, generates, shares, and acts on knowledge by revising its strategy as new knowledge becomes available. This type of firm is an **adaptive enterprise**—one that changes its strategy or business model as the conditions of the marketplace require. Philippe Kahn, cofounder of Borland and Lightsurf, described adaptivity as improvisation [Malone, 2002]: "I don't know what an entrepreneur is, but to me it's the difference between a jazz musician and a classical musician. I think a classical musician is a kind of guy who's gonna work for a big company. A jazz musician's gotta work in a small band and know how to improvise. I think that's really the analogy, that's really the difference."

In a small start-up consisting of 10 to 20 people with shared values and objectives, an informal process of developing renewed strategies will suffice.

As the firm grows, it needs to continue to renew its strategy as conditions in the competitive marketplace require. At that time, the ability to adapt becomes a required organizational capability. The leadership team needs to learn from its experiences, adjust its strategy, and execute those changes. Learning to deal with change, discontinuity, and uncertainty and adapt in a timely way needs to become a skill for leaders of new ventures [Buchanan, 2004].

The effective management of risk is critical to success. Assessing the risks associated with new initiatives can help leaders make adjustments to mitigate these risks. For example, as leader of Amazon, Jeff Bezos is always asking what could go wrong and considering ways to mitigate the downside. Table 20.5 lists the characteristics of successful CEOs of adaptive enterprises.

The learning organization, as described in Section 12.8, uses a learning process or cycle as shown in Figure 20.4. The goal of the learning process is to generate new strategies in a cycle of renewal. The first step consists of a situation analysis (see Section 4.4) of the competitive marketplace, industry dynamics, and the firm's strengths and weaknesses. The outcome of this step is insight into the issue and alternative responses. The second step is to redefine the vision, mission, strategy, and adjusted business model. The outcome of this step is a statement of the performance, resource, and capability gaps in terms of the desired strategy and the actual reality. The third step is to adjust the structure, process, people, and culture of the firm to work toward the new strategy. The outcome of this step is an adjusted business plan. The fourth step is to execute the newly adjusted business plan. The outcome of the fourth step is the actual adjusted performance. After a period of actual performance, the learning cycle starts again. The firm continues to learn and adjust as it repeats the strategic learning cycle.

Effective learning involves continuing to ask the key questions about customers, competitors, capabilities, resources, and profitability. Entrepreneurs should not rely on excessive overconfidence and overcommit up-front resources, hindering their willingness to learn and adjust. The best degree of confidence lies at the level of willingness to decide and move ahead with the expectation that new knowledge will help the firm learn and adjust [Simon and Houghton, 2003].

**TABLE 20.5  Characteristics of successful CEOs.**

| Prone to failure | More likely to succeed |
| --- | --- |
| 1.  Arrogant, hubris | 1.  Humble, open-minded |
| 2.  Overconfident with their answers to problems | 2.  Realistic, learning, always challenging answers |
| 3.  Underestimate major obstacles and risks | 3.  Examine carefully potential negative consequences and all risks |
| 4.  Rely on what worked in the past | 4.  Challenge every decision and look for wise changes and opportunities to learn |

*Source:* Finkelstein and Mooney, 2003.

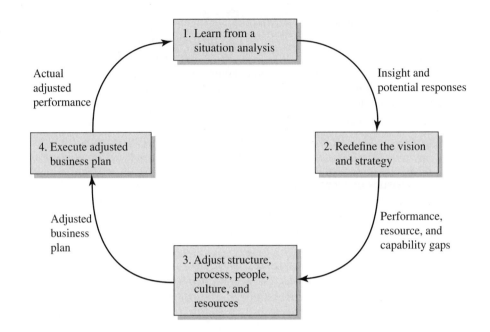

**FIGURE 20.4** Strategic learning cycle of a learning organization. (Adapted from Pietersen, 2002.)

Companies often conduct "postmortems" when they fail at a specific task, as with a failed product development effort, a failed product launch, or a quarter where revenue expectations were not met. These analyses enable the company to identify the factors responsible for the failure and to adjust accordingly. Fewer companies conduct a postmortem on their successes. Companies can learn a great deal, however, from analyzing the factors responsible for success and by determining the role of internal versus external factors. This analysis can keep a company from becoming overconfident [Gino and Pisano, 2011].

Most organizations face all kinds of unpredictable challenges that collectively place huge demands on people's creativity and imaginations. Resilience is the ability to recover quickly from setbacks. **Resilience** is a skill that can be learned and increased. A resilient organization acts on its learning and possesses a staunch acceptance of reality, a set of strongly held positive values, and a powerful ability to adapt [Coutu, 2002]. The firm has a clear, undistorted sense of reality about its competitive position. Then it needs the ability to make meaning out of difficult challenges. Resilient leaders build a new, improved vision of the future. Value systems of resilient firms evoke meaning and noble purpose. The third factor is the ability to make a future from what is available—a kind of inventiveness or ability to improvise a solution. Companies that survive regard improvisation as a core skill [Coutu, 2002]. Highly resilient organizations have the ability to read a weak signal of a problem and respond to it. Action tempered by reflection is the best method of responding to change and information [Coutu,

2003]. One way to identify threats and weaknesses is to bring in outsiders to test the resilience of the firm and its processes.

In the early 2000s, IBM was losing market share in the hardware business and its software business was not growing at the same pace of many of its competitors. Many doubted that IBM could continue to be a meaningful player in the information technology business. Under the leadership of its CEO, Sam Palmissano, the company divested non core growth businesses, including the PC manufacturing business, and invested heavily in building higher-margin software and cloud computing businesses. IBM made a number of key acquisitions, including the purchase of the consulting division of Price Waterhouse Coopers; large analytics companies, including Cognos, Netezza, and SPSS; and core system software companies, including Filenet and Rational Software. The company continued to grow and prosper, with IBM's share price growing 73% during the tenure of Sam Palmissano's leadership.

---

**Amyris Biotechnologies: Adapting to New Industry**

Amyris Biotechnologies is a noteworthy example of an adaptive company. Amyris was founded in 2003 by Dr. Jay Keasling, at the time a professor at U.C. Berkeley. In 2004, it received a $43MM grant from the Bill and Melinda Gates Foundation for the manufacture of a bacteria-produced antimalarial drug, artemisinin. Under the terms of the grant, Amyris would sell the drug at cost and make no profit.

By 2005, Amyris had succeeded in producing the drug microbially. The company now had both R&D funding and significant experience with the methodologies of synthetic biology. It was on the cutting edge of engineering microorganisms to become highly effective chemical factories.

In 2006, Amyris and its new CEO (John Melo) identified a new opportunity. With rising fuel prices, biofuels were increasingly looked at as a potential solution to the world's energy needs. With Amyris's unique competency in bioengineering, it felt well positioned to take a leadership position in the biofuels industry. Amyris executed an IPO in 2010 that raised $85MM for the company as it entered the biofuels business derived from sugarcane. When scaling became a difficult challenge, John Melo and Amyris adapted again to market needs and expanded into a third industry, cosmetics and fragrances, in 2012. The company effectively adapted to market needs and expanded into other industries.

---

It is wise to build an adaptive learning enterprise from a new venture's inception. Competitive advantage depends, in large part, on the ability of organizations to constantly change and reinvent themselves. They accomplish this through building and rebuilding a shared vision and team learning [Gabor, 2000]. Adaptive organizations strive to learn to manage within complex marketplaces. Complex, nonlinear, unpredictable markets challenge venture leaders to learn to manage continuous challenges. To a great extent, the iterative, adaptive process of Figure 20.4 is the only sustainable advantage for a firm in a dynamic economy.

## 20.4 Ethics

Life is filled with difficult ethical challenges. Ethics are a set of moral principles for good human behavior. Ethics provides the rules for conducting activities in a manner acceptable to society. Moral principles are concerned with goodness (or badness) of human behavior and usually are provided as rules and standards of human behavior. Thus, a common moral rule would be: do not lie. Of course, such rules are subject to interpretation, exemplified by the concept of the "white lie."

Ethics is concerned with doing the right (moral) thing. Society also establishes laws to guide actions. For example, U.S. law states that bribes and kickbacks are illegal. Laws are subject to interpretation, and paying a fee for sales help may be legal, while paying a bribe is illegal.

The success of new ventures, either profit or nonprofit, depends on winning against competitors. The competitive marketplace can put pressure on the entrepreneur to act unethically [Mishina et al., 2010]. The business leader finds it difficult to be fair to others without sacrificing customers or profits. The troubles of Enron and WorldCom show the poor practices that arise when competitive pressures win out over ethical principles.

Ethical conduct may reach beyond the law, since the law is inadequate for every task. Doing the right thing is an undefined but helpful standard. One moral goal would be to tell the truth. Thus, a businessperson would try to provide full and truthful information about his or her product or service. Telling the truth is a critical part of integrity, and integrity is the basis for reputation. Thus, firms, at the least, find it in their interest to be truthful. Fortunately, good ethics and self-interest usually coincide, since most firms want to develop and maintain a high reputation [Arnold et al., 2012].

**Integrity** can be defined as truthfulness, wholeness, and soundness. It can be described as a consistency of our words and our actions or our character and our conduct. A corporate model of integrity is based on ethical principles embedded in the corporate culture so that all stakeholders can conduct business to attain mutual benefits [Kaptein and Wempe, 2002]. A key task of a business leader is to establish an ethical culture [McCoy, 2007].

While firms have clear-cut business objectives such as profitability, they must consider them subordinate to ethical values. A firm's integrity cannot be sacrificed to short-term gain. The firm's moral compass points the way. The spotlight is on the CEO and his or her integrity.

People can have great values and still give way to error. One needs the competence and character to implement one's values. Often, major corruption begins with a single small misstep. As executives or employees take further actions that build upon or attempt to cover up this misstep, unethical behaviors can grow to

**TABLE 20.6  Obstacles to ethical decision making.**

*Complacency:* We believe that "it can't happen here."

*Self-Delusion:* We judge ourselves by our intentions, while others judge us by our actions

*Rationalizations:* We construct justifications and excuses for ethical missteps

*Survival Mentality:* We convince ourselves that ethical missteps are necessary

*Source:* McCoy, 2007.

massive proportions [McCoy, 2007]. Firms and individuals face a number of obstacles to ethical decision making, as indicated in Table 20.6.

As one becomes an entrepreneurial leader, the pressures to win at any cost will become powerful [Harris et al., 2009]. To be a good team player, one may be asked to cut corners. We know that the lack of truth and the collapse of integrity can lead to terrible outcomes, as illustrated in the Enron case of 2002. On Enron, Robert Bryce and M. Ivins write [2002]:

> Enron failed because its leadership was morally, ethically, and financially corrupt. Whether the question was accounting or marital fidelity, the executives who inhabited the 50th floor at Enron's headquarters became incapable of telling the truth, to the Securities and Exchange Commission, to their spouses, or to their employees. That corruption permeated everything they did, and it spread through the company like wildfire.

The challenges of Enron extended beyond the company, too. Enron's rise and fall was based on a partnership between its financial division and the investment bankers who put together the deals. "Enron loves these deals," wrote a Chase banker in 1998, "as they are able to hide funded debt from their equity analysts" [McLean and Elkind, 2003]. On Wall Street, investment bankers call their innovative structured-finance arrangements their "technology." In investment banking, the ethic for many was: "Can you get the deal? If you can, and you're not likely to be sued or jailed, it's a good deal."

### MiniScribe: Cooking the Books

MiniScribe was a Longmont, Colorado, producer of disk drives that found itself in trouble when IBM canceled major purchasing contracts. When actual sales failed to materialize, the CEO badgered and bullied MiniScribe executives to meet quarterly revenue targets, no matter what it took. Executives turned to cooking the books. Their "cooking" activities included counting raw inventory as finished goods, creating false inventory, and grossly overstating actual shipments.

As MiniScribe's sales and profits continued to tumble, the pressures on the firm increased. At one point, executives rented a private warehouse. Over a weekend, staffers and spouses packed bricks in disk drive shipping boxes, then shipped pallet-loads of them to "BW," a fake customer. To pull off the brick shipping plan, they created a custom computer program called "Cook Book." Over $4 million worth of "very hard" disks were shipped utilizing Cook Book.

Once the fraud was exposed, MiniScribe's stock plummeted, and investors lost hundreds of millions of dollars, but not before the same executives, in an example of insider trading, sold most of their shares at a healthy profit. The CEO and CFO served time in prison. Unethical behavior most often gets exposed with severe consequences.

Fortunately, there are a number of tools to help one act ethically. McLemore [2003] suggests two tests for difficult actions: (1) are things so questionable that you lose sleep over them and (2) could you live with the newspaper reporting your actions tomorrow? Ernest Hemingway wrote in his novel, *Death in the Afternoon*: "What is moral is what you feel good after, and what is immoral is what you feel bad after." When difficult issues arise, contemplating them and engaging in conversation with other people can serve to encourage an ethical decision [Gunia et al., 2012].

Several tools for acting ethically appear in Table 20.7. For individuals, it is helpful to have a "personal business plan" that records the importance of various activities and relationships and that sets goals for the coming year. Actions that deviate from this plan should be evaluated carefully. Individuals should also identify ethical "partners" who can serve as sounding boards and supporters. When confronted with unethical situations, McLemore [2003] wisely suggests you say, "I don't feel comfortable doing that." Of course, the risk of loss of

**TABLE 20.7  Tools for acting ethically in tough situations.**

- Maintain involvement in a variety of activities and with a variety of people. This will help avoid being pressured to go along with what everybody around you says is acceptable behavior.

- Create a personal board of directors consisting of people you admire and who possess admirable values. If difficult situations arise, call on them for advice.

- Keep a cash reserve of six months to a year of salary. This will allow you to escape further involvement with an unethical firm and seek other options.

- Increase your defenses to negative forces of influence and persuasion by reading a book on the subject such as Cialdini [2001].

- Apply the "front, left-hand-side page of the Wall Street Journal" test. Would your actions change if you knew they would be exposed publicly someday?

- Write down your personal core values for later use. They can provide a helpful reference in difficult times to remind you of what is important in your life.

- Take a break before making a decision. When feeling pressured, ask for time to leave the room and gather your thoughts.

your position is real, but you can repeat without judgment, "I am uncomfortable doing that."

For firms, it is important to have a "Credo" that makes values explicit to all employees. At J&J, for example, all employees must read, accept, and sign the Credo, which aligns everyone along a clear set of principles. When trouble does occur, an open and transparent approach is essential for maintaining trust and facilitating ethical responses [McCoy, 2007].

A person with an ethical mind asks herself, "What kind of a person, worker and citizen do I want to be?" [Gardner, 2006]. When circumstances are tempting you to drop your standards, life becomes difficult. With a firm belief in personal integrity, a person needs to be willing to resign or be fired for what he or she believes is right.

---

**Building an Ethical Venture the Intel Way**

Entrepreneurs should plan an ethical foundation for the firm so that they can build integrity and reputation. Maintenance of integrity is critical to the long-run success of any firm. Former Intel CEO Craig Barrett promotes a "3M" philosophy to help leaders make decisions with the right amount of integrity and ethics. The three Ms—manager, media, and mother—represent the three constituencies with whom leaders should be comfortable sharing their decision. Only if the leaders expect that their manager, the media, and their mother will all approve of their decision should they proceed with their chosen course of action.

*Source:* Barrett, 2003.

---

# 20.5 Spotlight on Intuitive Surgical

Intuitive Surgical Inc. is a company that designs, makes, and services surgical devices. The origin of the device called the da Vinci surgical system was a research project at the Stanford Research Institute (SRI) in the period 1990-95. In 1995, Dr. Frederick Moll acquired the rights to this research and founded Intuitive Surgical. The first da Vinci device was approved by the FDA in 2000. The device provides 3D vision and instruments that rotate and bend beyond the abilities of humans. These capabilities permit more effective procedures and smaller incisions that lead to shorter recovery times. The da Vinci system is used for prostrate, thoracic, and colorectal surgery.

Intuitive Surgical has grown to become a leader in surgical devices. Today, more than 2,600 da Vinci systems are installed in more than 2,000 hospitals worldwide. The system was used to perform approximately 450,000 procedures in the year 2012 alone. Intuitive Surgical continues to invest heavily in R&D and it recently developed a remote surgery device called Clearpoint, which has potential for use in urology, gynecology, and thoracic surgery.

In 2000, Intuitive Surgical (ISRG) issued an IPO, which raised $46 million. The company's 2013 market value was about $20 billion. Its 2012 revenue was nearly $2.2 billion.

Intuitive Surgical is a strong example of a company that has leveraged leading-edge technology to meet an important market need. By relentlessly executing and by making strategic adjustments along the way, Intuitive Surgical has grown to become a model technology venture.

## 20.6 Summary

The implementation of a creative, well-defined business plan is essential to the success of a new enterprise. Good execution depends on the logical alignment of the firm's strategy with its goals and the efforts of its people. Turning a concept into a successful reality depends on goals, deadlines, teamwork, and focus on achieving the desired outcomes. Choosing the right people for the right jobs and helping them see where to go and how to get there are critical elements of building a great company. With the right people, a great strategy, and a sound road map, a start-up can strive to achieve an outstanding execution of the plan.

A new business grows from fledgling start-up to growth to maturity in stages. Managing a new business through the stages requires varying skills and organizational arrangements. Start-ups need to have talented, multiskilled people in place as they move through the stages of growth.

Emerging firms are constantly subject to challenge and change. Organizing a firm for resilient response to these challenges calls for an adaptive corporation. The ability to adapt to change may be a firm's only sustainable advantage. Furthermore, a firm needs to sustain its ethical principles through difficult times.

---

**Principle 20**
The ability to continuously and ethically execute a business plan and adapt that plan to changing conditions provides long-term success.

---

## Video Resources

Visit http://techventures.stanford.edu to view experts discussing content from this chapter.

| | | |
|---|---|---|
| Obligation to Create Value | Tim O'Reilly | O'Reilly Media |
| The Fight for Corporate Control | Mark Pincus Bing Gordon | Zynga KPCB |
| Company Ethics | Karen Richardson | E.piphany |

## 20.7 Exercises

**20.1** In May 2003, Zipcar of Boston decided it was time to bring in new funding to reach profitability (www.zipcar.com). However, the willing investors insisted on replacing the CEO and the board of directors. Examine Zipcar's subsequent progress in terms of execution and need at different stages of the life of this company.

**20.2** The global financial crisis in 2007–2010 drastically impacted the spending patterns of many businesses and industries. Select an industry, list specific spending changes that occurred during this period, and describe how a start-up selling into this industry should adapt its strategies to account for the broader market conditions.

**20.3** Your emerging new company is selling a high-priced software system to the oil and gas industry. Each sale amounts to $100,000 or more. Your firm is scheduled to deliver a system next week to one of your best customers. However, your chief technical officer has just told you that they have found a major software error that will take two to three weeks to fix. You are counting on the sale within this month so that you can meet payroll and pay all your delinquent bills. Your CFO suggests you ship the system now and send in a team later to fix the error. Your CTO wants to fix it first and then ship. What should you do?

**20.4** Southeby's and Christie's are the two largest upscale auction houses. Both enjoyed a growing business in the boom years of the late 1990s. In 2000, both firms were accused of price fixing. The Sherman Antitrust Act was passed in 1890 to control the power of trusts and monopolists. In 1995, both firms announced they would charge a fixed, nonnegotiable sliding-scale commission on the sales price. Is this the age-old tactic of price fixing? What constitutes legal pricing policies versus illegal price fixing?

**20.5** Your cash-strapped company is bidding for a badly needed contract. As the bid deadline nears, an employee of your nearest competitor pays you a visit. He says he will provide details of your competitor's bid in return for the promise of a job in six months, after the dust has settled. You know your competitor can survive losing this contract, but you cannot. Unfortunately, hiring a new employee will mean someone who currently works for you will have to go. Even so, is this an offer you cannot refuse?

**20.6** Your new firm is considering offering one of two health benefit options. One is more complete but also more costly than the other. Should you ask your employees to accept the lower-cost option? Should you explain the benefits of both plans? If you do, most people will prefer the better plan. What should you do?

**20.7** You attend a critical partner meeting with your CEO. After the meeting, your CEO misrepresents the results of the meeting to the broader management team to further a different agenda. How do you handle this situation with your CEO? With your other team members?

**20.8** Select an example of a white-collar (business) crime in the technology industry and describe what happened. How could this crime have been avoided?

## VENTURE CHALLENGE

1.  Briefly describe your plan for executing your business plan after you receive the resources.

2.  Describe your venture's plans to act as an adaptive organization.

3.  What mechanisms will you use to instill ethical behavior in your venture?

# REFERENCES

Aaker, David. 2001. *Developing Business Strategies.* 6th ed. New York: Wiley and Sons.

Aaker, David, and E. Joachimsthaler. 2000. *Brand Leadership.* New York: Free Press.

Aaker, David, V. Kumar, and G. S. Day. 2001. *Marketing Research.* New York: Wiley & Sons.

Abate, Tom. 2008. "Who Is Doing What with Technology?" *San Francisco Chronicle* (28 July).

Adner, Ron, and D. A. Levinthal. 2002. "The Emergence of Emerging Technologies." *California Management Review* (Fall), pp. 50–66.

Agarwal, Rajshree, M. B. Sarkar, and R. Echambadi. 2002. "The Conditioning Effect of Time on Firm Survival." *Academy of Management Journal* 5:971–94.

Ahuja, Gautam, and C. M. Lampert. 2001. "Entrepreneurship in the Large Corporation." *Strategic Management Journal* 22:521–43.

Aiello, Robert, and M. Watkins. 2000. "The Fine Art of Friendly Acquisition." *Harvard Business Review* (December), pp. 101–16.

Albrinck, Jill, J. Hornery, D. Kletter, and G. Neilson. 2002. "Adventures in Corporate Venturing." *Strategy and Business* 22:119–29.

Allen, T. J. 2000. "Architecture and Communication Among Product Development Engineers." *Engineering Management Society. Proceedings of the 2000 IEEE* (pp. 153–158). IEEE.

Anand, Bharat, and A. Galetovic. 2004. "How Market Smarts Can Protect Property Rights." *Harvard Business Review* (December), pp. 73–79.

Anders, George. 2003. *Perfect Enough.* New York: Penguin Putnam.

Arnold, Denis G., Tom L. Beauchamp, and Norman L. Bowie. 2012. *Ethical Theory and Business*, 9th Edition. New York: Pearson.

Arthurs, Jonathan, and Lowell Busenitz. 2006. "Dynamic Capabilities and Venture Performance: The Effects of Venture Capitalists." *Journal of Business Venturing* 21:195–215.

Astebro, Thomas. 1998. "Basic Statistics on the Success Rate and Profits for Independent Inventors." *Entrepreneurship Theory and Practice* (Winter):41–48.

Audretsch, David, and Erik Lehmann. 2005. "Does the Knowledge Spillover Theory of Entrepreneurship Hold for Regions?" *Research Policy* 34:1191–1202.

Audretsch, David, Erik Lehmann, and Susanne Warning. 2005. "University Spillovers and New Firm Location." *Research Policy* 34:1113–22.

Austin, James, Roberto Gutierrez, Enrique Ogliastri, and Ezequiel Reficco. 2007. "Capitalizing on Convergence." *Stanford Social Innovation Review* (Winter): 24–31.

Baer, Markus. 2012. "Putting Creativity to Work: The Implementation of Creative Ideas in Organizations." *Academy of Management Journal* 55(5): 1102–1119.

Bagley, Constance, and Craig Dauchy. 2011. *The Entrepreneur's Guide to Business Law.* Cincinnati: West/Thomson.

Bailey, Jeff. 2003. "For Investors, Founders Are Short-Term CEOs." *Wall Street Journal* (21 October), p. A24.

Baker, Wayne. 2000. *Achieving Success Through Social Capital.* San Francisco: Jossey-Bass.

Bakke, Dennis W. 2005. *Joy at Work.* Seattle: PVG Publishers.

Balachandra, R., M. Goldschmitt, and J. Friar. 2004. "The Evolution of Technology Generations," IEEE Trans. on Engineering Management, February, 3–12.

Baldwin, Carliss and Eric von Hippel. 2011. "Modeling a Paradigm Shift: From Producer Innovation to User and Open Collaborative." *Organization Science* (November/December) 22:1399–1417.

Barkema, Harry, J. Baum, and E. Mannix. 2002. "Management Challenges in a New Time." *Academy of Management Journal* 5:916–30.

Barney, Jay. 2002. *Gaining and Sustaining Competitive Advantage,* 2d ed. Upper Saddle River, N.J.: Prentice-Hall.

Barney, Jay. 2001. "Is the Resource Based View a Useful Perspective for Strategic Management Research? Yes." *Academy of Management Review* (January): 41–56.

Baron, James, and M. T. Hannan. 2002. "Organizational Blueprints for Success in High-Tech Start-Ups." *California Management Review* 3:18–24.

Baron, Robert, and G. D. Markman. 2003. "Beyond Social Capital." *Journal of Business Venturing* 18:41–60.

Barrett, Craig. 2003. Address at the AEA/Stanford Executive Institute, Stanford University, Palo Alto, Calif. (August 14).

Barroso, L. A., J. Dean, and U. Holzle. 2003. "Web Search for a Planet: The Google Cluster Architecture." *Micro, IEEE* 23(2): 22–28.

Barthelemy, Jerome. 2003. "The Seven Deadly Sins of Outsourcing." *Academy of Management Executive* 2:87–100.

Bartlett, Christopher, and S. Ghoshal. 2002. "Building Competitive Advantage through People." *MIT Sloan Management Review* (Winter), pp. 34–41.

Batten, Frank. 2002. *The Weather Channel.* Boston: Harvard Business School Press.

Baumol, William. 2002. *The Free Market Innovation Machine.* Princeton, N.J.: Princeton University Press.

Baumol, William, Robert Litan, and Carl Schramm. 2007. *Good Capitalism, Bad Capitalism, and the Economics of Growth and Prosperity.* New Haven: Yale University Press.

Beatty, Jack. 2001. *Colossus—How the Corporation Changed America.* New York: Broadway Books.

Beauchamp, Tom, and N. E. Bowie. 2001. *Ethical Theory and Business,* 6th ed. Upper Saddle River, N.J.: Prentice-Hall.

Bechky, Beth, and S. O'Mahony. 2006. "Stretchwork: Managing the career progression paradox in external labor markets." *Academy of Management Journal* 49(5):918–941.

Beckman, Christine, M. Diane Burton, and Charles O'Reilly. 2007. "Early Teams: The Impact of Team Demography on VC Financing and Going Public." *Journal of Business Venturing* 22:147–73.

Bennis, Warren, and R. J. Thomas. 2002. "Crucibles of Leadership." *Harvard Business Review* (September), pp. 39–45.

Bergl, Skylar. 2013. "Most Innovative Companies 2013." *Fast Company* 11 February 2013.

Berkery, Dermot. 2008. *Raising Venture Capital for the Serious Entrepreneur.* New York: McGraw-Hill.

Bernstein, Peter. 1996. *Against the Gods.* New York: Wiley & Sons.

Bertoni, Fabio, Massimo Colombo, and Luca Grilli. 2011. "Venture Capital Financing and the Growth of High-Tech Start-ups: Disentangling Treatment from Selection Effects." *Research Policy* (September) 40(7):1028–1043.

Best, M.H. 2001. *The New Competitive Advantage: The Renewal of American Industry.* New York: Oxford UP.

Bhardwaj, Gaurab, John Camillus, and David Hounshell. 2006. "Continual Coporate Entrepreneurial Search for Long-Term Growth." *Management Science* 52(2):248–61.

Bhargava, Hement. 2003. "Contingency Pricing for Information Goods." *Journal of Management Information Systems* (Fall): 115–38.

Bhattacharya, C.B., Sankar Sen, and Daniel Korschun. 2008. "Using Corporate Social Responsibility to Win the War for Talent." *MIT Sloan Management Review* 49(2):37–44.

Bhidé, Amar. 2000. *The Origin and Evolution of New Business.* New York: Oxford University Press.

Bhidé, Amar. 2008. *The Venturesome Economy.* Princeton, N.J.: Princeton University Press.

Bingham, Christopher B. 2005. "Learning from Heterogeneous Experience: The Internationalization of Entrepreneurial Firms." Dissertation: Stanford University Department of Management Science and Engineering.

Bingham, Christopher B. 2009. "Oscillating Improvisation: How Entrepreneurial Firms Create Success in Foreign Market Entries Over Time." *Strategy Entrepreneurship Journal* 3:321–345.

Birley, Sue. 2002. "Universities, Academics, and Spinout Companies: Lessons from Imperial." *International Journal of Entrepreneurship Education* 1:133–53.

Black, J. Stewart, and H. B. Gregersen. 2002. *Leading Strategic Change.* Upper Saddle River, N.J: Prentice-Hall.

Blank, S., and B. Dorf. 2012. *The Startup Owner's Manual.* Pescadero, CA: K&S Ranch Publishers.

Blank, Steve. 2013. *The Lean LaunchPad Educators Teaching Handbook.* Pescadero, CA: K&S Ranch Publishers.

Boeker, Warren, and R. Karichalil. 2002. "Entrepreneurial Transitions." *Academy of Management Journal* 3:818–26.

Boeker, Warren, and Robert Wiltbank. 2005. "New Venture Evolution and Managerial Capabilities." *Organization Science* 16(2):123–33.

Boer, Peter. 2002. *The Real Options Solution.* New York: Wiley & Sons.

Bolino, Mark, W. Turnley, and J. Bloodgood. 2002. "Citizenship Behavior and the Creation of Social Capital." *Academy of Management Review* 4:505–22.

Bossidy, Larry, and Ram Charan. 2002. *Execution.* New York: Crown.

Bosworth, Michael. 1995. *Solution Selling.* New York: McGraw-Hill.

Boulding, William, and M. Christen. 2001. "First Mover Disadvantage." *Harvard Business Review* (October), pp. 20–21.

Bower, Joseph. 2001. "Not All M&As Are Alike and That Matters." *Harvard Business Review* (March), pp. 93–101.

Bradley, Bill, P. Jansen, and L. Silverman, 2003. "The Nonprofit Sector's $100 Billion Opportunity." *Harvard Business Review* (May), pp. 94–103.

Brandenburger, Adam, and Barry Nalebuff. 1997. *Co-opetition.* New York: Currency Doubleday.

Branscomb, Lewis M. 2008. "Research Alone Is Not Enough." *Science* (15 August), pp. 915–916.

Brinckmann, Jan, Dietmar Grichnik, and Diana Kapsa. 2010. "Should Entrepreneurs Plan or Just Storm the Castle? A Meta-Analysis on Contextual Factors Impacting the Business Planning—Performance Relationships in Small Firms." *Journal of Business Venturing* (January) 25:24–40.

Brown, David. 2002. *Inventing Modern America.* Cambridge: MIT Press.

Brown, Rex. 2005. *Rational Choice and Judgment.* New York: Wiley.

Brown, Shona, and K. M. Eisenhardt. 1998. *Competing on the Edge.* Boston: Harvard Business School Press.

Brown, Tim. 2008. "Design Thinking." *Harvard Business Review* (June), pp. 84–92.

Bryce, Robert, and M. Ivins. 2002. *Pipe Dreams: Greed, Ego and the Death of Enron.* New York: Public Affairs–Perseus.

Brynjolfsson, Erik, Yu (Jeffrey) Hu, and Mohammad S. Rahman. 2009. "Battle of the Retail Channels: How Product Selection and Geography Drive Cross-Channel Competition." *Management Science* 55:1755–1765.

Buchanan, Mark, 2004. "Power Laws and the New Science of Complexity Management," Strategy and Business, 34:71–79.

Bunnell, David. 2000. *Making the Cisco Connection.* New York: Wiley & Sons.

Burgelman, Robert A. 2002. *Strategy as Destiny.* New York: Free Press.

Burgelman, Robert A., and L. Valikangas. 2005. "Managing Internal Corporate Venturing Cycles." *Sloan Management Review* (Summer), pp. 26–34.

Burgelman, Robert A., C. M. Christensen, and S. C. Wheelwright. 2004. *Strategic Management of Technology and Innovation.* Burr Ridge, Ill.: McGraw-Hill Irwin.

Burnham, Daniel Hudson, 1909

Burton, Thomas. 2005. "Medtronic Settles Patent Fight." *Wall Street Journal* (25 April), p. B4.

Cao, Belind. 2013. "Alibaba M&A Fuels Gains as Sohu Jumps: China Overnight." *Bloomberg* (12 May).

Carl, Fred. 2007. "The Best Advice I Ever Got." *Harvard Business Review* (February), pp.____.

Carr, Austin, et al. 2013. "The World's Most Innovative Companies 2013." *Fast Company* (February).

Carr, Geoffrey. 2008. "The Power and the Glory." *Economist* (19 June).

Carr, Nicholas. 2002. "Unreal Options." *Harvard Business Review* (December), p. 22.

Chakravorti, Bhaskar, 2004. "The New Rules for Bringing Innovations to Market," *Harvard Business Review* (March), pp. 58–67.

Charan, Ram. 2007. *What the Customer Wants You to Know.* New York: Portfolio.

Chatman, Jennifer, and S. E. Cha. 2003. "Leading by Leveraging Culture." *California Management Review* (Summer), pp. 20–33.

Chatterjee, Sayan. 1998. "Delivering Desired Outcomes Efficiently." *California Management Review* (Winter), pp. 78–94.

Chatterji, Aaron. 2009. "Spawned with a silver spoon? Entrepreneurial performance and innovation in the medical device industry." *Strategic Management Journal* 30: 185–206.

Chatterji, Aaron, and Kira Fabrizio. 2012. "How Do Product Users Influence Corporate Invention?" *Organization Science* (July/August) 23:971–87.

Chesbrough, Henry. 2002. "Making Sense of Corporate Venture Capital." *Harvard Business Review* (March), pp. 90–99.

Choi, Young Rok, Moren Lévesque, and Dean Shepherd. 2008. "When Should Entrepreneurs Expedite or Delay Opportunity Exploitation?" *Journal of Business Venturing* 23:333–55.

Chrisman, James J., Ed McMullan, and J. Hall. 2004. "The Influence of Guided Preparation on the Long-term Performance of New Ventures." *Journal of Business Venturing* (Fall), pp. 18–26.

Christensen, Clayton. 1999. *Innovation and the General Manager.* Burr Ridge, Ill.: McGraw-Hill Irwin.

Christensen, Clayton. 2002. "The Rules of Innovation." *Technology Review* (June), pp. 21–28.

Christensen, Clayton, and M. E. Raynor. 2003. *The Innovator's Solution.* Boston: Harvard Business School Press.

Christensen, Clayton, and Richard Tedlow. 2000. "Patterns of Disruption in Retailing." *Harvard Business Review* (January), pp. 36–42.

Christensen, Clayton, M. Raynor, and M. Verlinden. 2001. "Skate to Where the Money Will Be." *Harvard Business Review* (November), pp. 73–81.

Christensen, Clayton, S. D. Anthony, and E. A. Roth. 2004. *Seeing What's Next: Using the Theories of Innovation to Predict Industry Change.* Boston: Harvard Business School Press.

Churchill, Neil, and J. W. Mullins. 2001. "How Fast Can Your Company Afford to Grow?" *Harvard Business Review* (May), pp. 135–42.

Cialdini, Robert. 2008. *Influence: Science and Practice.* Boston: Allyn and Bacon.

Cialdini, Robert. 1993. *Influence.* New York: Morrow.

Clark, Don. 2003a. "Intel's John Miner to Be President of Investment Unit." *Wall Street Journal* (18 April), p. C5.

Clark, Don. 2003b. "Renting Software Online." *Wall Street Journal* (3 June), p. B1.

Clemons, Eric, and J. A. Santamaria. 2002. "Maneuver Warfare." *Harvard Business Review* (April), pp. 57–65.

Coburn, Pip. 2006. *The Change Function.* New York: Portfolio.

Cohen, Adam. 2002. *The Perfect Store.* Boston: Little, Brown and Company.

Cohen, Wesley, and Dan Levinthal. 1990. "Absorptive Capacity: A New Perspective on Learning and Motivation." *Administrative Science Quarterly* 35:128–53.

Coleman, James. 1990. *Foundations of Social Theory.* Cambridge, MA: Belknap Press.

Collins, James, and William Lazier. 1992. *Beyond Entrepreneurship.* Upper Saddle River, N.J.: Prentice-Hall.

Collins, James, and J. Porras. 1996. "Building Your Company's Vision." *Harvard Business Review* (September), pp. 65–77.

Collins, James. 2001. *Good to Great.* New York: Harper Collins.

Conger, Jay. 1998. "The Necessary Art of Persuasion." *Harvard Business Review* (May), pp. 85–95.

Constable, Giff. 2011. "12 Tips for Customer Development Interviews." 29 July 2011 post at http://giffconstable.com.

Copeland, Tom, and Peter Tufano, 2004. "A Real-World Way to Manage Real Options," *Harvard Business Review* (March), pp. 90–99.

Corstjens, Marcel, and J. Merrihue. 2003. "Optimal Marketing." *Harvard Business Review* (October), pp. 114–21.

Courtney, Hugh. 2001. *20–20 Foresight.* Boston: Harvard Business School Press.

Coutu, Diane. 2003. "Sense and Reliability." *Harvard Business Review* (April), pp. 84–90.

Coutu, Diane. 2002. "How Resilience Works." *Harvard Business Review* (May), pp. 46–55.

Covey, Stephen. 2006. The Speed of Trust. New York: Free Press.

Covey, Stephen, and A. R. Merrill. 1996. First Things First. New York: Free Press.

Crawford, Fred, and Ryan Matthews. 2001. *The Myth of Excellence.* New York: Crown Business.

Cross, Rob, and L. Prusak. 2002. "The People Who Make Organizations Go—or Stop." *Harvard Business Review* (June), pp. 105–12.

Dahan, Ely, and V. Srinivasan. 2000. "The Predictive Power of Internet-Based Product Concept Testing." *Journal of Product Innovation Management* 17:99–109.

Dahl, Michael and Olav Sorenson. 2012. "Home Sweet Home: Entrepreneurs' Location Choices and the Performance of Their Ventures." *Management Science* (June) 58:1059–71.

Dahle, Cheryl. 2005. "The Change Masters." *Fast Company* (January).

Davidsson, Per. 2002. "What Entrepreneurship Research Can Do for Business and Policy Practice." *International Journal of Entrepreneurship Education* 1:5–24.

Davis, Jason and Kathleen Eisenhardt. 2011. "Rotating Leadership and Collaborative Innovation: Recombination Processes in Symbiotic Relationships." *Administrative Science Quarterly* (June) 56:159–201.

Davis, Julie L., and S. S. Harrison. 2001. *Edison in the Boardroom.* New York: Wiley & Sons.

Davis, Stan, and Christopher Meyer. 2000. *Future Wealth.* Boston: Harvard Business School Press.

Dean, Thomas, and Jeffrey McMullen. 2007. "Toward a Theory of Sustainable Entrepreneurship: Reducing Environmental Degradation Through Entrepreneurial Action." *Journal of Business Venturing* 22:50–76.

de Boer, Joop. 2012. "Trend 8: The Peer-to-Peer Economy." *www.popupcity.net.* (January 5).

Dees, J. Gregory, J. Emerson, and P. Economy. 2002. *Strategic Tools for Social Entrepreneurs.* New York: Wiley & Sons.

De Jong, Bart, and Tom Elfring. 2010. "How Does Trust Affect the Performance of Ongoing Teams? The Mediating Role of Reflexivity, Monitoring, and Effort." *Academy of Management Journal* (June) 53(3):535–49.

DeLong, Thomas, and V. Vijayaraghavan. 2003. "Let's Hear It for B Players." *Harvard Business Review* (June), pp. 96–102.

DeMeyer, Arnold, C. H. Loch, and M. T. Pich. 2002. "Managing Project Uncertainty." *MIT Sloan Management Review* (Winter), pp. 60–67.

Demos, Nick, S. Chung, and M. Beck. 2002. "The New Strategy and Why It Is New." *Strategy and Business* 25:15–19.

Dhar, Ravi. 2003. "Hedging Customers." *Harvard Business Review* (May), pp. 86–92.

Diamond, Jared. 2000. "The Ideal Form of Organization." *Wall Street Journal* (12 December), p. A17.

Di Stefano, Giada, Alfonso Gambardella, and Gianmario Verona. "Technology Push and Demand Pull Perspectives in Innovation Studies: Current Findings and Future Research Directions." *Research Policy* (October) 41(8):1283–95.

Dobrev, Stanislav, and William Barnett. 2005. "Organizational Roles and Transition to Entrepreneurship." *Academy of Management Journal* 48(3):433–49.

Dorf, Richard C. 2001. *Technology, Humans and Society: Toward a Sustainable World.* San Diego: Academic Press.

Dorf, R. C. (Ed.). 2004. *The Engineering Handbook, 2nd ed.* Boca Raton, FL: CRC Press.

Douglas, Evan, and Dean Shepherd. 1999. "Entrepreneurship as a Utility Maximizing Response." *Journal of Business Venturing* 15:231–51.

Douglas, Evan, and D. Shepherd. 2002. "Self-Employment as a Career Choice." *Entrepreneurship, Theory and Practice* (Spring): 81–89.

Downing, S. 2005. "The Social Construction of Entrepreneurship: Narrative and Dramatic Processes in the Co-Production of Organizations and Identities." *Entrepreneurship Theory and Practice,* 29(2), 185–204.

Drucker, Peter. 2002. *Managing in the Next Society.* New York: St. Martin's Press.

Dushnitsky, Gary, and Michael Lenox. 2005. "When Do Incumbents Learn from Entrepreneurial Ventures? Corporate Venture Capital and Investing Firm Innovation Rates." *Research Policy* 34:615–39.

Dyck, Bruno, et al. 2002. "Passing the Baton: The Importance of Sequence, Timing, Technique and Communication in Executive Succession." *Journal of Business Venturing* 17:143–62.

Dye, Renee. 2000. "The Buzz of Buzz." *Harvard Business Review* (November), pp. 139–46.

Dyer, Jeffrey, P. Kale, and H. Singh. 2004. "When to Ally and When to Acquire." *Harvard Business Review* (July), pp. 109–15.

Earley, P. Christopher, and Elaine Mosakowski. 2004. "Cultural Intelligence." *Harvard Business Review* (October), pp. 139–146.

*Economist.* 2006. "Behind the Bleeding Edge." (23 September), p. 16.

*Economist.* 2012a. "The Third Industrial Revolution." (April 21), pp 3–4.

*Economist.* 2012b. "Bad Medicine." (October 13), pp. 74–75.

*Economist.* 2012c. "Another Game of Thrones." (1 December).

Eesley, C. E. and Roberts, E. B. 2012. "Are You Experienced or Are You Talented?: When Does Innate Talent versus Experience Explain Entrepreneurial Performance?" Strategic *Entrepreneurship Journal* 6: 207–219

Eisenhardt, Kathleen, and D. N. Sull. 2001. "Strategy as Simple Rules." *Harvard Business Review* (January), pp. 106–16.

Eisenmann, T. 2012. "Business Model Analysis for Entrepreneurs." *Harvard Business School Entrepreneurial Management Case* (812–096).

El-Haik, Basem, and D. M. Roy. 2005. *Service Design for Six Sigma.* New York: Wiley.

Elias, Stephen, and R. Stim. 2003. *Patent, Copyright and Trademark,* 6th ed. Berkeley: Nolo Press.

Ellis, Kimberly, Taco Reus, Bruce Lamont, and Annette Ranft. 2011. "Transfer Effects in Large Acquisitions: How Size-Specific Experience Matters." *Academy of Management Journal* (December) 54(6):1261–76.

Elsbach, Kimberly. 2003. "How to Pitch a Brilliant Idea." *Harvard Business Review* (September), pp. 40–48.

Erikson, Truls. 2002. "Entrepreneurial Capital: The Emerging Venture's Most Important Asset and Competitive Advantage." *Journal of Business Venturing* 17:275–290.

Ertel, Danny. 2004. "Getting Past Yes." *Harvard Business Review* (November), pp. 60–68.

Estrin, Judy. 2009. *Closing the Innovation Gap.* New York: McGraw-Hill.

Ettenberg, Elliott. 2002. *The Next Economy.* New York: McGraw-Hill.

Ettenson, Richard, and J. Knowles. 2008. "Don't Confuse Reputation with Brand." *MIT Sloan Management Review* (Winter), pp. 19–21.

Fahey, Liam, and R. M. Randall. 1998. *Learning from the Future.* New York: Wiley & Sons.

Farrell, Diana. 2004. "Beyond Offshoring." *Harvard Business Review* (December), pp. 82–90.

Fast Company, 1997, "John Doerr's Startup Manual," (March), p. 82.

Ferguson, Glover, S. Mathur, and B. Shah. 2005. "Evolving from Information to Insight." *Sloan Management Review* (Winter), pp. 51–58.

Fernandez-Araoz, Claudio. 2005. "Getting the Right People at the Top." *Sloan Management Review* (Summer), pp. 67–72.

Fine, Charles, et al. 2002. "Rapid Response Capability in Value Chain Design." *MIT Sloan Management Review* (Winter), pp. 69–75.

Finkelstein, Sydney. 2003. *Why Smart Executives Fail.* New York: Portfolio Penguin.

Finkelstein, Sydney, and A. Mooney. 2003. "Not the Usual Suspects." *Academy of Management Executive,* 2:101–12.

Fischer, Bill, and Andy Boynton. 2005. "Virtuoso Teams." *Harvard Business Review* (July), pp. 117–123.

Fisher, Lawrence. 2002. "Yves Doz: The Thought Leader Interview." *Strategy and Business* 29:115–23.

Fisher, Robert, and W. Ury. 1991. *Getting to Yes.* New York: Penguin.

Fitza, M., S.F. Matusik, and E. Mosakowski. 2009. "Do VCs Matter? The Importance of Owners on Performance Variance in Start-Up Firms. *Strategic Management Journal,* 30(4): 387–404.

Fleming, John H., C. Coffman, and J. K. Harter. 2005. "Manage Your Human Sigma." *Harvard Business Review* (July), pp. 107–14.

Fleming, Lee, and Olar Sorenson. 2001. "The Dangers of Modularity." *Harvard Business Review* (September), pp. 20–21.

Florida, Richard. 2002. *The Rise of the Creative Class.* New York: Basic Books.

Florida, Richard, and J. Goodnight. 2005. "Managing for Creativity." *Harvard Business Review* (July), pp. 125–32.

Florin, Juan. 2005. "Is Venture Capital Worth It?" *Journal of Business Venturing* 20:113–35.

Folta, Timothy B., Frédéric Delmar, and Karl Wennberg. 2010. "Hybrid Entrepreneurship." *Management Science* 56: 253–269

Foster, William, and J. Bradach. 2005. "Should Non-Profits Seek Profits?" *Harvard Business Review* (February), pp. 92–100.

Freeman, John, and Jerome Engel. 2007. "Models of Innovation: Startups and Mature Corporations." *California Management Review* 50(1):94–119.

Freiberg, Kevin, and Jackie Freiberg. 1997. *Nuts!* New York: Broadway Books.

Friedman, Thomas L. 1999. *The Lexus and the Olive Tree.* New York: Farrar, Straus, and Giroux.

Friedman, Thomas. 2008. *Hot, Flat, and Crowded.* New York: Farrar, Straus, and Giroux.

Furr, Nathan R., Fabrice Cavarretta, and Sam Garg. 2012. "Who Changes Course? The Role of Domain Knowledge and Novel Framing in Making Technology Changes." *Strategy Entrepreneurship Journal* 6:236–56.

Gaba, Vibha, and Alan D. Meyer. 2008. "Crossing the Organizational Species Barrier: How Venture Capital Practices Infiltrated the Information Technology Sector." *Academy of Management Journal* 51(5):976–98.

Gabor, Andrea. 2000. *The Capitalist Philosophers.* New York: New York Times Books.

Gaffney, John. 2001. "How Do You Feel About a $44 Tooth Bleaching Kit?" *Business 2.0* (October), pp. 126–27.

Galor, Oden, and O. Moav. 2002. "National Selection and the Origin of Economic Growth." *Quarterly Journal of Economics* (November): 1133–91.

Gardner, Howard. 2006. *Five Minds for the Future.* Boston: Harvard Business School Press.

Garg, Sam. 2013. "Venture Boards: Distinctive Monitoring and Implications for Firm Performance." *Academy of Management Review* 28(1):90–108.

Gargiulo, Terrence, 2002. *Making Stories*, Westport, CT:Quorum.

Garrett, Alexander. 2012. "What's Your Margin?" *www.managementtoday.com* (May 31).

Garrett, E. M. 1992. "Branson the Bold." *Success* (November), p. 22.

Garvin, David. 1993. "Building a Learning Organization." *Harvard Business Review* (July), pp. 78–91.

Garvin, David, and Lynee Levesque. 2006. "Meeting the Challenge of Corporate Entrepreneurship." *Harvard Business Review* (October), pp. 102–12.

Gatewood, Elizabeth. 2001. "Busting the Stereotype." *Kelley School Business Magazine* (Summer), pp. 14–15.

Gavetti, Giovanni, and J. W. Rivkin. 2005. "How Strategists Really Think." *Harvard Business Review* (April), pp. 54–63.

Gawer, Annabelle, and M. A. Cusumano. 2008. "How Companies Become Platform Leaders." *MIT Sloan Management Review* (Winter), pp. 28–35.

Gershenfeld, Neil. 2005. *FAB: The Coming Revolution on Your Desktop.* New York: Basic Books.

Gerstner, Louis. 2002. *Who Says Elephants Can't Dance?* New York: Harper Collins.

Gilbert, C. G., and M.J. Eyring. 2010. "Beating the Odds When You Launch a New Venture." *Harvard Business Review* 88(5): 92–98.

Gilbert, Brett, Patricia McDougall, and David Audretsch. 2008. "Clusters, Knowledge Spillovers and New Venture Performance: An Empirical Examination." *Journal of Business Venturing* 23:405–22.

Gino, Francesca and Gary P. Pisano. 2011. "Why Leaders Don't Learn from Success." *Harvard Business Review* 89(4):68–74.

Girard, Kim. 2000. "Pandesic's Failed Union." *Business 2.0* (September), pp. 16–18.

Gittell, Jody H. 2003. *The Southwest Airlines Way.* New York: McGraw-Hill.

Gladwell, Malcolm. 2000. *The Tipping Point.* Boston: Little, Brown.

Goldenberg, Jacob, Roni Horowitz, Amnon Levav, and David Mazursky. 2003. "Finding Your Innovation Sweet Spot." *Harvard Business Review* (March), pp. 120–29.

Goleman, Daniel, R. Boyatzis, and A. McKee. 2002. *Primal Leadership.* Boston: Harvard Business School Press.

Goleman, Daniel, R. Boyatzis, and A. McKee. 2001. "Primal Leadership: The Hidden Driver of Great Performance." *Harvard Business Review* (December), pp. 44–51.

Gompers, Paul, and J. Lerner. 2001. *The Money of Invention.* Boston: Harvard Business School Press.

Gompers, Paul, and W. A. Sahlman. 2002. *Entrepreneurial Finance.* New York: Wiley & Sons.

Gosling, Jonathan, and H. Mintzberg. 2003. "The Five Minds of a Manager." *Harvard Business Review* (November), pp. 54–63.

Gottfredson, Mark, R. Puryear, and S. Phillips. 2005. "Strategic Sourcing." *Harvard Business Review* (February), pp. 135–39.

Govindarajan, Vijay, and Chris Trimble. 2005a. "Building Breakthrough Businesses Within Established Organizations." *Harvard Business Review* (May), pp. 58–68.

Graebner, Melissa. 2004. "Momentum and Serendipity: How Acquired Leaders Create Value in the Integration of Firms." 25:*Strategic Management Journal.*

Graham, Paul. 2005. "How to Start a Startup" www.paulgraham.com (March).

Graham, Paul. 2006. "Want to start a startup?" *www.paulgraham.com* (October).

Graham, Paul. 2008. *www.paulgraham.com blog* (October).

Green, Heather. 2003. "Companies That Really Get It." *Business Week* (25 August), p. 144.

Greene, Patricia, C. G. Brush, and M. M. Hart. 1999. "The Corporate Venture Champion." *Entrepreneurship Theory and Practice* (Spring): 103–22.

Grégoire, Denis A., and Dean A. Shepherd. 2012. "Technology-Market Combinations and the Identification of Entrepreneurial Opportunities: An Investigation of the Opportunity-Individual Nexus." *Academy of Management Journal* 55(4): 753–785.

Grove, Andy. 2003. "Churning Things Up." *Fortune* (11 August), pp. 115–18.

Gruber, Marc. 2007. "Uncovering the Value of Planning in New Venture Creation: A Process and Contingency Perspective." *Journal of Business Venturing* 22:782–807.

Gundry, Lisa, and H. Welsch. 2001. "The Ambitious Entrepreneur." *Journal of Business Venturing* 16:453–70.

Gunia, Brian, Long Want, Li Huang, Jiunwen Want, and J. Keith Murnighan. 2012. "Contemplation and Conservation: Subtle Influences on Moral Decision Making." *Academy of Management Journal* (February), 55(1):13–22.

Hall, Bronwyn, and Nathan Rosenberg. 2010. *Handbook of the Economics of Innovation* 1:3–730.

Hallen, Benjamin, and Kathleen Eisenhardt. 2012. "Catalyzing Strategies and Efficient Tie Formation: How Entrepreneurial Firms Obtain Investment Ties." *Academy of Management Journal* (February), 55(1): 35–70.

Hamel, Gary. 2001. "Revolution versus Evolution: You Need Both." *Harvard Business Review* (May), pp. 150–56.

Hamel, Gary. 2000. *Leading the Revolution.* Boston: Harvard Business School Press.

Hamermesh, Richard, Paul Marshall, and Taz Pirmohamed. 2002. "Note on Business Model Analysis for the Entrepreneur." *Harvard Business School Case* 9–802–048.

Hamm, John. 2002. "Why Entrepreneurs Don't Scale." *Harvard Business Review* (December), pp. 110–15.

Hammer, Michael. 2001. *The Agenda.* New York: Crown Business.

Hardy, Quentin. 2003. "All Eyes on Google." *Forbes* (26 May), pp. 100–10.

Hargadon, Andrew, and Y. Douglas. 2001. "When Innovations Meet Institutions." *Administrative Science Quarterly* 46 (September): 476–501.

Harris, Jared, Harry Sapienza, and Norman Bowie. 2009. "Ethics and Entrepreneurship." *Journal of Business Venturing* 24:407–18.

Harzberg, Friderick. 2003. "How Do You Motivate Employees?" *Harvard Business Review* (January), pp. 87–92.

Hastie, Reid, and R. M. Dawes. 2001. *Rational Choice in an Uncertain World.* Thousand Oaks, Calif.: Sage.

Hayward, Mathew, Dean Shepherd, and Dale Griffin. 2006. "A Hubris Theory of Entrepreneurship." *Management Science* 52(2):160–72.

Heath, Chip, and Dan Heath. 2007. *Made to Stick.* New York: Random House.

Heifetz, Ronald, and D. Laurie. 2001. "The Work of Leadership." *Harvard Business Review* (December), pp. 131–40.

Helft, Miguel. 2002. "Fashion Fast Forward." *Business 2.0* (May), pp. 61–66.

Henderson, James, Benoit Leleux, and Ian White. 2006. "Service-for-Equity Arrangements: Untangling Motives and Conflicts." *Journal of Business Venturing* 21:886–909.

Henderson, Rebecca, and Kim Clark. 1990. "Architectural Innovation." *Administrative Science Quarterly* 35:9–30.

Hill, Charles W., and Gareth R. Jones. 2001. *Strategic Management,* 5th ed. Boston: Houghton Mifflin.

Hill, Charles, and F. Rothaermel. 2003. "The Performance of Incumbent Firms in the Face of Radical Technological Innovation." *Academy of Management Review* 28:257–74.

Hill, Michael, R. Ireland, S. Camp, and D. Sexton. 2002. *Strategic Entrepreneurship.* Malden, Mass.: Blackwell.

Hinds, Pam, and Sara Kiesler. 2002. *Distributed Work.* Cambridge, MA: MIT Press.

Hirsh, Evan, S. Hedlund, and M. Schweizer. 2003. "Reality Is Perception—The Truth about Car Brands." *Strategy and Business* (Fall): 20–25.

Hitt, Michael A., R. D. Ireland, S. M. Camp, and D. L. Sexton. 2001. "Entrepreneurial Strategies for Wealth Creation." *Strategic Management Journal* 22:479–91.

Hmieleski, Keith M., and Robert A. Baron. 2009. "Entrepreneurs' Optimism And New Venture Performance: A Social Cognitive Perspective." *Academy of Management Journal* 52(3): 473–488.

Ho, Yew Kee, H. T. Keh, and J. M. Ong. 2005. "The Effects of R&D and Advertising on Firm Value." *IEEE Transactions on Engineering Management* (February), pp. 3–14.

Holt, Douglas. 2003. "What Becomes an Icon Most?" *Harvard Business Review* (March), pp. 43–49.

Hoover, Gary. 2001. *Hoover's Vision.* New York: Texere.

Howell Jane M., Christine M. Shea, and? Christopher A. Higgins. 2005. "Champions of product innovations: Defining, developing, and validating a measure of champion behavior." *Journal of Business Venturing* 20(5):641–661.

Hrebiniak, Lawrence G. 2005. *Making Strategy Work.* Upper Saddle River, N.J.: Pearson.

Hsu, David. 2004. "What Do Entrepreneurs Pay for Venture Capital Affiliation?" *Journal of Finance* (August), pp. 1805–36.

Hsu, David. 2006. "Venture Capitalists and Cooperative Start-up Commercialization Strategy." *Management Science* 52(2):204–19.

Hsu, David, Edward Roberts, and Charles Eesley. 2007. "Entrepreneurs from Technology-Based Universities: Evidence from MIT." *Research Policy* 36:768–88.

Hughes, Jonathan, and Jeff Weiss. 2007. "Simple Rules for Making Alliances Work." *Harvard Business Review* (November), pp. 122–31.

Huntington, Tom. 2003. "The Gimmick That Ate Hollywood." *Invention and Technology* (Spring): 34–45.

Huxley, Aldous. 1990. *The Perennial Philosophy.* New York: Harper Collins.

Hvide, Hans K., and Jarle Møen. 2010. "Lean and Hungry or Fat and Content? Entrepreneurs' Wealth and Start-up Performance." *Management Science* 56(8):1242–58.

Iansiti, Marco, and Roy Levien, 2004. "Strategy as Ecology."*Harvard Business Review* (March), pp. 68–78.

Ibarra, Hermina. 2002. "How to Stay Stuck in the Wrong Career." *Harvard Business Review* (December), pp. 40–47.

Ibarra, Hermina, and Kent Lineback. 2005. "What's Your Story?" *Harvard Business Review* (January), pp. 65–71.

Isenberg, Daniel. 2008. "The Global Entrepreneur." *Harvard Business Review* (December).

Jackman, Jay, and M. H. Strober. 2003. "Fear of Feedback." *Harvard Business Review* (April), pp. 101–7.

Jackson, Ira, and J. Nelson. 2004. *Profits with Principles.* New York: Doubleday.

Jakle, John A., K. A. Sculle, and J. S. Rogers. 1996. *The Motel in America.* Baltimore: Johns Hopkins University Press.

Jassawalla, Avan, and H. C. Sashittal. 2002. "Cultures that Support Product Innovation Processes." *Academy of Management Executive* (August): 42–54.

Jensen, Richard, and Marie Thursby. 2001. "Proofs and Prototypes for Sale: The Licensing of University Inventions." *American Economic Review* 91:240–259.

Jiang, Lin, Justin Tan, and Marie Thursby. 2011. "Incumbent Firm Invention in Emerging Fields: Evidence from the Semiconductor Industry." *Strategy Management Journal* 32:55–75.

Johnson, Mark, Claytom Christensen, and Henning Kagermann. 2008. "Reinventing Your Business Model." *Harvard Business Review* (December).

Joyce, William, N. Nohria, and B. Roberson. 2003. *What Really Works.* New York: Harper Collins.

Julien, Pierre Andre, and C. Pamangalahy. 2003. "Competitive Strategy and Performance of Exporting SMEs." *Entrepreneurship Theory and Practice* (Spring): 227–45.

Kacperczyk, Aleksandra. 2012. "Opportunity Structures in Established Firms: Entrepreneurship versus Intrapreneurship in Mutual Funds." *Administrative Science Quarterly* (September) 57:484–521.

Kalnins, Arturs, and Wilbur Chung. 2006. "Social Capital, Geography, and Survival: Gujarti Immigrant Entrepreneurs in the U.S. Lodging Industry." *Management Science* 52(2):233–47.

Kanter, Rosabeth Moss. 2003. "Leadership and the Psychology of Turnarounds." *Harvard Business Review* (June), pp. 58–67.

Kaplan, Robert S., and David P. Norton. 2004. *Strategy Maps.* Boston: Harvard Business School Press.

Kaptein, Muel, and J. Wempe. 2002. *The Balanced Company.* New York: Oxford University Press.

Karlgaard, Rich. 2005. "Flying Ellipse's Pocket Jet." *Forbes* (August 15), p. 27.

Katila, Riitta, Jeff Rosenberger, and Kathy Eisenhardt. 2008. "Swimming with Sharks: Technology Ventures, Defense Mechanisms, and Corporate Relationships." *Administrative Science Quarterly* 53(2):295–332.

Kawasaki, Guy. 2004. *The Art of the Start.* New York: Penguin.

Kelley, T. 2001. *The Art of Innovation: Lessons in Creativity from IDEO, America's Leading Design Firm.* New York: Doubleday.

Kelley, Thomas, and Jonathan Littman. 2005. *The Ten Faces of Innovation: IDEO's Strategies for Defeating the Devil's Advocate and Driving Creativity throughout Your Organization.* New York: Doubleday.

Kellner, Tomas. 2002. "One Man's Trash." *Forbes* (4 March), pp. 96–98.

Kessler, Eric, and Paul Bierly. 2002. "Is Faster Really Better? An Empirical Test of the Implications of Innovation Speed." *IEEE Transactions on Engineering Management* (February): 2–12.

Khosla, V. 2012. Gene Pool Engineering for Entrepreneurs. http://www.khoslaimpact.com/wp-content/uploads/2012/03/Gene_Pool_Engineering_1_31_2012.pdf

Khurana, Rakesh. 2002. *Searching for a Corporate Savior.* Princeton, N.J.: Princeton University Press.

Kim, W. Chan, and R. Mauborgne. 2005. "Blue Ocean Strategy. *"California Management Review* (Spring), pp. 105–21.

King, Jr., Martin Luther. 1963. *Strength to Love.* Minneapolis: Fortress Press.

Kingon, Angus. 2014. "University-Corporate Partnerships for Technology Commercialization: Classroom Insights and Implications." Working paper. Brown University.

Kirchhoff, Bruce A. 1994. Entrepreneurship and Dynamic Capitalism. New York: Praeger.

Kirkman, Bradley, et al. 2002. "Five Challenges to Virtual Team Success." *Academy of Management Executive* 3:67–79.

Kirkpatrick, David. 2003. "Brainstorm 2003." *Fortune* (27 October), pp. 187–90.

Klein, Alec. 2003. *Stealing Time.* New York: Simon and Schuster.

Klein, Mark, and A. Einstein. 2003. "The Myth of Customer Satisfaction." *Strategy and Business* 30:8–9.

Kleiner, Art. 2003. "Making Patient Capital Pay off." *Strategy and Business* (Fall): 26–30.

Knopper, Steve. 2009. *Appetite for Self Destruction.* New York: Free Press.

Koehn, Nancy. 2001. *Brand New.* Cambridge: Harvard Business School Press.

Koetsier, John. 2012. "Alibaba Reaches 1 Ttrillion RMB ($157B) in Sales to Become Biggest E-Commerce Company in the World." *Venture Beat* (3 Dec).

Komisar, Randy. 2000. *The Monk and the Riddle.* Boston: Harvard Business School Press.

Krajewski, Lee, and L. Ritzman. 2002. *Operations Management,* 6th ed. Upper Saddle River, N.J.: Prentice-Hall.

Krueger, Richard and Mary Anne Casey. 2008. *Focus Groups: A Practical Guide for Applied Research.* Thousand Oaks, CA: SAGE Publications, Inc.

Krupp, Fred, and M. Horn. 2008. *Earth: The Sequel.* New York: Norton.

Kuemmerle, Walter. 2005. "The Entrepreneur's Path to Global Expansion." *Sloan Management Review* (Winter), pp. 42–49.

Kuemmerle, Walter. 2002. "A Test for the Fainthearted." *Harvard Business Review* (May), pp. 4–8.

Landes, Nate. 2012. "Top Ten Personal Finance Start-ups" *Consumer* 2 February 2012.

Langerak, Fred, and E. J. Haltink. 2005. "The Impact of New Product Development Acceleration Approaches on Speed and Profitability." *IEEE Transactions on Engineering Management* (February), pp. 30–41.

Lapre, Michael, and L. N. Van Wassenhove. 2002. "Learning across Lines." *Harvard Business Review* (October), pp. 107–11.

LaSalle, Diana, and T. A. Britton. 2003. *Priceless.* Boston: Harvard Business School Press.

Lassiter, Joseph B., III. 2002. "Entrepreneurial Marketing: Learning from High-Potential Ventures." *Harvard Business School Case 9–803–036.*

Lawrence, Thomas B., E. A. Morse, and S. W. Fowler. 2005. "Managing Your Portfolio of Connections." *Sloan Management Review* (Winter), pp. 59–65.

Lax, David A., and J.K. Sebenius. 2003. "3-D Negotiation: Playing the Whole Game." *Harvard Business Review* (November), pp. 65–74.

Lax, Eric. 1985. "Banking on Biotech Business." *New York Times* (22 December).

Lechler, Thomas. 2001. "Social Interaction: A Determinant of Entrepreneurial Team Venture Success." *Small Business Economics* 16:263–78.

Lee, Jae-Nam, and Y. G. Kim. 2005. "Understanding Outsourcing Partnership." *IEEE Transactions on Engineering Management* (February), pp. 43–58.

Lee, Lena, Poh Kam Wong, Maw Der Foo, and Aegean Leung. 2011. "Entrepreneurial Intentions: The Influence of Organizational and Individual Factors." *Journal of Business Venturing* 26(1): 124–136.

Lee, Soo Hoon, P. K. Wong, and C. L. Chong. 2005. "Human and Social Capital Explanations for R&D Outcomes." *IEEE Transactions on Engineering Management* (February), pp. 59–67.

Leibovich, Mark. 2002. *The New Imperialists.* Paramus, N.J.: Prentice-Hall.

Leifer, Richard, Christopher M. McDermott, Gina Colarelli O'Connor, and Lois S. Peters. 2000. *Radical Innovation.* Boston: Harvard Business School Press.

Leonard-Barton, Dorothy. 1995. *Wellsprings of Knowledge.* Boston: Harvard Business School Press.

Leslie, Mark, and Charles Holloway. 2006. "The Sales Learning Curve." *Harvard Business Review* (July-August).

Lévesque, Moren, and Dean A. Shepherd. 2004. "Entrepreneurs' Choice of Entry Strategy in Emerging and Developed Markets." *Journal of Business Venturing* 19:29–54.

Lévesque, Moren, Dean Shepherd, and Evon Douglas. 2002. "Employment or Self-Employment: A Dynamic Utility-Maximizing Model." *Journal of Business Venturing* 17:189–210.

Lévesque, Moren, Maria Minniti, Dean Shepherd. 2009. "Entrepreneurs' Decisions on Timing of Entry: Learning from Participation and from the Experiences of Others." *Entrepreneurship: Theory & Practice* 33(2):547–70.

Levitt, Harold. 2003. "Why Hierarchies Thrive." *Harvard Business Review* (March), pp. 96–102.

Lewis, Michael. 2000. *The New, New Thing.* New York: Norton.

Liker, J. K. 2004. *The Toyota Way.* New York: McGraw-Hill.

Lojacono, Gabriella, and G. Zaccai. 2004. "The Evolution of the Design-Inspired Enterprise." *Sloan Management Review* (Spring), pp. 75–79.

Lord, Michael, S. W. Mandel, and J. D. Wager. 2002. "Spinning out a Star." *Harvard Business Review* (June), pp. 115–21.

Lounsbury, Michael, and M. Glynn. 2001. "Cultural Entrepreneurship: Stories, Legitimacy and the Acquisition of Resources." *Strategic Management Journal* 22:545–64.

Low, Murray B., and E. Abrahamson. 1997. "Movements, Bandwagons and Clones: Industry Evolution and the Entrepreneurial Process." *Journal of Business Venturing* 12:435–57.

Lowe, Robert A. 2001. "The Role and Experience of Start-ups in Commercializing University Inventions." *Entrepreneurial Inputs and Outcomes* 13:189–222.

Lu, Jane, and Paul Beamish. 2006. "Partnering Strategies and Performance of SMEs' International Joint Ventures." *Journal of Business Venturing* 21:461–86.

Lucier, Chuck, and J. Dyer. 2003. "Creating Chaos for Fun and Profit." *Strategy and Business* 30:14–20.

Luenberger, David. 2006. *Information Science.* Princeton: Princeton University Press.

Lumpkin, G. T., and B. B. Lichtenstein. 2005. "The Role of Organizational Learning in the Opportunity Recognition Process." *Entrepreneurship Theory and Practice* (July), pp. 451–72.

Lynn, Gary, and Richard Reilly. 2002. *Blockbusters.* New York: Harper Collins.

Magretta, Joan. 2002. *What Management Is.* New York: Free Press.

Maher, Michael, C. P. Stickney, and R. L. Weil. 2004. *Managerial Accounting,* 8th ed. Cincinnati: Southwestern.

Majumdar, Sumit. 1999. "Sluggish Giants, Sticky Cultures and Dynamic Capability Transformation." *Journal of Business Venturing* 15:59–78.

Malik, Om. 2003. *Broadbandits.* Hoboken, N.J.: Wiley & Sons.

Malone, Michael. 2002. *Betting It All: The Entrepreneurs of Technology.* New York: Wiley & Sons.

Mangalindan, Mylene, and S. L. Hwang. 2001. "Insular Culture Helped Yahoo! Grow, But Has Now Hurt It in the Long Run." *Wall Street Journal* (9 March), p. A1.

Mankins, Michael, and R. Steele. 2005. "Turning Strategy into Great Performance." *Harvard Business Review* (July), pp. 65–72.

Markides, Constantinos, and Paul Geroski. 2005. *Fast Second: How Smart Companies Bypass Radical Innovation to Enter and Dominate New Markets.* San Francisco: Josey-Bass.

Markman, Gideon, D. Balkin, and R. A. Baron. 2002. "Inventors and New Venture Formation." *Entrepreneurship Theory and Practice* (Winter): 149–65.

Marn, Michael, Eric Roegner, and Craig Zawada. 2004. *The Price Advantage.* New York: John Wiley & Sons.

Martens, Martin, Jennifer Jennings, and P. Devereaux Jennings. 2007. "Do the Stories They Tell Get Them the Money They Need? The Role of Entrepreneurial Narratives in Resource Acquisition." *Academy of Management Journal* 50(5):1107–32.

Martin, Roger L. 2002. "The Virtue Matrix." *Harvard Business Review* (March), pp. 69–75.

Martin, Roger, and Sally Osberg. 2007. "Social Entrepreneurship: The Case for Definition." *Stanford Social Innovation Review* (Spring), pp. 28–39.

Marvel, Matthew, Abbie Griffin, John Hebda, and Bruce Vojak. 2007. "Examining the Technical Corporate Entrepreneurs' Motivation: Voices from the Field." *Entrepreneurship Theory and Practice* (September), pp. 753–68.

Mason, Heidi, and T. Rohner. 2002. *The Venture Imperative.* Boston: Harvard Business School Press.

Maurer, Indre, and Mark Ebers. 2006. "Dynamics of Social Capital and Their Performance Implications: Lessons from Biotechnology Start-ups." *Administrative Science Quarterly* 51:262–92.

McCall, Morgan. 1998. *High Flyers.* Boston: Harvard Business School Press.

McCoy, Bowen H. 2007. *Living into Leadership.* Stanford: Stanford University Press.

McElroy, Mark. 2003. *The New Knowledge Management.* Boston: Elsevier.

McFarland, Keith. 2008. *The Breakthrough Company.* New York: Crown.

McGrath, James, F. Kroeger, M. Traem, and J. Rocken Haeuser. 2001. *The Value Growers.* New York: McGraw-Hill.

McGrath, Rita G. 2005. "Market Busting: Strategies for Exceptional Business Growth." *Harvard Business Review* (March), pp. 81–89.

McGrath, Rita Gunther, Thomas Keil, and Taina Tuki-ainen. 2006. "Extracting Value from Corporate Venturing." *MIT Sloan Management Review* 48(1):50–56.

McKee, Robert. 2003. "Storytelling That Moves People." *Harvard Business Review* (June), pp. 51–55.

McKenzie, Ray. 2001. *The Relationship-Based Enterprise.* New York: McGraw-Hill.

McLean, Bethany, and P. Elkind. 2003. *The Smartest Guys in the Room.* New York: Portfolio.

McLemore, Clinton. 2003. *Street-Smart Ethics.* Louisville, Ky.: Westminster John Knox Press.

McMullen, Jeffrey, and Dean Shepard. 2002. "Regulatory Focus and Entrepreneurial Intention." Presentation at the Academy of Management Meeting (August).

McNulty, James, et al. 2002. "What's Your Real Cost of Capital?" *Harvard Business Review* (October), pp. 114–21.

McQuivey, James. 2013. Digital Disruption: Unleashing the Next Wave of Innovation. Las Vegas, NV: Amazon Publishing.

Melnyk, Steven, and M. Swink. 2002. *Value-Driven Operations Management.* New York: McGraw-Hill.

Mezias, John, and W. H. Starbuck. 2003. "What Do Managers Know, Anyway?" *Harvard Business Review* (May), pp. 16–17.

Miles, Morgan, and J. Covin. 2002. "Exploring the Practice of Corporate Venturing." *Entrepreneurship, Theory and Practice* (Spring): 21–40.

Minniti, Maria, and W. Bygrave. 2001. "A Dynamic Model of Entrepreneurial Learning." *Entrepreneurship Theory and Practice* (Spring): 5–16.

Mintzberg, Henry, B. Ahlstrand, and J. Lampel. 1998. *Strategy Safari.* New York: Free Press.

Mishina, Y., B.J. Dykes, E.S. Block, and T.G. Pollock. 2010. "Why 'Good' Firms Do Bad Things: The Effects of High Aspirations, High Expectations, and Prominence on the Incidence of Corporate Illegality." *Academy of Management Journal* 53(4): 701–722.

Mittelstaedt, Robert. 2005. *Will Your Next Mistake Be Fatal?* Upper Saddle River, N.J.: Pearson.

Mokyr, Joel. 2003. *The Gifts of Athena—Historical Origins of the Knowledge Economy.* Princeton, N.J.: Princeton University Press.

Moore, Geoffrey. 2002. *Crossing the Chasm.* New York: Harper Collins.

Moore, Geoffrey. 2000. *Living on the Fault Line.* New York: Harper Collins.

Moore, Geoffrey A. 2004. "Darwin and the Demon: Innovating Within Established Enterprises." *Harvard Business Review* (July), pp. 86–92.

Mullins, John. 2006. *The New Business Road Test.* Harlow, England: Prentice Hall.

Mullins, John, and Randy Komisar. 2009. *Getting to Plan B: Breaking Through to a Better Business Model.* Boston: Harvard Business School Press.

Muna, Farid A., and Ned Mansour. 2005. "Leadership lessons from Canada geese." *Team Performance Management* 11:316–326.

Murray, F., & O'Mahony, S. 2007. "Exploring the Foundations of Cumulative Innovation: Implications for Organization Science." *Organization Science* 18(6), 1006–1021.

Nagle, Thomas, and John Hogan. 2006. *The Strategy and Tactics of Pricing.* Upper Saddle River, N.J.: Prentice-Hall.

Nalebuff, Barry, and Ian Ayres. 2003. *Why Not?* Boston: Harvard Business School Press.

Nambisan, Satish. 2002. "Designing Virtual Customer Environments for New Product Development." *Academy of Management Review* 27(3):392–413.

National Venture Capital Association (NVCA). 2013. 2013 *National Venture Capital Association Yearbook*. New York: Thomson Reuters.

Nelson, A. J. 2005. "Cacophony or Harmony? Multivocal Logics and Technology Licensing by the Stanford University Department of Music." *Industrial and Corporate Change* 14(1): 93–118.

Niraj, Rakesh, Mahendra Gupta, and Chakravarthi Narasimhan. 2001. "Customer Profitability in a Supply Chain." *Journal of Marketing* 65:1–16.

Northouse, Peter G. 2001. *Leadership,* 2d ed. Thousand Oaks, Calif.: Sage.

O'Farrell, John. 2011. "Building the Global Startup." *www.fastcompany.com* (June 11).

Ogawa, Susumu, and Frank Piller. 2006. "Reducing the Risks of New Product Development." *MIT Sloan Management Review* 47(2):65–71.

Ogden, Joseph, F. Jen, and P. O'Connor. 2003. *Advanced Corporate Finance.* Upper Saddle River, N.J.: Prentice-Hall.

Ogle, Sean. 2012. "7 Reasons Most People Should Build Business, Not Startups." *www.forbes.com* (September).

O'Reilly, Charles, and M. L. Tushman. 2004. "The Ambidextrous Organization." *Harvard Business Review* (April), pp. 22–30.

Osterwalder, A., and Y. Pigneur. 2010. *Business Model Generation: A Handbook for Visionaries, Game Changers, and Challengers.* New York: Wiley.

Oviatt, Benjamin, and Patricia McDougall. 2005. "Defining International Entrepreneurship and Modeling the Speed of Internationalization." *Entrepreneurship Theory and Practice* (September), pp. 537–53.

Ozcan, Pinar, and Kathleen M. Eisenhardt. 2009. "Origin of Alliance Portfolios: Entrepreneurs, Network Strategies, and Firm Performance." *Academy of Management Journal* 52(2): 246–279.

Ozgen, Eren, and Robert Baron. 2007. "Social Sources of Information in Opportunity Recognition: Effects of Mentors, Industry Networks, and Professional Forums." *Journal of Business Venturing* 22:174–92.

Packalen, Kelley. 2007. "Complementing Capital: The Role of Status, Demographic Features, and Social Capital in Founding Teams' Abilities to Obtain Resources." *Entrepreneurship Theory and Practice* (November), pp. 873–91.

Packard, David. 1995. *The HP Way.* New York: Harper Collins.

Park, Haemin Dennis, and H. Kevin Steensma. 2012. "When Does Corporate Venture Capital Add Value for New Ventures?" *Strategy Management Journal* 33:1–22.

Perlow, Leslie, and S. Williams. 2003. "Is Silence Killing Your Company?" *Harvard Business Review* (May), pp. 52–58.

Perlow, Leslie, G. Okhuysen, and N. P. Repenning. 2002. "The Speed Trap." *Academy of Management Journal* 5:931–55.

Peters, Tom, and R. Waterman. 1982. *In Search of Excellence.* New York: Harper & Row.

Petroski, Henry. 2003. *Small Things Considered: Why There Is No Perfect Design.* New York: Knopf.

Pfeffer, J., and R. Sutton. 2006. "Evidence-based Management." *Harvard Business Review* (January), pp. 51–60.

Pfeffer, Jeffrey, and R. Sutton. 2000. *The Knowing-Doing Gap.* Boston: Harvard Business School Press.

Phan, Phillip, Mike Wright, Deniz Ucbasaran, and Wee-Liang Tan. 2009. "Corporate Entrepreneurship: Current Research and Future Directions." *Journal of Business Venturing* 24(3):197–205.

Phelps, Corey. 2010. "A Longitudinal Study of the Influence of Alliance Network Structure and Composition on Firm Exploratory Innovation." *Academy of Management Journal* (August) 53(4):890–913.

Phinisee, Ivory, I. Elaine Allen, Edward Rogoff, Joseph Onochie, and Monica Dean. 2008. 2006–2007 *National Entrepreneurial Assessment for the United States of America.* Global Entrepreneurship Monitor.

Pietersen, Willie. 2002. *Reinventing Strategy.* New York: Wiley & Sons.

Popper, Ben. 2013. "Kickstarter Is on Pace to Collect $1 Billion in Pledges This Year." *www.theverge. com* (January 9).

Porter, Michael. 2001. "Strategy and the Internet." *Harvard Business Review* (March), pp. 63–78.

Porter, Michael E. 1998. *On Competition.* Boston: Harvard Business School Press.

Post, James, L. E. Preston, and S. Sachs. 2002. "Managing the Extended Enterprise." *California Management Review* (Fall), pp. 6–20.

Prahalad, C. K. 2005. *The Fortune at the Bottom of the Pyramid.* Upper Saddle River, N.J.: Pearson.

Prahalad, C. K., and A. Hammond. 2002. "Serving the World's Poor, Profitably." *Harvard Business Review* (September), pp. 48–57.

Prahalad, C. K., and V. Ramaswamy. 2000. "Co-opting Customer Competence." *Harvard Business Review* (January), pp. 79–87.

Prasad, Dev, G. Vozikis, and G. Bruton. 2001. "Commitment Signals in the Interaction between Business Angels and Entrepreneurs." *Entrepreneurial Inputs and Outcomes* 13:45–69.

Prestowitz, Clyde. 2005. *Three Billion New Capitalists.* New York: Basic Books.

Prusak, Laurence, and D. Cohen. 2001. "How to Invest in Social Capital." *Harvard Business Review* (June), pp. 86–94.

Quadrini, Vincerizo. 2001. "Entrepreneurial Financing, Savings and Mobility." *Entrepreneurial Inputs and Outcomes* 13:71–94.

Quinn, J. B., P. Anderson, and S. Finkelstein. 1996. "Leveraging Intellect." *The Academy of Management Executive* 10(3): 7–27.

Rappaport, Alfred, Michael J. Mauboussin, and Peter L. Bernstein. 2001. *Expectations Investing.* Boston: Harvard Business School Press.

Reeves, Marin, and Mike Deimler. 2011. "Adaptability: The New Competitive Advantage." *Harvard Business Review* (July).

Reichheld, Frederick. 2001. "Lead for Loyalty." *Harvard Business Review* (July), pp. 76–84.

Ridgway, Nicole. 2003. "Something to Sneeze at." *Forbes* (21 July), pp. 102–4.

Ries, Al, and Jack Trout. 2001. *Positioning: The Battle for Your Mind.* New York: McGraw-Hill.

Ries, Eric. 2011. *The Lean Startup.* New York: Crown Business.

Rigby, Darrell. 2001. "Moving Upward in a Downturn." *Harvard Business Review* (June), pp. 99–105.

Rigby, Darrell, and C. Zook. 2002. "Open Market Innovation." *Harvard Business Review* (October), pp. 80–89.

Riggs, Henry. 2006. *Understanding the Financial Score.* San Rafael, CA: Morgan and Claypool.

Riggs, Henry. 2004. *Financial and Economic Analysis for Engineering and Technology Management,* 2d ed. Hoboken, N.J.: Wiley & Sons.

Robbins-Roth, Cynthia. 2000. *From Alchemy to IPO.* Cambridge, Mass.: Perseus.

Roberts, Jeff. 2012. "iLegal: As Apple's Products Evolved, So Did a Strategy to Protect Them." *GigaOM* (1 August).

Roberts, John. 2004. *The Modern Firm.* New York: Oxford University Press.

Roberts, Michael. 2003. "Managing the Growing Venture." *Harvard Business School Note 9–803–137.*

Roberts, Michael, and Diana Gardner, 2000. Advanced Inhalation Research. Harvard Business School, Case 899292.

Rogers, Everett. 2003. *Diffusion of Innovations,* 5th ed. New York: Free Press.

Rohlfs, Jeffrey. 2001. *Bandwagon Effects in High Technology Industries.* Cambridge: MIT Press.

Roman, Kenneth. 2003. *How to Advertise,* 3rd ed. New York: Thomas Dunne.

Rosen, Corey, J. Case, and M. Staubus. 2005. "Every Employee an Owner. Really." *Harvard Business Review* (June), pp. 122–30.

Ross, Stephen, R. Westerfield, and J. Jaffe. 2012. *Corporate Finance.* New York: McGraw-Hill Irwin.

Rothaermel, Frank, and D. Deeds. 2004. "Exploration and Exploitation Alliances in Biotechnology." *Strategic Management Journal* (Winter): 100–21.

Rothaermel, Frank, and David Deeds. 2006. "Alliance Type, Alliance Experience and Alliance Management Capability in High-Technology Ventures." *Journal of Business Venturing* 21:429–60.

Russo, Michael. 2010. *Companies on a Mission: Entrepreneurial Strategies for Growing*

*Sustainability, Responsibility, and Profitability.* Stanford: Stanford University Press.

Ryans, Adrian, R. More, D. Barclay, and T. Deutscher. 2000. *Winning Market Leadership.* New York: Wiley & Sons.

Sachs, Jeffrey. 2008. *Common Wealth.* New York: Penguin Press.

Sahlman, William. 1999. *The Entrepreneurial Venture.* Boston: Harvard Business School Press.

Sanchez, José, Tania Carballo, and Andrea Gutiérrez. 2011. "The Entrepreneur from a Cognitive Approach." *Psicothema* 23(3):433–38.

Sandner, Philipp G., and Joern Block. 2011. "The Market Value of R&D, Patents, and Trademarks." *Research Policy* 40(7): 969–985.

Santos, Filipe, and Kathy Eisenhardt. 2009. "Constructing Markets and Shaping Boundaries: Entrepreneurial Action in Nascent Markets." *Academy of Management Journal* 52:643–671.

Sarasvathy, Saras D., and Sankaran Venkataraman. 2011. "Entrepreneurship as Method: Open Questions for an Entrepreneurial Future." *Entrepreneurship: Theory & Practice* 35(1): 113–135.

Sathe, Vijay. 2003. *Corporate Entrepreneurship.* New York: Cambridge University Press.

Sawhney, Mohan, and J. Zabin. 2001. *The Seven Steps to Nirvana.* New York: McGraw-Hill.

Schrage, Michael. 2002. "Ease of Learning." *Technology Review* (December), p. 23.

Schramm, Carl. 2004. "Building Entrepreneurial Economies." *Foreign Affairs* (July), pp. 104–15.

Schultz, Howard. 1997. *Pour Your Heart Into It: How Starbucks Built a Company One Cup at a Time* New York: Hyperion.

Schumpeter, Joseph. 1984. *Capitalism, Socialism and Democracy.* New York: Harper Torchbooks.

Schwartz, Evan. 2002. *The Last Lone Inventor.* New York: Harper Collins.

Schwartz, Barry. 2004. *The Paradox of Choice.* New York: Echo Press.

Sebenius, James. 2001. "Six Habits of Merely Effective Negotiators." *Harvard Business Review* (April), pp. 87–95.

Seelig, Tina. 2012. *inGenius: A Crash Course on Creativity.* NewYork: HarperCollins Publishers.

Selden, Larry, and G. Colvin. 2003. "What Customers Want." *Fortune* (7 July), pp. 122–25.

Selden, Larry, and G. Colvin. 2002. "Will This Customer Sink Your Stock?" *Fortune* (30 September), pp. 127–32.

Shah, Sonali K. 2003. Innovation & Product Development within User Communities: Findings from Open Source Software And Consumer Sporting *Goods, PhD Thesis,* MIT, Boston, MA, May 2003.

Shah, Sonali K., and Mary Tripsas. 2007. The Accidental Entrepreneur: The Emergent & Collective Process of User Entrepreneurship. *Strategic Entrepreneurship Journal.* 1(1), pp. 123–140.

Shane, Scott, and S. Venkataraman. 2000. "The Promise of Entrepreneurship as a Field of Research." *Academy of Management Review* 25(1):217–26.

Shane, Scott A. 2005. *Finding Fertile Ground.* Upper Saddle River, N.J.: Pearson.

Shapiro, Hal, and H. Varian. 1998. *Innovation Rules.* Cambridge, Mass: Harvard Business School Press.

Shaw, Gordon, R. Brown, and P. Bromiley. 1998. "Strategic Stories: How 3M Is Rewriting Business Planning." *Harvard Business Review* (May), pp. 41–50.

Shepherd, Dean. 2003. "Learning from Business Failure." *Academy of Management Review* 2:318–28.

Shepherd, Dean, Evan Douglas, and Mark Shanley. 2000. "New Venture Survival." *Journal of Business Venturing* 15:393–410.

Shepherd, Dean, R. Ettenson, and A. Crouch. 2000. "New Venture Strategy and Profitability." *Journal of Business Venturing* 15:449–67.

Shepherd, Dean, and N. F. Krueger. 2002. "An Intentions-Based Model of Entrepreneurial Teams." *Entrepreneurship Theory and Practice* (Winter): 167–85.

Shepherd, Dean, and Mark Shanley. 1998. *New Venture Strategy.* Thousand Oaks, Calif.: Sage Publications.

Sheth, Jagdish, and R. Sisodia. 2002. *The Rule of Three.* New York: Free Press.

Shrader, Rodney, and Mark Simon. 1997. "Corporate versus Independent New Ventures." *Journal of Business Venturing* 12:47–66.

Shu, Catherine. 2013. "China's E-Commerce Market Grew to $190B in 2012, Driven by Mobile Users and Social Media, Says CNNIC." *TechCrunch* (16 April).

Shu, Catherine, and Ingrid Lunden. 2013. "Confirmed: Alibaba Group Takes $294M, 28% Stake in China's Navigation/Location Services Specialist AutoNavi, As it Gets Down to Business on Mobile." *TechCrunch* (9 May).

Simon, Mark, and S. M. Houghton. 2003. "The Relationship between Overconfidence and the Introduction of Risky Products." *Academy of Management Journal* 2:139–49.

Simons, Robert. 2005. "Designing High-Performance Jobs." *Harvard Business Review* (July), pp. 55–62.

Sine, Wesley, Heather Haveman, and Pamela Tolbert. 2005. "Risky Business? Entrepreneurship in the New Independent-Power Sector." *Administrative Science Quarterly* 50:200–232.

Singh, Jasjit, and Lee Fleming. 2010. "Lone Inventors as Sources of Breakthroughs: Myth or Reality?" *Management Science* 56:41–56.

Sloan, Paul. 2005. "What's Next for Apple?" *Business 2.0* (April), pp. 69–78.

Slywotzky, Adrian. 2002. *The Art of Profitability.* New York: Warner Books.

Slywotzky, Adrian. 2007. *The Upside.* New York: Crown.

Slywotzky, Adrian, and J. Drzik. 2005. "Countering the Biggest Risk of All." *Harvard Business Review* (April), pp. 78–88.

Slywotzky, Adrian, and David Morrison. 2000. *How Digital Is Your Business?* New York: Crown Business.

Slywotzsky, Adrian, David J. Morrison, Ted Moser, Kevin A. Mundt, and James A. Quella. 1999. *Profit Patterns.* New York: Random House.

Smith, Janet, and R. L. Smith. 2004. *Entrepreneurial Finance,* 3d ed. Hoboken, N.J.: Wiley & Sons.

Sonnenfeld, Jeffrey. 2002. "What Makes Boards Great." *Harvard Business Review* (September), pp. 106–12.

Sørensen, Jesper. 2007. "Bureaucracy and Entrepreneurship: Workplace Effects on Entrepreneurial Entry." *Administrative Science Quarterly* 52:387–412.

Sørensen, Jesper B., and Magali A. Fassiotto. 2011. "Organizations as Fonts of Entrepreneurship." *Organization Science* 22(5):1322–31.

Spekman, Robert, and L. Isabella. 2000. *Alliance Competence.* New York: Wiley & Sons.

Spulker, Daniel. 2004. *Management Strategy.* Burr Ridge, Ill.: McGraw-Hill.

Sternberg, Robert, L. A. O'Hara, and T. I. Lubart. 1997. "Creativity as Investment." *California Management Review* (Fall), pp. 8–21.

Stevenson, Howard, et al. 1999. *New Business Ventures and the Entrepreneur,* 5th ed. Burr Ridge, Ill.: McGraw-Hill Irwin.

Stim, Richard. 2012. *Patent, Copyright & Trademark: An Intellectual Property Desk Reference*, 12th ed. Berkeley, CA: Nolo.

Stringer, Kortney. 2003. "How Do You Change Consumer Behavior?" *Wall Street Journal* (17 March), p. R6.

Stuart, Toby, and Waverly Ding. 2006. "When Do Scientists Become Entrepreneurs? The Social Structural Antecedents of Commercial Activity in the Academic Life Sciences." *American Journal of Sociology* 112(1):97–144.

Suarez, Fernando, and G. Lanzolla. 2005. "The Half Truth of First Mover Advantage." *Harvard Business Review* (April), pp. 121–27.

Sull, Don. 2004. "Disciplined Entrepreneurship." *MIT Sloan Management Review* (Fall), pp. 71–77.

Sull, Donald, and C. Spinosa. 2005. "Using Commitments to Manage Across Units." *MIT Sloan Management Review* (Fall), pp. 73–81.

Sutton, Robert. 2002. *Weird Ideas That Work.* New York: Free Press.

Szulanski, Gabriel, and S. Winter. 2002. "Getting It Right the Second Time." *Harvard Business Review* (January), pp. 62–69.

Tallman, Stephen, and K. Fladmoe-Lindquist. 2002. "Internationalization Globalization and Capability Strategy." *California Management Review* (Fall), pp. 110–34.

Tapscott, Don, and A. D. Williams. 2008. *Wikinomics.* New York: Portfolio.

Tapscott, Don, D. Ticoll, and A. Lowy. 2000. *Digital Capital.* Boston: Harvard Business School Press.

Taylor, Suzanne, and K. Schroeder. 2003. *Inside Intuit.* Boston: Harvard Business School Press.

Tedeschi, Bob. 2003. "End of the Paper Chase." *Business 2.0* (March), p. 64.

Tedlow, Richard S. 2001. *Giants of Enterprise.* New York: Harper Collins.

Teitelman, Robert. 1989. *Gene Dreams.* New York: Basic Books.

Thomas, Paulette. 2003. "Entrepreneur's Biggest Problems." *Wall Street Journal* (17 March), p. R1.

Thompke, Stefan. 2001. "Enlightened Experimentation." *Harvard Business Review* (February), pp. 48–52.

Thompke, Stefan, and D. Reinertsen. 1998. "Agile Product Development." *California Management Review* (Fall), pp. 8–28.

Thompke, Stefan, and Eric von Hippel. 2002. "Customers as Innovators." *Harvard Business Review* (April), pp. 74–81.

Thurm, Scott. 2003. "A Go-Go Giant of the Internet Age, Cisco Is Learning to Go Slow." *Wall Street Journal* (7 May), p. A1.

Thurm, Scott. 2002. "Cisco Details the Financing for Its Start-Up." *Wall Street Journal* (12 March), p. A3.

Thursby, Jerry G., and Marie C. Thursby. 2004. "Are Faculty Critical? Their Role in University-Industry Licensing." *Contemporary Economic Policy* 22:162–178.

Tichy, Noel, and Warren Bennis. 2007. *Judgment.* New York: Portfolio.

Tiwana, Amrit, and A. A. Bush. 2005. "Continuance in Expertise-Sharing Networks." *IEEE Transactions on Engineering Management* (February), pp. 85–100.

Treacy, Michael. 2004. "Innovation as a Last Resort." *Harvard Business Review* (July), pp. 29–30.

Tuggle, Christoper, Karen Schnatterly, and Richard Johnson. 2010. "Attention Patterns in the Boardroom: How Board Composition Processes Affect Discussion of Entrepreneurial Issues." *Academy of Management Journal* (June) 53(3):550–71.

U.S. Patent and Trademark Office . 2013. www.uspto.gov

Ullman, David. 2003. *The Mechanical Design Process.* New York: McGraw-Hill.

Ulrich, Karl, and S. Eppinger. 2004. *Product Design and Development,* 3rd ed. New York: McGraw-Hill.

Van den Ende, Jan, and N. Wijaberg. 2003. "The Organization of Innovation and Market Dynamics." *IEEE Transactions on Engineering Management* (August): 374–82.

Van Praag, Miriam. 2006. *Successful Entrepreneurship: Confronting Economic Theory with Empirical Practice.* Cheltenham: Edward Elgar.

Venkataraman 2001

Vermeulen, Freek. 2005. "How Acquisitions Can Revitalize Companies." *Sloan Management Review* (Summer), pp. 45–51.

Vogelstein, Fred. 2003. "24/7 Customer." *Fortune* (24 November), p. 212.

Vogelstein, Fred. 2005. "Yahoo's Brilliant Solution." *Fortune* (8 August), pp. 42–54.

Vonderembse, Mark, and G. White. 2004. *Operations Management.* Hoboken, N.J.: Wiley & Sons.

Von Hippel, Eric. 2005. *Democratizing Innovation.* Cambridge, MA: MIT Press.

Von Hippel, E. 2009. "Democratizing Innovation: The Evolving Phenomenon of User Innovation." *International Journal of Innovation Science* 1(1): 29–40.

Waaser, Ernest et al., 2004. "How You Slice It: Smarter Segmentation for Your Sales Force." *Harvard Business Review* (March), pp. 105–10.

Wadhwa, Anu, and Suresh Kotha. 2006. "Knowledge Creation Through External Venturing: Evidence from the Telecommunications Equipment Manufacturing Industry." *Academy of Management Journal* 49(4):819–35.

Wang, T., and P. Bansal. 2012. "Social Responsibility in New Ventures: Profiting from a Long-Term Orientation." *Strategic Management Journal* 33: 1135–1153.

Wasserman, Noam. 2008. "The Founder's Dilemma." *Harvard Business Review* (February), pp. 103–109.

Wasserman, N. 2012. *The Founder's Dilemmas: Anticipating and Avoiding the Pitfalls that Can Sink a Startup.* Princeton, NJ: Princeton University Press.

Welch, Jack. 2002. *Jack: Straight from the Gut.* New York: Warner.

Wennekers, Sander, L. M. Uhlaner, and R. Thurik. 2002. "Entrepreneurship and Its Conditions." *International Journal of Entrepreneurship Education* 1:25–64.

West, Joel, and S. O'Mahony. 2008. "The role of participation architecture in growing sponsored open source communities." *Industry and Innovation* 15(2):145–168.

Weterings, Anet, and Ron Boschma. 2009. "Does Spatial Proximity to Customers Matter for Innovative Performance?: Evidence from the Dutch Software Sector." *Research Policy* (June) 38(5):746–55.

Winborg, Joakim, and Hans Landström. 2000. "Financial Bootstrapping in Small Businesses: Examining Small Business Managers' Resource Acquisition Behaviors." *Journal of Business Venturing* 16:235–54.

Winer, Russell. 2000. *Marketing Management.* Upper Saddle River, N.J.: Prentice-Hall.

Winer, Russell. 2001. "A Framework for Customer Relationship Management." *California Management Review* (Summer), pp. 89–104.

Wolcott, Robert, and Michael Lippitz. 2007. "The Four Models of Corporate Entrepreneurship." *MIT Sloan Management Review* 49(1):75–82.

Wood, Robert C., and G. Hamel. 2002. "The World Bank's Innovation Market." *Harvard Business Review* (November), pp. 104–12.

Wright, Randall. 2008. "How to Get the Most from University Relationships." *MIT Sloan Management Review* 49(3):75–80.

Yakura, Elaine. 2002. "Charting Time: Timelines as Temporal Boundary Objects." *Academy of Management Journal* 5:956–70.

Yankelovich, Daniel, and David Meer. 2006. "Rediscovering Market Segmentation." *Harvard Business Review* (February): 122–31.

York, Jeffrey G., and S. Venkataraman. 2010. "The Entrepreneur-Environment Nexus: Uncertainty, Innovation, and Allocation." *Journal of Business Venturing* 25(5): 449–463.

Young, Jeffrey S. 1988. *Steve Jobs—The Journey Is the Reward.* Glenview, Ill.: Scott-Foresman.

Yu, Jifeng, Brett Anitra Gilbert, and Benjamin M. Oviatt. 2011. "Effects of Alliances, Time, and Network Cohesion on the Initiation of Foreign Sales by New Ventures." *Strategy Management Journal* 32:424–46.

Zacharakis, Andrew, G. D. Meyer, and J. DeCastro. 1999. "Differing Perceptions of New Venture Failure." *Journal of Small Business Management* (July):1–14.

Zahra, Shaker, D. O. Neubaum, and M. Huse. 2000. "Entrepreneurship in Medium Size Companies." *Journal of Management* 5:947–76.

Zider, Bob. 1998. "How Venture Capital Works." *Harvard Business Review* (December), pp. 131–39.

Zimmerman, Monica, and G. J. Zeitz. 2002. "Beyond Survival: Achieving New Venture Growth by Building Legitimacy." *Academy of Management Review* 3:414–31.

Zott, Christopher, and Raphael Amit. 2007. "Business Model Design and the Performance of Entrepreneurial Firms." *Organization Science* 18(2):181–99.

# Sample Business Plan

## EXECUTIVE SUMMARY

**David Gal**
**Buzz Bonneau**

*Note:* With permission by Calcula founders for use in educational only settings and environments.

Each year, kidney stones account for 1% of all emergency room visits and $5.9B in health care costs[1]. Current kidney stone therapies are indicated only for stones larger than 10mm despite the fact that 85% of patients suffer from stones smaller than 10mm. These patients are prescribed narcotics and told to go home to pass the stone naturally over the course of several weeks. During this time patients revisit the ER, miss an average of 2.5 days of work, and suffer through acute pain onset that has been described as "the worst imaginable pain a person can experience."

Calcula is developing products that solve this unmet need by removing kidney stones smaller than 10mm. Our minimally invasive, patent-pending device will render patients stone free after a single visit to the urologist without the use of general anesthesia. This opportunity represents a dramatic disruption in the field of urology.

Extracorporeal shockwave lithotripsy (ESWL) and Ureteroscopy (URS) are the two existing minimally invasive procedures to treat kidney stones. Both procedures fracture stones into smaller pieces and require specialized facilities, which include an operating room, general anesthesia, capital equipment, and fluoroscopy. Both take an hour or more to perform. The anesthesia risk and facility requirements of ESWL and URS render urological guidelines to recommend that patients with stones smaller than 10mm wait 4–6 weeks in the hopes of spontaneous passage before intervening. In addition, both require capital equipment investments between $250K and $750K.

Calcula has developed a precise method of stone removal requiring only local anesthetic. The product is based on early clinical testing from the 1980s, which proved the technique feasible and safe, but was never commercialized

---

[1] Pearle MS, Calhoun EA, Curhan GC; Urologic Diseases of America Project. J Urol. 2005 Mar;173(3):848–57. PubMed PMID: 15711292.

due to a lack of precision control and a fear of anatomical damage due to requisite large instrumentation. Calcula has solved both of these issues with its novel device design, which includes precision software control and miniaturized design feasible only with modern manufacturing techniques. Calcula has verified the concept in animal models, and IP has been filed on the technology.

Calcula's solution will fit under an existing CPT reimbursement code. The device is a planned 510(k) submission with both technology and intended use predicates. Clinical data is not required though Calcula plans to perform a two-phased trial to drive adoption. This data may be submitted with the 510(k) with with a clear endpoint (stone removal) and short follow up time (30 days). This regulatory path has been validated by Calcula's advisors.

In 2011, 1.45M patients had kidney stones treated only with pain medication because their stones were smaller than 10mm$^2$. Calcula plans to focus initially on those visiting the urologist with stones smaller than 10mm. Calcula plans to fully subsidize its $10,000 control system (compared to capital equipment cost of ESWL and URS - $750k and $225k respectively) and its disposable device at $275 (under existing CPT codes).

Rapid adoption of any new medical device is always a risk. Urologists may be hesitant to remove stones in the office setting because it is outside of current standard of care. This risk is mitigated by existing preliminary clinical data that proves the feasibility and safety of the technique. Additionally, thought leaders in urology have strongly validated the Calcula approach. Calcula's product will be easily deployable during an existing procedure, considered a "bread and butter" procedure in every urology practice. Calcula's stone removal will be among the most lucrative procedures urologists perform due to the short procedure time and high reimbursement. Clinical safety, ease of implementation, and fiscal incentive ensure that adoption of the Calcula device will be rapid and widespread.

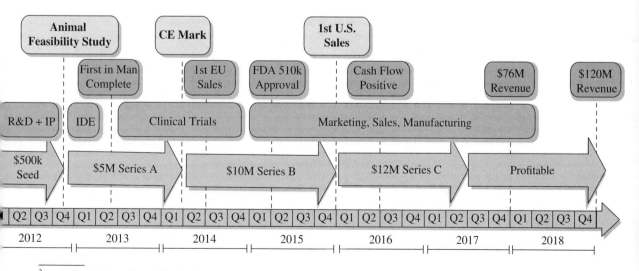

---

$^2$ Internal calculations based off of reference 1.

Calcula plans to fully subsidize the capital equipment cost for early adopters to drive adoption. Considering the disposable cost alone and conservatively assuming that 50% of urologist visits require intervention the initial addressable market is $400M. The total addressable market surpasses $1B when including the European market and console revenue. By treating patients in a single urologic visit Calcula will save the health care system $1.1B/year[3]. Calcula anticipates a requisite $24M cash burn in order to break even by year 5 following FDA 510(k) approval, which is expected in late year 4. Calcula is currently seeking $500K in seed financing to complete an in vivo animal trial in preparation for a $5M series A. We anticipate growing to 103 employees by 2018 in order to realize the execution plan. See business plan for additional details.

The core team includes:

- Buzz Bonneau has an MS from Stanford and MBA from Berkeley. He spent 6 years designing robotics at Intuitive Surgical and worked for Mohr Davidow Ventures in their life sciences group.
- David Gal is an electrical engineer and computer scientist with an MS from Stanford. He spent 4 years designing innovative products for the consumer, medical, and defense industries.
- John Leppert, MD is an attending urologist at the Palo Alto V.A. and author of 35+ papers. Dr. Leppert treats upwards of a dozen patients a week at the "unofficial stone center of the west."

In addition, Calcula has secured an experienced board of advisors to guide the development of this innovative solution:

- Paul Yock, MD – Stanford Professor and inventor of the Rapid Exchange catheter.
- Josh Makower, MD, MBA – Serial medical device entrepreneur. Founder of Acclarent.
- Tom Krummel, MD – Chief of Surgery at Stanford Hospitals and Clinics.
- Beverly Huss, MBA – Founder and CEO of Vibrynt Corporation.

---

[3] Internal calculation based off of department of labor statistics.

# Cases

# METHOD: ENTREPRENEURIAL INNOVATION, HEALTH, ENVIRONMENT, AND SUSTAINABLE BUSINESS DESIGN

It was spring 2007, and Method cofounder Adam Lowry sat deep in thought over enchiladas at Mercedes, a restaurant a block from his company's office on Commercial Street in San Francisco. He began to sketch out ideas on a piece of paper to sort through the issues troubling him. As a company known for environmentally healthy household products with designer brand appeal, Method was eager to develop a biodegradable cleaning cloth. Sourcing polylactic acid (PLA) cloth from China had not been in his plans, but every U.S. PLA manufacturer Lowry had talked to told him it was impossible for them to create the dry floor dusting cloth he wanted. And then there was the genetic modification issue. U.S. PLA producers did not screen their corn plant feedstock to determine whether it came from genetically modified organisms (GMOs). In any case, however, weren't any bio-based and biodegradable materials a better alternative than oil-based polyester, the material used by the competition? Yet certain major retailers were unwilling to stock products that weren't certifiably GMO-free. It was hard enough to manage a fast-growing new company, but why were there some people who seemed willing to stop progress while they held out for perfection on the environmental front? The naysayers made him have to think carefully about what it meant to be true to the environmental philosophy that formed the backbone of his business. He had often said that Method's business was to change the way business was conducted. But where did you draw the line?

As a hot new company that had received widespread publicity for its dedication to environmental values and healthy, clean production, use, and disposal of all its products, Method had set high standards. In a relatively short time, it had created a model for excellence in integrating health and environmental concerns into corporate strategy. From only a germ of an idea in 1999, Method had experienced explosive growth during the intervening years. The company proved that home cleaning products could evolve from toxic substances that had to be locked away from children and hidden in cupboards to nice-smelling, stylishly packaged, biodegradable, benign products that you proudly displayed on your countertop. In 2006, *Inc.* magazine listed Method Products at number seven of the 500 fastest and most successfully growing firms in the United States. Method stood out in many ways from the typical entrepreneurial firm.

---

This case was prepared by Andrea Larson, Associate Professor of Business Administration. It was written as a basis for class discussion rather than to illustrate effective or ineffective handling of an administrative situation. Copyright © 2007 by the University of Virginia Darden School Foundation, Charlottesville, VA. All rights reserved. *To order copies, send an e-mail to sales@dardenbusiness-publishing.com. No part of this publication may be reproduced, stored in a retrieval system, used in a spreadsheet, or transmitted in any form or by any means—electronic, mechanical, photocopying, recording, or otherwise—without the permission of the Darden School Foundation.*

Leveraging only $300,000 in startup capital, twentysomethings Adam Lowry and Eric Ryan caused small-scale "creative destruction" across a $17 billion industry in the United States by emphasizing the health, environmental, and emotional aspects of the most mundane of products: household cleaners. The company's differentiating characteristic? Lowry and Ryan assumed from the start that incorporating ecological and human health concerns into corporate strategy was simply good business. By 2007, Method was growing rapidly and was profitable with 45 employees and annual revenues of more than $50 million. Its products were available in well-known distribution channels (drugstores, department stores, supermarkets, and other retail outlets) in the United States, Canada, Australia, and the United Kingdom. Customers embraced Method's products giving the company live feedback on the Web site, praising the firm and providing tips for the future. They were a loyal crowd and a signal that the time was right for this kind of business model. They even requested T-shirts featuring the Method brand; the company responded by offering two different shirts, one that said "Cleans like a mother" and another that simply said "Method," both with the company slogan—"People against dirty"—on the back; a baseball cap was also available.

Indeed, "People against dirty" was Method's stated mission. From the company Web site: "Dirty means the toxic chemicals that make up many household products, it means polluting our land with nonrecyclable materials, it means testing products on innocent animals . . . these things are dirty and we're against that." Under Lowry and Ryan's leadership, Method shook up the monolithic and staid cleaning products markets by delivering high-performance products that appealed to consumers from a price, design, health, and ecological perspective—simultaneously. From the original offering of a clear cleaning spray, Method's product line had expanded by 2007 to a 125-product line of home solutions including dishwashing liquids and hand and body soaps. The "aircare" line, an array of air fresheners housed in innovatively designed dispensers, extended the product offerings in 2006, and the O-mop™ was added in 2007.

All products were made in alignment with Method's strategy. They had to be biodegradable, contain no propellants, aerosols, phosphates, or chlorine bleach, and had to be packaged in minimal and recyclable materials. Method used its product formulation, eye-catching design, and a lean outsourcing network of 50 suppliers to remain nimble and quick-to-market while building significant brand loyalty.

Method sold its products in the United States through several national and regional groceries, but one of the company's key relationships was with Target, the nation's number-two mass retailer in 2007. Through Target's 1,400 stores in 47 states, Method reached consumers across the United States. International sales were expanding, and the firm was in discussion with new distribution channels regularly.

## An Upstart Innovator in an Industry of Giants

The U.S. market for soaps and cleaning products did not seem a likely industry for innovation and environmental consciousness. It was dominated by corporate giants, many of which were integral to its founding. Although the soap and cleaning product industry was fragmented around the edges, with a typical supermarket stocking up to 40 brands, market share was dominated by companies such as SC Johnson, Procter & Gamble (P&G), Unilever, and Colgate-Palmolive.

To put Method's position in perspective, its total annual sales were approximately 10% of Procter & Gamble's sales in dish detergent alone ($317.6 million) (2006). P&G's total annual sales in the category were more than $1 billion. Furthermore, the market for cleaning products was under steady cost pressure from private label brands, increasing raw materials prices, and consumers' view of these products as commodities. Those companies that reported positive numbers in the segment between 2000 and 2006 did so by cutting costs and consolidating operations. Startups such as Seventh Generation and others attempted to penetrate the mass market with "natural" products, but those products were largely relegated to health food stores and chains such as Whole Foods. For Method to have obtained any foothold in this heavily consolidated segment dominated by market giants seemed improbable at best. But for Method founders Lowry and Ryan, the massive scale and cost focus of their competitors offered an opportunity.

## Method to Their Madness

"You have all your domestic experiences in that house or wherever you live," Ryan explained. And so, "from the furniture you buy to your kitchenware, you put a lot of thought and emotion into what you put in that space. Yet the commodity products that you use to maintain this very important space tend to be uninteresting, ugly, and toxic—and you hide them away." Lowry and Ryan didn't understand why it had to be that way.

They decided to take the opposite approach; if they could create products that were harmless to humans and the natural environment and were attractively designed with interesting colors and aromas, they could disrupt an industry populated with dinosaurs. By differentiating themselves from the competition in a significant and meaningful way, Lowry and Ryan hoped to offer an attractive alternative that also reduced the company's ecological footprint and had a positive environmental impact. "It's green clean for the mainstream," said Lowry, "which wouldn't happen if it wasn't *cool*."

To make green cool, Method took a two-pronged approach. First, it formulated new product mixtures that performed as well as leading brands while minimizing environmental and health impacts. Cleaning product manufacturers had been the target of environmental complaints since the 1950s, when the federal government enacted the Federal Water Pollution Control Act, in part to address the foaming of streams due to the use of surfactants, chemicals used in soaps and detergents to increase cleaning power. In addition to surfactants, household cleaners often contained phosphates, chemicals used as water softeners, which also acted as a plant nutrient, providing an abundant food source for algae. Fast-growing algae resulted in algal blooms, which depleted oxygen levels and starved aquatic life. Water sources contaminated with phosphates were also toxic for animals to drink. Another environmentally problematic compound in cleaning products was chlorine bleach, which when released into the environment could react with other substances to create toxic compounds. According to the Method Web site:

> A major problem with most household cleaners is that they biodegrade slowly, leading to an accumulation of toxins in the environment. The higher the concentration of toxins, the more dangerous they are to humans, animals, and plant life. The key is to create products that biodegrade into their natural components quickly and safely.

With a degree in chemical engineering from Stanford University, and experience researching "green" plastics and at a climate-change think tank, Lowry saw these issues as opportunities.

Method counted on the competition's seeing environmental and health issues as "problems." Doing so allowed Method to seize competitive advantage through designing out human health threats and ecological impacts from the start, while their larger competitors struggled to deal with increasing legislative and public image pressures. Method products sold at a slight premium to compensate for the extra effort. "I knew as a chemical engineer that there was no reason we couldn't design products that were nontoxic and used natural ingredients," Lowry said. "It would be more expensive to do it that way. But that was okay as long as we created a brand that had a 'premiumness' about it, where our margins would support our extra investments in product development and high-quality ingredients."

The second prong of Method's attack on the entrenched cleaning products industry was to utilize design and brand to appeal to consumers tired of the same old products. In an industry rife with destructive price competition, Method realized it would have to be different. The founders believed that their competition was so focused on price that "they weren't able to invest in fragrance or interesting packaging or design." Lowry explained:

> Our idea was to turn that reality on its head and come up with products that absolutely could connect with the emotion of the home. We wanted to make these products more like "home accessories." We believed there was an opportunity to really reinvent, and in the end, change the competitive landscape.

By focusing their marketing and packaging as the solution "against dirty," they tapped into consumers' disquiet with the ingredients in their household cleaners. Through packaging that stood out from the rest, they created the opportunity to deliver the environmental and health message of the products ingredients.

Design of packaging to deliver that message was integral to Method's success from its first sale. Method's home-brewed cleaning formulas for kitchen, shower, bath, and glass surfaces were all originally packaged in clear bottles that stood out on a shelf. "The manager of the store just liked the way the packaging looked," said David Bennett, the co-owner of Mollie Stones, a San Francisco Bay–area grocer that was Method's first retail customer. "It looked like an upscale product that would meet our consumer demands, so we went with it."

Design continued to be a key element of Method's appeal, with the recruiting of Karim Rashid, a renowned industrial designer who had worked with Prada and Armani. Rashid was responsible for bringing a heightened sense of style to Method's packaging, while continuing to focus on environmental impact. This led to the use of easily recycled number-one and number-two plastics (the types of plastic most commonly accepted by municipal recycling centers). Method's approach seemed to represent a younger generation's more holistic mental model. This small firm seemed to provide a window into a future where health, environmental, and what were increasingly called "sustainability issues" would be assumed as part of business strategy and product design.

### Wipes, the O-mop™, and PLA Material

PLA was an innovative and relatively new plastic material derived from plants such as corn, rice, beets, and other starch-based agricultural crops. PLA biodegraded at the high temperatures and humidity levels found in most composting processes. NatureWorks was the first large-scale plant in the United States to produce PLA in resin (pellet) form, based on milled material made from farm-supplied corn and corn waste. The resin pellets went to a fiber manufacturer who made bales; those bales of PLA material went next to the nonwoven cloth manufacturer, who converted it into giant rolls of nonwoven cloth. Next, a converter took the bulk nonwoven cloth, cut it into shapes, and packaged it according to the specifications of a customer such as Method. When NatureWorks first began operations, demand was limited. That picture changed quickly between 2004 and 2006, and by 2007 the plant could not produce its PLA feedstock resins fast enough to meet worldwide demand. PLA came out of the facility in pellet form and was melted, extruded, spun, and otherwise manipulated by converters at different steps down the supply chain into a virtually endless spectrum of materials for different applications across a wide range of product categories.

As a replacement for ubiquitous oil-based plastic feedstock, PLA promised a departure from the petroleum-based plastic materials that had come to dominate since synthetic plastics were first developed in volume after World War II. PLA had proved itself a particularly high-performing and cost-effective raw material that was well suited as a substitute for polyethylene terephthalate (PET) in many applications. PET was the oil-based polymer known generically as polyester and used extensively in packaging, films, and fibers for textiles and clothing.

The competition's wipes and mop heads were made of petroleum-based non-biodegradable plastic material, typically polyester or polypropylene. Although microfiber was quickly becoming commonplace, microfiber and the denier unit of measurement were first associated with material in women's hosiery. Technology advances permitted polyester microfiber production for very fine fiber applications, and just as microfiber had become common in clothing lines, it was also used as a more effective wiping and cleaning product.

Microfiber was fiber with strands measured at less than one denier, a unit of weight used to describe extremely fine filaments and equal to a yarn weighing one gram per 9,000 meters. Whether made from corn or oil, microfiber material, used by most companies selling residential cleaning wipes by 2006, made an excellent cleaning cloth. Its structure enabled the fiber surface to more effectively pick up dirt and dust compared with conventional materials and methods. The microfiber wipes could be washed and reused, providing greater durability than alternative products that were typically thrown away immediately after use.

Consistent with Method's environmental and sustainability philosophy, Lowry wanted to use biobased materials, specifically PLA nonwoven cloth, for the dry floor dusting product. Ultimately he wanted PLA to be the basis for all fibers used, both nonwoven disposable cloth and reusable woven microfiber. If customers weren't grabbed by the marketing message that the mop was sexy and hip (a message consistent with Method's playful tone) they might be pulled in by the ergonomic O-mop's more effective, biotech-based, and nontoxic floor cleaning.

Lowry knew most disposable wipes ended up in landfills, not compost piles, even with their extended life. So the company supported municipal recycling and composting infrastructure development in an effort to encourage cradle-to-cradle[1] resource use, or at least raise awareness and encourage behavior in that direction. Method estimated that 83,000 tons of "wipe" material made of polyester or polypropylene plastic was ending up in landfills annually, enough to fill 9,000 tractor-trailers. If using PLA could reduce oil feedstock use even a little, he reasoned, it was an improvement. Even if the PLA fiber went to landfills, where temperature and humidity never reached the ideal composting levels that would quickly and thoroughly break it down, it would still decompose safely, perhaps after one to two months, unlike oil-based fibers, which could remain in landfill disposal sites in the same condition for thousands of years.

The market for biobased plastic materials had taken off by 2007, but Lowry had had no luck finding a U.S. manufacturer to create a PLA-based fabric suitable for the white nonwoven dry floor duster cloth used with the O-mop. He had just talked with the last on his list of PLA manufacturers, and the answer was no. They had all told him it couldn't be done. The material was too brittle, they couldn't process it, it wouldn't run on their machines, and the strands were too weak: PLA nonwoven cloth for this application was technologically impossible.

He picked up the phone and placed a call to a company he knew in China—a departure from business as usual given that 90% of Method's inputs were

---

[1] "Cradle-to-cradle" was an increasingly popular term that referred to a product cycle in which materials could be manufactured, used, then broken down and used again with no loss of quality; for more information on this concept, see William McDonough and Michael Braungart's *Cradle-to-Cradle: Remaking the Way We Make Things* (North Point Press, 2002).

sourced in the United States. Chinese suppliers often were excellent, but domestic sourcing was preferable to avoid the high transportation costs of moving product long distances. Typically the farther the transport requirement, the greater the fossil fuel use, so the choice seemed inconsistent with the firm's sustainability approach. But Lowry was sure the dry floor dusting cloth could be made with PLA resins, and the Chinese manufacturer confirmed it. He placed the order. A Taiwanese fiber manufacturer would make the bales, and send them to the Chinese nonwoven cloth manufacturer that would pass the cloth on to a converter close by that would in turn cut and package to meet Method's needs. He knew the suppliers were good and reliable, and that the product would arrive promptly. Perhaps all Method's PLA products would need to come from China. But was sourcing from the other side of the world "sustainable" in the sense that he and Ryan tried to apply sustainability principles to the company's operations?

The other issue on his mind was that Method's products could be deemed unacceptable in certain distribution channels that would not tolerate any genetically engineered organisms in their products. PLA was produced from agricultural material (often corn or cornfield waste material) that was brought by farmers to a centrally located milling plant that converted it and separated out the components from which PLA was made. There was no monitoring of the corn coming into the milling facility, thus there was no guarantee that all inputs to the PLA resin-producing process were free of GMOs. If Lowry used PLA, it meant certain large and reputable buyers would refuse to put Method products on their shelves. Even so, to Lowry, it seemed preferable to substitute PLA for petroleum-derived products and compromise on the GMO issue for the time being. After a particularly discouraging conversation with a company that declined to do business with Method until it agreed to stop using GMO agricultural inputs, he decided to write out his thoughts in an essay, both to sort them out for himself and to draft a position paper that he could later post on the Method Web site.

## Questions

1. What was the original opportunity that these entrepreneurs identified?
2. Why and how has this company been successful?
3. How have Adam Lowry and Eric Ryan built environmental and sustainability issues into their company?
4. Should Method source PLA cloth from China? Why or why not?

# METHOD PRODUCTS: SUSTAINABILITY INNOVATION AS ENTREPRE-NEURIAL STRATEGY

## Introduction

Method Products, Inc., had hit a sweet spot for its buyers by 2010. Since its founding in 2000, the privately held Method had had a clear mission: Make good-smelling, high-performing household cleaners that were healthy through-out their material life spans and packaged in attractive, eye-catching, and eco-designed containers. Said Adam Lowry, cofounder and chief "greenskeeper":

> We wanted to change the way people view home cleaning. There is a dis-connect between the way people feel about and care for their homes and the design of the products they use to clean them. We set out to evolve the household cleaner from an object that lived under the sink to a countertop accessory and must-have item by providing cool-looking, effective, non-toxic products that are healthy for both the environment and the home.

As scientific studies revealed growing health problems associated with chemi-cal exposure and regulation of chemicals steadily rose around the world, more informed customers were seeking effective but healthier cleaning product options on retail shelves.

Despite the company's small size and entrepreneurially disruptive approach, Method's stylish cleaning products had quickly become state of the art in an industry moving toward sustainable business thinking. "We want to be thought leaders and we want to evoke change," said cofounder and design guru Eric Ryan. Method had in fact altered the once staid market for cleaning products in which large competitors traditionally fought over shelf facings, thin margins, and fractional market share points. Method introduced a three-times-more-concentrated laundry detergent in 2004 that contributed to Wal-Mart's requiring all detergent suppliers to concentrate their products to save resources on packag-ing and shipping. In 2006, Method began getting many of its products Cradle-to-Cradle-certified by McDonough Braungart Design Chemistry. This certification meant the products were nontoxic and used fewer resources throughout their life cycles. The next year, Method became one of the founding B Corporation (http://www.bcorporation.net [accessed August 30, 2010]), a form of company

---

that built environmental and social goals into its charter and passed third-party standards for sustainable practices. Method also worked through its public communications and Web site to express the goals and values that were an integral part of its company culture and products—that is, protecting health, children, and pets through eco-friendly and socially conscious products designed from a full life-cycle perspective.

Yet Method wanted to go further. "Our business model is to try to catalyze the next chapter of innovation within our product categories," Lowry said.[1] He and the company aspired to launch two major products a year. In 2008, Method turned to laundry detergents, a several billion-dollar market in the United States alone. It devised an eight-times-concentrated detergent in an encapsulated tablet, or monodose format, which would further save packaging and product materials and drastically reduce manufacturing and distribution of energy usage. The consumer could toss the tablet in with a load of laundry. It was convenient, efficient, less messy, and prevented using excess soap in each laundry load.

At a critical point in the product development process, everything about the monodose was working except one thing: The gel that encapsulated the detergent didn't dissolve entirely in cold water, a result of the company's decision not to use animal-derived ingredients such as the gelatins most often used for capsules. A bit of the plant-based gel casing could remain in the washing machine during the rinse cycle. Could Method and its loyal customers accept the residue as the price for a much more concentrated and environmentally compatible detergent formulation? Method could turn to petroleum-derived or gelatin capsules rather than plant-based materials for the capsule in order to solve the problem. Alternatively, the company could abandon the monodose concept and its inherent benefits. Lowry and his team contemplated what to do.

## The Emergence of Method

Lowry and Ryan knew each other growing up in Michigan, where both their families had entrepreneurial companies that became significant suppliers to the automobile industry. Lowry earned his bachelor of science in chemical engineering from Stanford University and worked on climate change policy at the Carnegie Institution for Science, an organization that focused on innovation and discovery. There, he helped develop software tools for the study of global climate change. In his post-college work experience, Lowry honed his unique approach to commercial environmentalism, which would form the basis of Method's success. And through his combined education and employment, Lowry became convinced that business was "the most powerful agent for positive change on the planet. But it's not business as we know it today. It is fundamentally and profoundly different. It is business redesigned."[2]

---

[1] Case writer interview with Adam Lowry, San Francisco, January 20, 2010; unless otherwise indicated, all subsequent attributions derive from this interview.

[2] "Behind the Bottle," http://www.methodhome.com/behind-the-bottle/ (accessed August 24, 2010).

In 2006, Ryan was named one of *Time* magazine's eco-leaders and received similar accolades from *Vanity Fair*. Ryan attended the University of Rhode Island and went into marketing, eventually doing work for the Gap and Saturn. The old high school classmates ran into each other on an airplane in 1997 and realized they lived on the same block of Pine Street in San Francisco. They soon thereafter became roommates, helping to maintain a house full of college fraternity brothers. As the story went, no one liked to clean. The two spent time discussing what was cool and what was not in commercial markets—and hence ripe for innovation. The pair settled on cleaning products, a bastion of typically harsh, dangerous chemicals—definitely not cool.

Lowry and Ryan used their bathtub to mix their own cleaners from natural, fragrant, and gradually more benign and renewable ingredients. Initial funding of $300,000 was provided through convertible debt from friends and family. They filled one beer pitcher at a time with their cleaners and made their first sale in February 2001 to a local grocery store—an order they then scrambled to fill from their bathtub. The next day, Lowry and Ryan hired their new boss, an entrepreneurial CEO with an MBA and years of experience in the consumer packaged goods industry.

The founders knew presentation would matter when it came to making an impression on customers. Lowry said:

> [We want to] inspire a happy, healthy home revolution. . .We want it to be happy because, in our opinion, way too much of the green movement has been heavy handed in sort of an education-based instead of inspiration-based [way]. That's one of the reasons that you see us concentrating so much on making our products not only work great and be green but be beautiful as well.[3]

They cold-pitched New York-based industrial designer Karim Rashid, whose aesthetically appealing work had appeared in museums, and he accepted the offer to design for Method. Soon, his designs would appear on countertops across the country. Rashid's name and Ryan's contacts got Method a pilot deal with Target to sell its products at 200 stores around Chicago and San Francisco. Method found a contract manufacturer to scale up production. Method's commitment to excellence and attention to detail impressed Target, especially after a problem with leaky containers led Lowry and other Method employees to walk through Target and pull the leaky bottles off the shelves themselves; their container supplier quickly corrected the problem.

Method's growth accelerated even as the company stayed true to its core values of style and social and environmental soundness. Target decided to carry Method products in all its stores, and Method went from having $16 in cash, credit card debt, and arrears to vendors in 2001 to profitability in 2005.[4] In 2006,

---

[3] Jacob Gordon, "Adam Lowry, the Man Behind the Method (Cleaning Products)," April 30, 2009, http://www.treehugger.com/files/2009/04/th-radio-adam-lowry.php (accessed August 24, 2010).
[4] Rich Whittle, "How Two Friends Built a $100 Million Company," *Inc.*, March 23, 2010.

Method experienced rapid growth, ending the year with about 45 employees, 50 vendors and suppliers, and a foothold in the United Kingdom. The next year pushed sales to $71 million.[5] Method expanded rapidly from hand soaps and countertop cleaners to body washes, floor cleaners, dish soaps, and laundry detergents. These products were carried by large retail chains such as Costco, Target, Lowe's, and Whole Foods and generated over $100 million in revenue by 2010. Method's former CEO distilled the company's approach for success in 2006: "Method has to enter a category with a huge disruption. The story cannot be copied overnight or eroded by competitors. It has to have disruptive packaging, ingredients, and fragrance."[6]

The company also continued to use naturally occurring or naturally derived ingredients as much as possible. If synthetic ingredients were needed, they were screened for biodegradability and toxicity to humans and the environment but without the use of animal testing. People for the Ethical Treatment of Animals (PETA) gave Lowry and Ryan its 2006 award for Person of the Year. The founding duo wrote a guidebook in 2008 titled *Squeaky Green: The Method Guide to Detoxing Your Home*. Their company, meanwhile, strove to reduce its carbon footprint through efficiency, switching to biodiesel trucks, or buying offsets such as methane digesters for manure at three Pennsylvania dairy farms. It also became the first company to introduce a custom-made bottle manufactured from 100% post-consumer recycled (PCR) polyethylene terephthalate (PET), whose recycling number, as part of the resin identification code, was one.

Despite its innovation, growth, and sterling public image, Method remained tiny relative to the competition. While Seventh Generation, an established producer of green cleaners and a fellow B Corporation, had sales comparable to Method's, generating $93 million in revenue in 2007 and over $120 million the following year, the makers of conventional cleaning products were orders of magnitude larger. One of the largest companies in the world, Procter & Gamble (P&G), had a market capitalization of $180 billion in April 2010, and its Household Care business unit alone had sales of $37.3 billion in 180 countries in 2009. P&G's laundry detergents in 2009 included Tide, the first synthetic heavy-duty detergent launched in 1946 and a now billion-dollar brand; Gain, another billion-dollar brand; Ace and Dash, each of which generated over $500 million in sales; and Cheer.[7] In short, P&G's laundry detergents dominated that market and by themselves generated more than 30 times Method's revenue.

Other giants, with broad product portfolios, also operated in the laundry detergent market. In 2009, Unilever, also a food producer, had total sales of about $55 billion. Colgate-Palmolive, known for toothpastes and dish soaps, had sales of $15 billion. Clorox, best known for its chlorine bleach, had sales of $5.5 billion. Clorox also had proven particularly adroit at moving into the green

[5] "No. 7 Method Products," *Inc.*, September 1, 2007.
[6] Stephanie Clifford, "Running Through the Legs of Goliath," *Inc.*, February 1, 2006.
[7] Jeffrey Hollender, "How I Did It: Giving Up the CEO Seat," *Harvard Business Review*, March 2010; the Procter & Gamble Company annual report, 2009.

cleaning market. It launched its Green Works line in the United States in 2008, rapidly expanded into 14 countries, and, according to the company, captured 47% of the natural cleaner market from mid-2008 to mid-2009, more than doubling the closest competitor's share. Church & Dwight Co., makers of Arm & Hammer brand baking soda, pulled in another $2.5 billion in sales in 2009 and marketed a series of baking-soda-based green cleaners and laundry detergents under its Arm & Hammer Essentials line.[8]

As Method mulled its own new laundry detergent, it remained very much aware of its smaller stature. As Eric Ryan told Inc. magazine in 2007, "When you run through the legs of Goliath, you need to spend a lot of time thinking about how to act so you don't put yourself in a place you can be stepped on."[9] Josh Handy, the lead Method designer, made a similar point:

> Where we've gone awry sometimes is when we've forgotten how small we are and therefore while we talked about ourselves as being the biggest green brand in the world, which typically we were, that's the wrong mindset for Method. What we are is the 35th-smallest cleaning products brand in the world.

Handy understood that Method's work environment had to support the creativity required for David-esque innovation. After Handy came to Method, he actively encouraged people to break rules in order to innovate, at one point literally drawing on a piece of furniture. Other employees followed his lead, and soon a room of once-uniform and uncomfortable white furniture was thus decorated and dubbed "the Wiggle Room." Commitment to giving people and ideas room to "wiggle" was serious. The mission was stated as "Keep it weird, keep it real, keep it different," and, as Eric Ryan commented:

> We don't build rockets over here; we build soap. And it's hard to be different in soap. So ideas have to be flowing. We have to have an environment where people are comfortable sharing ideas. We do everything we can to make people as connected as possible. We have to have every brain in the game. The more different an idea, the more fragile it is and the more likely it is beaten down and doesn't go anywhere. We have to cultivate our ability to be different, to be open to ideas. It means putting as much work into the culture as the product you are creating.

## Concentrating the Formula and Pushing the Monodose to Failure

When Method decided to pursue an improved detergent in early 2008, it turned the matter over to its team of "green chefs," including Fred Holzhauer, whom Lowry characterized as being "as close to a true mad scientist as anyone I've

---

[8] Unilever annual report, 2009; Colgate-Palmolive Company annual report, 2009; the Clorox Company, "Financial Overview," http://investors.thecloroxcompany.com/financials.cfm (accessed August 24, 2010); The Clorox Company annual report, 2009; "Church & Dwight Reports 2009 Earnings per Share of $3.41," *Business Wire*, February 9, 2010.
[9] Whittle.

ever met." Lowry gave the green chefs the mission of creating a better detergent and trusted them to figure out the details. "What we do is set up a system,"
Lowry explained, "a way of working, an environment that allows the innovation
to occur within the boundaries that we want." Drummond Lawson, Method's
"Green Giant" (or director of sustainability) seconded that notion. Method's
strategy was to hire creative people and then get out of their way. In the case of
Holzhauer, Lawson said:

> He has this opportunity to really play with everything in the lab, bring it
> through, and get some prototypes up to the point where we can take them
> and put them in other people's hands. Whereas if we mandate—go make this
> formulation with these characteristics—he'd be checked out and bored and
> gone. You'd end up with what you ask for instead of myriad opportunities.[10]

Method's green chefs decided to build upon the success of Method's three-
times-concentrated detergent. Further concentration would decrease the water
in the product, thereby decreasing volume, mass, packaging, storage space, and
freight costs. Method also wanted to encapsulate the detergent in tablet form for
the user's convenience and to reduce detergent wasted by inaccurate measurement. The green chefs listed their goals and all the tools at their disposal, including the conventional harsh, artificial tools. They focused initially on getting the
detergent formula right. As Holzhauer explained:

> The first thing you do is you build it the old way. You build it with all the
> nasty stuff that a competitor would and say, "What would be the highest per-
> forming biggest-payoff thing that we could build?" And you build it and then
> you say, "This stuff rocks. This is my benchmark. How close can I get using
> materials that are available with green?" Then there's the whole process of draw-
> ing lines through a whole bunch of stuff you wish you could use. And you're left
> with some holes. You're definitely left with some holes.[11]

Method, in collaboration with the Environmental Protection Encouragement
Agency based in Hamburg, Germany, had amassed a list of safe, biodegradable
chemicals to use as starting points for its products. Holzhauer had to ensure no
unwanted interactions among those ingredients, but he also had to fill the holes
in his toolkit in order to get the results he wanted. Method prized its products'
effectiveness above all else. If the products didn't clean, it didn't matter that they
were natural, nontoxic, and beautiful. But other constraints aside from performance, rather than limiting Method, forced the company to be more innovative.

Lowry had once called tradeoffs among these various qualities "just a symp-
tom of poor design." In addition, Lowry considered the following to be essential:

> . . .to make sure that what you're doing is really compelling for reasons
> other than being green. It has to be great in its own right, and green has to

---

[10] Case writer interview with Drummond Lawson, San Francisco, January 20, 2010; unless otherwise
indicated all subsequent attributions derive from this interview.
[11] Case writer interview with Fred Holzhauer, San Francisco, January 20, 2010; unless otherwise indicated all subsequent attributions derive from this interview.

be just another part of its quality. The whole idea of eco-entrepreneur should become the standard for entrepreneurship in general.[12]

Therefore, the chefs pushed further. They began to consult their networks. Holzhauer said:

> This is where collaboration and innovation really pay off. You start asking people who make detergents, [who] know you're handy, and you leverage your relationships and you say, "Hey, would you guys entertain this thought? You guys can make sodium laurel sulfate, but nobody's making MIPA sulfate, and that's the kind of tool that would really make a difference in what I'm doing, and it's just not available commercially. How about you whip me up a lab sample, you know?" And [from that] you get a new tool and you try it.

The chefs continued to use their contacts, get new tools, and test and alter them. They eventually refined the formula, dubbed "smartclean," to be as effective as "the nasty stuff" yet naturally based and eight times as concentrated. They found their detergent showed nonlinear improvement; doubling the concentration more than doubled its effectiveness.

Unexpectedly, they had moved into a new realm of chemistry in which few people had any experience working with such concentrated liquids. They kept testing the formula until they understood at the molecular level what exactly was happening (**Exhibit 1**). They also realized that the increased effectiveness meant they needed far less of the detergent to do the job, which meant the product could now compete on cost.[13]

The gel capsule, however, continued to be a problem. Holzhauer talked to people in the paintball industry to get a sense of how big he could make a glycerin or gelatin capsule to hold the detergent. He wanted something that could fit in your hand, and the paintball people thought that could be done. Holzhauer concentrated the detergent enough so the capsule size would contain all the detergent needed. Yet the detergent was so effective that it dissolved the capsules as well unless they were made sufficiently thick, at which point they would not fully dissolve in cold-water wash conditions. Holzhauer tinkered with the formula. He knew petroleum-based and animal-derived ingredients could make it all work perfectly, but that violated Method's premise of sustainability and ethics. "We tried and tried and tried. We just never got where we wanted to go," Holzhauer said.

## Surviving Failure: Collaboration and the Container

Ever optimistic that a tablet/capsule solution would be found, the chefs still wanted to test the "smartclean" detergent itself. They thought of a diaper cream

---

[12] Gordon; Susanna Schick, "Interview with OppGreen 2009 Speaker, Adam Lowry, Chief Greenskeeper of Method," *Opportunity Green*, November 3, 2009.

[13] On April 23, 2010, special sales excluded, Method's detergent sold on Amazon.com for about $0.31 per load, the same price as Tide with Febreeze, while Seventh Generation sold for about $0.27 per load and Gain for $0.19 per load.

Method already sold and got a simple idea: Put the detergent in the pump dispenser used for the diaper cream. Let people squirt the detergent straight into the washing machine instead of dissolving a tablet. Suddenly, the critical solution had taken hold in the chef group: a pump, not a tablet for this detergent. Holzhauer talked with Josh Handy to refine the pump bottle further. "If you get a bright idea," Holzhauer said, "you walk over to Josh's desk and you say, 'Hey, dude. I need one of these. Could you whip it up for me?' And he's like, 'Sure.' He'll ask a few questions about it and make sure it's worth the time, but he'll do it." Handy began working on Holzhauer's pump and posted his drawings outside the bathroom so that employees from all over could see them and provide feedback.

The switch from gel tablet to pump sent Holzhauer back to the lab to refine the formula. The detergent was incredibly viscous and had to be tweaked to work in a pump. It also had to be uniformly mixed so that each squirt dispensed exactly the same proportion of ingredients, whereas in the tablet the ingredients would all mix eventually once the tablet dissolved and thus could start off unevenly dispersed. Holzhauer wanted to keep tweaking the formula, which he had already gotten to 95% effectiveness using all benign and renewable ingredients, but Method was preparing to launch the product. Holzhauer patented his work to date and continued to work on a revised version for future release.

The new detergent formula appeared to work in the pump, so Method shifted emphasis to making sure it could get exactly the container it needed. Handy's final design featured a pump mechanism that was easy to depress without Herculean strength (**Exhibit 2** and **Exhibit 3**). A standard cap and bottle made measurement a two-handed task. In contrast, Method's customer could hold a laundry basket or child in one arm and dispense the necessary smartclean detergent—four short squirts—with the other hand. Whereas a full bottle of typical two-times concentrate could easily weigh seven pounds or more, Method's 50-load smartclean container, when full, weighed less than two pounds.

Handy's design next had to be mass-produced. That task fell largely to the packaging engineering and project management teams. Collaboration was the key for these groups; said one packaging engineer: "I can literally turn my chair around and help contribute." Method employees worked at common long tables, rather than in cubicles, so they could overhear discussions and add to them. The project manager believed innovation was more important than strict adherence to procedures. The packaging engineers, known internally as plastic surgeons, found they could use a stock engine (the internal workings of the pump) for the detergent but needed a custom top to match Method's aesthetic and operating needs. They reached out to various suppliers and found only one willing to collaborate on the custom design. Method agreed to pay for the tooling necessary to produce the top.

Method wanted the bottle itself to have transparent windows to show the contents and the newly designed angled dip tube that sucked in the detergent. That transparency would allow customers to ensure the tube always rested at the bottom of the container to extract all the detergent rather than leave some behind. Method selected an independent, California-based plastics manufacturer that produced almost 200 million containers annually and had experience using post-consumer recycled number-two plastic (HDPE). Method pushed the recycled content as high as it could before it started sacrificing transparency. Hence, the company ultimately settled on 50% virgin HDPE and 50% PCR HDPE.

Simultaneously, Method worked to produce the detergent in sufficient volume. Method's operations department began working with the contract manufacturer it selected to make the product, a supplier for almost all the biggest personal care and cleaner companies, including P&G, Unilever, and others. Method's ops team ushered the smartclean detergent from batches in Method's labs to identical, but much more massive, batches in the factory. A pilot manufacturing run revealed new problems—and a new opportunity. The first factory batch was contaminated by dirt from the bottling equipment because the Method detergent was so powerful, it cleaned out the lines as it moved through the system. Although Method had to nix the batch, it stumbled into a potential market: industrial cleaners. Indeed, the manufacturer began using the smartclean laundry detergent as its default factory equipment cleaner.

Finally, Method needed new shrink-wrapping equipment to put a label around the bottle's unique shape and to keep the pump locked even while allowing customers to unscrew the cap and sniff the detergent, a design requirement responding to the desire of many buyers to smell the contents. Method worked with the manufacturer to get the results it needed and invested in the new equipment.

## Quantifying Sustainability

Moving through the development and production of the smartclean laundry detergent, Method wanted to assess the product's environmental impact. The first step was Cradle-to-Cradle certification, which smartclean became the first laundry detergent ever to obtain. Smartclean also was recognized by the U.S. Environmental Protection Agency's Design for Environment program because of its nontoxic, biodegradable formulation.

In addition, Method wanted to calculate the detergent's overall carbon footprint. It collaborated with Planet Metrics, a Silicon Valley start-up founded in 2008 with $2.3 million in venture capital Series A funding. The young company was eager to collaborate with a company such as Method to build its reputation and beta-test its Rapid Carbon Modeling software. The ultimate goal of the software was to give companies a quick way to calculate returns on investment for various sustainability options. It did so by measuring life-cycle carbon emissions from a product throughout Scopes 1, 2, and 3. Those scopes are defined by the Greenhouse Gas (GHG) Protocol accounting method as follows:

- Scope 1: all direct GHG emissions
- Scope 2: indirect GHG emissions from consumption of purchased electricity, heat, or steam

■ Scope 3: other indirect emissions, such as the extraction and production of purchased materials and fuels, transport-related activities in vehicles not owned or controlled by the reporting entity, electricity-related activities (e.g., transmission losses) not covered in Scope 2, outsourced activities, waste disposal, etc.[14]

Planet Metrics analyzed Method's detergent from cradle-to-gate: all the activities and material needed to produce the product and get it ready to ship to retailers. Of course, later carbon dioxide emissions would presumably be lower for shipment to retailers, recycling old bottles, and so on, because the bottles used less material overall, which meant less energy and less mass to be moved around for an equivalent cleaning capacity. Looking just at cradle-to-gate, however, Method's smartclean detergent per load had a carbon footprint 35% smaller than the average two-times concentrated laundry detergent. It also used 36% less plastic and 33% less oil and energy. Finally, consumers would be more likely to use the appropriate amount of detergent, making the actual reductions even greater.

## Conclusion

Reflecting on the path of the smartclean laundry detergent, Lowry considered it a success born of failure. People were given space to create and collaborate within Method and across its supply chain. They didn't shut down after encountering obstacles; they got more creative and talked to each other. Lowry noted some factory workers where the detergent was made volunteered to work extra, unpaid shifts because they believed they were part of something bigger: a social change, not just another way to make money off people's dirty laundry. He said:

> Cultures are the only sustainable competitive advantage. We don't see the innovation per se as the competitive advantage. We see the ability to innovate as the competitive advantage. If you're going to do that you've got to build a different type of company where you literally are built—the people and the culture of the place—around the ability to bring the best ideas forward and let them live and let them thrive. Each innovation gives us license to innovate again.

By summer 2010, it appeared the new laundry detergent was a successful launch. What was the next focus for this young, fast-moving company? Typical growth challenges faced the firm and its entrepreneurial founders. As it grew from start-up to midsize, could its innovative output be maintained? How should management's attention be allocated across its innovation imperatives and its proliferation of product offerings and growth demands? And what was the end point? Was the goal to grow Method indefinitely?

---

[14] The Greenhouse Gas Protocol Initiative, "FAQ," 2010, http://www.ghgprotocol.orgicalculation-tools/faq (accessed August 24, 2010).

**Exhibit 1**

# METHOD PRODUCTS: SUSTAINABILITY INNOVATION AS ENTREPRENEURIAL STRATEGY

Conventional Detergent Micelle

Detergents typically consisted of two parts: a tail that could stick to oils (and other grime) and a head that stuck to water, so the dirt could be rinsed away in the wash. This structure led detergent molecules to clump into spheres, called micelles, in water because the hydrophilic heads (circle) faced out, while the hydrophobic tails (line), consisting of chemicals similar to fat, faced in.

Conventional detergents worked by breaking open the micelles in the washing machines so the tails could grab dirt and then getting the micelles to clump with the dirt in the middle so it could be washed away in the water. Breaking open micelles required agitation and thermal energy.

Method's smartclean laundry detergent kept hydrophobic tails on the outside, available to interact with dirt immediately. This inverted micelle made the detergent more efficient: Cleaning was improved while less detergent was wasted because it interacted more readily with dirt and less energy was needed to agitate and heat the laundry. This inverted property also reduced the amount of water needed in the liquid, thereby concentrating the detergent and reducing its mass and volume.

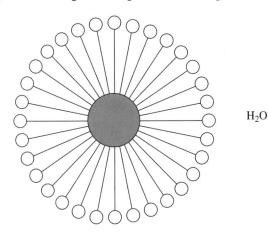

$H_2O$

Source: Created by case writer.

Source: Created by case writer.

**Exhibit 2**

# METHOD PRODUCTS: SUSTAINABILITY INNOVATION AS ENTREPRENEURIAL STRATEGY

The Method Laundry Detergent Bottle's Workings

**The Method Laundry Detergent Bottle's Workings**

**no-mess projectile squirt**

**custom pump**
for maximum grippability
+ easy dispensing

**screw-off cap**
enables sniffing at shelf

**precise dosing pump**
2 pumps = small load
4 pumps = medium load
6 pumps = large or heavily soiled load

**iconic method shape**

Source: Method.

## Exhibit 3

# METHOD PRODUCTS: SUSTAINABILITY INNOVATION AS ENTREPRENEURIAL STRATEGY

Method Laundry Detergent Advertisement

This ad emphasized the small size of the bottle and easy, one-handed use relative to detergent for an equivalent number of loads.

Company cofounders Adam Lowry and Eric Ryan

Source: Method.

# BIODIESEL INCORPORATED

Joshua Maxwell shut down his laptop and looked out the window. From the second floor of the Graduate School of Management's new building, he could see a number of cars driving on the nearby freeway and sitting in the adjacent parking lot.

Josh was in his last term of the full-time MBA program at UC Davis. He would soon be graduating and entering a new chapter of his life. While he had the luxury of having several management-level job offers from which to choose, he was unsure whether he wanted to follow such a traditional route. There was one opportunity in particular that had recently come across his path which gave him pause.

## Background

The previous term, Josh had been enrolled in Professor Dorf's class on Business and Sustainability. While the class was offered at the GSM, it was open to the entire university. In this class, he met Hannah Long, who was in her final year of her undergraduate studies in Agricultural Economics, and Matthew Hammond, who was a senior in the Mechanical Engineering department.

The three began working on a class project, which would ultimately turn into a formidable business opportunity. The impetus for their collaboration began with a lecture-discussion regarding the challenges and opportunities in the emerging renewable energy industry.

## The Challenge

Dependence on energy is a worldwide reality. Energy powers the machines and equipment around us in order to make life more convenient and efficient. In our everyday lives, energy is synonymous with the forms that it can assume. The major generation sources—petroleum, coal, natural gas and nuclear—are non-renewable resources and have detrimental effects on the environment. In our daily lives, the two most common forms of this energy are liquid fuel (refined from petroleum) and electricity.[1]

Increasingly, developed and developing countries alike are consuming liquid fuel for the purposes of mobility, food production, and the facilitation of trade. All of these functions essentially provide a substitute for human effort. Due to the widespread consumption of petrol-based liquid fuel, an incredibly large global infrastructure and set of surrounding institutions have grown around the support of such consumption. The petrochemical fuel industry manifests itself in the form of oil fields and reserves, pipelines, transport ships, and fueling stations.

---

Prepared by MBA candidate Benjamin Finkelor; Assistance from MBA candidate Sonja Yates and Paul Yu-Yang under the supervision of Professor Richard C. Dorf, Graduate School of Management of UC Davis.

[1] Technically speaking, liquid fuel is a form of energy, and electricity is considered a carrier of energy. For the purposes of this proposal, the distinction is not significant.

The way energy is used worldwide is not sustainable. It is well-documented that the use of these fuels is depleting the world's natural resource reserves, harming communities in terms of health and displacement, and polluting the air and water in local environments. The drilling, refining, and transporting of oil leads to spills on land and in oceans, and when petrol-based fuels are used to power machines and automobiles, the air is polluted with greenhouse gases and particulate matter such as carbon dioxide, carbon monoxide, sulfur, and nitrous oxide emissions.

In spite of the drawbacks, the current energy industry is committed to the continuation of these ways, primarily because of considerable assets and investment in the existing form of infrastructure.

The challenge, which became clear to the team from class discussion and further brainstorming, is to find a form of fuel or technology that can mitigate the current negative affects on the environment of petrol-based fuel while utilizing the existing infrastructure. The urgency of this challenge is heightened by the astounding projected growth in the global population and per-capita consumption of liquid fuels.

## The Concept

Matthew's coursework in engineering coupled with a bit of networking with fellow engineers suggested the emerging technology of biodiesel as a possible solution to this challenge. As the group explored the environmental benefits and the viability of the diesel fuel substitute, the three began to realize the potential of the biodiesel market.

Biodiesel is a vegetable- and/or animal-based product that serves as a substitute for traditional diesel fuel. Although its chemical composition is dissimilar from the petrol-based diesel, biodiesel will still work in diesel engines built in and after 1996 with no modification. For engines made before that time, modifications can be made to allow for the use of biodiesel fuel. The choice of biodiesel as a product of biomass is an intentional one. Producing a product that can be utilized by the existing infrastructure and social patterns of use[2] increases the likelihood of its adoption. "Entrepreneurs must locate their ideas within the set of existing understandings and actions that constitute the institutional environment yet set their innovations apart from what already exists."[3] This economic viability is coupled with a significant potential to the environment: biodiesel showcases an innovation that is a step in the right direction for air quality.

Biodiesel's greatest promise to sustainability as a renewable energy source is its lower emissions over conventional diesel. Compared to traditional diesel, biodiesel achieves significant reductions in harmful emissions. Additionally, the ozone-forming impact of biodiesel is nearly half of that of petroleum fuel. Further benefits can be counted when looking at lifecycle effects. If biodiesel is

---

[2] M. Ahmad, *Practical Handbook on Biodiesel Production and Properties*, CRC Press. 2012.
[3] A. Hargadon and Y. Douglas, "When Innovations Meet Institutions," *Administrative Science Quarterly*, 46(2001): 476.

obtained using soybeans as an example, the amount of $CO_2$ taken up by soybeans and released upon burning the fuel, is a near zero sum balance. Contrast this with petroleum products where release of $CO_2$ is unidirectional into the atmosphere.

Because biodiesel is biodegradable and dependent on organic material as opposed to fossil fuels, the energy source is considered renewable. Production of biodiesel begins with feedstock, preferably in the form of oils or fats. Oils can be processed from oleic varieties of plants such as soy, canula, sunflower and saf-flower. Fats can come directly from grease such as and tallow/lard and recycled cooking grease from restaurants. The oils or fats are mixed with alcohol and a catalyst in a process that forms esters resulting in biodiesel defined as mono-alkyl esters of long chain fatty acids and glycerin.

Ultimately, the large-scale production of biodiesel would generate a dramatic impact on the economic value of the feedstocks involved. For example, according to one study, if biodiesel demand over the next ten years were to increase to 200 million gallons, a commensurate amount of soy oil would be required and net average farm income would increase by $300 million per year. A bushel of soybeans would increase by an average of 17 cents over the ten-year period.[4] The potential economic benefit to farmers seems considerable.

Even with such economies of scale, however, the wholesale price of 100 percent biodiesel would rarely be lower, and therefore cost-competitive, with traditional diesel fuel. Barring some crisis that would drive up the price of crude oil or reduce the capacity of diesel refineries, the current regulatory structure and assets devoted to petrol-diesel will more often than not yield a lower price with petrol-based diesel. Biodiesel as a fuel additive however, does provide a cost-competitive potential. Studies have shown that splash-mixing even 1 percent biodiesel with traditional diesel "can increase the lubricity of petroleum diesel by up to 65 percent."[5] This is not to mention the sulfur- and other emissions-reducing benefits that splash-mixing provides. As more consumer and regulatory pressure is placed on traditional diesel users, biodiesel producers will be able to charge the premium necessary to offset higher relative costs. Markets for 100 percent biodiesel will grow as well in such specialty markets as the marine industry, railroads, electricity generators, and even agriculture.

### Biodiesel Incorporated

Josh, Hannah, and Matthew presented a compelling business case for their final class project: Biodiesel Incorporated. This new venture would enlist and develop a series of local producer's cooperatives in an effort to capitalize on the emerging biodiesel market as described in the following list:

- Members would grow feedstock crops and gather crop residues with high fat content.

---

[4] G. Knothe and J. V. Gerpen, *The Biodiesel Handbook*, 2nd ed., AOCS Publishers, 2010.
[5] "Biodiesel Carries New Weight in Premium Diesel Market," Biodiesel Bulletin, Sept. 2002. http://www.biodiesel.org/news/bulletin/2003/080403.pdf.

■ Capital equipment costs would be shared and spread over membership. Oils would be extracted from the collected biomass and biodiesel would be produced using these oils.

■ Biodiesel Incorporated would distribute the biodiesel locally using the existing petroleum-based infrastructure.

## Advantages of the Cooperative Business Form

The cooperative model has been successfully used to allow small farmers to maintain a competitive edge against the larger corporate farming organizations. "Today, there are more than 4,000 agricultural cooperatives in the U.S., with a total net income of nearly \$2 billion and net business volume of more than \$89 billion."[6] A coop is owned and controlled by the members, with self-reliance and self-help being key characteristics—ideal for the implementation of emerging and disruptive innovations such as biodiesel.

Biodiesel Incorporated will:

■ Utilize the collective purchase power of the coop to obtain necessary capital-intensive equipment and to gain economies of scale.

■ Increase negotiating power, allowing it to
  ■ Stabilize crop prices and biodiesel output
  ■ Gain access to higher-volume contracts

■ Serve to unite rural communities and preserve agricultural economy

Biodiesel Incorporated offers the unique service of both the bargaining and manufacturing of biodiesel on behalf of its farmer members. It will serve to control the production of agricultural products (i.e., the biomass feedstock), the price and terms set for members' production, and price and terms for biodiesel output.[7]

## Questions

1. What are the key factors in determining if this is a viable business opportunity for Josh, Hannah, and Matthew?
2. What market drivers should they research and be aware of?
3. What are the flaws in the current business strategy?
4. What type of financing should they use if they choose to go forward with this?
5. What types of distribution channels should they go into?
6. How can they improve their chances for success?
7. What is the next step?

---

[6] http://co-operatives.ucdavis.edu/Agricultural Co-operatives.htm.
[7] "New Biodiesel Hub Sets Up in Portland," Sustainable Business Oregon, April 10, 2012.

# YAHOO!

*Guess, three and a half years ago, if we were looking to start a business and
make a lot of money, we wouldn't have done this.*
**—Jerry Yang, 1997**

It was April of 1995—a key decision point for Jerry Yang and David Filo. These
two Stanford School of Engineering graduate students were the founders of
Yahoo!, the most popular Internet search site on the World Wide Web. Yang and
Filo had decided that they could transform their Internet hobby into a viable business. While trying to decide between several different financing and partnering
options that were available to them, they attended a meeting with Michael Moritz,
a partner at Sequoia Capital. Sequoia, one of the leading venture capital firms in
Silicon Valley, had been discussing the possibility of investing in Yahoo!.

Michael Moritz leaned forward in his chair. As he looked across the conference table at Jerry and Dave, he laid out Sequoia's offer to fund Yahoo!:

> As you know, we have been working together on this for some time now.
> We have done a lot of hard work and research to come up with a fair value
> for Yahoo!, and we have decided on a $4 million valuation. We at Sequoia
> Capital are prepared to offer you $1 million in venture funding in exchange
> for a 25 percent share in your company. We think that with our help, you
> have a real chance to make Yahoo! something special. Our first order of
> business will be to help you assemble a complete management team, after
> which we should be able to really start helping you to develop and manage
> your site's vast amount of content.
>
> Right now, the biggest risk that you guys run is *not* making a decision.
> You *have* to make a decision, because if you don't, someone else is going
> to run you over. You might get run over by Netscape. You might get run

This case was prepared by Michael K. Chang and Matthew Garman, graduate students at Stanford
University's School of Engineering, and Thomas J. Kosnik, consulting professor, Stanford School of
Engineering, as basis for class discussion rather than to illustrate either effective or ineffective handling
of an administrative situation. Some facts have been disguised. Dialogue between case actors has been
reconstructed from multiple sources based on their recollection of past events, and is not intended as a
verbatim quotation at the time of the meeting.
Revised August 22, 2001.

over by AOL. You might get run over by one of these other venture-backed start-ups. It is imperative that you make a decision now if you are going to survive. To help you make a decision, I am going to give you a deadline: tomorrow. If you don't want to do business with Sequoia, that's OK. I'll be disappointed, but that's OK. But you are going to have to call me by 10 A.M. tomorrow morning to tell me yes or no.[1]

Yang and Filo gazed around the Sequoia conference room and noticed the many posters of companies such as Cisco, Oracle, and Apple that were hung from the walls—all success stories from past Sequoia investments. They wondered if Yahoo!'s poster would someday join that group. The two were excited at the possibilities; however, they still had some decisions to make. There were several other financing options available, and they were still not sure if they wanted to accept Sequoia's funding. Yang responded:

That sounds like a pretty fair offer, Mike. Let us talk this over tonight, and we will get back to you by tomorrow after we weigh all of our options. However you have to realize that we're still grad students, and we don't even usually wake up by 10 A.M., so can you give us until noon?

## Yahoo!

Yahoo! was an Internet site that provided a hierarchically organized list of links to sites on the World Wide Web. It offered a way for the general public to easily navigate and explore the Web. Users could click through multiple topic and category headings until they found a list of direct links to Web sites related to their interest. In addition, Yahoo! offered a central place where people could go to just to see what was out there. This made it easy for people with little previous exposure to the Web to start searching through Yahoo!'s lists of links, often just to see if they could find something of interest. In a little over a year since its inception, it had become one of the most heavily visited sites on the Web.

But Yang and Filo believed Yahoo! had the potential to be much more than a way for Web surfers to find what they were looking for. In 1995, John Taysom, a vice president of marketing of Reuters, a London-based provider of news and financial data, called Jerry Yang to explore the idea of a Yahoo!-Reuters partnership. It seemed to Taysom that affiliating with Yahoo! could help Reuters to build a distribution network on the Web.

"The first thing Jerry said to me," Taysom remembers, "was 'if you hadn't called me, I would have called you.'" Jerry *got* the news feed vision. He had been thinking about it for months. He further surprised Taysom by informing him that as far as he was concerned, Yahoo! was "not just a directory but a media property."[2]

---

[1] Michael Moritz, personal interview, November 10, 1998.
[2] Rob Reid (1997), *Architects of the Web,* John Wiley and Sons, New York, p. 253.

Yang further believed that: "Primarily we're a brand. We're trying to promote the brand and build the product so that it has reliability, pizzazz, and credibility. The focus of the business deals we are doing right now is not on revenues but on our brand."[3]

## Dave and Jerry at Stanford

David Filo, a native of Moss Bluff, Louisiana, attended Tulane University's undergraduate program in computer engineering. In 1988, Filo finished his undergraduate work and enrolled in Stanford's master's program in electrical engineering. Completing his master's degree, he opted to stay at Stanford and try for his PhD in electrical engineering. Extremely competent in the technical arena, Filo had been described by many as a quiet and reserved individual.

Jerry Yang was a Taiwanese native who moved to California at the age of 10. Yang was raised by his widowed mother and grew up in San Jose with his younger brother, Ken. Yang was a member of the Stanford class of 1990 and completed both his bachelor's and master's degrees in electrical engineering. Yang also opted to stay at Stanford for a PhD in electrical engineering. Also technically competent, Yang was considered much more outgoing than Filo.

Yang and Filo met each other in the electrical engineering department at Stanford; Filo was Yang's teaching assistant for one of his classes. They also both worked in the same design automation software research group. They became close friends while teaching at the Stanford campus in Kyoto, Japan. Upon returning to the Stanford campus, they moved into adjacent cubicles in the same trailer to conduct their graduate research. They both enjoyed working together, as their individual personalities perfectly complemented each other, forming a unique combination.

Their office was not much to look at, but it served as a place for them to work on their research as well as a place from which they could run their website. "The launching pad (for Yahoo!) was an oxygen-depleted, double-wide trailer, stocked by the university with computer workstations and by the students with life's necessities . . . that prompted a friend to call the scene 'a cockroach's picture of Christmas'."[4] Michael Moritz remembered his early visits to Jerry and Dave's cube:

> With the shades drawn tight, the Sun servers generating a ferocious amount of heat, the answering machine going on and off every couple of minutes, golf clubs stashed against the walls, pizza cartons on the floor, and unwashed clothes strewn around . . . it was every mother's idea of the bedroom she wished her sons never had.[5]

[3] Jerry Yang (1995) interview in *Red Herring* (October), online back issue, p. 9.
[4] Randall E. Stross (1998), "How Yahoo! Won the Search War," *Fortune,* http://www.pathfinder.com.fortune/1998/980302/yah.html, p. 2.
[5] Rob Reid (1997), *Architects of the Web*, John Wiley and Sons, New York, p. 254.

## Mosaic and the World Wide Web

In 1993, the University of Illinois-Urbana Champaign's National Center for Supercomputing Applications (NCSA) revolutionized the growth and popularity of the World Wide Web by introducing a Web browser they had developed called Mosaic. Mosaic made the Web "an ideal distribution vehicle for all kinds of information in the professional and academic circles in which it was known."[6] It provided an easy-to-use graphical interface that allowed users to travel from site to site simply by clicking on specified links. This led to the widespread practice of surfing the Web, as people spent hours trying to find new and interesting sites. This easy-to-use browser for navigating the Internet was estimated to have 2 million users worldwide in just over one year.

## Creating Jerry's Guide to the World Wide Web

With Mosaic's introduction in late 1993, Filo and Yang, along with thousands of other students, began devoting large amounts of time to surfing the Web and exploring the vast content available. As they discovered interesting sites, they made bookmarks of the sites. The Mosaic Web browser had an option to store a bookmark list of your favorite sites. This feature allowed users to return directly to a page that they had visited, without having to navigate through several different links. As the popularity of the Web quickly increased, so did the total number of sites created, which in turn led to an increase in the number of interesting sites that Filo and Yang wanted to bookmark. Eventually, their personal list of favorite Web sites grew large and unwieldy, due to the fact that the earliest versions of Mosaic were unable to sort bookmarks in any convenient manner.

To address this problem, Filo and Yang wrote software using Tcl/TK and Perl scripts that allowed them to group their bookmarks into subject areas. They named their list of sites "Jerry's Guide to the World Wide Web" and developed a Web interface for their list. People from all over the world started sending Jerry and Dave e-mail, saying how much they appreciated the effort. Yang explained: "We just wanted to avoid doing our dissertations."[7]

The two set out to cover the entire Web. They tried to visit and categorize at least 1,000 sites a day. When a subject category grew too large, subcategories were created, and then sub-subcategories. The hierarchy made it easy for even novices to find websites quickly. "Jerry's Guide" was a labor of love—lots of labor, since no software program could evaluate and categorize sites. Filo persuaded Yang to resist the engineer's first impulse to try to automate the process. "No technology could beat human filtering," Filo argued.[8]

---

[6] Rob Reid (1997), *Architects of the Web,* John Wiley and Sons, New York, p. 11.

[7] Randall E. Stross (1998), "How Yahoo! Won the Search War," *Fortune,* http://www.pathfinder.com. fortune/1998/980302/yah.html, p. 2.

[8] Randall E. Stross (1998), "How Yahoo! Won the Search War," *Fortune,* http://www.pathfinder.com. fortune/1998/980302/yah.html, p. 3.

Though engineers, Yang and Filo had a great sense of what real people wanted. Consider their choice of name. Jerry hated "Jerry's Guide," so he and Filo opted for "Yahoo!," a memorable parody of the tech community's obsession with acronyms (this one stood for "Yet Another Hierarchical Officious Oracle"). Why the exclamation point? Said Yang: "Pure marketing hype."[9]

### Yahoo!'s Growing Popularity

At first, Yahoo! was only accessible by the two engineering students. Eventually, they created a Web interface that allowed other people access to their guide. As knowledge of Yahoo!'s existence spread by word of mouth and e-mail, more people began using their site, and Yahoo!'s network resource requirements increased exponentially. Stanford provided them with sufficient bandwidth to the Internet, but bottlenecks came from limitations in the number of TCP/IP connections that could be made to the two students' workstations.[10] Additionally, the time required to maintain the site was becoming unmanageable, as Yang and Filo found themselves continually updating their Web site with new links. Classes and research fell behind as Yang and Filo devoted more and more time to their ever-expanding hobby.

### Competing Services

A number of businesses already existed in the Internet search space. While none offered the same service that Yahoo! did, these companies could definitely provide potential competition to any new business that Yahoo! would start. Among the competitors were Architext, soon to be renamed Excite, Webcrawler at the University of Washington, Lycos at Carnegie Mellon, the World Wide Web Worm, and Infoseek, founded by Steven Kirsh. AOL and Microsoft in 1995 represented larger competitors who could enter the market either by building their own capability or acquiring one of the other start-ups.

Yahoo!'s human-crafted hierarchical approach to organizing the information for intuitive searches was a key component of its value proposition. Rob Reid, a Venture Capitalist with 21st Century Internet Venture Partners, explained how this made Yahoo! unique among Internet search providers.

> The Yahoo! hierarchy is a handcrafted tool in that all of its . . . categories were designated by people, not computers. The sites that they link to are likewise deliberately chosen, not assigned by software algorithms. In this, Yahoo! is a very labor intensive product. But it is also a guide with human discretion and judgment built into it—and this can at times make it almost uncannily effective. . . .

---

[9] Randall E. Stross (1998), "How Yahoo! Won the Search War," *Fortune,* http://www.pathfinder.com. fortune/1998/980302/yah.html, p. 3.

[10] Mark Holt and Marc Sacoolas (1995), "Chief Yahoos: David Filo and Jerry Yang," *Mark & Marc Interviews,* May, http://www.sun.com/950523/yahoostory.html.

This is the essence of Yahoo!'s uniqueness and (let's say it) genius. It isn't especially interesting to point to information that many people are known to find interesting. *TV Guide* does this. So do phone books, and countless Web sites that cater to well-defined interest groups. . . . But Yahoo! is able to build intuitive paths that might be singularly, or even temporarily important to the people seeking it. And it does this in a way that no other service has truly replicated.[11]

However, if Yahoo!, as a business, was to survive and flourish in the face of increasingly well-funded competition, it would quickly need to find some outside capital.

## Leaving Stanford and Starting the Business

Yang and Filo had been in Silicon Valley long enough to realize that what they really wanted to do was to start their own business. They split much of their free time between their Internet hobby and sitting around thinking up possible business ideas.

"A considerable period of time passed before it occurred to them that the most promising idea was sitting under their noses, and some of the credit for their eventual illumination belongs to their PhD adviser, Giovanni De Micheli. Toward the end of 1994, De Micheli noted that inquiries to Yahoo! were rising at an alarming rate. In a single month, the number of hits jumped from thousands to hundreds of thousands daily. With their workstations maxed out, and the university's computer system beginning to feel the load, De Micheli told them that they would have to move their hobby off campus if they wanted to keep it going."[12]

By fall of 1994, the two received over two million hits a day on their site. It was then that Jerry and Dave commenced the search for outside backing to help them continue to build up Yahoo!, but with only modest hopes. Yang thought they might be able to bootstrap a workable system, using personal savings to buy a computer and negotiating the use of a network and a Web server in return for thank-you banners. Unexpected overtures from AOL and Netscape caused them to raise their sights, although both companies wanted to turn Filo and Yang into employees.

If they were going to abandon their academic careers (as they soon did, six months shy of their doctorates), they reasoned that they should hold out for some control. Filo and Yang had three main potential options to explore: (1) sell Yahoo! outright; (2) partner with a corporate sponsor; (3) start an independent business using venture capital financing.

---

[11] Rob Reid (1997), *Architects of the Web,* John Wiley and Sons, New York, pp. 243–244.

[12] James, Lardner, (1998), "Yahoo! Rising," *U.S. News,* May 18, http://www.usnews.com.usnews/issue/80518/18yaho.html, p. 3.

## The Search for Funding

Looking to receive funding and create a credible business out of Yahoo!, Filo and Yang began preliminary discussions with potential partners in October 1994. One of the first people who contacted them was John Taysom, a vice-president of marketing at Reuters, the London-based media service. Taysom was interested in integrating Reuters' news service into Yahoo!'s Web pages. Yahoo! would gain the advantage of being able to provide news services from a well-known source, while Reuters would be able to begin developing its own presence on the Internet. Unfortunately, since Yahoo! did not generate revenues, it was in a poor negotiating position. Talks between the two were cordial, but they also progressed very slowly.

Yahoo! also talked to Randy Adams, founder of the Internet Shopping Network (ISN), a company that styled itself as "the first online retailer in the world." ISN, funded by Draper Fisher Jurvetson, was one of the first venture funded Internet companies. It had recently been purchased by the Home Shopping Network, in order to expand its possible exposure. ISN was interested in being a host site for Yahoo!, offering them the chance to finally generate some revenue. However, there were also definite possible disadvantages that came from being associated with a shopping network.

Another company that approached Yahoo! was Netscape Communications Corporation. Founded in April 1994 by Jim Clark, who also founded Silicon Graphics, and Marc Andressen, who created the NCSA Mosaic browser with a team of other UIUC students and staff, Netscape was a hot private company developing an improved browser based on the old Mosaic technology. Andressen contacted Yang and Filo over e-mail and, in Yang's words said, "Well, I heard you guys were looking for some space. Why don't you come on into the Netscape network? We'll host you for free and you can give us some recognition for it."[13] This was a fortuitous contact that allowed Yahoo! to move itself off of Stanford's campus. By early 1995, Yahoo! was running on four Netscape workstations.

Soon after, Netscape offered to purchase Yahoo! outright in exchange for Netscape stock. The advantage of this option was that Netscape was already planning its initial public offering and had tremendous publicity and momentum behind it. Coupled with high profile founders and backers like Clark, James Barksdale, former president and CEO of AT&T Wireless Services, and the venture capital firm Kleiner Perkins Caufield & Byers, this offer was a potentially lucrative one for the two Yahoo! founders. Additionally, Netscape's company culture was more in tune with what the two students were looking for, in comparison to some of the more established market players.

---

[13] Jeff Ubois (1996), "One Thing Leads to Another," *Internet World,* January, http://www.Internetworld.com/print/monthly/1996/01/yahoo.html, p. 1.

## Corporate Partnerships

Yahoo! was also feeling tremendous pressure to partner or accept corporate sponsorship from other large content companies and online service providers like America Online (AOL), Prodigy, and Compuserve. These companies offered the carrot of money, stock, and/or possible management positions. They argued that if Yahoo! did not partner with them, as large players they could develop their own competing services that would cause Yahoo! to fail. One potential disadvantage with corporate funding was the potential taint that came with such sponsorship. Yahoo! had started as a grass-roots effort, free of commercialization. A second disadvantage was the lack of control that the two Yahoo! founders would have over their creation. "Building Yahoo! was fun, particularly without adult supervision. (Dave) and Jerry were also worried that selling to AOL would have 'most likely killed' Yahoo! in the end."[14]

With partner discussions beginning to heat up, Yang requested help from Tim Brady, a friend and second-year Harvard Business School student. As a class project, Brady generated a business plan for Yahoo! during the 1994–1995 Christmas vacation. (See the Appendix for excerpts of the business plan circa 1995.)

With Brady's business plan in hand, Filo and Yang began to approach different venture capital firms on nearby Sand Hill Road. Venture capital firms brought experience, valuable contacts in the Silicon Valley, and most importantly, money. However, they also required substantial ownership in return for their services. One venture firm that the Yahoo! founders approached was Kleiner Perkins Caufield & Byers. KPCB had an excellent reputation as one of the most prestigious VC firms in the Silicon Valley, and their list of successful investments included Sun Microsystems and Netscape. KPCB showed a definite interest in Yahoo!; however, Vinod Khosla of KPCB and Geoffrey Yang of Institutional Venture Partners had just invested $0.5M in Architext (later renamed Excite), another company started by Stanford engineering students that was developing a search-and-retrieval text engine. Architext was receiving increased press coverage, with a March 1995 *Red Herring* magazine spotlighting the company and its venture capital partners. KPCB proposed to fund Yahoo!, but only if they agreed to merge with Architext.

## Sequoia Capital

Another venture capital firm that Yahoo! approached was Sequoia Capital. It was during partnership discussions with Adams at the Internet Shopping Network that Yang and Filo were first introduced to Michael Moritz, a partner at Sequoia Capital. Moritz went to visit Jerry and Dave, who were at the time still operating out of their tiny Stanford trailer. Said Yang, "The first time we sat down with Sequoia, Mike (Moritz) asked, 'So, how much are you going to charge subscribers?' Dave and I looked at each other and said, 'Well, it's going

---

[14] Rob Reid (1997), *Architects of the Web,* John Wiley and Sons, New York, p. 256.

to be a long conversation."[15] Fortunately, Moritz, who came from a journalistic background at *Time* was flexible in his thinking. Some of the major advantages that Moritz brought to the negotiating table were his contacts with publications and knowledge about how to manage content. Moritz talked about the roots of Sequoia's interest in working with Yang and Filo. "I think we are always enamored with people that seem to be on to something, even if they can't define that something. They had a real passion and a real spark."[16]

Sequoia Capital had a long tradition of success in the venture capital market, citing that the total market capitalization for Sequoia backed companies exceeded that of any other venture capital firm. Sequoia's trademark *modus operandi* was funding successful companies using only a small amount of capital. Its list of successful investments included Apple Computer, Oracle, Electronic Arts, Cisco Systems, Atari, and LSI Logic. Said Moritz, Sequoia preferred "to start wicked infernos with a single match rather than 10 million gallons of kerosene."[17]

In February 1995, Filo and Yang were weighing a number of possibilities and in no hurry to accept any of them, when Michael Moritz made them an offer. Sequoia Capital would fund Yahoo! for $1 million and would help them to assemble a top management team. In return, Sequoia would receive a 25 percent share of the company. Additionally, Moritz gave them only 24 hours to accept the deal before it was pulled off the table. "I felt a need to deliver them from the agony of indecision," claimed Moritz. With the deadline quickly approaching, Yang and Filo sat down to weigh their options. The decisions that they made that night would determine the direction of their careers as well and the future of Yahoo!

## The Decision

Sitting in their tiny office on the Stanford campus, Jerry and Dave shared a late-night pepperoni and mushroom pizza as they explored their options and tried to come to a decision. It was already getting pretty late, and they only had until noon the next day to make their decision.

Yang took a bite from his pizza as he looked over the terms sheet that Sequoia had given them.

> We have some pretty tough decisions to make, and Michael has really forced the issue now with this 24-hour deadline. As I see it, we have a couple of options. The first is to accept Sequoia's offer and launch Yahoo! as our own company. We would be giving up a significant percentage of Yahoo!, but we really need the money if we are going to survive. Moritz and the rest of the resources at Sequoia could also prove to be invaluable as we try to assemble the rest of our management team.

---

[15] Jerry Yang, "Found You on Yahoo!" *Red Herring,* October 1995, p. 3.

[16] Michael Moritz, personal interview, November 10, 1998.

[17] Anthony Perkins, *Red Herring,* June 1996.

Our second option is to accept corporate sponsorship. This would allow us to get the funding we need and still retain 100 percent ownership of Yahoo!. However, I am worried about selling out to corporate America. We were fortunate to be able to develop our site in an educational setting as a noncommercial free site. I am afraid if we accept the corporate sponsorship, it will taint Yahoo!'s image.

Finally, we could agree to merge with an existing corporation. The word is that Netscape is pretty close to their IPO, and Architext has some really big time investors behind it. If we merge with Netscape or Architext in exchange for stock options, it could mean a lot of money for us in the next couple of years.

Filo got up from his seat and kicked aside some of the empty pizza boxes that had started to accumulate. He walked over to Yahoo!'s tiny office window and stared at Stanford's Hoover Tower, which was barely visible in the distance.

It's true that we could make some money if we sell to Netscape or Architext, but we would have to give up primary control of Yahoo! if we did. We would never know what we could have done if we would have maintained control of the site ourselves.

There is also a fourth option you forgot to mention. I'm excited by Sequoia's offer, but I'm wondering if maybe we are giving up too much of our company. A fourth option could be to not decide tonight and look for better terms with another VC firm. I know Michael said that we should decide quickly, but I would hate to give up 25 percent of our company, only to find out in a week that another firm would have offered us $3 million for the same percentage. I know that time is really important, and we like working with Michael Moritz. On the other hand, I don't want to be regretting our decision two months from now.

As they grappled with the alternatives facing them, Filo and Yang began to envision life outside of the Stanford trailer in which Yahoo! was born. It was well past 2 A.M., and they had to make a decision in less that ten hours. What should they do?

## Questions

1. What makes Yahoo! an attractive opportunity (and not just a good idea)?
2. How will Yahoo! make money (i.e., business model)?
3. Identify the major risks in each of these categories: technology, market, team, and financial. Rank order them.
4. What are the advantages and disadvantages of each of the funding options they could pursue? Which one do you recommend?

## Video Resources

Visit http://techventures.stanford.edu to view a video of the founders of Yahoo! and others discussing the outcome of the case.

**EXHIBIT 1  Yahoo!** Founders and Potential Investor

*Jerry Yang*

Jerry Yang was a Taiwanese native who was raised in San Jose, California. He co-created the Yahoo! online guide in April of 1994. Jerry took a leave of absence from Stanford University's electrical engineering PhD program after earning both his BS and MS degrees in electrical engineering from Stanford University.

*David Filo*

David Filo, a native from Moss Bluff, Louisiana, co-created the Yahoo! online guide in April 1994 and took a leave of absence from Stanford University's electrical engineering PhD program in April 1995 to co-found Yahoo!, Inc. Filo received a BS degree in computer engineering from Tulane University and a MS degree in electrical engineering from Stanford University.

*Michael Moritz, Partner, Sequoia Capital*

Moritz was a general partner at Sequoia Capital since 1988 and focused on information technology investments. Moritz served as a director of Flextronics International and Global Village Communication, as well as several private companies. Between 1979 and 1984, Moritz was employed in a variety of positions by Time, Inc. Moritz had an MA degree in history from Oxford University and an MBA from the Wharton School.

**Appendix** Selected Excerpts from the Yahoo! Business Plan.

Yahoo!'s first business plan was developed by Tim Brady as part of a course project at the Harvard Business School. The plan was continuing to evolve during discussions between Jerry Yang and David Filo at Yahoo! and Michael Moritz of Sequoia Capital. For this case, the company has provided excerpts of this business plan that are not proprietary.

The case writers thank Mr. J. J. Healy, director of corporate development, and others at Yahoo! for their efforts in providing this original archival information to enhance the learning experience of future entrepreneurs.

**Business Strategy**

Yahoo!'s goal is to remain the most popular and widely used guide to information on the Internet. The Internet is in a period of market development characterized by extremely high rates of both user traffic growth and entry of new companies focused on various products and services. By virtue of its early entry, Yahoo! has developed its current position as the leader in this segment. Yahoo!'s ability to expand its position and develop long-term, sustainable advantages will depend on a number of things. Some of these relate to its current position and others relate to its future strategy.

Today, Yahoo! solves the main problem facing all Internet users. It is next to impossible for users, faced with millions of pieces of information scattered globally on the Internet, to easily find that what is relevant to them without a guide like Yahoo! Not only is the amount of information huge, it is expanding almost exponentially.

All enhancements to Yahoo! will be governed by the goal of making useful information easy to find for individuals.

We believe that Yahoo's enormous following has been generated by the following list:

- Yahoo! was the first company to create a fast, comprehensive and enjoyable guide to the Internet, and in so doing, built a strong brand early and created momentum.

- The unique interest-area based structure of Yahoo! makes it an easier and more enjoyable way for the user to find relevant information than the classic search engine approach where key words and phrases are used as the starting point.

- Through its editorial efforts, Yahoo! has continually built a guide which is noticeably better than its competition through a combination of comprehensiveness and high quality.

The company will focus on the directory and the guide business and generate revenue from advertising and sponsorship.

Yahoo!'s strategy is to:

- **Continue to build user traffic and brand strength** on the primary server site through product enhancements and extensions as well as through an aggressive marketing communications program.

- **Develop and integrate the leading technology** required to maintain a leadership position. Underlying the extremely appealing guide is Yahoo!'s scaleable core technology in search engine, database structure, and communication software. These core technologies are relevant to the user's experience to the extent that it enables Yahoo! customers' access to a broader array of high quality information in an intuitive way, faster than any competitors product. Yahoo! is discussing a full license to advanced web-wide search engine technologies, web-wide index data, and crawler services with Open Text of Waterloo, Canada. Yahoo! will be the first guide with a seamless integrated directory/web-wide search product. The proposed agreement with Open Text also includes ongoing joint development of advanced search and database technologies leveraging the strengths of both companies. All jointly developed products will be distributed by Yahoo! allowing the company to continue to introduce advanced features on a regular and aggressive basis.

- **Extend the reach to a broader audience** through establishment of contractual relationships with Internet access providers such as MSN, America Online, and Compuserve and very popular web sites.

- **Extend the reach and appeal to international users** through partnerships with international access providers who can operate foreign mirror sites for Yahoo and add localization in the form of foreign language, local advertisers, and local content.
- **Retain the users ("readership") of Yahoo!** through constant enhancements to the content and interface of the guide.
- **Rapidly extend the product line** by introducing regional guides, vertical market guides, and more importantly, individually personalizeable guides. Our intention is to be the first to market in all or most of these categories and outrun our competition by constantly "changing the competitive rules and targets." Our introduction of personalized guides will be a first in the market and will leverage core technology owned both internally as well as through our license with Open Text.

## Market Analysis

The Internet, whose roots trace back almost 20 years, is experiencing a period of incredibly rapid growth in the area of online access base and user population. According to IDC and a recent report by Montgomery Securities, there are approximately 40 million users of the Internet, a majority using it only for email. However, it is estimated that about 8 million people have access to the Internet and World Wide Web. Most of these access the Web from the workplace because of the availability of high bandwidth hardware and communications ports there. It is expected that over the next two to four years as higher bandwidth modems, home-based ISDN lines and cable modems are adopted, that both the growth and penetration of Web access into the home will increase dramatically. IDC estimates that by 2000, 40 percent of the homes and 70 percent of all businesses in the United States will have access to the Internet. In the Western European and Japan markets, the comparable penetration rates might be as high as 25 percent and 40 percent respectively. If this holds true, there will be as many as 200 million users on the Internet and Web by the year 2000.

## Market Segmentation and Development

We believe that between now and the year 2000 there will be three principal user groups driving the growth of the Web:

- Large businesses using the Internet for both internal wide area information management and communication as well as intrabusiness communication and commerce.
- Small home based businesses using it for retrieval of information relevant to the business as well as for vendor communication and commerce.
- The individual user/consumer using it initially to find and access information which is relevant to their personal entertainment and learning and later to make purchases of products and services.

We also believe that the evolution of the Internet will include three stages of market development:

- Availability and proliferation of enabling technology.
- Establishment of widespread access and communication services.
- Widespread distribution of high value content.

We are currently in the first stage of market development consisting primarily of infrastructure building and including rapid growth in the adoption and sale of computer, network, and communication products and entering into the second stage involving the initial establishment of "access" service based businesses.

**Internet Market Size**

Estimates of the amount of current and projected revenue for Internet related business vary. However, primary research conducted by both Montgomery Securities as well as Goldman Sachs indicate that the total served market for Internet hardware, software, and services will total approximately $1B in 1995, up from approximately $300M in 1994. Projections are that these categories might grow to a total of $10B by the year 2000. Several research firms including Forrester and Alex Brown & Sons have estimated the revenues to be produced by Web-based advertising at approximately $20M in 1995, $200M in 1996, and over $2B by the year 2000.

**Market Trends**

During the current, rapidly expanding stages of market and industry development, the following trends are clear:

- There is large scale adoption of enabling technology in the areas of network hardware and software, as well as communication hardware and software. The World Wide Web with its inherent support of multimedia begs for the adoption of higher and higher bandwidth platform and communication hardware and software.
- Telecommunication companies and newly entering Internet access providers are rushing to put in place basic "hook-ups" in high bandwidth form.
- The price for high-speed computer and communication "port" hardware and software of adequate bandwidth to support acceptable levels of transport and display is still somewhat high. Partly for this reason, the adoption of fully capable ports onto the Web is still principally occurring at businesses.
- With the availability of 28.8K baud modems, ISDN lines and high performance/low price personal computers, home adoption of Internet access is on the rise and slated to have extremely high growth over the next five years. Adoption of cable modems could accelerate this trend.

- Formerly closed network online services such as America Online, Compuserve, and Prodigy are now offering Internet access and opening up their services. Other companies such as Microsoft as well as divisions of MCI, AT&T, and others are attempting to put in place Internet online services in which a range of programming content is presented.
- Companies such as Yahoo! which provide means to navigate the Web are growing rapidly as measured by amount of end user traffic.
- These high traffic sites already provide a high volume platform for delivering electronic advertising.

During this stage, and sustainably for all stages to come, there is one fundamental need which users have: The location of meaningful information easily and quickly on this large and exponentially growing source called the Internet.

### Competition

Yahoo! intends to effectively beat any emerging competitors by:

- Establishing broader distribution earlier than any other competitor in order to maintain the Yahoo! guide as the most widely used in its class.
- Broadening the product line faster than the competition through the introduction of vertical market focused guides and personalizeable editions of the guide.
- Staying ahead of the competition with regular core product updates which continue to make it faster, easier to use, and more effective.
- Delivering high quality audiences and compelling results to advertisers.

### Risks

The main risks facing Yahoo! are:

- *The ability to increase traffic and enhance the Yahoo! brand.* Management believes it can achieve both these goals.
- *Ability to introduce key new products faster and better than the competition.* We believe that our current core technologies and platform will allow us to do this if supplemented by funded expansion of product development and marketing functions.
- *Ability to develop an international presence and leading brand internationally before the competition.* At the present time, Yahoo! is being pursued by a number of very high visibility and capable international affiliates. The funded addition of limited marketing and business development resources will allow us to respond to these opportunities in a timely way.
- *The introduction of competitive products internally developed by access providers.* While there is no assurance that this will not happen, we have secured relationships with several of the leading providers already in which the Yahoo! product is featured and are in advanced discussions with others.

We believe that many of the access providers already respect Yahoo!'s strong brand, comprehensive guide and focus and are concluding that they will not be inclined to reinvent this late in lieu of a mutually favorable affiliate business relationship with Yahoo!.

- *Ability to scale our support of both the traffic through our main site as well as mirror sites of our affiliates.* If the demands of traffic outgrow the bandwidth of servers we install, then response rates might go down and lead to customer dissatisfaction. Yahoo! has successfully scaled and operated its server site. We believe we will be able to support the needed growth.

- *That the growth of the Internet industry as a whole slows significantly, or that the adoption of the Web as a significant platform for advertising does not grow as projected.* These are both out of Yahoo!'s control. However, the company believes that the industry is in a secure phase of adoption which should fuel growth.

## Yahoo!'s sustainable advantages

The Internet is in a period of market development characterized by extremely high rates of both user traffic growth and entry of new companies focused on various products and services. By virtue of its early entry, Yahoo! has developed its current position as the leader in its segment. Yahoo!'s ability to sustain and grow its position will depend on a number of things. Some of these relate to its current core advantages and others relate to future execution of its strategy.

At present, Yahoo!'s core strategic advantages include:

- *It's strong brand.* The company executed early and well with its unique, context focused, quick and intuitive guide and benefited from the widespread adoption of the Yahoo! product. The guide is the standard in the world of Web navigation.

- *Yahoo!'s scalable core technology in search engine, database structure, and communication software.* These core technologies are relevant to the user's experience to the extent that it enables the Yahoo! customer's access to a broader array of high quality information in an intuitive way, faster than any competitor's product.

# BARBARA'S OPTIONS

## Introduction

Barbara Arneson strolled through the campus of the University of Maryland at College Park on a Spring evening in 2014. She often came to the quad at the end of a day for some quiet reflective time. Tonight she was mulling over her career options and the path her life would take in the next few years. Graduation was only five days away, and tomorrow Barbara would pick up her parents at the airport for a short visit and the ceremony. She hoped to be able to share her career decision with them and then relax over the next few days.

Having completed an undergraduate degree in biology a year prior, Barbara would soon be receiving her masters degree in computer science from Maryland before beginning a career in high technology. Barbara felt lucky that she had been offered a number of career options, thanks to the strong high-tech economy and growth of investment in the new field of bioinformatics, which applied software and Internet technology to the process of identifying and using genetic information for the life sciences industries. Barbara's personal objective was to work in product development and eventually move into management and maybe someday start her own company. As an interim step, she thought she might return to graduate school for an MBA in a few years.

On this beautiful evening, however, Barbara had to make a decision between two attractive job offers.

## Barbara's Dilemma

Barbara had interviewed extensively with high-technology companies, and had decided BioGene Systems, Inc., and InterWeb Genetics Corp. were her top choices. She had hoped to receive an offer from at least one of them, but had received offers from both. Now she had a tough decision.

BioGene was a 7-year-old company that was successful and had grown rapidly. Its product-development group was highly regarded technically, and Rasha Motwani, to whom Barbara would report, had more than 10 years of development experience in several product areas in which Barbara was interested. She liked Rasha and felt she would learn a lot from her.

InterWeb was a start-up that had been funded about a year ago by two venture capital investors, both of whom had successfully funded technology companies. The InterWeb team was hard at work on its first product, which would launch in about a year. Barbara would be joining a team of about 10 engineers, most of whom had extensive experience in the areas relating to the product they were developing. The technical team leader was Robert Jackson, one of the company founders, who was only a few years older than Barbara and had a reputation as a technical "visionary."

Barbara had been trying to decide for several days. As she strolled toward her dorm, she reflected on her thoughts. "Okay, I've been trying to make a decision on the basis of my key priorities, namely the types of projects I would be working on, the quality of the people I would be working with, and the opportunities for personal growth. Although BioGene and InterWeb are not directly comparable—each has potential strengths and weaknesses—the fact is that I think I would be equally happy with either one. For me, the decision is a toss-up. I guess the only way to determine which is better is to evaluate the financial offers. Because both proposed similar salaries and benefits, that means analyzing the stock option offers."

## The Stock Option Packages

Not all companies offer stock options to new college graduates. Because of Barbara's success at school and a hot job market in bioinformatics, both BioGene and InterWeb had included a stock option package in their offers.

A stock option gives an individual the right to purchase, during a fixed time period called the "term," a certain number of shares of stock from the company at a fixed price, called the "exercise price." The option expires at the end of the term, but it can be "exercised" (or bought) all or partially during the term, usually subject to certain conditions such as "vesting." Options have no financial risk to the employee—if the value of the stock remains below the exercise price, he or she need not ever exercise the option.

BioGene had offered Barbara options for 6,000 shares at an exercise price of $16.00 per share. BioGene had gone public in June 2008 at $10 per share, and the stock was currently selling for roughly $16. In extending the offer to Barbara, Karen Hershfield, manager of recruiting for BioGene, had said, "We have a proven record of rapid, profitable growth, and we expect that kind of growth to continue. You should receive a handsome return on this option package."

InterWeb had offered 60,000 shares at an exercise price of $0.10 per share. Because the company was private, this price reflected a favorable pricing decision at the time of the venture capital investment received by the company. Robert Jackson, in discussing the offer with Barbara, had commented, "The great thing about going with a start-up is that if it is successful, everybody gets rich. Our business plan shows that we should be in a position to do our IPO (initial public offering) in 3 or 4 years, and because companies usually go public at $10 to $15 per share, you can see that this option could be worth almost a million dollars!"

Both options had identical restrictions. Vesting was 25% per year with a term of 4 years. This meant that at the end of one year of employment, Barbara had the right to exercise 25% of the shares at her discretion any time. At the end of two years of employment, she would vest for another 25%, and so on. If at any time she left the company, she could, within 90 days, exercise any options for which she was vested, but any unvested options would terminate. Any unexercised portion of the option would expire at the end of 10 years from the start of employment.

Having decided that she would be equally satisfied with joining either company, Barbara was understandably excited by the prospect of big financial gains on stock options. She had even begun day-dreaming about what she could do with a financial windfall—travel abroad for a year, buy a new car for her parents, and pay for an MBA without worrying about 2 years of no salary and huge loan payments. Because she had learned a lot about financial analysis in her entrepreneurship courses, she had obtained and analyzed financial data from both companies, as shown in Exhibits 1 and 2. She also had analyzed the opportunity and strategies of both companies, and felt each had excellent prospects of achieving its objectives. She had examined stock market data for public high-tech companies and knew that the price/earnings (PE) ratios of bioinformatics companies averaged 25. InterWeb had offered a lot more shares than BioGene, but there was a higher element of risk. She knew that many start-ups failed to be successful.

She recalled that autumn day 5 years ago when her parents had dropped her off at school for the start of her freshman year. When she picked them up at the airport tomorrow, she wanted to share her career decision with them and make them proud of her.

## Questions

1. What is the number of shares outstanding at BioGene as of May 31, 2014? What is its current PE ratio? Why do you think it is higher than the current average of other bioinformatics companies (Hint: consider the recent annual growth rates of revenues and profits)?

2. What is Barbara's potential percentage ownership in each firm?

3. Compare the firms in 4 years (i.e., 2018) when the stock options will be fully vested. Assuming Barbara remains employed until that time, which stock option offer is better? Make sure to include the cost of the stock options and state all critical assumptions.

4. In addition to compensation matters, what other factors would you suggest Barbara consider in making her decision?

**EXHIBIT 1** BioGene Systems profit and loss history.

|  | FY10 | FY11 | FY12 | FY13 | FY14 |
|---|---|---|---|---|---|
| Revenue: | 10.1 | 17.1 | 25.6 | 42.4 | 74.6 |
| Cost of revenue | 4.0 | 6.8 | 10.2 | 17.0 | 29.8 |
| Gross margin | 6.1 | 10.3 | 15.4 | 25.4 | 44.8 |
| Expenses: | | | | | |
| Engineering | 1.2 | 2.1 | 3.1 | 5.1 | 9.0 |
| Marketing | 2.5 | 4.3 | 6.4 | 10.6 | 18.7 |
| G&A | 0.6 | 1.0 | 1.5 | 2.5 | 4.5 |
| Total expenses | 4.3 | 7.4 | 11.0 | 18.2 | 32.2 |
| Pretax profit | 1.7 | 2.9 | 4.4 | 7.2 | 12.7 |
| Taxes | 0.7 | 1.2 | 1.7 | 2.9 | 5.1 |
| After-tax profit | 1.0 | 1.7 | 2.6 | 4.3 | 7.6 |
| EPS (earnings per share) | $0.06 | $0.08 | $0.12 | $0.19 | $0.33 |

*Note:* (1) All numbers in millions except EPS. (2) Stock is traded on NASDAQ. Closing price on 5/31/14 was $16.25. (3) End of fiscal year for 2014 is June 30. FY14 numbers are estimates by stock market analysts and consistent with guidance by management.

**EXHIBIT 2** InterWeb proforma profit and loss projections from business plan.

|  | FY13 | FY14 | FY15 | FY16 | FY17 | FY18 |
|---|---|---|---|---|---|---|
| Sales | 0.0 | 0.0 | 5.0 | 20.0 | 41.0 | 62.0 |
| Cost of sales | 0.0 | 0.0 | 2.0 | 8.0 | 16.4 | 24.8 |
| Engineering | 0.7 | 1.0 | 1.5 | 2.4 | 4.9 | 7.4 |
| Marketing | 0.3 | 0.5 | 1.3 | 5.0 | 10.3 | 15.5 |
| G&A | 0.1 | 0.2 | 0.3 | 1.2 | 2.5 | 3.7 |
| Total expenses | 1.1 | 1.7 | 3.1 | 8.6 | 17.7 | 26.6 |
| Pretax profit | −1.1 | −1.7 | 0.0 | 3.4 | 7.0 | 10.5 |
| Taxes | 0.0 | 0.0 | 0.0 | 0.2 | 2.8 | 4.2 |
| After-tax profit | −1.1 | −1.7 | 0.0 | 3.2 | 4.2 | 6.3 |

*Note:* (1) 118.6 million shares outstanding as of 5/31/14. Management's business plan requires no additional venture capital or other funding. (2) End of fiscal year for 2014 is June 30.

# ARTEMIS IMAGES

Christine Nazarenus tried to retain her optimism. Thirteen had always been a lucky number for her, but Friday, the thirteenth of July, 2001, had the earmarks of being the unluckiest day of her life. She was more than disappointed. She was shattered. Yet she knew that she had hard facts, not just gut feel, that offering images and products on the World Wide Web was the wave of the future. She was sure that the management team she had put together had the creativity and skills to turn her vision into reality. Managing her own company had seemed the obvious solution, but she hadn't counted on how overwhelming the start-up process would be. Now, two years later, she was trying to figure out what went wrong and if the company could survive.

It had been so clear on day one. Archived photographs and images had tremendous value if they could be efficiently digitized and catalogued. Sports promoters and publishers had stores of archived information, most of it inaccessible to those who wanted it. Owners and fans represented only part of the untapped markets that the Internet and digital technology could serve. She had conceived a simple business model: digitize documents using the latest technology, tag them with easy-to-read labels, and link them to search engines for easy retrieval and widespread use. But over the ensuing months, so many factors affected the look, feel and substance of the company that Artemis Images would become.

So many things seemed outside her control that she wondered how she could have been so sure of herself back in February of 1999. Enthusiastically, Chris had approached a number of friends and acquaintances to help in the formation of a new "dot.com" company that seemed a sure bet. Frank Costanzo, a former colleague from Applied Graphics Technologies (AGT), shared Chris's enthusiasm, as did long-time friend George Dickert. George, in turn, contacted Greg Hughes, who was enrolled in a Business Planning course. Grateful for the opportunity to help launch a real company, Greg took the idea and honed it as part of a class assignment. The plan was a confirmation of Chris's confidence in the venture. But as she looked over the original plan, she knew there was a lot of work yet to do. Greg understood the business idea, but he didn't understand the work involved to actually run a business. George and Frank understood digital technology and project management, but, like Chris, had never launched, much less worked for, a start-up company. Chris knew that she had the technology and talent she needed

© 2002 by Joseph R. Bell, University of Northern Colorado, and Joan Winn, University of Denver. Published in *Entrepreneurship Theory & Practice,* 28(2) Winter 2003. The authors wish to thank Chris Nazarenus and the staff of Artemis Images for their cooperation in the preparation of this case. Special thanks to Herbert Sherman, Southampton College, Long Island University, and Dan Rowley, University of Northern Colorado. This case is intended to stimulate class discussion rather than to illustrate the effective or ineffective handling of a managerial situation. *All events and individuals in this case are real, but some names may have been disguised.*

and felt confident that the four friends could construct a business model that would put Artemis ahead of the current image providers. Greg's business plan looked like the perfect vehicle to appeal to investors for the funds they needed to proceed.

## The Business Idea

In 1999, Chris had been working for three years as VP-Sales out of the Colorado office of AGT, a media management company that provided digital imaging management and archiving services for some of the largest publishers and advertisers in the world. AGT had sent Chris to Indianapolis to present a content management technology solution to the Indianapolis Motor Speedway Corporation (IMSC) as it prepared marketing materials for the 2001 Indy 500. IMSC is the host of the 80-plus-year-old Indy 500, the largest single-day sporting event in the world, NASCAR's Brickyard 400, the second-largest single-day sporting event in the world, and other events staged at the track. Chris's original assignment was a clear one: IMSC needed to protect its archive of photographs, many of which had begun to decay with age. The archive included five million to seven million photographs and dynamically rich multimedia formats of video, audio, and in-car camera footage.

Chris discovered that the photo archives at IMSC were deluged with requests (personally or via letters) from fans requesting images. She was amazed that a relatively unknown archive had generated nearly $500,000 in revenues in 1999 alone. Further discussions with IMSC researchers revealed that requests often took up to two weeks to research and resulted in a sale of only $60 to $100. However, IMSC was not in a position, strategically or financially, to acquire a system to digitize and preserve these archives. Not willing to leave the opportunity on the table, Chris asked herself, "What is the value of these assets for e-commerce and retail opportunities?" Without a doubt, IMSC and some of her other clients (Conde Nast, BBC, National Motor Museum) would be prime customers for digitization and content management of their collections.

Chris knew that selling photos on the Internet could generate substantial revenue. She conceived of a business model where the system would be financed through revenue-sharing, rather than the standard model where the organization paid for the system up front. IMSC was interested in this arrangement, but it was outside the normal business practices of AGT. AGT wanted to sell systems, not give them away. They couldn't see the value of managing other organizations' content.

As Chris told the story, her visit to the archives at IMSC was her *Jerry Maguire* experience. In the movie, Jerry is sitting on the bed when everything suddenly becomes clear and now he must pursue his dream. Like Jerry, Chris believed so passionately that her idea would bear fruit that when AGT turned down Chris's request for the third time, she quit her job to start Artemis Images on her own.

## Building a Team

When AGT was not interested in Chris's idea of on-site digitization and sale of IMSC's photo archives, Chris was not willing to walk away from what she saw as a gold mine. She contacted her friends and colleagues from AGT. Swept up in the dot.com mania, Chris named her company "e-Catalyst." e-Catalyst was incorporated as an S-corporation on May 3, 1999, by a team of four people: Christine Nazarenus, George Dickert, Frank Costanzo, and Greg Hughes. (See Exhibit 1 for profiles of these partners.) Expecting that they would each contribute equally, each partner was given a 25 percent interest in the company. Chris fully expected them to work as a team, so no formal titles were assigned, largely as a statement to investors that key additions to the team might be needed and welcomed. As another appeal to potential investors—and to broaden the team's expertise—Chris and George put together a roster of experts with content management, systems and technology experience as their first advisory board. Greg's professor and several local business professionals agreed to serve on the board of advisors, along with an Indy 500 winning driver-turned-entrepreneur, and Krista Elliott Riley, president of Elliott Riley, the marketing and public relations agency that represented Indy 500 and Le Mans Sports Car teams and

**EXHIBIT 1**  Artemis Images management team 1999–2000.

Christine Nazarenus, 34, was formerly vice president of national accounts for AGT, one of the top three content management system providers in the world securing million dollar deals for this $500 million company. She is an expert in creating digital workflow strategies and has designed and implemented content management solutions for some of the largest corporations in the world including Sears, Conde Nast, Spiegel, Vio, State Farm, and Pillsbury. Ms. Nazarenus has extensive general management experience and has managed a division of over one hundred people. Chris holds a BA in communications from the University of Puget Sound.

George Dickert, 32, most recently worked as a project manager for the Hibbert Group, a marketing materials distribution company. He has experience with e-commerce, Web-enabled fulfillment, domestic and international shipping, call centers and CD-ROM. He has overseen the implementation of a million-dollar account, has managed over $20 million in sales, and has worked with large companies including Hitachi, Motorola, ON Semiconductor, and Lucent Technologies. Mr. Dickert has an MBA from the University of Colorado. George and Christine have been friends since high school.

Frank Costanzo, 40, is currently a senior vice president at Petersons.com. Petersons.com has consistently been ranked as one of the top one hundred sites worldwide. Mr. Costanzo is an expert in content management technology and strategy and was previously a vice president at AGT. Mr. Costanzo has done in-depth business analysis and created on-site service solutions in the content management industry. He has worked on content management solutions for the world's top corporations including General Motors, Hasbro, Bristol-Meyers Squibb, and Sears.

Greg Hughes, 32, is currently a senior sales executive with one of the largest commercial printers in the world. Mr. Hughes has 10 years' sales experience and has sold million-dollar projects to companies like US West, AT&T, R. R. Donnelly, and MCI. His functional expertise includes financial and operational analysis, strategic marketing, fulfillment strategies and the evaluation of start to finish marketing campaigns. Mr. Hughes has an MBA from the University of Colorado.

drivers. Chris felt confident that her team had the expertise she needed to launch a truly world-class company.

Chris and George quit their jobs and took the challenge of building a company seriously. They contacted one of the Rocky Mountain region's oldest and most respected law firms for legal advice. They worked with two lawyers, one who specialized in representing Internet companies as general counsel and one who specialized in intellectual property rights. With leads from her many contacts at AGT, Chris contacted venture capitalists to raise money for the hardware, software licensing, and personnel costs of launching the business.

The dot.com bust of 2000 did not make things easy. Not wanting to look like "yet another dot.com" in search of money to throw to the wind, Chris and her team changed their name to Artemis Images. Artemis, the Greek goddess of the hunt, had been the name of Chris's first horse as well as her first company, Artemis Graphics Greeting Cards, her first entrepreneurial dabble at the age of 16. Chris had always been enthralled with beautiful images.

## Artemis Images's Niche

In her work at AGT, Chris had observed that many organizations had vast stores of intellectual property (photos, videos, sounds and text), valuable assets often underutilized because they exist in analog form and may deteriorate over time. Chris's vision was to preserve and enable the past using digital technology and the transportability of the World Wide Web. Chris envisioned a company that would create a digitized collection of image, audio and video content that she could sell to companies interested in turning their intellectual property into a source of revenue.

Publishers and sports promoters were among the many organizations with large collections of archived photos and videos. Companies like Boeing, General Motors, and IMSC are in the business of producing planes, cars, or sporting events, not selling memorabilia. However, airplane, car, and sports fans are a ready market for photos of their favorite vehicle or videos of their favorite sports event.

Proper storage and categorization of archived photos and videos is complex and expensive. In 2000, the two common solutions were to sell the assets outright or to set up an in-house division devoted to managing and marketing them. Most organizations were unwilling to sell their assets, as they represented their priceless brand and heritage. Purchasing software and hiring specialized personnel to digitize and properly archive their assets was a costly proposition that lay beyond the core competence of most companies. Chris's work with AGT convinced her that there were literally thousands of companies with millions of assets that would be interested in a company that would digitize and manage their photo and video archives.

Chris understood a company's resistance to selling its archives, and the high cost of obtaining and scanning select images for sale. However, she also understood the value to an organization of having its entire inventory digitized, thus creating a permanent history for the organization. She proposed a revenue sharing model whereby Artemis Images would digitize a client's archives but would not take ownership. Instead, her company would secure exclusive license to the archive, with 85 percent of all revenue retained by Artemis Images and 15 percent paid to the archive owner. She expected that the presence of viewable archives on the Artemis Images website would lure buyers to the site for subsequent purchases.

The original business model was a "B2C" (business-to-consumer) model. Starting with the IMSC contract, Artemis Images would work with IMSC to promote the Indy 500 and draw the Indy race fans to the Artemis Images website. Photos of the current-year Indy 500 participants—and historical photos including past Indy participants, winners, entertainers, celebrities (e.g., Arnold Palmer on the Indy golf course)—would be added to IMSC's archived images and sold for $20 to $150 apiece to loyal fans. A customer could review a variety of photo options on the Artemis website, then select and order a high-resolution image. The order would be secured through the Web with a credit card, the image transferred to the fulfillment provider, and a hard copy mailed to the eager recipient. The website was sure to generate revenue easier than IMSC's traditional sales model of the past.

Having established the model with IMSC content in the auto racing market, Chris and George built the business plan around obvious market possibilities that might appeal to a wider range of consumers and create a comprehensive resource for stock photography. Since the Artemis Images team had prior business dealings with two of the three largest publishers in the world, publishing was the obvious target for future contracts. Future markets would be chosen similarly, where the Artemis team had established relationships. These markets would be able to build on the archive already created and would bring both consumer-oriented content and saleable stock images. Greg made a list of examples of some industries and the content that they owned:

- Sports: images of wrestling, soccer, basketball, bodybuilding, football, extreme sports
- Entertainment: recording artists, the art from their CDs, movie stars, pictures of events, pictures from movie sets
- Museums: paintings, images of sculpture, photos, events
- Corporations: images of food, fishing, planes, trains, automobiles
- Government: coins, stamps, galaxies, satellite imaging

As Chris and George worked with Greg to put together the business plan, they began to see other revenue-generating opportunities for their virtual-archive company. Customers going to IMSC or any other Artemis client's website would be linked to Artemis Images's website for purchase of photos or videos. Customer satisfaction with image sales would provide opportunities to sell merchandise targeted to specific markets and to syndicate content to other websites. For motor sports, obvious merchandise opportunities would include T-shirts, hats, and model cars. For landscapes, it might be travel packages or hiking gear.

Corporate customers might be interested in software, design services, or office supplies. Unique content on Artemis Images's website could be used to draw traffic to other companies' sites. Chris and her team planned to license the content on an annual basis to these sites, creating reach and revenues for Artemis Images.

Another potential market for Artemis Images lay in the unrealized value of the billions of images kept by consumers worldwide in their closets and drawers. These images were treasured family heirlooms which typically sat unprotected and underutilized. Consumers could offer their photographs for sale or simply pay for digitization services for their own use. If just 10% of the U.S. population were to allow Artemis Images to digitize their archive and half of these people ordered just one 8" × 10" print, Artemis Images could create a list of 25 million consumers and generate revenues of approximately $250 million. Because images suffer no language barriers, the worldwide reach of the Internet and the popularity of photography suggested potential revenues in the billions.

Working together on the business plan, the Artemis team brainstormed ways they could attract customers to the Artemis Images site by providing unique content and customer experiences. A study by Forrester Research analyzed the key factors driving repeat site visits and found that high-quality content was cited by 75 percent of consumers as the number one reason they would return to a site. The Artemis team wanted to create a community of loyal customers through additional unique content created by the customers themselves. This would include the critical chats and bulletin boards that are the cornerstone of any community-building program. Artemis Images could continuously monitor this portion of the site to add new fan experiences to keep the experience "fresh." Communities would be developed based on customer interests.

As the company gained clients and rights to sell their archived photos and videos, Artemis would move toward a "B2B" (business-to-business) model. Chris and George knew marketing managers at *National Geographic,* CMG World Wide, the BBC, Haymarket Publishing (includes the Formula 1 archive), Conde Nast, and International Publishing Corporation. These large publishers controlled and solicited a wide range of subject matter (fashion, nature, travel, hobbies, etc.) yet often had little idea of what existed in their own archives or had difficulty in getting access to it. Finding new images was usually an expensive and time-consuming proposition. Artemis Images could provide the solution. For example, Conde Nast (publisher of *Vogue, Bon Appetit, Conde Nast Traveler, House & Garden,* and *Vanity Fair*) might like a photo for its travel magazine from the *National Geographic* archives. They would be willing to pay top dollar for classic stock images, given the number of viewers who would see the image. Price-per-image was typically calculated on circulation volume, much like royalty fees on copyrighted materials. Similarly, advertising agencies use hundreds of images in customer mockups. For example, an agency may desire an image of a Pacific island. If Artemis Images held the rights to Conde Nast and *National Geographic,* there might be hundreds of Pacific island photos from which to choose. As with the B2C concept, a copy of the image would be transferred through the Web with a credit card or on account, if adequate bandwidth were available (only low-resolution images would be available to view initially), or via overnight mail in hard copy or on disk.

The transition from B2C to B2B seemed a logical progression, one that would amass a large inventory of saleable prints and, at the same time, draw in larger per-unit sales. The basic business model was the same. Artemis would archive photos and videos that could be sold to other companies for publication and promotion brochures. Chris and George expected that this model could be replicated for other vertical markets including other sports, nature, entertainment, and education.

While the refocus on the B2B market seemed a surer long-term revenue stream for the company, both B2B and B2C were losing favor with the investing community. Chris and George refocused the business plan as an application service provider (ASP). With the ASP designation, Artemis Images could position itself as a software company, generating revenue from the licensing of its software processes. In 2000, ASPs were still in favor with investors.

Artemis Images's revenue would come from three streams: (1) sales of images to businesses and consumers, (2) syndication of content, and (3) sales of merchandise. Projected sales were expected to exceed $100 million within the first four years, with breakeven occurring in year three. (See Exhibits 2, 3, and 4 for projected volume and revenues.)

To implement this strategy, Artemis Images, Inc., needed an initial investment of $500,000 to begin operations, hire the team, and sign four additional content agreements. A second round of $1.5 million and a third round of $3 million to $8 million (depending on number of contracts) were planned, to scale the concept to 28 archives and over $100 million in assets by 2004. (See Exhibit 5 for funding and ownership plan.)

## The Content Management Industry

According to GISTICS, the trade organization for digital asset management, the content management market (including the labor, software, hardware, and physical assets necessary to manage the billions of digital images) was projected to be a $2 trillion market worldwide in the year 2000 (1999 Market Report). Content could include images, video, text and sound. Artemis Images intended to pursue two subsets of the content management market. The first was the existing stock photo market, a business-to-business market where rights to images were sold for limited use in publications such as magazines, books, and websites. Deutsche Bank's Alex Brown estimated this to be a $1.5 billion market in 2000. Corbis, one of the two major competitors in the digital imaging industry, estimated it to be a $5 billion market by 2000.

**EXHIBIT 2** Anticipated sales volume and on-site operations.

**Volumes 2001**

|  | Jan-01 | Feb-01 | Mar-01 | Apr-01 | May-01 | Jun-01 | Jul-01 | Aug-01 | Sep-01 | Oct-01 | Nov-01 | Dec-01 | Total |
|---|---|---|---|---|---|---|---|---|---|---|---|---|---|
| Consumer Photos | 0 | 0 | 7,500 | 7,500 | 27,000 | 9,000 | 9,000 | 22,500 | 18,000 | 9,000 | 4,500 | 22,500 | 136,500 |
| Stock Photos | 0 | 0 | 0 | 3,750 | 4,500 | 5,250 | 6,750 | 7,500 | 9,000 | 9,750 | 10,500 | 11,250 | 68,250 |
| Subtotal | 0 | 0 | 7,500 | 11,250 | 31,500 | 14,250 | 15,750 | 30,000 | 27,000 | 18,750 | 15,000 | 33,750 | 204,750 |
| Licensing Deals | 0 | 0 | 0 | 0 | 0 | 1 | 1 | 2 | 4 | 5 | 6 | 7 | 26 |
| Merchandise Orders | 0 | 0 | 6,000 | 6,000 | 21,600 | 7,200 | 7,200 | 18,000 | 14,400 | 7,200 | 3,600 | 18,000 | 109,200 |

**Volumes 2002**

|  | Jan-02 | Feb-02 | Mar-02 | Apr-02 | May-02 | Jun-02 | Jul-02 | Aug-02 | Sep-02 | Oct-02 | Nov-02 | Dec-02 | Total |
|---|---|---|---|---|---|---|---|---|---|---|---|---|---|
| Consumer Photos | 12,000 | 15,000 | 15,000 | 15,000 | 54,000 | 18,000 | 18,000 | 45,000 | 36,000 | 18,000 | 9,000 | 45,000 | 300,000 |
| Stock Photos | 6,000 | 7,500 | 9,000 | 10,500 | 12,000 | 15,000 | 15,000 | 15,000 | 15,000 | 15,000 | 15,000 | 15,000 | 150,000 |
| Subtotal | 18,000 | 22,500 | 24,000 | 25,500 | 66,000 | 33,000 | 33,000 | 60,000 | 51,000 | 33,000 | 24,000 | 60,000 | 450,000 |
| Licensing Deals | 8 | 9 | 10 | 11 | 12 | 13 | 14 | 15 | 16 | 16 | 16 | 16 | 156 |
| Merchandise Orders | 9,600 | 12,000 | 12,000 | 12,000 | 43,200 | 14,400 | 14,400 | 36,000 | 28,800 | 14,400 | 7,200 | 36,000 | 240,000 |

**Volumes 2003**

|  | Jan-03 | Feb-03 | Mar-03 | Apr-03 | May-03 | Jun-03 | Jul-03 | Aug-03 | Sep-03 | Oct-03 | Nov-03 | Dec-03 | Total |
|---|---|---|---|---|---|---|---|---|---|---|---|---|---|
| Consumer Photos | 24,800 | 31,000 | 31,000 | 31,000 | 111,600 | 37,200 | 37,200 | 93,000 | 74,400 | 37,200 | 18,600 | 93,000 | 620,000 |
| Stock Photos | 17,000 | 21,250 | 25,500 | 29,750 | 34,000 | 42,500 | 42,500 | 42,500 | 42,500 | 42,500 | 42,500 | 42,500 | 425,000 |
| Subtotal | 41,800 | 52,250 | 56,500 | 60,750 | 145,600 | 79,700 | 79,700 | 135,500 | 116,900 | 79,700 | 61,100 | 135,500 | 1,045,000 |
| Licensing Deals | 16 | 16 | 16 | 16 | 16 | 16 | 16 | 16 | 16 | 16 | 16 | 16 | 192 |
| Merchandise Orders | 15,600 | 19,500 | 19,500 | 19,500 | 70,200 | 23,400 | 23,400 | 58,500 | 46,800 | 23,400 | 11,700 | 58,500 | 390,000 |

**ONSITE OPERATIONS** (by quarters)

| Year | 2000 | 2001 | | | | 2002 | | | | 2003 | | | |
|---|---|---|---|---|---|---|---|---|---|---|---|---|---|
| Quarter | Qtr 4 | Qtr 1 | Qtr 2 | Qtr 3 | Qtr 4 | Qtr 1 | Qtr 2 | Qtr 3 | Qtr 4 | Qtr 1 | Qtr 2 | Qtr 3 | Qtr 4 |
| Onsites (cumulative) | 1 | 4 | 4 | 7 | 10 | 13 | 13 | 16 | 19 | 22 | 25 | 28 | 28 |

Source: e-Catalyst Business Plan, February 28, 2000.

**EXHIBIT 3** Projected monthly revenue stream.

**Revenues 2001**

| | Jan-01 | Feb-01 | Mar-01 | Apr-01 | May-01 | Jun-01 | Jul-01 | Aug-01 | Sep-01 | Oct-01 | Nov-01 | Dec-01 | Total |
|---|---|---|---|---|---|---|---|---|---|---|---|---|---|
| Consumer Photos | $ 0 | $ 0 | $ 149,925 | $ 149,925 | $ 539,730 | $ 179,910 | $ 179,910 | $ 449,775 | $ 359,820 | $ 179,910 | $ 89,955 | $ 449,775 | $ 2,728,635 |
| Stock Photos | $ 0 | $ 0 | $ 0 | $ 562,500 | $ 675,000 | $ 787,500 | $ 1,012,500 | $ 1,125,000 | $ 1,350,000 | $ 1,462,500 | $ 1,575,000 | $ 1,687,500 | $ 10,237,500 |
| Subtotal | $ 0 | $ 0 | $ 149,925 | $ 712,425 | $ 1,214,730 | $ 967,410 | $ 1,192,410 | $ 1,574,775 | $ 1,709,820 | $ 1,642,410 | $ 1,664,955 | $ 2,137,275 | $ 12,966,135 |
| Syndication | $ 0 | $ 0 | $ 0 | $ 0 | 0 | $ 8,333 | $ 16,667 | $ 33,333 | $ 66,667 | 108,333 | $ 158,333 | $ 216,667 | $ 608,333 |
| Merchandise | $ 0 | $ 0 | $ 45,000 | $ 45,000 | $ 162,000 | $ 54,000 | $ 54,000 | $ 135,000 | $ 108,000 | $ 54,000 | $ 27,000 | $ 135,000 | $ 819,000 |
| Total | $ 0 | $ 0 | $ 194,925 | $ 757,425 | $ 1,376,730 | $1,029,743 | $1,263,077 | $ 1,743,108 | $ 1,884,487 | $ 1,804,743 | $1,850,288 | $ 2,488,942 | $ 14,393,468 |

**Revenues 2002**

| | Jan-02 | Feb-02 | Mar-02 | Apr-02 | May-02 | Jun-02 | Jul-02 | Aug-02 | Sep-02 | Oct-02 | Nov-02 | Dec-02 | Total |
|---|---|---|---|---|---|---|---|---|---|---|---|---|---|
| Consumer Photos | $ 239,880 | $ 299,850 | $ 299,850 | $ 299,850 | $ 1,079,460 | $ 359,820 | $ 359,820 | $ 899,550 | $ 719,640 | $ 359,820 | $ 179,910 | $ 899,550 | $ 5,997,000 |
| Stock Photos | $ 900,000 | $1,125,000 | $1,350,000 | $1,575,000 | $ 1,800,000 | $2,250,000 | $2,250,000 | $ 2,250,000 | $ 2,250,000 | $ 2,250,000 | $2,250,000 | $ 2,250,000 | $ 22,500,000 |
| Subtotal | $1,139,880 | $1,424,850 | $1,649,850 | $1,874,850 | $ 2,879,460 | $2,609,820 | $2,609,820 | $ 3,149,550 | $ 2,969,640 | $ 2,609,820 | $2,429,910 | $ 3,149,550 | $ 28,497,000 |
| Syndication | $ 283,333 | $ 358,333 | $ 441,667 | $ 533,333 | $ 633,333 | $ 741,667 | $ 858,333 | $ 983,333 | $ 1,116,667 | $ 1,250,000 | $1,383,333 | $ 1,516,667 | $ 10,100,000 |
| Merchandise | $ 72,000 | $ 90,000 | $ 90,000 | $ 90,000 | $ 324,000 | $ 108,000 | $ 108,000 | $ 270,000 | $ 216,000 | $ 108,000 | $ 54,000 | $ 270,000 | $ 1,800,000 |
| Total | $1,495,213 | $1,873,183 | $2,181,517 | $2,498,183 | $ 3,836,793 | $3,459,487 | $3,576,153 | $ 4,402,883 | $ 4,302,307 | $ 3,967,820 | $3,867,243 | $ 4,936,217 | $ 40,397,000 |

**Revenues 2003**

| | Jan-03 | Feb-03 | Mar-03 | Apr-03 | May-03 | Jun-03 | Jul-03 | Aug-03 | Sep-03 | Oct-03 | Nov-03 | Dec-03 | Total |
|---|---|---|---|---|---|---|---|---|---|---|---|---|---|
| Consumer Photos | $ 495,752 | $ 619,690 | $ 619,690 | $ 619,690 | $ 2,230,884 | $ 743,628 | $ 743,628 | $ 1,859,070 | $ 1,487,256 | $ 743,628 | $ 371,814 | $ 1,859,070 | $ 12,393,80 |
| Stock Photos | $2,550,000 | $3,187,500 | $3,825,000 | $4,462,500 | $ 5,100,000 | $6,375,000 | $6,375,000 | $ 6,375,000 | $ 6,375,000 | $ 6,375,000 | $6,375,000 | $ 6,375,000 | $ 63,750,000 |
| Subtotal | $3,045,752 | $3,807,190 | $4,444,690 | $5,082,190 | $ 7,330,884 | $7,118,628 | $7,118,628 | $ 8,234,070 | $ 7,862,256 | $ 7,118,628 | $6,746,814 | $ 8,234,070 | $ 76,143,800 |
| Syndication | $1,650,000 | $1,783,333 | $1,916,667 | $2,050,000 | $ 2,183,333 | $2,316,667 | $2,450,000 | $ 2,583,333 | $ 2,716,667 | $ 2,850,000 | $2,983,333 | $ 3,116,667 | $ 28,600,000 |
| Merchandise | $ 117,000 | $ 146,250 | $ 146,250 | $ 146,250 | $ 526,500 | $ 175,500 | $ 175,500 | $ 438,750 | $ 351,000 | $ 175,500 | $ 87,750 | $ 438,750 | $ 2,925,000 |
| Total | $4,812,752 | $5,736,773 | $6,507,607 | $7,278,440 | $10,040,717 | $9,610,795 | $9,744,128 | $11,256,153 | $10,929,923 | $10,144,128 | $9,817,897 | $11,789,487 | $107,668,800 |

Source: e-Catalyst Business Plan, February 28, 2000.

**EXHIBIT 4**  Pro forma financial summary 2000.

### Summary profit and loss statement

| | 2000 | 2001 | 2002 | 2003 | Total |
|---|---|---|---|---|---|
| Revenues | $ 0 | $14,393,468 | $40,397,000 | $107,668,800 | $162,459,268 |
| Cost of sales | $ 0 | $ 5,186,454 | $11,398,800 | $ 30,457,520 | $ 47,042,774 |
| Gross profit | $ 0 | $ 9,207,014 | $28,998,200 | $ 77,211,280 | $115,416,494 |
| Operations | $ 439,847 | $13,623,571 | $27,109,143 | $ 47,078,657 | $ 88,251,217 |
| Net income before tax | ($ 439,847) | ($ 4,416,556) | $ 1,889,057 | $ 30,132,623 | $ 27,165,277 |
| Taxes (38%) | $ 0 | $ 0 | $ 0 | $ 10,322,805 | $ 10,322,805 |
| Net income | ($ 439,847) | ($ 4,416,556) | $ 1,889,057 | $ 19,809,818 | $ 16,842,472 |

### Summary balance sheet

| Assets | 2000 | 2001 | 2002 | 2003 |
|---|---|---|---|---|
| Cash and equivalents | $ 428,020 | $ 4,490,768 | $ 4,958,270 | $ 21,508,477 |
| Accounts receivable | $ 0 | $ 2,488,942 | $ 4,936,217 | $ 11,789,487 |
| Inventories | $ 0 | $ 0 | $ 0 | $ 0 |
| Prepaid expenses | $ 0 | $ 0 | $ 0 | $ 0 |
| Depreciable assets | $ 0 | $ 0 | $ 0 | $ 0 |
| Other depreciable assets | $ 0 | $ 0 | $ 0 | $ 0 |
| Depreciation | $ 0 | $ 0 | $ 0 | $ 0 |
| Net depreciable assets | $ 0 | $ 0 | $ 0 | $ 0 |
| Total assets | $ 428,020 | $ 6,979,710 | $ 9,894,487 | $ 33,297,964 |

| Liabilities and capital | 2000 | 2001 | 2002 | 2003 |
|---|---|---|---|---|
| Accounts payable | $ 367,867 | $ 1,836,113 | $ 2,861,833 | $ 5,589,379 |
| Accrued income taxes | $ 0 | $ 0 | $ 0 | $ 866,113 |
| Accrued payroll taxes | $ 0 | $ 0 | $ 0 | $ 0 |
| Total liabilities | $ 367,867 | $ 1,836,113 | $ 2,861,833 | $ 6,455,492 |
| Capital contribution | $ 500,000 | $10,000,000 | $10,000,000 | $ 10,000,000 |
| Stockholders' equity | $ 0 | $ 0 | $ 0 | $ 0 |
| Retained earnings | ($ 439,847) | ($ 4,856,403) | ($ 2,967,346) | $ 16,842,472 |
| Net capital | $ 60,153 | $ 5,143,597 | $ 7,032,654 | $ 26,842,472 |
| Total liabilities and capital | $ 428,020 | $ 6,979,710 | $ 9,894,487 | $ 33,297,964 |

Source: e-Catalyst Business Plan, February 28, 2000.

**EXHIBIT 5** Artemis images original funding plan.

| Projected Plan | Round 1 | Round 2 | Round 3 | Round 4 | Exit |
|---|---|---|---|---|---|
| **Financing assumptions:** | | | | | |
| 2003 Revenues | $110,000,000 | | | | |
| 2003 EBITDA | $ 30,000,000 | | | | |
| 2003 Revenue growth rate | 40% | | | | |
| 2003 Valuation | $440,000,000 | | | | |
|   Valuation/revenue | 4 | | | | |
|   Valuation/EBITDA | 14.67 | | | | |
| Round 1 Financing | $ 500,000 | | | | |
| Round 2 Financing | $ 1,500,000 | | | | |
| Round 3 Financing | $ 3,000,000 | | | | |
| Round 4 Financing | $ 5,000,000 | | | | |

| | Round 1<br>Oct-00 | Round 2<br>1-Jan | Round 3<br>1-Mar | Round 4<br>1-Jun | Exit<br>3-Dec |
|---|---|---|---|---|---|
| **Number of shares outstanding** | | | | | |
| Total number of shares outstanding prior to financing | 6,000,000 | 7,200,000 | 9,000,000 | 11,250,000 | 11,250,000 |
| Shares issues this round | 1,200,000 | 1,800,000 | 2,250,000 | 1,406,250 | 1,406,250 |
| Total number shares outstanding after financing | 7,200,000 | 9,000,000 | 11,250,000 | 12,656,250 | 12,656,250 |
| **Valuations** | | | | | |
| Premoney valuation | $2,500,000 | $6,000,000 | $12,000,000 | $40,000,000 | $440,000,000 |
| Amount of financing | $ 500,000 | $1,500,000 | $ 3,000,000 | $ 5,000,000 | 0 |
| Postmoney valuation | $3,000,000 | $7,500,000 | $15,000,000 | $45,000,000 | $440,000,000 |
| **Price per share** | $0.42 | $0.83 | $1.33 | $3.56 | $34.77 |
| **Resulting ownership** | | | | | |
| Founders | 83.33% | 66.67% | 53.33% | 47.41% | 47.41% |
| Round 1 investors | 16.67% | 13.33% | 10.67% | 9.48% | 9.48% |
| Round 2 investors | 0.00% | 20.00% | 16.00% | 14.22% | 14.22% |
| Round 3 investors | 0.00% | 0.00% | 20.00% | 17.78% | 17.78% |
| Round 4 investors | 0.00% | 0.00% | 0.00% | 11.11% | 11.11% |
| Total | 100.00% | 100.00% | 100.00% | 100.00% | 100.00% |
| **Value of ownership** | | | | | |
| Founders | $2,500,000 | $5,000,000 | $ 8,000,000 | $21,333,333 | $208,592,593 |
| Round 1 investors | $ 500,000 | $1,000,000 | $ 1,600,000 | $ 4,266,667 | $ 41,718,519 |
| Round 2 investors | $ 0 | $1,500,000 | $ 2,400,000 | $ 6,400,000 | $ 62,577,778 |
| Round 3 investors | $ 0 | $ 0 | $ 3,000,000 | $ 8,000,000 | $ 78,222,222 |
| Round 4 investors | $ 0 | $ 0 | $ 0 | $ 5,000,000 | $ 48,888,889 |
| Total | $3,000,000 | $7,500,000 | $15,000,000 | $40,000,000 | $440,000,000 |
| **Payback to investors** | **Round 1** | **Round 2** | **Round 3** | **Round 4** | |
|   Holding period (years) | 3.25 | 3 | 2.75 | 2.5 | |
|   Times money back | 83.44 | 41.72 | 26.07 | 9.78 | |
|   Internal rate of return (IRR) | 290% | 247% | 227% | 149% | |

Source: e-Catalyst Business Plan, February 28, 2000.

Commercially produced images were also in demand by consumers. Industry insiders believed that this market was poised for explosive growth in 2000, as Web-enabled technology facilitated display and transmission of images directly from their owners to individual consumers. The archives from the Indianapolis Motor Speedway was an example of this business-to-consumer model. Historically, consumers who bought from the archive had to visit the museum at IMSC or write a letter to the staff. Retrieval and fulfillment of images then required a manual search of a physical inventory, a process which could take as long as two weeks. Web-based digitization and search engines would reduce the search time and personnel needed for order fulfillment and allow customers the convenience of selecting products and placing orders on-line. The *Daily Mirror,* a newspaper in London, had displayed its archived images on its own website and had generated over $30,000 in sales to consumers in its first month of availability. IMG, a sports marketing group, placed a value of $10 million on the IMSC contract.

## Competition

There were a variety of stock and consumer photo sites ranging from those that served only the business-to-business stock photo market to amateur photographers posting their pictures. Most sites did not offer a "community," the Internet vehicle for consumer comments and discussion, a powerful search engine, and ways to repurpose the content (e-greeting cards, prints, photo mugs, calendars, etc.). In addition, the archives available in digital form were limited because other content providers worked from the virtual world to the physical world versus the Artemis Images model of working from the physical world to the virtual world. Competitors had problems with integrated digital workflows and knowing where the original asset resided due to the distributed nature of their archives. They scanned images on demand, which severely limited the content available to be searched on their websites.

Chris and Greg evaluated the five major competitors for their business plan:

**www.corbis.com:** Owned by Bill Gates with an archive of over 65 million images, only 650,000 were available on the Web to be accessed by consumers for Web distribution (e-greeting cards, screen savers, etc.). Only 350,000 images were available to be purchased as prints. The site was well designed and the search features were good, but there was no community on the site. The niche Corbis pursued was outright ownership of archives and scanning on demand. Corbis had recently acquired the Louvre archive, for a reported purchase price of over $30 million.

**www.getty-images.com:** An archive of over 70 million images. In 1999, this site was only a source to link to their other wholly owned subsidiaries, including art.com. There were no search capabilities, no community. This website functioned only as a brochure for the company. Like Corbis, Getty was focused on owning content and then scanning on demand.

**www.art.com:** A good site in design and navigation, this site was a wholly owned subsidiary of Getty and was positioned as the consumer window to a portion of the Getty archive. Similar to Corbis, customers were able to buy prints, send e-greeting cards, etc. Despite the breadth of the Getty archive, this site had a limited number of digitized images available.

**www.mediaexchange.com:** Strictly a stock photo site targeted toward news sources, the site was largely reliant on text. It was difficult to navigate and had an unattractive graphical user interface.

**www.thepicturecollection.com:** Strictly a stock site offering the *Time* photo archive, this site was well designed with good search capabilities. Searches yielded not only a thumbnail image but a display of the attached locator tags, or metadata.

**www.ditto.com:** The world's leading visual search engine, ditto.com enabled people to navigate the Web through pictures. The premise was two-fold: deliver highly relevant thumbnail images and link to relevant Web sites underlying these images. By 2000, they had developed the largest searchable index of visual content on the Internet.

Exhibit 6 shows a comparison of Artemis Images to the two major players in the stock photography market, Getty and Corbis. This table illustrates only revenues from stock photo sales and does not include potential revenue from consumer sales, merchandise, advertising or other potential revenue sources.

According to its marketing director, Corbis intended to digitize its entire archive, and was in the process of converting analog images into digital images, with 63 million images yet to be converted. While Getty and Corbis were established players in the content industry, they were just recently feeling the effects of e-commerce:

- In 1999, Corbis generated 80 percent of its revenues from the Web versus none in 1996.
- Getty's e-commerce sales were up 160 percent between 1998 and 1999.
- 34 percent of Getty's 1999 revenues came from e-commerce versus 17 percent in 1998.

## Strategy

Artemis Images intended to provide digitization and archive management by employing a professional staff who would work within each client-company's organization, rather than in an off-site facility of its own. Chris's model was to provide digitized archive services in exchange for (1) exclusive rights to market the content on the Internet, (2) merchandising rights, and (3) promotion of Artemis Images's URL, effectively co-branding Artemis Images with each client-partner. Chris envisioned a software process that would be owned or licensed by Artemis, and which could be used for digitizing different archive media, such as photos, videos, and text.

Chris and George expected Artemis Images to partner with existing sellers of stock photography and trade digitizing services for promotion through their sales channels. Artemis Images would pursue these relationships with traditional

**EXHIBIT 6** Anticipated sales volume comparisons.

| Stock photo market | | | | | | |
|---|---|---|---|---|---|---|
| | **Artemis Images** | **Artemis Images** | **Artemis Images** | **Artemis Images** | **Getty** | **Corbis** |
| | **Indy Archive 2000\*** | **2000** | **2001** | **2002** | **1999** | **1999** |
| Archive size | 5,000,000 | 5,000,000 | 50,000,000 | 95,000,000 | 70,000,000 | 65,000,000 |
| Cumulative number of images digitized | 345,600 | 345,600 | 6,796,800 | 21,542,400 | 1,200,000 | 2,100,000 |
| % digitized\*\* | 7% | 7% | 14% | 23% | 1.71% | 3.2% |
| # of image sales needed to hit revenue target | 0 | 0 | 151,484 | 623,493 | 1,646,667 | 666,666 |
| % of archive that must be sold to hit revenue target\*\*\* | 0 | 0 | 0.30% | 0.16% | 2.35% | 1.00% |
| Revenues\*\*\*\* | 0 | 0 | $22,722,600 | $93,523,950 | $247,000,000 | $100,000,000 |
| Revenue per image in archive | 0 | 0 | $0.45 | $0.98 | $ 3.53 | $ 1.54 |
| Revenue per digitized image | 0 | 0 | $3.25 | $4.30 | $205.83 | $47.62 |

\*Artemis Images had already secured an exclusive content agreement from the Indianapolis Motor Speedway Corporation.
\*\*Estimates based on scanning 1,920 images a day per scanner, 2 scanners per archive. As scanning technologies improve, the throughput numbers were expected to go up.
\*\*\*The percentage of the Artemis Images archive that needed to be sold to hit revenues projections varied between 0.03% and 0.22%, as compared to an actual 2.35% for Getty and to 0.6% for Corbis.
\*\*\*\*The Artemis Images revenue numbers were based on selling a certain number of images at $150 per image; $150 was the minimum average price paid for stock photographs. Corbis was privately held; this figure was an estimate.

sales and marketing techniques. Sales people would call on the major players and targeted direct mail, trade magazine advertising and PR would be used to reach the huge audience of smaller players. In addition, content partners were expected to become customers, as they were all users of stock photography.

As Artemis Images gained clients, the company would have access to some of the finest and most desirable content in the world. Chris knew that the workflow expertise of the management team would put them in a good position to provide better quality more consistently than either Corbis or Getty. This same expertise would allow Artemis to have a much larger digital selection, with a website design that would be easily navigable for customers to find what they needed.

Using on-site equipment, the client's content would be digitized, annotated (by attaching digital information tags, or metadata) and uploaded to the corporate hub site. Metadata would allow the content to be located by the search engine and thus viewed by the consumer. For example, a photo of Eddie Cheever winning the Indy 500 would have tags like Indy 500, Eddie Cheever, win photo, 1998, etc. Therefore, a customer going to the website and searching for "Eddie Cheever" would find this specific photo, along with the hundreds of other photos associated with him. The Artemis corporate database was intended to serve as the repository for search and retrieval from the website.

The traditional content management strategy forced organizations to purchase technology and expertise. Artemis Images's model intended to alleviate this burden by exchanging technology and expertise for exclusive web distribution rights and a share of revenues. The operational strategy was to create an infrastructure based on installing and operating digital asset management systems at their customers' facilities to create a global digital archive of images, video, sound and text. This would serve to lock Artemis Images into long-term relationships with these organizations and ensure that Artemis Images would have both the historical and the most up-to-date content. Artemis Images would own and operate the content management technology, with all other operational needs outsourced including Web development, Web hosting, consumer data collection, and warehousing and fulfillment of merchandise (printing and mailing posters or prints). Artemis Images would scan thousands of images per day, driving down the cost per image to less than $2.00, versus the Corbis and Getty model of scan-on-demand, where the cost per image was approximately $40.00. The equipment needed for both the content management and photo production would be leased to minimize start-up costs and ensure greater flexibility in the system's configuration.

The original plan was to purchase and install software and hardware at their main office in Denver, Colorado, contract with a Web development partner, and set up the first on-site facility at Indianapolis Motor Speedway Corporation. The Denver facility would serve as a development lab, to create a standard set of metadata to be used by all of their partners' content. This consistency of annotation information was intended to allow for consistent search and retrieval of content. Artemis Images's goal was to build a world-class infrastructure to handle content management, consumer data collection, and e-commerce. This infrastructure would allow them to amass a large content and transaction volume by expanding to other market segments. Developing their own structure would ensure standardization of content and reduced implementation time. Outreach for news coverage and the development of community features would be negotiated concurrently. The time line in Exhibit 7 illustrates the Artemis Images development plan.

### Financial Projections

Revenues were expected to come from four primary sources:

*Consumer photos:* IMSC's archive sold approximately 53,000 photos in 1999 to a market limited to consumers who visited the archive or wrote to its staff. Artemis Images based its projected sales on an average of

## Three phases of development

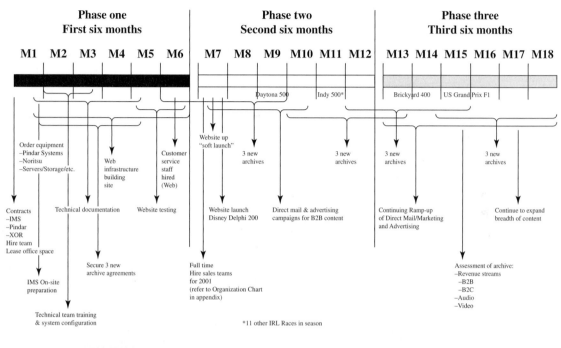

Phase One    ██████████

This phase was intended to take Artemis Images from initial funding to operationally being ready to sell images and take orders. The three main components included establishing the on-site facility at IMSC, construction and testing of the website and establishing the fulfillment operations. Phase one assumed money was in the bank.

Phase Two    ▭▭▭▭▭▭

This phase assumed that three additional archives had been secured and implemented, at least one of which would include breadth of content. Focus would be sales and ramping up revenues. B2B and B2C marketing strategies were to be executed and evaluated. Toward the end of phase two, three more archives would be secured.

Phase Three    ▭▭▭▭▭▭

Phase three  continued to build more archives and breadth of content. Marketing and sales would continue to be core focus for revenue development. Audio and video content assessed based on the state of market and technology (e.g., bandwidths) and a decision would be made on timing to enter this market.

**EXHIBIT 7**  Artemis Images development time line.

Source: e-Catalyst Business Plan, February 28, 2000.

15,000 images sold per archive in 2001, increasing to 20,000 images per archive in 2003. Price: $19.99 (8" × 10").

*Stock photos:* Stock photos ranged in price from $150 to $100,000, depending on the uniqueness of the photo. Competitors Getty and Corbis, two of the leaders in this market, sold 2.35 percent and 0.6 percent of their archive, respectively. Based on an average selling price of $150, Getty generated approximately $6.00 in revenue for each image in its archive; Corbis generated approximately $1.85. Artemis Images constructed financial projections based on sales of 0.30 percent of its archive in 2001 and 0.16 percent of its archive in 2002. Artemis Images's margin was based on a return of $0.20 per image in its archive for 2001, increasing to $0.60 per image in 2003.

*Syndication:* The team's dot.com experience led them to believe that websites with exclusive content were able to syndicate their content to other websites. They anticipated that Artemis Images would generate revenues of $100,000 per year from each contract for content supplied as marketing tools on websites. Existing companies with strong content had been able to negotiate five new agreements per week for potential annual revenues of $5 million.

*Merchandise:* According to America Online/Roper Starch Worldwide, approximately 30 percent of Internet users regularly make purchases. Artemis Images used a more conservative assumption that only 1 percent of unique visitors would make a purchase. Estimates of the average purchase online varied widely, ranging from Wharton's estimate of $86.13 to eMarketers's estimate of $219. The Artemis Images team viewed $50 per purchase as a conservative figure.

Chris and George felt confident that Artemis Images would be able to reach the revenue projections for number of photos sold. IMSC's archive had sold approximately 53,000 photos in 1999, an increase of 33 percent over 1998. These sales had been generated solely by consumers who had visited the archive in person, estimated at 1 million people. In other words, one out of every 28 possible consumers actually purchased an image. Chris and George assumed that if even one out of 160 unique visitors to the website purchased a photo, the Artemis website would generate 42 percent more than IMSC's 1999 figures (see Exhibit 6 for projected sales volume). Chris and George believed that this projection was reasonable in light of the fact that IMSC did not market its archive and significant publicity and advertising would accompany Artemis Images's handling of the archive. As breadth of content and reach of the Web increased, 2002 revenues should easily be double those of 2001.

Since the team previously had configured and sold content management systems, they were familiar with the costs associated with this process, including both equipment and personnel. They carefully conducted research to stay abreast of recent improvements in technology and intended to be on the lookout for cost reductions and process improvements.

## The Launch: Problems from the Start

Chris dove into the Artemis Images project with a vengeance. Having secured a five-year contract for exclusive rights and access to the IMSC archive, she found a dependable technician who was eager to relocate to Indianapolis to start the scanning and digitizing process. A reputable, independent photo lab agreed to handle printing and order fulfillment. Chris's visit to the Indy 500 in May 2000 was a wonderful networking opportunity. She met executives from large companies and got leads for investors and clients. She secured an agreement with a Web design company to build the Artemis Images site, careful to retain ownership of the design. She contacted over 100 potential venture capitalists and angel investors.

Personally, she was on a roll. Financially, she was rapidly going into debt. Frank and Greg, legal owners of the company, had long since contributed ideas, contacts, or legwork to the Artemis Images launch. While confident that his work on the business plan would appeal to investors, Greg viewed the start-up company as a risk to which he was unwilling to commit. Likewise, Frank decided to hold onto his job at Petersons.com, a unit of Thompson Learning, until the first round of investor funding had been secured. Frank continued to offer advice, but he had a wife and two pre-school-age children to support.

Each meeting with a potential funder resulted in a suggestion on how to make the business more attractive for investment. Sometimes they helped, sometimes they just added to Chris's and George's frustration. Beating the bushes for money over two years was exhausting, to say the least. The lack of funds impacted the look and feel of the business and severely strained relationships among the founding partners. Heated discussions ensued as to the roles that each was expected to play, the reallocation of equity ownership in the company, and the immediate cash needed to maintain the Indianapolis apartment and pay the scanning technician and Web developers, not to mention out-of-pocket expenses needed to manage and market the business.

Chris and George appealed to their families for help. George's father contributed $5,000. Chris's mother tapped into her retirement, mostly to pay Chris's mortgage and to fund Chris's trips to potential clients and investors in London, New York, and Boston. By May 2001, Chris's mother's contribution had exceeded $200,000. A $50,000 loan from a supportive racing enthusiast provided the impetus for Artemis Images to reorganize as a C-corporation. All four original partners had stock in the new company, but Chris held the majority share (66 percent), George held 30 percent, and Frank and Greg's shares were each reduced to 2 percent. Financial projections were revised downward (see Exhibit 8).

The site was officially launched on May 18, 2001. It was beautiful. Chris held her breath as she put in her credit card late that evening when the site went live. The shopping cart failed and the order could not be processed. Chris knew she was in trouble.

**EXHIBIT 8**   Revised pro forma financial summary 2001.

| Summary Profit and Loss Statement | | | | | |
|---|---|---|---|---|---|
| | 2001 | 2002 | 2003 | 2004 | Total |
| Revenues | $ 5,312* | $373,779 | $2,294,116 | $4,735,400 | $7,408,607 |
| Cost of sales | $ 1,700 | $ 43,368 | $ 265,312 | $ 564,480 | $ 874,860 |
| Gross profit | $ 3,612 | $330,411 | $2,028,804 | $4,170,920 | $6,533,747 |
| Operations | $ 52,499 | $328,550 | $1,235,363 | $2,035,430 | $3,651,842 |
| Net income before tax | ($48,887) | $ 1,861 | $ 793,441 | $2,135,490 | $2,881,905 |
| Taxes (38%) | $ 0 | $ 0 | $ 283,638 | $ 811,486 | $1,095,124 |
| Net income | ($48,887) | $ 1,861 | $ 509,803 | $1,324,004 | $1,786,781 |

**Summary Balance Sheet**

| Assets | 2001 | 2002 | 2003 | 2004 |
|---|---|---|---|---|
| Cash and equivalents | $ 45,113 | $ 78,260 | $ 675.347 | $2,615,573 |
| Accounts receivable | $ 0 | $ 13,610 | $ 222,950 | $ 462,200 |
| Inventories | $ 0 | $ 0 | $ 0 | $ 0 |
| Prepaid expenses | $ 0 | $ 0 | $ 0 | $ 0 |
| Depreciable assets | $ 0 | $ 0 | $ 0 | $ 0 |
| Other depreciable assets | $ 0 | $ 0 | $ 0 | $ 0 |
| Depreciation | $ 0 | $ 0 | $ 0 | $ 0 |
| Net depreciable assets | $ 0 | $ 0 | $ 0 | $ 0 |
| Total assets | $ 45,113 | $ 40,574 | $ 898,297 | $3,077,773 |

| Liabilities and capital | | | | |
|---|---|---|---|---|
| Accounts payable | $ 4,000 | $ 12,355 | $ 61,882 | $ 105,868 |
| Accrued income taxes | $ 0 | $ 0 | $ 283,638 | $1,095,124 |
| Accrued payroll taxes | $ 0 | $ 0 | $ 0 | $ 0 |
| Total liabilities | $ 4,000 | $ 12,355 | $ 345,520 | $1,200,992 |
| Capital contribution | $90,000† | $ 90,000 | $ 90,000 | $ 90,000 |
| Stockholders' equity | $ 0 | $ 0 | $ 0 | $ 0 |
| Retained earnings | ($48,887) | ($61,781) | $ 462,777 | $1,786,781 |
| Net capital | $ 41,113 | $ 28,219 | $ 552,777 | $1,876,781 |
| Total liabilities and capital | $ 45,113 | $ 40,574 | $ 898,287 | $3,077,773 |

*Notes:* *Approximately two-thirds of these transactions were executed by Artemis staff and friends to test the website.
† Chris's mother's contribution to her daughter for mortgage and living expenses is not included.

## The Crash

From the first, the website had problems. The Web development contract stipulated that the website for the Indy 500 would go live by May 8, 2001, to coincide with the month-long series of events held at the Indianapolis Motor Speedway leading up to the Indy 500 on May 27. However, the Web development took longer than anticipated, and the site was first operational on May 18. Having neglected to test the Web interface properly, serious failures were encountered when the site was activated. The site went down for 24 hours, only to face similar problems throughout the following week, again shutting down on May 27. More technical difficulties delayed the reactivation of the site until May 31, after the Indy racing series had ended.

Throughout June, consumer traffic was far less than originally anticipated. The site was not easily navigable. The shopping cart didn't work. Yet the Web builder demanded more money. Fearful of a possible lawsuit, investors stayed away. The crash of the dot.coms added kindling to the woodpile. Chris and George started to rethink their original business model. They were held hostage, as they owned no tangible assets.

Website tracking data indicated that between May and July there had been at least $40,000 worth of attempted purchases. Chris read through hundreds of angry e-mails, and tried manually to process orders. Orders which were successfully executed resulted in spotty fulfillment. Many photos ordered were never shipped, were duplicated, or were incorrectly billed. At the same time, she tried to negotiate with the software developers' demand for payment and keep alive a $250,000 investment prospect.

On July 9, 2001, the Web development company threatened an all-or-nothing settlement. They wanted payment in full for the balance of the contract even though the sites didn't work. Absent full payment, they would shut down the sites within the week. The investor offered to put up 80 percent of the balance owed on the full contract to acquire the code to fix it. The company refused. On Friday, July 13, Chris had to tell IMSC that in less than 48 hours the sites would be shut down. The investor took his $250,000 elsewhere.

On Tuesday, July 17, Chris called an emergency meeting with George. George had had enough. The stress was affecting his health, his relationships, and his lifestyle. He believed that his family had already contributed more money than he had a right to ask. He was putting in long hours with no money to show for his efforts. His girlfriend had been putting pressure on George to quit for some time. Now he had run out of reasons to stay.

Chris was devastated. How could she face the people in Indianapolis? It was hard for her to come to grips with having let them down. Having put so much of herself into this venture, she wasn't sure she could let go. At the same time, she wasn't sure how to go on.

Chris reflected, "At one time, I defined success by my title, my salary, and my possessions. Working for AGT, I had it all. I started Artemis Images because I really cared about IMSC and making the Indy motorsports images available to its fans. Now, I realize that there is a profound satisfaction in building a company. I can see my future so clearly, but living day to day now is so hard. And I'm still enthralled with beautiful images."

## Questions

1. Discuss why Chris started her company. What was the opportunity?
2. What is your evaluation of the team's qualifications for this business?
3. Discuss the division of ownership among the team.
4. Evaluate the business model for Artemis. Is it strong and will the firm be profitable?

# SIRTRIS PHARMACEUTICALS: LIVING HEALTHIER, LONGER (ABRIDGED)

One Saturday in February 2007, Dr. David Sinclair and Dr. Christoph Westphal, co-founders of Sirtris Pharmaceuticals, a Cambridge, Massachusetts-based life sciences firm, were about to start their weekly strategic planning meeting. Sirtris conducted research into age-related diseases such as diabetes, cancer, and Alzheimer's. What attracted the attention of investors and the media, however, was that its research might lead to another result: extending human lifespans.

Sirtris had established a link between resveratrol (a compound found in red grapes) and sirtuins (a newly discovered family of enzymes with links to improved longevity and health in living organisms as diverse as yeast, mice and humans). Sinclair and Westphal were building Sirtris around the development of sirtuin-activating drugs for the diabetes market. The Sirtris team had developed its own proprietary formulation of resveratrol, called SRT501, and was developing new chemical entities (NCEs) that were up to 1,000 times more potent activators of sirtuins than resveratrol.

In the strategy meeting, CEO Westphal, and Sinclair, the co-chair of Sirtris's scientific advisory board, were meeting with Dr. Michelle Dipp, senior director of corporate development, and Garen Bohlin, the company's chief operating officer, to discuss three strategic options for Sirtris:

**In-licensing.** Should Sirtris in-license compounds to diversify its drug development platform beyond its narrow focus on SIRT1, one of seven sirtuin variants in the human body?

**Partnership with Pharma.** Should Sirtris partner with a large pharmaceutical firm? What would it mean for the organization to become tied to a pharmaceutical company at this stage of its development? Should it postpone any deal until it had more clinical data?

**Nutraceuticals.** Should Sirtris sell its proprietary version of resveratrol, SRT501, as a nutraceutical, an off-the-shelf health supplement that would not require FDA approval?

## Anti-Aging Science

The quest for long life has spurred the imagination of people in virtually every era in human history. Ancient Greeks imagined immortal gods, the sixteenth century Spanish adventurer Ponce de Leon searched for the fountain of youth, and 21st century scientists tested rodents for life-extending biological compounds. Until the 1990s, the only proven means of increasing lifespan in any animal was to reduce its calorie intake; however, for decades scientists did not understand

Professor Toby Stuart and Senior Researcher David Kiron, Global Research Group, prepared this case, with advice and contributions from Alexander Crisses. HBS cases are developed solely as basis for class discussion rather than to illustrate either effective or ineffective handling of an administrative situation. Reprinted by permission of Harvard Business School.

the biological mechanism triggered by calorie reduction. In the 1990s, research indicated that calorie reduction was a biological stressor that activated a defensive metabolic response. Few scientists paid much attention to this research as few serious scientists focused their careers on anti-aging studies.

Longevity research gained traction as a credible field of study in the 1990s after Massachusetts Institute of Technology professor Leonard Guarente traced the molecular pathway of calorie reduction in yeast to sirtuins, which are proteins (enzymes) found in all cells. In humans, there are seven types of sirtuins. Sirtris was focusing 90% of its R&D on one sirtuin, called SIRT1 in humans.

### David Sinclair

Sinclair earned his Ph.D. from the University of New South Wales in 1995 and then joined Guarente in his MIT lab. In 1997, he published an article with Guarente that described how the yeast equivalent of SIRT1 increased the longevity of yeast cells by 30%. In 1999, Sinclair left for a tenure track position at Harvard Medical School, but continued to collaborate with Guarente. They found that extra copies of the sirtuin gene extended the life span of roundworms — a very different organism than yeast—by as much as 50%. The search was on for sirtuin activating compounds. This was a high-stakes search. "No chemical or drug had ever increased the activity of sirtuins" said Dr. Dipp. "A compound that could activate sirtuins would increase the speed of cellular metabolism. It could have far reaching implications for human healthcare."

In 2003, Sinclair discovered that resveratrol, a compound found in red wine, activated sirtuins in yeast cells, a discovery which indicated that it might be possible to develop a drug that could activate the sirtuin enzyme. One way to activate the enzyme was to optimize the effects of resveratrol by giving it to patients in a highly purified form. Another was to mimic the effects of resveratrol using an entirely new and more potent chemical structure. Sinclair pursued both approaches. He described his work in a September 2003 article in Nature magazine that drew the attention of Westphal who had recently been promoted to general partner at Boston's Polaris Venture Partners.

## Christoph Westphal

In four years at Polaris, Westphal had had several successful investments. He co-founded five companies and served as the original CEO at four of them. In all cases, Westphal recruited a CEO to replace himself while he remained on the board as lead investor once he got the company off the starting blocks. Two went public and in early 2007 had a combined market value of $1.4 billion. Philip Sharp, a Nobel Prize winning biologist and Sirtris advisor, described Westphal's business and science acumen: "Christoph's combination of skills is very rare. I haven't seen his equivalent in 30 years of working in biotech."[1] In 2002, MIT's Technology Review recognized Westphal as one of the country's top 100 Young Innovators Under 35.

---

[1] David Stipp, "Drink Wine and Live Longer," *Fortune, February* 12, 2007.

The son of two doctors, Westphal, a former McKinsey consultant, sped through an MD/Ph.D. program at the Harvard Medical School in less than six years. He spoke four languages and was an accomplished musician. Several Sirtris board members described him as having "extraordinary energy," and a "rock star" reputation in the biotechnology world. One stated, "He has an unusual combination of abilities—to understand the science and its commercial potential, and explain it all clearly in an understated way that resonates with investors."

Westphal had a distinctive approach to building biotech companies. His major successes both held initial public offerings (IPO) before many market watchers believed them to be ready. In both cases, Westphal teamed with world-renowned authorities (with Paul Schimmel, a prominent scientist at the Scripps Institute, and biologist Philip Sharp at one; with Robert Langer, an MIT professor and one of the world's most prolific scientist/entrepreneurs at the other). Westphal described the elements he looked for and the approach he took in starting and building companies: "You need fantastic science. Second, you need a great story. Third, you need great venture capital (VC) support and lots of money. I am a big believer in raising as much money up front as possible."

Applying this model and exploiting an ever-growing network among the industry's prominent players, Westphal had developed a successful approach to launching companies. In 2003, he read Sinclair's paper in Nature and phoned Sinclair.

## Launching Sirtris

Sinclair met Westphal and came away impressed. Six months later, Westphal expressed an interest in starting a company with Sinclair and Sinclair, who had been exploring such options with other investors, agreed. After an agonizing decision process, Westphal chose to join Sirtris as its full- time CEO. In contrast to his other start-ups, he planned to remain with the company this time, leaving behind the VC life and a high six-figure salary. He explained his decision:

> Many people thought it was too risky to leave a successful VC career. I was taking a $500,000 paycut and my wife and I had just purchased an expensive house close to Fenway Park in Boston. At the time, David had no data that showed resveratrol activated sirtuins in mammals or could affect mammalian glucose levels or insulin, although we hoped all that would prove true. My VC friends were telling me that I was not being rational. In some ways they were right, but I was excited about Sirtris in a way I had not been at my other companies.

### *Sirtris's Scientific Advisory Board*

Virtually all early-stage biomedical companies created scientific advisory boards (SABs). Among other roles, SAB members advised their companies on scientific strategies; they sometimes assisted in securing access to intellectual property produced by SAB members; and they served as portals that kept the company abreast of developments in the field. Sirtris's goal was to form a world-class

SAB and attract the brightest scientists in the field of sirtuin research. Sinclair described the formation of the Sirtris SAB: "Christoph said, 'Give me a list of the top 10 people in your field.' Within two weeks, we were having conference calls with all of these people. In one case, an academic was going to start a rival company, and Christoph flew out to St. Louis and convinced him to join us instead." One observer noted:

> Since combining forces with Sinclair, Westphal has organized what is arguably the most pedigree-rich scientific advisory board in biotech, including MIT's Sharp; Robert Langer, one of medicine's most prolific inventors, also of MIT; Harvard gene-cloning pioneer Thomas Maniatis; and Thomas Salzmann, formerly executive vice president of Merck's research arm. The group now numbers 27, among them many of the world's leading experts on sirtuins.

> Westphal also assembled an impressive list of directors — they include Alkermes's [founder Richard] Pops; [Rich]Aldrich, the Boston hedge fund manager and biotech veteran; and Paul Schimmel, a prominent scientist at the Scripps Research Institute in La Jolla, Calif., who has cofounded half-a-dozen biotech concerns. Westphal's right arm at Sirtris is chief operating officer Garen Bohlin, formerly a senior executive at Genetics Institute, a biotech now owned by Wyeth.[2]

### Growing Sirtris

In mid-2004, Westphal and Sinclair went on a road show to local Boston-area VC groups. In August, they obtained a $5 million seed (Series A) round of financing from Polaris Ventures and three other VC investors. Sinclair explained: "Christoph was very good at getting us in to talk with the majority of VCs in the Boston area over a short period of time. Although a lot of people said "no" to us, Christoph set a small window of time to invest and more than a few people started getting nervous about being left out. The short timeframe built its own momentum and helped drive interest in the company."

In November 2004, Westphal and Sinclair secured another $13 million in a Series A-prime round. (Exhibit 1 details Sirtris's five investment rounds. Exhibit 2 details pre-IPO financing at three comparable biotechnology firms.) Regarding the first two funding rounds, Westphal explained their ability to raise funds before the firm had any mammalian data:

> We were very early in terms of the science. We raised $18 million without any mammalian data, something that is almost unheard of in today's biotechnology. Part of our success was getting people bit by the Sirtris bug. We had a long-term vision of where we could go with the biology and the anti-aging message is extremely powerful, especially when you are talking

---

[2] David Stipp, "Live Forever?" *Fortune, February* 5, 2007.

to a bunch of aging, overweight guys who are prime targets for the drugs you want to develop.

Sirtris had to decide how to focus its drug development efforts. Sinclair had theorized that sirtuins played a role in diabetes, cancer, heart disease, neurodegenerative diseases and other disorders. One thing was clear: there would be no effort to claim anti-aging effects from any drug the firm produced since the U.S. Food and Drug Administration (FDA), which regulated such claims, did not recognize aging as a disease. The firm ultimately decided to develop a drug for diabetes: a major disease afflicting millions of individuals worldwide.

Sinclair conducted experiments that sought a connection between resveratrol and sirtuins in mice. Westphal recalled, "By early 2005, David started getting data in his lab that showed resveratrol was going to extend lifespan in a mammal. We had evidence you could lower glucose and insulin in mice. All of a sudden, we were getting real proof that this actually could be a drug; that this could actually be a very valuable company." Sirtris began hiring its R&D team, successfully staffing leadership positions with executives who had had long stints at large pharmaceutical and biotech companies identifying small molecules, developing drugs, and advancing drugs through clinical trials. With their mammalian data Sinclair and Westphal sought additional capital to finance Sirtris's R&D efforts. In March 2005, Sirtris closed on a $27 million Series B-round of financing.

In the next year, Sirtris made three significant advances. First, the company created a formulation for SRT501 that kept resveratrol in its active form and increased its absorption into the bloodstream. The result was that SRT501 could deliver five times more resveratrol into the blood stream than the best other formulations currently available.

Second, Sirtris began conducting human clinical trials in India, assessing SRT501 as a diabetes therapy. The transition from an R&D-only company to a clinical-stage company was, as Westphal remarked, "an important milestone in our plan to develop a rich pipeline of therapeutics to treat diseases of aging." Third, and perhaps most significantly, Sinclair showed that middle-age mice fed high-calorie diets with resveratrol could run further, were leaner, and lived 30% longer than control-group mice without resveratrol.

Sirtris researchers were exhilarated by these findings, in part because the data suggested that sirtuins could play a role in managing or even delaying the onset of type 2 diabetes. They were also excited because in the past experimental data with the same rat model had tended to foreshadow a higher probability of success for drugs in human trials. In April 2006, Sirtris closed on a $22 million Series C round of financing, and obtained a $15 million line of venture debt.

Throughout 2006, Sirtris gained momentum as Sinclair's research made its way into high profile academic journals, newspapers, and other media. In June, Sirtris announced that it had successfully completed a Phase 1 safety trial of SRT501 with 85 human subjects. In October, an article about Sinclair's work appeared on the front page of the Wall Street Journal. The following month, papers published in Cell and Nature demonstrated that resveratrol increased the stamina of mice two-fold and significantly extended their lifespans. The 2006 Nature article also demonstrated that sirtuin activation increased within the cell the number of functional mitochondria, the powerhouses that sustain a range of cellular activities including glucose metabolism.

In late fall of 2006, Sinclair received an unsolicited email from John Henry, owner of the Boston Red Sox baseball team, requesting a meeting. Westphal described the meeting:

> After we present him the company, he says, "How can I be helpful to you?" I say, "I think you could invest $50 million in the company." And he says, "I don't think I can do $50 million, but I think I can do $20 million." And I said, "Can we close in two weeks?" John teamed up with Peter Lynch, the legendary former fund manager of Fidelity's Magellan Fund, who did extensive due diligence, and they said, "OK." Everyone wanted in after that.

By February 2007, Westphal had closed on another round of financing, raising $35.9 million from company executives, venture firms from previous rounds, as well as from John Henry and Peter Lynch. Westphal explained, "We've always been able to raise money, I think because we had money. We weren't desperate. Investors knew that they couldn't get away with trying to hammer us on the valuation, because if we don't get the valuation we want we just won't raise the money."

## Moving Forward

### *Nutraceuticals*

Westphal and Sinclair had a long-running debate over the commercial opportunity represented by their proprietary formulation of resveratrol (SRT501). Sinclair believed that there would be great public demand for SRT501 long before the clinical studies were completed, and that Sirtris, one way or another, should start selling SRT501 to the public as a nutraceutical. Sinclair received 30 to 50 emails a day from people looking to obtain resveratrol.

For a biotech company focused on drug development, a Sirtris foray into nutraceuticals would not be unprecedented. A few other biotechs had taken this step. Depending on its definition, market size estimates ranged from tens of billions to hundreds of billions of dollars in yearly sales.[3] Sales of one joint-health supplement exceeded $1 billion.[4] The economics of nutraceuticals were also compelling. It might cost $0.25 to manufacture each capsule. Current vendors of resveratrol that were far less well absorbed by the body than SRT501 were charging prices in the vicinity of $1 per capsule.

Westphal, however, had doubts about the nutraceutical option. There were several issues. The nutraceuticals market was unpredictable and largely

---

[3] "The Global Demand for Essential Minerals is Worth $6.9 Billion," Business Wire, November 2, 2006, via Factiva, accessed April 2012.

[4] Eric Nagourney, "Study Sees Little Benefit in Chondroitin for Arthritis," New York Times, April 17, 2007.

unregulated — no FDA approval was necessary to sell supplements and no evidence was necessary to prove a product's effectiveness or even its composition. Sinclair had tested resveratrol pills and found that some brands on the market had no active resveratrol at all and other brands had far too little to have any meaningful effects on humans. Even if Sirtris had a scientifically proven product, it was not clear that science alone would be enough to differentiate its offering from those of the dozen other resveratrol providers. How would the company distinguish itself in this field? Another concern was safety. "The potential that someone could attribute a death to SRT501 could easily derail a nutraceuticals business," said Sinclair.

Another concern was identity. Was the company a scientifically rigorous enterprise focused on developing FDA-approved medicines that physicians would prescribe to improve the health of patients with aging-related disorders? Would Sirtris's brand suffer if it started selling nutraceuticals? Its office walls were plastered with pictures of Sinclair and Westphal with Nobel-prize winning biologists and luminaries from the VC world next to pictures of them with celebrities. Rich Aldrich, one of the company's original investors and current board member, explained: "The Sirtris story is a balance. It's carefully constructed from the core science and Christoph and David's public outreach. It's not clear if a nutraceutical approach will taint that story or extend it."

Next, what was the business case for entering the nutraceutical market? How much of a return would warrant Sirtris's participation and how would market entry be achieved? Should the company set up a subsidiary or create a spin-off company devoted to SRT501? Would the nutraceutical subsidiary have the Sirtris name or a different name? Some Sirtris executives favored a wholly- owned subsidiary with a different name in order to make clear that the Sirtris brand was focused on drugs based on their NCE development platform. "Our long-term investors like this option because they are in this company for the NCEs. They're not in it for the 501 data," said Westphal.

Yet another issue was retail format. Sirtris could sell SRT501 through the Internet or by partnering with supplement stores such as GNC. It could also open its own retail stores. Would partnering with a retailer reduce control over the brand? Would developing its own retail distribution take away from its drug development?

### *Pharmaceutical Deal*

Throughout 2006, new evidence from laboratories around the world confirmed Sinclair's hypothesis of a connection between resveratrol, sirtuins and metabolic activity in mammals. Several pharmaceutical companies began talking with Westphal about a possible drug development partnership. He anticipated that the terms of a deal could include a significant upfront cash payment, an equity purchase agreement, guaranteed R&D support that would likely step up over a four to five year period, and payments tied to clinical development milestones. Sirtris would also receive royalties on sales of any SIRT drugs. (See Exhibit 3 for details on the potential terms of a deal and Exhibit 4 for details of two deals between big pharma and other private biotechnology firms.) As an alternative, the team also believed that they could ink a deal with terms similar to those in Exhibit 3 but in which the pharmaceutical partner would take a 51% equity stake in the company.

Another reason to contemplate a deal was that it might be a relatively inexpensive source of additional capital. Pharmaceutical partners were often willing to purchase an equity stake at a premium to VC investors. They might also be willing to finance Sirtris's R&D programs.

Westphal, Sinclair, and Dipp also considered that a drug development deal with a pharmaceutical company would help Sirtris should it decide to conduct an IPO, which they hoped to do. A successful IPO could deliver some of the funding needed to move drug development from the laboratory through clinical trials. Westphal stated, "The typical biotech playbook says to get a partnership deal done, then file for an IPO. Public investors are often reassured by the prior involvement of a pharmaceutical company." One close advisor noted that a typical biotechnology company would not be a strong IPO candidate until it had developed some clinical data, successfully made headway on two different research programs, established intellectual protection around its discovery of one or more new chemical entities, and signed at least one significant deal with a pharmaceutical company.

There were reasons not to do a deal. Westphal and Sinclair were building a company that they hoped would have great impact. This was the first company that Westphal had actively managed beyond the first two years. It was Sinclair's first company. How would a deal affect their control over the company and its future?

Another reason to defer a deal was Sirtris's promising development of NCEs. Resveratrol was a naturally occurring substance and because it was already on the market intellectual property protection related to it was limited. Sirtris had recently made strides in synthesizing new compounds that could be 1,000 times more potent than resveratrol and these would be patentable. Partnering with a pharma company would require out-licensing these compounds without knowing their full value. If Sirtris waited for more data, and that data was positive, it might be able to arrange a more lucrative deal. If the new data did not show the results Sirtris was expecting, the pharma deal terms would be substantially worse, assuming these companies remained interested at all.

### In-Licensing or Expanding the Scientific Base

Sirtris executives had long debated how much of the company's resources to focus on SIRT1 versus alternative sirtuin targets. The biotechnology industry was littered with failed companies that had stalled in moving from mouse studies and toxicity screens to human trials. Several board members thought that the firm should diversify its product development platform. There were two main alternatives: investigate the six other human sirtuins or in-license a compound that had a better known mechanism of operation from another biotechnology firm.

The study of sirtuins was still in its infancy, so investigating the other sirtuins would require a great deal of basic research, financing and time. (See

Exhibit 5 for Sirtris financial data, 2004-2006.) Even so, the clinical role of the other human sirtuins offered tantalizing commercial options.

Several members of the SAB and the board of directors considered the in-licensing option to be appealing, while others disagreed. It would balance their investment in SIRT1, which was absorbing 90% of the firm's R&D expenditures. Dr. Dipp explained that, after investigating more than one hundred potential compounds to in-license, Sirtris had found a few anti-diabetic compounds that had better characterized effects than sirtuin-activating compounds. "It would be what we call "me-too" drugs. We know how they work, and if we could get them on the market they would get at least a small percentage of market share. It's something to have in your back pocket."

In 2004, Sinclair and Westphal debated, but decided against, in-licensing a compound. Sinclair stated, "I was the only person at the company who thought that SIRT1 activation was the right bet to make. I told Christoph, 'don't stop it until you know it's wrong.' If I'm wrong, find out sooner than later, and then in-license something." Westphal offered another view, "For the first eight months of this company, I was sitting there like a venture guy thinking that resveratrol will not be a great drug. It's a great story, but we'll have to bring in other stuff to build the company around."

Sinclair, Westphal, Bohlin and Dipp continued to debate whether to focus the firm's time, money, and other resources on one target or divert more of the firm's resources to additional targets, including non-sirtuins. They still did not have evidence that SIRT1 had the effects in humans that Sinclair believed they would one day see.

## Conclusion

After discussing these three issues for several hours, Sinclair and Westphal decided to summarize their views on the decisions they needed to make. At a general level, they remained convinced that if sirtuin-activating drugs could be successfully developed, they would have a revolutionary impact on human disease. However, Sirtris was many years from completing development of these drugs, much less manufacturing and selling them. Success would require navigating technical and regulatory hurdles that had stymied the majority of other biotechnology companies at similar points in their history. According to a 2007 pharmaceutical industry association report, only one in five compounds entering clinical trials gained FDA approval and, only 30% of approved drugs recovered the average development cost of a new medicine.[5] Given all of the unknowns about what could happen, Sinclair and Westphal described several options for addressing the risks they faced.

One approach would be to fully "hedge their bets." Sirtris could try to complete a pharmaceutical deal, in-license a compound with better known effects, and enter the nutraceuticals business.

---

[5] PhRMA Pharmaceutical Profile Industry Report, 2007, http://www.phrma.org/sites/default/files/539/profile2007.pdf Accessed April, 2012.

Another approach would be to "swing for the fences." Sirtris could continue to focus on a sirtuin-activating drug development platform. If successful, Sirtris executives believed the sky was the limit. Under this option, however, an IPO would be less likely since markets often preferred that biotech firms had a validating deal with a large pharmaceutical company.

A third approach would be a "middle of the road" path that incorporated some hedging, such as pursuing an in-licensing deal, but also accepted some risk, such as deferring a potential pharmaceutical deal. Alternatively, Sirtris could try to complete a pharmaceutical deal now, but forego in-licensing and the nutraceuticals project.

**EXHIBIT 1** Sirtris Investment Rounds ($ millions)

| Round | Date | Investors | Investment | Pre-$ Valuation |
|-------|------|-----------|------------|-----------------|
| A | Aug. 2004 | Polaris Venture Partners and others | $5 | $2.8 |
| Al | Oct. 2004 | Polaris Venture Partners and others | $13 | $10.9 |
| B | Mar. 2005 | Three Arch Partners and others | $27 | $32.5 |
| C | Apr. 2006 | Bessemer Venture Partners and others | $22 | $95.4 |
| Cl | Feb. 2007 | John Henry Trust, Peter Lynch and others | $35.9 | $184.0 |

Source: Company document.

**EXHIBIT 2** Pre-IPO Financing of Comparable Biotechnology Companies

| Company | Round/Date | Investors | Investment |
|---------|-----------|-----------|------------|
| Anesiva | 2Q2001 | InterWest Partners and others | $13 million |
| (IPO 2004) | 2Q/2002 | Bear Stearns Health Innoventures and others | $50 million |
| | 4Q/2003 | Bristol Myers Squibb | $45 million |
| Cytokinetics | 2Q/1998 | Mayfield and others | $5 million |
| (IPO 2004) | 3Q/1999 | International BM Biomedicine Holdings and others | $20 million |
| | 4Q/2000 | Credit Suisse and others | $50 million |
| | 2Q/2001 | GlaxoSmithKline | $14 million |
| Momenta | 2Q/2002 | Cardinal Partners, Polaris Venture Partners | $4 million |
| (IPO 2004) | 2Q2003 | Atlas Venture and others | $19 million |
| | 1Q2004 | Mithra Group and others | $20 million |

Source: Adapted from Growthink Research.

## EXHIBIT 3  Potential Pharma Deal Terms

- Five year term.

- 19.9% OR 51% equity stake at a 50% premium to most recent share price.[a]

- $75 million upfront for an exclusive option to join Sirtris in developing and marketing compounds from its SIRT1 Activator Program.

- 5 years of guaranteed R&D support totaling $100 million. Payments to step up over time. $10 million in year one.

- A combination of royalties, possibly manufacturing profits, and co-promotion fees that equate to approximately a 50/50 split of profits in the U.S. This is a significant point for Sirtris since the SIRT1 activator program is a core program, and the one that represents 90% plus of the firm's value.

- The pharmaceutical company will lead marketing and country specific development ex-U.S., Sirtris to receive substantial, double digit royalties on ex-U.S. sales.

- Roughly $200 million in milestones concurrent with risk reducing progress. Roughly 15% upon successful completion of safety/PK of a SIRT1 activator NCE in humans; Roughly 25% based on observation of glucose effects in phase lb of NCE in man; Roughly 30% upon successful completion of a phase 2a efficacy study for an NCE in man; and roughly 30% upon completion of phase 2b studies.

Source: Company document.

## EXHIBIT 4  Comparison Pre-IPO Agreements between Biotechs and Large Pharma Companies

| Biotech Company | Agreement | Large Pharma/Terms |
|---|---|---|
| **Corgentech** is a late-stage biotech that seeks to be the leader in the development and commercialization of transcription factor decoys. The Company has four drug candidates in clinical development for multiple potential indications. | Jointly develop and commercialize Corgentech's E2F Decoy, a first-of-its-kind treatment currently in Phase III development for the prevention of vein graft failure following coronary artery bypass graft (CABG) and peripheral artery (i.e., leg) bypass graft surgery. | **Bristol-Myers Squibb** will make an initial payment to Corgentech of $45 million comprising cash and an equity investment in Corgentech, with the potential for an additional $205 million in clinical and regulatory milestone payments. Bristol-Myers Squibb and Corgentech will share development costs in the U.S. and Europe based on a pre-agreed percentage allocation. |
| **Cytokinetics** is a biotech dedicated to the discovery, development and commercialization of novel classes of small-molecule therapeutics, particularly in the field of cytoskeletal pharmacology. | A broad strategic collaboration to discover, develop and commercialize novel small-molecule therapeutics targeting mitotic kinesins for applications in the treatment of cancer and other diseases. | **GSK** has committed funding of $50 million over 5-year research term, including a $14 million upfront cash payment and a $14 million purchase of Cytokinetics preferred stock. In addition, GSK could make milestone payments to Cytokinetics ranging from $30-50 million per target for products directed to each of over 10 mitotic kinesins that will be the subject of collaborative activities. |

Source: Adapted from press release, "Bristol-Myers Squibb and Corgentech Enter Global Agreement to Develop and Commercialize Novel Cardiovascular Therapy," PR Newswire, October 13, 2003; and "Cytokinetics and GlaxoSmithKline Form Major Strategic Alliance in Cancer Drug Discovery," Business Wire, June 25, 2001, both via Factiva, accessed April 2012.

---

[a] Two different equity stakes were under discussion: 19.9% or 51% of Sirtris's equity. All other terms would remain the same.

**EXHIBIT 5** Sirtris Financial Data 2004-2006 (in thousands, except share and per share amounts)

| | Period from March 28, 2004 (date of inception) through December 31, 2004 | Year Ended December 31, | | Period from March 25, 2004 (date of inception) through December 31, 2006 |
|---|---|---|---|---|
| | | **2005** | **2006** | |
| Revenue | $ — | $ 68 | $ — | $ 68 |
| Operating expenses: | | | | |
| Research and development | 1,190 | 7,062 | 14,242 | 22,494 |
| General and administrative | 699 | 3,865 | 4,340 | 8,904 |
| Total operating expenses | 1,889 | 10,927 | 18,582 | 31,398 |
| Loss from operations | (1,889) | (10,859) | (18,582) | (31,330) |
| Interest income | 45 | 1,143 | 2,447 | 3,635 |
| Interest expense | — | — | (878) | (878) |
| Net loss | $(1,844) | $(9,716) | $(17,013) | $(28,573) |

Source: Company document.

## Questions

1. What made Sirtris an attractive opportunity (and not just a good idea) at the time that Christoph Westphal joined as CEO? How will Sirtris make money?

2. Identify the major risks in each of these categories: technology, market, team, and financial. What is the most important category and risk?

3. Should Sirtris do the deal with the pharmaceutical company? Why or why not? Should Sirtris launch a neutraceuticals business? Why or why not?

4. Assume that Christoph Westphal and his team have just been told that J.P. Morgan Securities, which is a reputable investment bank, is eager to help Sirtris go public soon by filing for an Initial Public Offering (IPO). Furthermore, assume that Glaxo Smith Kline (GSK) has also just contacted them to discuss an offer to buy Sirtris entirely. What are the advantages and disadvantages of each of these three options that they could pursue in order to finance and operate the venture (e.g., stay private for now and fund the company with more venture capital and corporate partnerships, take the company public via an IPO, or agree to be acquired to become an operating division of an established company)?

# COOLIRIS: BUILDING AN A+ TEAM

## Introduction

It was well past 2AM on a warm evening in July 2007 at the Kleiner Perkins Caufield Byers (KPCB) incubator in California. Josh Schwarzapel, the young and energetic co-founder of Cooliris, checked his email again to see if their top recruit had accepted the Cooliris employment offer. Across the gray cubicle wall, the technical team, Austin and Kyan, were coding steadily on the next product release. Although Austin and Kyan had been working feverishly for months, they urgently needed more engineers if they were to make the next release deadline.

Three months earlier, when the Cooliris co-founders received their first round of funding from KPCB, Josh had accepted the challenge to help Cooliris expand by building a world-class technical team. At the time Josh had felt full of confidence: how difficult could it be for a young company to attract great talent when it had the backing of one of the world's most successful venture capital firms, an incredible technical vision, an early product with great traction, and another product in the pipeline?

Much to Josh's surprise, however, it had been challenging to build the team. Josh understood the importance of an outstanding team to the success of a new venture. As a student at Stanford studying entrepreneurship, Josh had heard luminaries in the field highlight the necessity of a world-class team for eventual success. But, as he struggled to find and attract such a team, Josh began to realize that his courses had not prepared him well for *how* to build such a team. Nonetheless, in Cooliris' high-accountability culture, pointing the finger at an issue from his formal education would not excuse a failure. Soujanya Bhumkar, the company's CEO, had been supportive and helpful along the way, but Josh could sense the pressure: from the investors, from Soujanya, and worst of all from the team who had been forced to work long hours as they waited for the much needed new hires.

Now, in the early hours of the morning, Josh once again began to ask himself difficult questions. Why had he struggled to recruit a great team? In the past he had always been successful, both as a student at Stanford and a collegiate volleyball player. What had gone wrong? Inevitably, Josh began to ask himself . . . "Is it me? Have I failed?"

# COOLIRIS BACKGROUND AND HISTORY

At the time of Cooliris' founding, the Internet underwent a transformation frequently referred to as Web 2.0. During the first major Internet wave spanning the late 1990s, entrepreneurs and corporations focused primarily on establishing a presence on the Internet using standard interface design to communicate with their audience. However, as the Internet evolved, a second wave emerged during the early 2000s that shifted the focus from corporate-generated content to user-generated content and from proprietary interfaces to modular interfaces that could be recombined (sometimes referred to as "mashable"). As part of this movement, a host of new websites, many of them funded by venture capitalists, sprung up to empower everyday users to become content publishers by posting blogs, making profiles, publishing videos, and sharing photos. Social networks such as MySpace and Facebook became household names and social networking, or the connecting of users to each other in online communities, became a common buzzword for many online ventures.

With the democratization of content creation and publishing, however, the sheer volume of content became a very real problem for most web users. As millions of participants posted billions of new photos, videos, and blogs, the quantity and type of information expanded exponentially. Narrow search terms returned hundreds of thousands of results, and information coming from one's social graph on sites like Facebook and MySpace was becoming too lengthy to consume. In short, a growing challenge for Internet users was how to interact with and process the exploding volume of new content. It was this challenge that led to the founding of Cooliris.

Soujanya, Josh and Austin Shoemaker founded Cooliris in January 2006 around the idea that the Internet had indeed become a fundamental element in the lives of billions of people, but the user interaction metaphors had changed very little since the first browsers. In their view, the Internet had always been characterized by a clunky, non-intuitive navigation experience based on the table format inherent in HTML code. Although flash animations and video content had enriched basic text, the way people interacted with the Internet was still constrained within the 2D framework of the original browsers.

The initial idea for Cooliris had emerged earlier in 2005 during a conversation between Soujanya and a friend, Mayank Mehta. As the two talked about how to make the Internet a more rich experience, Mayank suggested the idea of creating a mouse-over preview of embedded links to other web pages. The preview would allow the user to see the content underneath the link in a contextual window without leaving the original page, thereby creating a more multi-dimensional media experience. Soujanya, a former engineer turned serial entrepreneur, was struck by the insight and began discussing it with colleagues to get their feedback.

During this period, Josh and Soujanya met for coffee to discuss ideas. Josh caught the vision of a better Internet experience and connected Soujanya with Austin, a fellow student with exceptional technical capability. Together they laid out a plan for Cooliris and their first product: Cooliris Previews. The team began development right away, self-funded on Soujanya's credit card and working part time while Josh and Austin finished school. By September of 2006, the team had released Cooliris Previews as a free Internet browser plug-in.

Over the course of a few months, the product began to get significant traction, becoming a featured plug-in for Firefox and attracting thousands of users who desired a richer online experience. The success of Previews validated the primary Cooliris hypothesis that users desire a richer interaction with content, and so the team developed the concept for a second product: PicLens. The new PicLens product would allow users to view online photos from photo sharing sites or online image searches as full-screen slideshows rather than as low-quality thumbnail images or larger files that had to be downloaded individually. PicLens felt like a natural next step for the young company, but as the team built the product, their vision of the company began to evolve into something bigger. The team realized that the challenge of improving web navigation extended beyond a browser add-in to the fundamental way in which users access, discover, and navigate information. The team began soliciting feedback from investors and industry friends on how to take their ideas to the next level. In one such meeting, Randy Komisar, a partner at KPCB, suggested that his firm might be able to provide some funding and incubate the company in their adjacent offices. After several weeks of follow-on meetings with the KPCB partnership, Cooliris received and signed a term sheet for investment, taking up residence in the KPCB incubator next door to offices of such famous venture capitalists as John Doerr and Brook Byers.

## Cooliris Hiring Process

After receiving funding and moving into their new offices on Sand Hill Road, the Cooliris team identified several critical next steps for the company. The team and investors decided that one of the most critical action items should be to hire a top-notch technical team to execute on the Cooliris vision. As Josh and Soujanya defined their recruiting strategy, they both agreed that they needed to recruit people who were both entrepreneurial and technically brilliant. Furthermore, they firmly believed in the wisdom of hiring great team members right from the beginning. As Guy Kawasaki of Garage Technology Ventures said, "A players hire A players, B players hire C players, and C players hire bozos."

As the team discussed recruiting, hiring great team members seemed a comparatively easy task given their recent round of prestigious funding and their exciting technical vision. In Josh's eyes, new recruits would get in on the ground floor of an exciting opportunity, operate in a highly supportive setting, and be mentored by industry-leading venture capitalists. In this positive light, the recruiting task seemed easy.

Excited about his new role and freshly graduated from college, Josh took the lead on recruiting the new team. Initially, Josh anticipated that recruiting should take about half of his time, leaving the other half of his time open for business development. Conventional entrepreneurial wisdom suggested that the Cooliris team should start the recruiting process by tapping their social networks for potential hires. Aware of the competition for great technical talent, however, Josh developed an incentive to motivate his extended social network to proffer the best technical candidates: Cooliris would pay $1,000 for a candidate recommendation that led to a hire. Josh soon flooded his own extensive social network with the news about the exciting opportunity to join the Cooliris team. Similarly, KPCB and the entire Cooliris team reached out to their social networks to find the best potential recruits available. At the same time, Josh recognized that the Cooliris team might not know or have links to all the best technical people. To fill this gap, Josh also searched online databases such as LinkedIn and Google, looking for technical talent at similar companies using search terms such as "3D graphics engineer." Finally, Josh posted advertisements on LinkedIn and the San Francisco Bay Area edition of Craigslist.

Over the course of the next few weeks, Josh's search effort yielded mixed results. Not surprisingly, tapping the team's social networks produced the best leads. Searching for candidates on LinkedIn and Google also produced candidates, but not at the caliber the team desired. In contrast to searches, advertisements on LinkedIn and Craigslist typically produced subpar candidates. Later, Josh discovered that the reason the ads proved so disappointing was because candidates with significant talent ignored ads in general; these candidates were already receiving good offers through other channels.

In the end, after an exhaustive initial search, Josh reviewed over 1,200 resumes in search of the ideal team. Of the resumes browsed, Josh reached out to potential hires who he estimated had at least a 50% chance of being an "A player." Once Josh had filtered through the initial list of resumes, he then reached out to candidates via an introductory email explaining the Cooliris opportunity and the team's interest in the candidate (see EXHIBIT 1 for the text of a sample email). In total, Josh contacted 400 candidates via email to invite them to talk more with Cooliris.

In the end, the Cooliris team brought in 50 candidates for a first round interview. Because the incubator was located behind the main KPCB office, it was a little difficult to find. Furthermore, the doors to the incubator were always locked. To solve this problem, Josh gave candidates instructions on how to drive around to the back of the building to where the incubator was located. He told candidates to call once they arrived so that he could personally meet them and show them into the Cooliris offices.

At the beginning of the interview, a candidate would sign a non-disclosure agreement after which either Josh or Soujanya would give the candidate a ten-minute outline of Cooliris's vision and the products that had already been developed. At the same time, because the company was in stealth mode and had some very high potential ideas, Josh and Soujanya were careful not to reveal too much about the future direction of the company or some of their upcoming products. Therefore, in a typical interview, after a brief presentation about the company, the interviewer spent the rest of the hour screening the candidate for his or her technical ability. Josh and Soujanya also experimented with interviewing candidates by themselves or with the entire team.

After a thorough examination of the candidates' abilities during the first round, the Cooliris team decided to invite nine candidates back for a final round interview. The final interview lasted at least two hours and focused on a deeper technical discussion. Although the Cooliris team was still evaluating the candidate for fit and talent, in the final round interview the Cooliris team revealed a little more about the exciting future of the company. Lastly, the interview always included a long chat with Soujanya about expectations. In particular, Soujanya believed that it was important to have open and clear communication about the potential upsides as well as the risks involved; otherwise, both the candidate and the Cooliris team would be entering into a relationship under false pretences—a bad start to any relationship.

Of the nine candidates who received final round interviews, the team decided to extend offers to five high-potential candidates. Josh and Soujanya carefully crafted the offers to be as financially competitive as possible, benchmarking against what Google might pay for a similar position as well as giving candidates potential upside through equity in the company. It now seemed that Cooliris could finally add needed resources to the skeleton technical team who had already been stretched to the maximum.

## Two Unanswered Questions

Then came the surprise. Despite the apparent excitement that candidates exhibited during the interviews, of the five offers extended, four offers were turned down. Fortunately, the fifth candidate verbally accepted the offer before departing for a long-planned trip to Europe. Although the yield for his efforts seemed slim, Josh felt that if they could at least hire one candidate then the last few months would not be wasted effort. Indeed, recruiting had taken an immense amount of time—much more than the 50% of his time that he expected. He had worked late nights, weekends, and holidays, all in an effort to succeed in building a technical team.

In the end, though, even the fifth candidate decided not to join Cooliris. Shortly after returning from Europe, the candidate emailed to say he had second thoughts and decided he would pass on the opportunity. The candidate's retraction came as a shock and led Josh to reflect seriously on the failure of the recruiting process. What had the last two and a half months been about? Why had he failed to build a great technical team? What could he have changed to make the process successful? And finally, the most challenging question of all, could it be him? Was the problem that he lacked what was needed to be an entrepreneur? These questions plagued him, but, on another level, Josh realized their challenges boiled down to two key questions. First, *who* is actually an "A player" and second, *how* do you attract those people to join your team? Soujanya peeked over the rim of Josh's cubicle and with his usual earnestness suggested that Josh get some rest: "Hey Josh, it's okay. Let's talk about it in the morning."

Josh nodded in agreement, grabbed his bag and headed for the door. The entire weekend was blocked out for the team to meet and discuss what had gone wrong in the recruiting process and what, if anything, could be changed. As he headed home, Josh wondered what he should suggest at the meeting. Were the team's standards too high? Should they hire whomever they could find to help? What was the problem? Why couldn't they recruit world-class team members?

## Questions

1. You have just raised your first round of financing and want to build a team that can innovate in a completely new area: how would you go about identifying the right candidates and striking a balance between ingenuity and experience? Where exactly would you search for them? How do you successfully attract your top choices to join your venture?

2. Make a list of what Cooliris is doing right and doing wrong, if anything, with its current recruiting process? How should they improve it?

3. How should recruiting processes differ for hiring various functional positions in this venture? For example, do the same rules apply to hiring engineers as sales and business development talent?

### EXHIBIT 1: Email to Potential Candidates

Hi (candidate name here),

I took a look at your profile and you are definitely the kind of guy that we would like to work with for our startup, Cooliris. To give you context, we're leveraging 3D graphics to build an immersive media environment for browsing web content (check out our downloadable app at www.cooliris.com). We've recently raised Series A investment from Kleiner Perkins (the same investors as Google, Amazon, Intuit etc.) and are working with people like Bill Joy (Chief Scientist at Sun Microsystems) and Randy Komisar (former CEO of Lucas Arts).

We'd be willing to explore both full time and contracting options with you, although we would greatly prefer people open to full time. Would you be interested in chatting further?

Sincerely,

Josh Schwarzapel
Cofounder and VP of Business Development
www.cooliris.com

**absorptive capacity** A firm's ability to assimilate new knowledge for the production of innovations. The more related knowledge a firm has, the easier it is for it to assimilate the new knowledge.

**acquisition** When one firm purchases another, the acquired company gives up its independence and the surviving firm assumes all assets and liabilities.

**adaptive enterprise** An enterprise that changes its strategy or business model, as the conditions of the marketplace require.

**advertising revenue model** Selling firm, usually media companies, provides space or time for advertisements and collects revenues for each use.

**affiliate revenue model** A model based on steering business to an affiliate firm and receiving a referral fee or percentage of revenues.

**alliance** See *partnership*.

**angels** Wealthy individuals, usually experienced entrepreneurs, who invest in business start-ups in exchange for equity in the new ventures.

**architectural innovation** A change in how components of a product are linked together while core design concepts are left untouched.

**asset** Something of monetary value that is owned by a firm or an individual.

**asset velocity** Ratio of sales to net assets.

**balanced scorecard** A strategy formulation device as well as a report of performance.

**barriers to entry** Whatever keeps a firm from entering an industry or market.

**base case** The calculation of cash flows based on a set of assumptions that portray outcomes that are most likely to happen.

**best customer** One who values your brand, buys it regularly whether your product is on sale or not, tells his or her friends about your product, and will not readily switch to a competitor.

**board of advisors** A group constituted to provide advice and contacts to a venture. The members have extensive skills and knowledge and provide good advice.

**board of directors** A group composed of key officers of a corporation and outside members responsible for the general oversight of the affairs of the entity.

**book value** The net worth (equity) of the firm, calculated by total assets minus intangible assets (patents, goodwill) and liabilities.

**bootstrap financing** Launching a start-up with modest funds from the entrepreneurial team, friends, and family.

**brand** A combination of name, sign, or symbol that identifies the goods sold by a firm.

**brand equity** The brand assets linked to a brand's name and symbol that add value to a product.

**breakeven** The point at which the total sales equal the total costs.

**burn rate** Defined as cash in minus cash out on a monthly basis.

**business design** A design that incorporates the venture's selection of customers, its offerings, the tasks it will do itself and those it will outsource, and how it will capture profits.

**business method patent** A type of a utility patent that involves the classification of a process.

**business model** A set of planned assumptions about how a firm will create value for all its stakeholders; the resulting outcome of the business design process.

**business plan** A document that describes the opportunity, product, context, strategy, team, required resources, financial return, and harvest of a business venture.

**cannibalization** The act of introducing products that compete with a firm's existing product line.

**capacity** The ability to act or do something. A firm has processes and assets that need to be expanded as the firm grows its sales volume.

**cash flow**   The sum of retained earnings plus the depreciation provision made by a firm. The cash coming into a firm minus the cash going out over a predetermined time period.

**C-corporation**   A business that provides limited liability, unlimited life, the ability to accept investments from other corporations, and the ability to merge with other corporations.

**certain**   An outcome resulting from an action in that it will definitely happen.

**challenge**   A call to respond to a difficult task and the commitment to undertake the required enterprise.

**champion**   An executive or leader in the parent company who advocates or provides support and resources as well as protection of the venture when parent company routines are breached. The champion helps describe and defend the venture and secure the necessary resources.

**chasm**   A large gap between visionaries and pragmatists in the adoption process.

**cluster**   A geographic concentration of interconnected companies in a particular field. Clusters can include companies, suppliers, trade associations, financial institutions, and universities active in a field.

**collaborative structure**   Primarily consists of teams with few underlying functional departments. In a collaborative structure, the operating unit is a team that may consist of 5 to 10 members.

**competitive intelligence**   The process of legally gathering data about competitors.

**complement**   A product that improves or perfects another product.

**complementors**   Companies that sell complements to another enterprise's product offerings.

**conjoint analysis**   A quantitative measure of the relative importance of one attribute as opposed to another.

**convergence**   The coming together or merging of several technologies or industries thought to be different or separate.

**convertible note**   A type of bond that the investor can convert to stock in the new venture or cash of equal value at an agreed-upon price.

**copyright**   An exclusive right granted by the federal government to the owner to publish and sell literary, musical, or other artistic materials. A copyright is honored for 70 years after the death of the author.

**corporate culture**   The basic style of a company and how people work with each other.

**corporate new venture (or corporate venture)**   A new venture started by an existing corporation for the purpose of initiating and building an important new business unit or organization, solely owned subsidiary, or spin-off as a new public company.

**corporate venture capital**   An initiative by a corporation to invest in either young firms outside the corporation or units formerly part of the corporation. These are often organized as corporate subsidiaries, not as limited partnerships.

**corporation**   A legal entity separate from its owners. A body of owners granted a charter to act as a separate entity distinct from its owners.

**cost driver**   Any expense that impacts total firm costs: fixed, variable, semi-variable, and nonrecurring.

**creative destruction**   The creation of new industrial structures and companies and the destruction of older structures.

**creativity**   The ability to use the imagination to develop new ideas, new things, or new solutions.

**crowdfunding**   The joining together of a collection of individuals—the "crowd"—each of whom contributes a small amount to help fund a business.

**customer relationship management**   A set of conversations that consist of (1) economic exchanges, (2) the product offering that is the subject of the exchange, (3) the space in which the exchange takes place, and (4) the context of the exchange with the customers.

**customization**   Provision of a product designed to meet a user's preferences.

**debt capital**   Money that a business has borrowed and must repay in a specified time with interest.

**design**   The activity leading to the arrangement of concrete details that embodies a new product idea or concept.

**design patents**   Grants of exclusive right of use for new original, ornamental, and nonobvious designs for articles of manufacture for a period of 14 years.

**diffusion of innovations**   The process by which innovations spread through a population of potential adopters.

**diffusion period**   The time required to move from 10 percent to 90 percent of the potential adopters.

**discount rate**   The rate (r) at which future earnings or cash flow is discounted because of the time value of money.

**disruptive innovation**　An innovation that uses new modules and new architecture to create new products.

**dominant design**　A design whose major components and underlying core concepts do not vary substantially from one product model to the other and that commands a high percentage of the market share for the product.

**due diligence**　Gathering and verifying facts and data provided in a business plan before making a commitment to the terms of an investment deal.

**dynamic capitalism**　The process of wealth creation characterized by the dynamics of new, creative firms forming and growing and old, large firms declining and failing.

**dynamic disequilibrium**　The constant change of factors in an economy.

**economic capital**　The value of an economy and the associated standard of living.

**economics**　The study of production, consumption, and distribution of goods and services. Economics can also be defined as the study of how society manages its scarce resources.

**economies of scale**　The concept that larger volumes sold reduce per-unit costs.

**economies of scope**　Economies obtained by sharing of resources by multiple products or business units.

**elevator pitch**　A short version of the venture story that quickly demonstrates that the entrepreneurs know their business and can communicate it effectively.

**emergent industries**　Newly created or re-created industries formed by product, customer, or context changes.

**emotional intelligence**　A bundle of four psychological capabilities that leaders exhibit: self-awareness, self-management, social awareness, and relationship management.

**entrepreneur**　(1) A person who undertakes an enterprise or business with the chance of profit or loss (or success or failure); or (2) a person or group that engages in the initiation and growth of a purposeful enterprise for the production of goods and services.

**entrepreneurial capital**　A combination of entrepreneurial competence and entrepreneurial commitment.

**entrepreneurial commitment**　A dedication of the time and energy necessary to bring the enterprise to initiation and fruition.

**entrepreneurial competence**　The ability (1) to recognize and envision taking advantage of opportunity and (2) to access and manage the necessary resources to take advantage of the opportunity.

**entrepreneurial intensity**　The degree of commitment of the entrepreneurial team to the growth of the firm.

**entrepreneurship**　The identification and exploitation of previously unexploited opportunities by enterprising individuals.

**equity**　The ownership of the firm, usually divided into certificates called common or preferred shares of stock.

**equity capital**　The investment by a person in ownership through purchase of the stock of the firm.

**ergonomics**　The science of making a physical task easier and less stressful to accomplish.

**ethics**　A set of moral principles for good human behavior. Ethics provides the rules for conducting activities in a manner acceptable to society.

**execution**　A discipline for meshing strategy with reality, aligning the firm's people with goals, and achieving the results promised.

**exit strategy**　The way entrepreneurs or investors get their money out of a venture.

**family-owned business**　A firm that includes two or more members of a family who hold control of the firm.

**financial capital**　Financial assets such as money, bonds, securities, land, patents, and trademarks.

**first-mover advantage**　Gain accruing to the first to enter a market.

**flexibility**　A measure of a firm's ability to react to a customer's needs quickly.

**flow-through entities**　Firms where all profits flow to the owners free of prior taxation.

**focus group**　A small group of people who are brought together to discuss a product or service.

**follower strategy**　An entrant to a market that follows the initial one and attempts an improvement of the first mover's strategy.

**founders**　The people responsible for starting a firm, usually all the members of the initial leadership team.

**global** A strategy that emphasizes worldwide creation of new products, sales, and marketing.

**globalization** The integration of markets, nation-states, and technologies enabling people and companies to offer and sell their products in any country in the world.

**growing industries** Industries that exhibit moderate revenue growth and have moderate stability and uncertainty.

**harvest plan** A plan that defines how and when the owners and investors expect to realize or attain an actual cash return on their investment.

**high growth business** Characterizes a business corporation that aims to build an important new business and requires a significant initial investment to start up.

**horizontal merger** A merger between firms that make and sell similar products in a similar market.

**human capital** Combined knowledge, skill, and ability of the people in the enterprise.

**hybrid model** This model, sometimes called "bricks and clicks," utilizes the best of the Internet as well as other channels. A hybrid model can extend a company's reach to new market segments as well as globally.

**increasing returns** When the marginal benefits of a good or of an activity are growing with the total quantity of the good or the activity consumed or produced.

**incremental innovation** An innovation that is a faster, better, and/or cheaper version of an existing product.

**independent venture** A new venture that is not owned or controlled by an established corporation. An independent venture is typically unconstrained in its choice of a potential opportunity yet is usually constrained by limited resources.

**industry** A group of firms producing products that are close substitutes for each other and serve the same customers.

**initial public offering** The first public equity issue of stock by a company.

**innovation** Invention that has produced economic value in the marketplace. It is the commercialization of new technology.

**installed base profit model** One of the most powerful profit models; the supplier builds a large installed base of users who then buy the supplier's consumable products.

**integrity** Truthfulness, wholeness, and soundness; the consistency of our words and our actions or our character and our conduct.

**intellectual capital** The sum of knowledge assets of an organization. The sources of intellectual capital are threefold: human capital, organizational capital, and social capital.

**intellectual property** The valuable intangible property owned by persons or companies.

**international** A strategy that aims to create value by transferring products and capabilities from the home market to other nations using export or licensing arrangements.

**Internet** A worldwide network of computer networks linking businesses, organizations, and individuals.

**intrapreneurship** The entrepreneurial process within the confines of an established corporation.

**IPO** See *initial public offering.*

**joint venture** A short-lived partnership with each partner sharing in the costs and rewards of the project; common in research, investment banking, and the health care industry.

**just-in-time** A method that focuses on reducing unnecessary inventory and removing non-value-added activities by receiving items only when needed.

**knowledge** The awareness and possession of information, facts, ideas, truths, and principles in an area of expertise.

**knowledge management** The practice of collecting, organizing, and disseminating the intellectual knowledge of a firm for the purpose of enhancing its competitive advantages.

**layout** The arrangement of a facility to provide a productive workplace. This can be accomplished by aligning the form of the space with its use or function.

**leadership** The ability to create change or transform organizations. A real measure of leadership is the ability to acquire needed new skills as the situation changes.

**lean startup** A way of developing enterprises based upon quick product cycles with intense hypothesis-driven experimentation and adaptive learning.

**lean systems**   Operations systems that are designed to create efficient processes by using a total systems perspective.

**learning organization**   A firm that captures, generates, shares, acts, and uses its corporate experiences to improve and adapt.

**license**   A grant to another firm to make use of the rights of the licensor's intellectual property.

**licensing**   Occurs when a firm (the licensor) grants the right to produce its product, use its production processes, or use its brand name or trademark to another firm (the licensee). In return for giving the licensee these rights, the licensor collects a royalty fee on every unit the licensee sells.

**limited liability company**   A way of legally organizing a firm to limit liability of the participants.

**local**   A strategy focusing all efforts locally (or regionally) since that is the venture's pathway to a competitive advantage.

**logistics**   The organization of moving, storing, and tracking parts, materials, and equipment. Logistic systems usually are based on electronic networks such as a supply-chain intranet.

**loyalty**   A measure of a customer's commitment to a company's product or product line.

**management**   A set of processes such as planning, budgeting, organizing, staffing, and controlling that keep the organization running well.

**marketing**   A set of activities with the objective of securing, serving, and retaining customers for the product offerings of the firm. Marketing is getting the right message to the right customer segment via the appropriate media and methods.

**marketing objectives statement**   A clear description of the key objectives of the marketing program.

**marketing plan**   A written document serving as a section of the business plan and containing the necessary steps required to achieve the marketing objective.

**market potential**   A prospective of the maximum sales under expected conditions.

**market segment**   A group with similar needs or wants and may include geographical location, purchasing power, and buying attitudes.

**market segmentation**   The division of the market into segments that have different buying needs, wants, and habits.

**mature industries**   Industries that have slow revenue growth, high stability, and intense competitiveness.

**merger**   The combining together of two companies.

**metanational**   A company that possesses three core capabilities: (1) being the first to identify and capture new knowledge emerging all over the world; (2) mobilizing this globally scattered knowledge to out-innovate competitors; and (3) turning this innovation into value by producing, marketing, and delivering efficiently on a global scale.

**modular innovation**   An innovation that uses new components and modules, but does not disrupt the linkages between modules.

**module**   An independent, interchangeable unit that can be combined with others to form a larger system.

**moral principles**   Tenets concerned with goodness (or badness) of human behavior and usually provided as rules and standards of human behavior.

**multidomestic**   A strategy that calls for a presence in more than one nation as resources permit.

**natural capital**   Those features of nature, such as minerals, fuels, energy, biological yield, or pollution absorption capacity, that are directly or indirectly utilized or potentially utilizable in human social and economic systems.

**negotiation**   A decision-making process among interdependent parties who do not share identical preferences.

**net present value (NPV)**   The present value of the future cash flow of a venture discounted at an appropriate rate (r).

**network economies**   Observed effects in industries where a network of complementary products is a determinant of demand.

**new venture team**   A small group of individuals who possess expertise, management, and leadership skills in the requisite areas.

**new venture valuation rule**   Uses the projected sales, profit, and cash flow in a target year (N) and the projected earning growth rate (g) for five years after year N to calculate value of a firm.

**niche business**   A firm that seeks to exploit a limited opportunity or market to provide the entrepreneurs with independence and a slow-growth buildup of the business.

**nonprofit organization**   A corporation or a member association initiated to serve a social or charitable purpose.

**oligopoly**   An industry characterized by just a few seller firms.

**on-time speed**   A measure of lead-time, on-time delivery, and product development speed.

**open source innovation**   An innovation that is the product of many firms and individuals working together under a common goal and an agreed-to governance system.

**operation**   A series of actions.

**operations management**   The supervising, monitoring, and coordinating of the activities of a firm carried out along the value chain. Operations management deals with processes that produce goods and services.

**opportunity**   A timely and favorable juncture of circumstances providing a good chance for a successful venture or progress; an auspicious chance of an action occurring at a favorable time.

**opportunity cost**   The value (cost) of the forgone action.

**option**   The right to purchase an asset at some future date and at a predetermined price.

**organic growth**   Growth enabled by internally generated funds.

**organic organizations**   Organizations that are flexible and effectively adapt to change.

**organizational capital**   An enterprise's management processes, work procedures, information technologies, and communication methods.

**organizational culture**   The bundle of values, norms, and rituals that are shared by people in an organization and govern the way they interact with each other and with other stakeholders.

**organizational design**   The design of an organization in terms of its leadership and management arrangements, its selection, training, and compensation of its talent (people), its shared values and culture, and its structure and style.

**organizational norms**   The guidelines and expectations that impose appropriate kinds of behavior for members of the organization.

**organizational rituals**   The rites, ceremonies, and observances that serve to bind together members of the organization.

**organizational values**   The beliefs and ideas of an organization about what goals should be pursued and what behavior standards should be used to achieve these goals. Values include entrepreneurship, creativity, honesty and openness.

**outsourcing**   Purchasing services or goods from suppliers rather than doing or producing them within the firm.

**partnership**   Business association of two or more people or firms who agree to cooperate with one another to achieve mutually compatible goals that would be difficult for each to accomplish alone. There are two types of partnerships: the general partnership and the limited partnership.

**patent**   A grant by the U.S. government to an inventor giving exclusive rights to an invention or process for 20 years. A U.S. patent does not always grant rights in foreign countries.

**PE ratio**   The ratio of the price of a stock to the company's earnings.

**performance**   The degree to which a product meets or exceeds certain operating characteristics.

**personalization**   The provision of content specific to a user's preferences and interests.

**pessimistic case**   When the outcome of the calculation of cash flows, based on a set of assumptions, is less than expected.

**place**   The channels for distribution of the product and, when appropriate, the physical location of the stores.

**plant patent**   A grant of exclusive right of use for a term of 20 years for certain new varieties of plants that have been asexually reproduced.

**positioning**   The act of designing the product offering and image to occupy a distinctive place in the target customer's mind.

**post-money value**   The valuation accorded a company after investment by venture capitalists or angels.

**preferred stock**   Stock with preferences or claims on dividends and assets before common stock owners.

**pre-money value**   The value accorded a company before investment from venture capitalists or angels.

**pricing policies**   Methods for setting prices for various customer categories and volume discount plans.

**private placements**   The sale of stocks or bonds to wealthy individuals, pension funds, insurance companies, or other investors without a public offering or any oversight from the Securities and Exchange Commission.

**process**   Any activity or set of activities that takes one or more inputs, transforms and adds value to them, and provides one or more outputs.

**product**   The item or service that serves the needs of the customer.

**productivity**   The quantity of goods and services produced from the sum of all inputs, such as hours worked and fuels used.

**product offering**   Communicates the key values of the product and describes the benefits to the customer.

**product platform**   A set of modules and interfaces that form a common architecture from which a stream of derivative products can be efficiently developed and produced.

**product sales model**   Sales of a product in units to a customer. Selling items for a price.

**profit**   The net return after subtracting the costs from the revenues.

**profit margin**   The ratio of profit divided by sales revenues.

**profit model**   The mechanism a firm uses to reap profits from its revenues.

**pro forma**   Provided in advance of actual data. Pro forma statements are forecasts of financial outcomes.

**promotion**   The communication of an initial product message using public relations, advertising, and sales methods to attract customers.

**proprietary**   That which is owned, such as a patent, formula, brand name, or trademark associated with the product or service, and not usable by another without permission.

**prototype**   A model that has the essential features of the proposed product or service but remains open to modification.

**public offering**   The sale of a company's shares of stock to the public by the company or its major stockholders.

**quality**   A measure of a product that usually includes performance and reliability.

**radical innovation**   An important new development that leads to a new industry or way of operating.

**rapid prototyping**   The fast development of a useful prototype that can be used for collaborative review and modification.

**real option**   The right to invest in (or purchase) a real asset (the start-up firm) at a future date.

**regional**   See *local.*

**regret**   The amount of loss that a person can tolerate.

**regular taxable corporation**   An enterprise subject to taxes on its reported profits.

**relational coordination**   Describes how people act as well as how they see themselves in relationship to one another.

**reliability**   A measure of how long a product performs before it fails.

**resilience**   The ability to recover quickly from setbacks. It is a skill that can be learned and increased.

**restricted stock**   Stock issued in an employee's name and reserved for his or her purchase at a specified price after a period of time.

**return on capital**   The ratio of profit to the total invested capital of a firm.

**return on equity**   The ratio of net income divided by owner's equity.

**return on invested capital**   The ratio of net income to investment.

**return on investment (ROI)**   The ratio of net income divided by invested capital.

**revenue model**   Describes how the firm will generate revenue.

**revenues**   A firm's revenues are its sales in dollars expressed after deducting for all returns, rebates, and discounts.

**risk**   The chance or possibility of loss, which could pertain to finance, physicality, or reputation. Risk is a measure of the potential variability that will be experienced in the future.

**robust product**   A product that is relatively insensitive to aging, deterioration, component variations, and environmental conditions.

**sales cycle**   The length from the first contact with a customer until a sale transaction is made.

**sales forecast**   An estimate of the amount of sales to be achieved under a set of assumed conditions within a specified period of time.

**scalability**   The extent to which a firm can grow in various dimensions to provide more service.

**scale of a firm**   The extent of the activity of a firm as described by its size. The scale of a firm's activity can be described by its revenues, units sold, or some other measure of size.

**scenario**   An imagined sequence of possible events or outcomes, sometimes called a mental model.

**scope of a firm**   The range of products offered or distribution channels utilized.

**S-corporation**   A firm that has elected to be taxed as a partnership under the subchapter S provision of the Internal Revenue Code.

**seed capital**   The first funds used to launch the new firm.

**self-organizing organization**   Teams of individuals that benefit from the diversity of the individuals and the robustness of their network of interactions.

**selling**   The transfer of products from one person or entity to another through an exchange mechanism.

**six forces model**   A model for evaluating the competitive forces in an industry: (1) firm rivalry, (2) threat of entry by new competitors, (3) threat of substitute products, (4) bargaining power of customers, (5) bargaining power of complementors, and (6) bargaining power of suppliers.

**small business**   A firm with fewer than 50 employees operating as a sole proprietorship, a partnership, or a corporation owned by a few people.

**social capital**   The quality of relationships with a firm's suppliers, allies, partners, and customers. Social capital refers to the resources available in and through personal and organizational networks.

**social entrepreneur**   A person or team that acts to form a new venture in response to an opportunity to deliver social benefits while satisfying environmental and economic values.

**sole proprietorship**   A firm owned by only one person and operated for his or her profit.

**sources of innovation**   For new ventures, these include universities, research laboratories, and independent inventors.

**spin-off unit**   An organization that is first established within an existing company and then sent off on its own.

**staged financings**   The provision of capital to entrepreneurs in multiple installments, with each financing conditional on meeting particular business targets. This helps ensure that the money is not squandered on unprofitable projects.

**stock**   The owner's shares of the corporation.

**stock options**   An offer in a plan under which employees can purchase shares of the company at a fixed price. Stock options take on value once the market price of a company's stock exceeds the exercise price. A stock option gives employees the right to buy the company's stock in the future at a preset price.

**strategic control**   The process used by firms to monitor their activities and evaluate the efficiency and performance of these activities and to take corrective action to improve performance, if necessary.

**strategic learning**   A cyclical process of adaptive learning using four steps: learn, focus, align, and execute.

**strategy**   A plan or road map of the actions that a firm or organization will take to achieve its mission and goals.

**subscription revenue model**   A type of business that offers content or a membership to its customers or members and charges a fee permitting access to the information or participation for a certain period of time.

**sunk costs**   A cost that has already incurred cannot be affected by any present or future decisions. In other words, funds and time invested on a new venture are already spent, regardless of any action taken today or later.

**supply-chain management**   A firm's processes and those of its suppliers that enable the flow of materials, resources, and information to meet customer demand.

**switching costs**   The costs to the customer to switch from the product of an incumbent company to the product of the new entrant.

**synergy**   The increased effectiveness and achievement produced as a result of the combined action of two or more firms.

**talent**   The people, often called employees, of an organization.

**team**   A small number of people with complementary capabilities and skills who are committed to a common objective, goals, and tasks for which they hold themselves mutually accountable.

**technology**   Devices, artifacts, processes, tools, methods, and materials applied to industrial and commercial purposes.

**term sheet**   A summary of the principal conditions for a proposed investment by a venture capital firm in a company.

**throughput**   The amount of units processed within a given time.

**throughput efficiency**   The ratio of value-adding time and the sum of value-adding time and non-value-adding time. The goal is to reduce non-value-adding time.

**tipping point**   The moment of critical mass or threshold that results in a jump in adoption of a product or service.

**trademark**   Any distinctive word, name, symbol, slogan, shape, sound, or logo a firm uses to designate its product.

**trade secret**   An intellectual asset protected by confidentiality, nondisclosure, and assignment of inventions agreements as well as physical barriers such as safes and limited access.

**transaction fee revenue model**   A type of business that provides a transaction source or activity for a fee.

**transnational**   A strategy resting on a flow of product offerings created in any one of the countries of operation and transferred between countries.

**triple bottom line**   The three factors of a product or business: economic, environmental, and social equity.

**trust**   A firm belief in the reliability or truth of a person or an organization.

**uncertain**   An outcome resulting from an action in that the outcome is not known or is likely to be variable.

**unique selling proposition**   A short version of a firm's value proposition often used as a slogan or summary phrase to explain the key benefits of the firm's offering versus that of a key competitor.

**usability**   A measure of the quality of a user's experience when interacting with a product.

**utility patents**   Rights of exclusive use issued for the protection of new, useful, nonobvious, and adequately specified processes, machines, and manufacturing processes for a period of 20 years.

**valuation rule**   The algorithm by which an investor, such as an angel or venture capitalist, assigns a monetary value to a new venture.

**value**   The worth, importance, or usefulness to the customer. In business terms, value is the worth in monetary terms of the social and economic benefits a customer pays for a product or service.

**value chain**   The sequence of steps or subprocesses that a firm uses to produce its product or service.

**value proposition**   Summarizes the values offered to the customer.

**value web**   Consists of the extended enterprise within a network of interrelated stakeholders that create, sustain, and enhance its value-creating capacity. It is usually based on an Internet infrastructure to manage operations dispersed in many firms.

**venture capital**   A source of funds for new ventures that is managed by investment professionals on behalf of the investors in the venture capital fund.

**venture capitalists**   Professional managers of investment funds.

**versioning**   The creation of multiple versions of a products and selling their modified versions to different market segments at different prices.

**vertical integration**   The extension of a firm's activities into adjacent stages of productions (i.e., those providing the firm's inputs or those that purchase the firm's outputs).

**vertical merger**   The merger of two firms at different places on the value chain.

**viral marketing**   Building knowledge of a product through word of month.

**virtual organization**   A venture that manages a set of partners and suppliers linked by the Internet, fax, and telephone to provide a source or product.

**vision**   An informed and forward-looking statement of purpose in response to an opportunity.

**working capital**   The amount of funds available to support a firm's normal operations, such as unexpected or out-of-the-ordinary, one-time-only expenses. Working capital is a firm's current assets minus its current liabilities.